COMPUTATIONAL METHODS IN PHYSICS AND ENGINEERING

COMPUTATIONAL METHODS IN PHYSICS AND ENGINEERING

2nd edition

Samuel S M Wong
University of Toronto

World Scientific
Singapore • New Jersey • London • Hong Kong

Published by

World Scientific Publishing Co. Pte. Ltd.
5 Toh Tuck Link, Singapore 596224
USA office: Suite 202, 1060 Main Street, River Edge, NJ 07661
UK office: 57 Shelton Street, Covent Garden, London WC2H 9HE

Library of Congress Cataloging-in-Publication Data
Wong, S. S. M. (Samuel Shaw Ming)
Computational methods in physics and engineering / Samuel S. M. Wong, -- 2nd ed.
p. cm.
Includes bibliographical references and index.
ISBN 9810230176 ISBN 9810230435 (pbk)
1. Physics -- Data processing. 2. Mathematical physics. 3. Engineering -- Data processing. 4. Engineering mathematics.
I. Title.
QC52.W66 1997
530'.0285--dc21 96-51106
CIP

British Library Cataloguing-in-Publication Data
A catalogue record for this book is available from the British Library.

First edition published in 1992 by Prentice-Hall, Inc.
This edition © 1997 by World Scientific Publishing Co. Pte. Ltd.

Reprinted 2003

This book is printed on acid-free paper.

Printed in Singapore by Uto-Print

Contents

Preface to the First Edition

Computational methods form an increasingly important part of the undergraduate curriculum in physics and engineering these days. This book is mainly concerned with the ways that computers may be used to advance a student's understanding of physics. A large part of the material is common to engineering as well.

The subject matter covered in this volume may be classified also under the title of "computational physics." There are several ways to organize the material that should be included. The choice made here is to follow the traditional approach of mathematical physics. That is, the chapters and sections are grouped around methods, with physical problems used as the motivation and examples. One attractive alternative is to group around physical phenomena. The difficulty of following this way of organization is the heavy reliance on the physics background of the readers, thus making it harder to follow for students at early stages of their education. For this reason, such an approach is rejected.

The intimate relation between physics and mathematics may be seen by the way that physics is usually taught in the undergraduate curriculum. With a knowledge of calculus, for example, the subject of mechanics is discussed in a more rigorous manner. By the time the student is introduced to differential equations, topics such as harmonic oscillators and alternating current circuits, are brought in. Long experience in the community has shown this way of teaching to be very successful. The major problem here is the delay in introducing certain other basic concepts, because of the need to acquire first a certain maturity in mathematics. The fault is not with the mathematics required but the way it is used. For example, discussions on a pendulum are usually limited to small amplitudes at the early years. For finite amplitudes, the differential equation is nonlinear and the necessary skill to solve such equations by analytical methods comes only in later years. On the other hand, it is possible to use the same numerical methods to solve both types of differential equations for the pendulum problem and they are no more difficult than analytical methods. In this way, the discussion on pendulum does not have to be limited to small amplitude oscillations.

Until quite recently, the mathematics required in undergraduate physics and engineering, to a large extent, consists of analytical techniques to manipulate algebraic equations, to carry out integrals, and to solve simple differential equations. In addition to these "algebraic" methods, there is also a large class of numerical approaches that can be used to solve physical problems. While it is true that computers have made numerical calculations popular, many of the methods have their origins with the same group of mathematicians as the algebraic methods, such as Gauss and Newton. In the intervening years since the introduction of these mathematical techniques, numerical calculations have lost out to "algebraic" ones, perhaps because of the tedium of carrying out numerical calculations by hand. This reason is certainly no longer true, as attested by the explosion of numerical solutions in research papers. In spite of its popularity in research, the introduction of computational physics to the undergraduate syllabus is only starting.

To carry out numerical calculations on a computer, the usual practice is to write one or more programs in one of the high-level languages, such as FORTRAN or C. To simplify the process, one may make use of standard subroutine libraries to take over specific tasks, such as inverting a determinant or diagonalizing a matrix. In contrast, the general approach to algebraic computation is to make use of one of the symbolic manipulation packages. Most of these packages are very powerful and are able to carry out a large variety of complicated calculations. Although a substantial amount of "programming" can be done in most cases, the symbolic manipulation instructions are not programming languages on the same level as, for example, FORTRAN or C. Furthermore, there has been little attempt to standardize the instructions between different packages. As a result, any discussions of algebraic calculations in a volume on computational methods must either be very abstract or very specific in terms of one of these packages. The choice made here is the former, as it is not clear what is available to the average reader.

The results obtained from computer calculations often appear in the form of a large table of numbers. For human beings to comprehend such huge quantities of information, graphical presentations are essential. For this reason, graphical techniques are essential parts of computational methods. On the other hand, it is not possible to cover all three aspects, numerical, symbolic, and graphical, in a single volume. The choice made here is to select a few of the standard topics in physics covered in the undergraduate curriculum and present them in ways that computers may be useful for their solutions. Even this is too big a task. The compromise is to put the emphasis on numerical techniques. For reasons mentioned in the previous paragraph, only an introduction is made to symbolic manipulation techniques. Computer graphics is perhaps one of the fastest growing areas in computing. For this reason also, it is best to leave everything beyond an introduction to volumes specializing on the subject.

There is quite a bit of interest in the physics community in developing courses in computational physics, as has already been done in many places. Such courses should be regarded as in parallel with the more traditional ones in mathematical physics and experimental physics. It is one of the aims of this volume to serve as a textbook or major reference for such a course. At the same time, many graduate students and senior undergraduates may not have benefited from such a course. It is also the intention here to serve this group of physicists. Engineers and other professional people who make use of computational techniques in physics may also find it useful to examine some of the background involved in solving some of the physical problems on computers.

Although a set of computer program is present here, it is not the primary intention of the author to provide a library of subroutines for common problems in physics. The computer programs used as examples and included in the accompanying diskette are intended as illustrations for some of the materials discussed. They can be modified for other applications. However, before one does that, as with any computer program, the programs should be thoroughly tested first for the intended purpose.

S. S. M. Wong

Preface to the Second Edition

The excellent reception of the first edition by the reader community worldwide makes it imperative for a new edition. Many subtle changes have taken place in scientific computation in the time interval. Most of these are driven by the tremendous increase in computational power available to the individual user, making it possible to carry out work almost unthinkable a little while ago and there is every indication that the trend will continue in the near future. As a result, computational techniques become even more indispensable to engineers and scientists than ever before. At the same time, the convenience of communication through internet is also having an impact on how computers are used and how some of the computations are carried out these days.

The main changes in the second edition include the addition of a chapter on finite element methods. The growth of available computing power makes it more and more attractive to solve many complicated differential equations numerically and a different approach from the traditional finite difference methods is getting increasing attention, especially in the physics community. In order to keep the volume from expanding into unmanageable size, some sacrifices have to be made. These include introductory chapters on graphics and computer algebra and a couple of sections on topics that are less popular. Some derivations that can be left to references are also omitted. In order to accommodate these changes, most of chapters have been substantially rewritten. The end result is a volume that is more ideal in size both for the classroom and on the desk.

Many valuable suggestions have been received from readers and they have been incorporated, as far as possible, into the new edition. The author is greatly indebted to colleagues for caring to communicate and share their thoughts. The strong encouragement and support from World Scientific also form an essential ingredient in making the decision to finish on the second edition.

Samuel S.M. Wong
Toronto

Chapter 1

Computational Methods

Modern electronic computers owe their origin, to a large extent, to the needs in science and engineering. In the 50 years or so since their appearance, computers have out-performed their original goals of solving numerical problems and keeping tracks of information. They are now an essential tool in almost every aspect of the daily routines of engineers and scientists, from data collection to writing technical reports. The high speed of computation available to us these days opens up not only new ways of carrying out traditional tasks but also new areas of endeavor that have implications going well beyond what we can realize at the moment. Our concern here is limited to a small, albeit important, corner of the role of modern computers in science and engineering, namely some of the general techniques to solve common problems encountered in physics and engineering.

1-1 Numerical calculations and beyond

When we use a computer to solve a problem in science, the general assumption is that it is done numerically. Indeed, the proper name of most computers in the market is "digital computer," reminding us of the fact that numbers are being manipulated. However, in addition to mathematical operations, such as addition and multiplication, the central processor of a computer is also capable of logical operations, that is, making decisions depending on whether a particular condition is true or false. Furthermore, a binary digit, or "bit," of the computer memory may be regarded as a logical unit, representing the value "true" if it is on (= 1) and "false" if it is off (= 0). In this way, a computer can be programmed equally well to carry out logical decisions or, more generally, symbolic manipulations.

At the same time, the computer screen is made of lines of horizontal dots or "pixels." On a monochrome screen, each dot can be turned on or off. For a color monitor, there is the further capability of displaying different colors at each dot. As a result, very effective graphical images can be displayed. The same is also true for paper output. For this reason, in addition to numerical calculation and symbolic manipulation, computers are used extensively for graphics.

In this volume, we shall be mainly concerned with numerical calculations. Before we get totally immersed in the topic, it is useful to remind ourselves that both symbolic manipulation and graphic presentation are also important in scientific applications of computers. We shall give a brief description of both topics before getting on to numerical calculations.

Computer algebra Among the various possibilities of using computers for "artificial intelligence" applications, symbolic manipulation, more commonly referred to as computer algebra, is an important tool in scientific endeavors. However, we shall not go into the subject in this volume. The main reason for this choice is that, most computer algebra are carried out using one of the available "packages," such as Maple (Symbolic Computation Group, University of Waterloo), Mathematica (Wolfram Research, Inc), and Reduce (Rand Corporation). Although the Lisp language, for example, is designed for the purpose of symbolic manipulation, most of computer algebra applications these days are carried without explicitly going to the level of actually programming a computer using one of the general purpose programming languages.

In most computer applications, it is desirable for us to give the instructions in a language that is as close as possible to the working language of the subject we are involved with. In the case of algebraic calculations, it is natural for us to want to work in terms of algebraic equations. For example, if we wish to solve the set of equations:

$$\begin{aligned} ax + by &= c \\ dx + ey &= f \end{aligned} \tag{1-1}$$

it would be nice if all we need to say to the computer is something like

SOLVE

$$ax + by = c$$
$$dx + ey = f$$

FOR x AND y.

However, to solve Eq. (1-1) requires some knowledge of linear algebra that is beyond the basic mathematical and logical operations a computer is designed for.

As we shall see later in §5-1, the solution may be expressed in terms of ratios of determinants. In general, we cannot expect a computer or any general purpose programming language to possess such advanced knowledge of algebra. On the other hand, for commonly encountered applications, one can think of developing a set of codes that specialize in these problems. In this way, we can always call up the codes whenever we want to solve, for example, a set of linear equations as in the example above. In fact, we can put together a number of such codes that carry out related functions, such as evaluating determinants, inverting matrices, finding the roots of linear equations, and write a "driver" to manage them. Such a collection is often loosely referred to as a "package." The development of computer algebra, to a large extent, has followed this route. As a result, algebraic calculations are usually carried on computers in terms of one of the existing packages. In fact, some of these symbolic manipulation packages are so versatile that they can be regarded as programming languages for a large class of calculations.

Because of this tendency, discussions on algebraic calculations are often based on one of the popular packages. Since most textbooks in physics and engineering have already done a good job in presenting the algebraic aspects of various topics, the main work we need to do is to cast the problems in terms of the language of one of the packages. We shall not do this here. Instead, we shall give an example of algebraic calculation later in §1-4, just to provide some introduction to those who have not been exposed to the wonders of computer algebra.

The development of numerical methods for physics and engineering is somewhat different. Although there are many excellent "packages" available to carry out specific calculations, such as eigenvalue problems and matrices,[2] they tend to exist in the form of subprogram libraries. In this case, the user often has to write a "calling" program to transform the problem in hand into one that can take advantage of the library to perform some of the calculations. Similar to other tools, a basic knowledge of the numerical methods used in these subprogram is essential in this case.

Computer graphics An important market for computers these days is in graphics. This is, in part, due to the success in computer animation and computer aided design. Such applications clearly belong to totally different treatments from what we intend here. However, there is a small area of computer graphics that is important to numerical work, namely graphical representation of results.

In numerical calculations, the results often appear as a table of numbers and, for a complicated problem, such a table can be an extensive one. A good way to gain an overall feeling for a large set of numbers is to view them in the form of a plot. Before computers, plots are usually done on a sheet of graph paper. For simplicity, let us consider the problem of making a linear plot for some function y of a single independent variable x. The graph paper we shall use in this case is nothing but a sheet of paper with $(N_x + 1)$ evenly spaced horizontal lines and $(N_y + 1)$ evenly spaced vertical lines. The plotting area of our graph paper may therefore be regarded as made up of $N_x \times N_y$ squares. We can select one of the horizontal lines as our x-axis and a vertical line as the y-axis. The scales of our two axes are set by the ranges of values we wish to display. The actual plotting is carried out by putting on the graph paper a symbol for the value of y corresponding to each one of the values of x we are interested in. For a continuous function, the plot of y versus x is a continuous curve. However, for our purpose here, such a continuous curve may be regarded as a collection of closely spaced points, one for each possible values of the independent variable x.

We can follow basically the same steps to plot the graph on a computer screen or a sheet of output. The reason is that both types of device are made of a number of dots or pixels, very similar to the squares on our graph paper. For example, many monitors are said to have a resolution of 1024×746. For our present purpose, it may be regarded as a graph paper with 1024 horizontal lines and 746 vertical lines. The only difference is that, for a variety of technical reasons which we do not need to go into here, the spacings between the horizontal lines and vertical lines are not necessarily equal. In fact, the aspect ratio, i.e., the ratio of the horizontal and vertical size of each element, on a computer screen is usually less than one. As a result, each one of the basic elements on our "graph paper" is, often, a rectangle instead of a square as on a normal sheet of graph paper. While this difference causes some nuisance in displaying a graph, it does not impose any fundamental problem and we shall ignore it here. For output on paper, it is not difficult to obtain resolutions of 300 to 600 dots per inch. This means that we can easily have the equivalent of 300^2 to 600^2 elements on each square inch of a sheet of output.

The computer screen (often paper output as well) has the advantage of color. As a result, we have an additional dimension to express our "graphs" that is not easily available on graph papers. Furthermore, it is possible to generate many frames of a "graph" in a short time and, as a result, one can have a "movie" of our calculated results to show, for example, the time development of a process.

Although the basic principles of computer graphics are simple, the actual applications require some preparations. Most computers come with machine-level instructions to access and to manipulate the pixels. However, it will be extremely tedious to plot a graph using such low-level instructions. In fact, what we prefer is that, for example, once two arrays, say X_ARRAY and Y_ARRAY, are generated, we can issue an instruction like

PLOT Y_ARRAY versus X_ARRAY.

The computer will then go and find the best axes and scales to represent y as a function of x and display the results on the screen. Another instruction,

OUTPUT PLOT

will produce a printed version of the plot on a sheet of paper. For more complicated plots, such as histograms, log plots, contours, and surfaces, we can think of, at least in principle, developing similar conversation-like instructions.

Needless to say, we are not close to this ideal level of graphics programming on computers. Partly because of the fact that the standardization of computer graphics hardware and software is still in the development stage, there are only, at the time of writing, the beginnings of common high-level graphics programming languages and "interfaces" that are portable between different types of computers. As a result, we must once again resort to "packages." Similar to computer algebra, there are many fairly extensive plotting packages available, both public domain and commercial, and most plottings are done using one of these packages.

1-2 Integers and floating numbers

Most computers make a clear distinction between integers and floating numbers. An integer is a number, such as 5, −213, and 0, without a part that must be represented by a decimal point. A floating or *real* number is any other type of number, such as 3.1415926..., 3.0×10^{23}, and −9.9, that requires a decimal point to specify its value. The reason for differentiating between these two types of numbers comes from the structure of the computer memory.

The basic unit for storing a number in a computer is a *bit*, the state of an electronic component that is either on or off, as we saw earlier. The two possible states of a bit may be used to represent two numerical values 0 (off) and 1 (on). Since a single bit is too small for most interests, 8 bits are grouped together into a *byte*. The status of a byte may be represented by an eight-digit binary number $b\sqcup\sqcup\sqcup\sqcup\sqcup\sqcup\sqcup\sqcup$, where we have added a prefix b in front of the number to indicate that it is in the binary representation. For example, the integer 5 is shown as b00000101 or simply as b101. The largest integer that can be represented in one byte of storage is then $b11111111 = 2^8 - 1 = 255$.

Before we leave the subject of internal representation of integers in the computer memory, we shall define two other representations. The binary representation used above is inconvenient in many cases because of the large number of digits required to express most of the integers of interest to us. The hexadecimal representation is based on powers of 16 (in an analogous way as the decimal system is based on powers of 10). Each hexadecimal digit can take on values 0 through 15, and they are usually written as z0, z1, z2, z3, z4, z5, z6, z7, z8, z9, zA, zB, zC, zD, zE, and zF, where we have added a prefix z to indicate that they are given in hexadecimal representation. In terms of computer memory, each hexadecimal digit represents one of the possible values stored in four binary digits ($2^4 = 16$). The value that can be stored in a byte is then represented by two hexadecimal digits. Examples of numbers in the hexadecimal representation and their corresponding values in decimal and binary representations are given in Table 1-1.

Another way of displaying binary-based numbers is the octal representation. To distinguish it from others, we shall prefix a number in the octal representation with the letter o. (The lowercase o is used here instead of the uppercase so as to make it easy to differentiate it from the number zero.) Each octal digit represents the value given by 3 bits. This is convenient for some models of computer whose memories are made of multiples of 3 bits, such as 36 and 60. In addition, octal representation is also a convenient way for carrying out many types of manipulations.

In many calculations, a byte, which can only store integers up to 255, is still too small. For this reason, each integer is often assigned either two or four bytes of memory. Such a grouping of bytes is sometimes called a computer word, or just *word* for short. Before we go into the question of the range of integer values a computer word can store, we must recall that most numbers we are interested in have a ± sign associated with them. It is common practice to designate the first bit of an integer

Table 1-1: Decimal, hexadecimal, octal, and binary representations of numbers.

Decimal	Hexa-decimal	Octal	Binary	Decimal	Hexa-decimal	Octal	Binary
0	*z*0	*o*0	*b*0	10	*z*A	*o*12	*b*1010
1	*z*1	*o*1	*b*1	11	*z*B	*o*13	*b*1011
2	*z*2	*o*2	*b*10	12	*z*C	*o*14	*b*1100
3	*z*3	*o*3	*b*11	13	*z*D	*o*15	*b*1101
4	*z*4	*o*4	*b*100	14	*z*E	*o*16	*b*1110
5	*z*5	*o*5	*b*101	15	*z*F	*o*17	*b*1111
6	*z*6	*o*6	*b*110	16	*z*10	*o*20	*b*1 0000
7	*z*7	*o*7	*b*111	17	*z*11	*o*21	*b*1 0001
8	*z*8	*o*10	*b*1000	18	*z*12	*o*22	*b*1 0010
9	*z*9	*o*11	*b*1001	19	*z*13	*o*23	*b*1 0011
255	*z*FF	*o*377	*b*1111 1111				
256	*z*100	*o*400	*b*1 0000 0000				
257	*z*101	*o*401	*b*1 0000 0001				

word as the sign bit. Thus, for a two-byte integer I_2, only 15 bits are available to store the magnitude of the number. The possible values that can be represented by such a word is then

$$-32,768 \leq I_2 \leq +32,767.$$

That is, -2^{15} through $(2^{15}-1)$. The reason that the maximum positive integer value is one less than 2^{15} comes from the fact that +0 must also be considered as one of the integers. (Why, then, is the maximum absolute value of a negative integer one larger?) For a four-byte word, the possible value of an integer is then

$$-2,147,483,648 \leq I_4 \leq +2,147,483,647$$

corresponding to -2^{31} to $(2^{31}-1)$. Although the allowed range of integer values for I_4 may be large, it is still not adequate for many purposes. For example, the upper limit of I_4 is less than $13! = 6,227,020,800$. As a result, it may not be possible to carry out certain types of calculations that involve factorials. We shall see later ways to circumvent this difficulty.

For most calculations in science and engineering, the use of integers alone is too restrictive. Floating numbers broaden the range of values that can be kept in the computer memory by allocating a part of a word to store the exponent of each number. That is, each number now has a sign, a fraction part or mantissa, and an exponent. Since the computer memory is made of bits, some arrangements must be made to represent a number in this way. The usual case is to use four bytes for a single-precision number. Among the 32 bits in such a word, 1 bit is devoted to the sign, 8 bits are assigned to the exponent, and the remaining 23 bits are left for the mantissa. In this way, numbers with absolute values in the range from approximately

1.2×10^{-38} to 3.4×10^{38} can be represented. In double precision, eight bytes are usually used for each floating number. Among the 64 bits here, 1 bit is for the sign, 11 bits for the exponent, and 52 bits for the mantissa. Floating numbers with absolute values approximately in the range from 2.2×10^{-308} to 1.8×10^{308} may be represented by such an arrangement.

If we get a floating number whose absolute value is too small to be represented in a computer, an *underflow* condition is created. This happens, for example, when we get a single precision floating number with absolute value less than 10^{-38}. The usual course of action on such occasions is to replace the number by zero. If a number is produced with an absolute value much larger than 10^{38} in single precision or 10^{308} in double precision, an *overflow* condition is created. In this case, the usual response of the computer is to suspend the calculation, as an error will be introduced if we replace the number by anything else. Generally, one should check for possible underflows and, in particular, overflows in a program where such conditions are likely to occur and take the appropriate response.

The maximum number of significant figures that can be achieved in a floating number calculation is limited, in the first place, by the number of bits assigned to the mantissa. In single precision, the maximum is roughly seven significant figures and in double precision it is possible to reach 15 significant figures. Since the maximum number of significant figures is limited, *truncation* errors become a part of any floating number calculations. One important consideration in numerical work is to choose an algorithm that minimizes the truncation errors. If this is not done, the numerical errors may accumulate and become so large that no significant figures are left in the final results. At the same time, there is no use in trying to design an algorithm with a precision beyond the limitations imposed by truncation errors. We shall see different aspects of these considerations in many of the examples following.

1-3 Programming language and program library

In addition to hardware, such as central processor, memory and disk, software is also an integral part of any computer. As users, we do not wish to be overly involved in all the aspects that make the machine work for us; however, some background knowledge is useful. In getting a computer to solve our problem, the pieces of the software of most direct concern to us are the programming language and the subroutines to carry out some of the tasks to solve our problem.

High-level programming language For a computer to carry out a task, it is necessary to give it a set of specific instructions. This is generally referred as programming the computer. From a user point of view, it is desirable to give these instructions in terms of a language that is close to how we think of the problem, for example, in terms of equations. On the other hand, since the computer hardware is designed to carry out a very limited number of basic operations, we are still a long way from the possibility of such a direct communication. Furthermore, equations, by themselves, are sometimes incomplete in describing a problem and, as a result, inadequate for a machine to carry out the calculations implied. For this reason, computer languages are developed as the interface between the user and the machine.

For our interest here, we shall be concerned with what are normally called high-level languages. That is, languages that are close to how users think of the problem and yet can be translated, or *compiled* into computer instructions. In scientific computations, C, Fortran, and Lisp are the three examples of high-level languages that come to mind most readily. Fortran is fairly easy to learn and has the support of a large number of subroutine libraries for scientific calculations developed over the years. Furthermore, because of its simple structure, it is relatively easy to develop highly optimized compilers. However, Fortran lacks the flexibility of the more modern C language. With the development of libraries accessible to both Fortran and C (as well as several other high-level languages) C is quickly becoming the language of choice in scientific computations. Lisp is more widely used in symbolic manipulation and is also essential for those interested in some of the finer details of computer algebra. Most of the discussions in this book are independent of computer languages. However, since Fortran remains to be the more popular medium of communication in numerical applications, the examples are based on Fortran.

Subroutine library Even with a high-level computer language, it will be quite tedious in most cases if we have to program everything needed in the project. Fortunately, among the software normally accompanying a computer these days, there is a large number of subroutines and utilities to carry out quite a few of the "routine" tasks in computation. For our purpose, we can divide these supporting programs into four categories, intrinsic library, system subroutines, utilities, and application library.

Many parts of a calculation are often common to a variety of problems. One obvious example is the trigonometry functions. For this reason, each high-level language compiler is usually associated with a standardized *intrinsic library* containing many of the functions generally needed in calculations. As an example, the functions in the intrinsic library of Fortran is given in Table 1-2.

Besides the intrinsic library, computers are equipped with a number of system routines that are special to the operating system, either in the way they are used or because they are specially adapted for the particular computer. One example is the routine to read the system clock so that one can, among others, print out the time when a calculation is done. Since the time is related to the master clock that, in a sense, sets the pace for the computer, the system clock routine depends very much on how the computer is built. Another example is the routine for us to ask whether an overflow or underflow condition has occurred so that we can take appropriate actions.

In using a computer, we need also to carry out a number of tasks not directly related to the problem to be solved. For example, to write a program, a text editor is required to enter each line of the code and to make corrections. We need also utilities to maintain files on the disk and to check on the status of our job. It will be impossible to make use of modern computers without such tools.

Table 1-2: Intrinsic Fortran functions.

Name	Function	Name	Function
Type conversion		**Trigonometric functions**	
INT	To integer	COS	Cosine
REAL	To real	SIN	Sine
DBLE	To double precision	TAN	Tangent
CMPLX	To complex	ACOS	Arc cosine
AINT	Truncation to nearest integer	ASIN	Arc sine
ANINT	Nearest whole number	ATAN	Arc tangent (1 argument)
NINT	Nearest integer	ATAN2	Arc tangent (2 arguments)
Numeric functions		**Hyperbolic functions**	
ABS	Absolute value	COSH	Hyperbolic cosine
MOD	Modulo	SINH	Hyperbolic sine
SIGN	Transfer of sign	TANH	Hyperbolic tangent
DIM	Positive difference		
MAX	Maximum	**Character functions**	
MIN	Minimum	LGE	Lexical $\geq$
DPROD	Double precision product	LGT	Lexical $>$
AIMAG	Imaginary part	LLE	Lexical $\leq$
CONJG	Conjugate	LLT	Lexical $<$
Transcendental functions		CHAR	Integer to character
SQRT	Square root	ICHAR	Character to integer
EXP	Exponential	INDEX	Location of string
LOG	Natural logarithm	LEN	Length of string
LOG10	Logarithm base 10		

Application libraries, as the name implies, are more specific to the particular type of work we wish to carry out. For example, for calculations involving matrices, one of the important library is lapack[2] which contains a large number of routines to perform matrix operations. Each user may also have a collection of routines accumulated over time that are useful on a personal basis. There are also packages to make plots, to carry out algebraic manipulations, and to do the word processing for publication. All these also constitute an indispensable part of making the computer work for us.

A particular group of routines that merits special mentioning is the basic linear algebra subprograms, or BLAS, both for the functions they perform and as an example of application library. In linear algebra, as well as many other types of calculations, we often need to work with vectors and matrices. As an example, we can think of taking the scalar or dot product of two vectors (dot)

$$\alpha = \boldsymbol{x} \cdot \boldsymbol{y}$$

or adding a constant times one vector to another vector (axpy)

$$\boldsymbol{y} = \alpha \boldsymbol{x} + \boldsymbol{y}$$

where α is a scalar quantity and $\boldsymbol{x}$ and $\boldsymbol{y}$ are vector quantities. Since the operations are fairly basic it is not difficult to write a few lines of code to carry out the calculations.

One of the reasons for using routines in BLAS to carry out such basic steps lies in the potential for optimization. The work that can be taken over by BLAS are often some of the more time consuming parts in many calculations. As a result, it is useful to make them as efficient as possible. This means writing the codes in a language closer to the operations of the computer itself so that we can take advantage of the particular architecture of the machine we have in hand. For this reason, computers designed for intensive numerical work are often supplied with BLAS optimized for the particular system. As a result, it is possible to achieve performance far superior than what is possible with high-level languages alone. The usefulness of BLAS actually goes one step beyond this. Since these optimized routines are, by necessity, system dependent, it may be difficult to take programs depending on them to a computer with a different architecture. This is the question of portability of computer programs, an increasingly important issue as we are getting more and more into the situation that our codes must be used on a variety of "platforms." By standardizing BLAS and equipping each machine with its own optimized version, our code can be both efficiency and portable.

More detailed information on BLAS can be found in lapack[2] and references cited there. For computers without specially optimized BLAS library, a set of the routines in Fortran are available from netlib.[55] They are also useful for certain debugging purposes even on computers equipped with the optimized version.

Special libraries In addition to subprogram libraries that are useful for a large variety of calculations, there are also a number of application libraries written for specific tasks, such as matrix calculations, symbolic manipulations, and plotting. In general, they may be divided into two categories, public domain and commercial. Public domain software is usually available without charge and is "donated" by the authors for the benefit of users in general. With the spread of internet, it has become quite easy to obtain these codes once they are located. However, there are no depositories similar to central libraries for books for all the software available. In addition to papers published by the authors in such journals as *Computers in Physics* by the American Institute of Physics and *Computer Physics Communications* by North Holland, a good place to start is netlib.[55] Commercial software, on the other hand, is usually well advertised on the relevant journals to attract business.

In generally, the use of existing software, public domain, commercial, or one's own collection, tends to save the development time for a project. However, to make intelligent use of the these tools, the user must be knowledgeable in the general methods behind the packages so as to be able to judge if they are suitable for the job in hand. Furthermore, the onus is still on the user to test the packages thoroughly to ensure that they are used correctly and that they are right for the job.

1-4 Examples of algebraic, integer and floating number calculations

We shall give in this section an example each of algebraic, integer number, and floating number calculations to illustrate some of the points discussed earlier. By necessity, they are trivial examples. However, since we will not be directly involved with the first case from now on, it may be of interest for some readers to see an example. Although the rest of the book will be mainly concerned with floating number calculations, the other two examples here attempt to single out certain points that are important but cannot be easily emphasized later. Integer calculations are always present in any nontrivial numerical work and we shall not belittle their importance.

Elementary algebra operations In numerical calculations, the expression

$$C = A + B$$

means that the value of C is the sum of those of A and B. If either A or B is not given a numerical value, the expression becomes meaningless. In fact, a good compiler will often warn the programmer if any quantity on the right side of an equation is "undefined," that is, not associated with a numerical value. In algebraic computations, A and B are symbols whose meanings can be anything the user wishes to assign. Our interest here is to see how algebraic results may be obtained by manipulating these symbols on a computer according to some well-defined set of rules. Very often, these rules are quite simple to write down, but the work involved in applying them can be tedious. The use of a computer to carry out the operations is therefore very attractive from the point of view of time as well as accuracy (meaning not making any mistakes here).

A simple illustration is provided by the problem of proving the identity

$$\sum_{k=1}^{n} k = \frac{n(n+1)}{2}. \tag{1-2}$$

One way to achieve this is to use induction. It is obvious that the relation holds for $n = 1$. The next step in the argument is to assume that the relation is true up to $n = (m-1)$ for some $m > 0$. In other words, we take it for granted that

$$\sum_{k=1}^{m-1} k = \frac{(m-1)m}{2}.$$

For the convenience of discussion, we shall use the symbols $S_L(m-1)$ and $S_R(m-1)$ for, respectively, the left and right sides of the equation. That is,

$$S_L(m-1) \equiv \sum_{k=1}^{m-1} k \qquad S_R(m-1) \equiv \frac{(m-1)m}{2}. \tag{1-3}$$

Now let us test the relation Eq. (1-2) for $n = m$, using the fact that $S_L(m-1) = S_R(m-1)$ from our assumption. The expression for neither $S_L(m-1)$ nor $S_L(m)$ is known for arbitrary m. However, their difference is well defined and has the value

$$S_L(m) - S_L(m-1) = m.$$

The proof of the identity Eq. (1-2) is complete once we show that the difference between $S_R(m)$ and $S_R(m-1)$ is also m.

The calculations involved may be carried out using a computer. Let us define S_R by the algebraic relation

$$S_R = N * (N+1)/2 \tag{1-4}$$

where N is an algebraic symbol. If we let $N = (m-1)$, then S_R takes on the value specified by Eq. (1-3) with N replaced by $(m-1)$. This gives us the value of $S_R(m-1)$. If we reassign the value of N to be m, the value of S_R becomes that of $S_R(m)$. To prove the identity of Eq. (1-2), we need to check if it is true that

$$S_R(N=m) - S_R(N=m-1) - m = 0.$$

Since this condition is met, we can be satisfied that Eq. (1-2) holds for all $n \geq 1$, as the value of m is arbitrary here (as long as $m > 0$).

It is also possible to demonstrate the identity Eq. (1-2) using numerical calculations. For example, we can calculate the sum $S_L(m)$ numerically to some large value of n, such as 10, and we find that the result is 55. For $n = 10$, the right side of Eq. (1-2) is $S_R(5) = 10 * 11/2 = 55$. This verifies that the identity Eq. (1-2) holds for $n = 10$. We can repeat the calculation for other values of n for which we have sufficient numerical accuracy. The net result is that we can be fairly well convinced of the correctness of Eq. (1-2). On the other hand, from the point of view of logic, all we have done by numerical calculations is to show that there is no contradiction to Eq. (1-2) within certain limits. It will be difficult to convince any mathematician that the demonstration constitutes a proof of the identity.

The example clearly shows the power of algebraic calculations, a point that hardly needs to be emphasized. On the other hand, it is perhaps too simple an example to demonstrate that there are calculations complicated enough for a computer to be useful. As an exercise, the interested reader can try to prove the following identity by induction:

$$\sum_{k=1}^{n} k^7 = n^2(n+1)^2(3n^4 + 6n^3 - n^2 - 4n + 2)/24. \tag{1-5}$$

The tedium involved should convince anyone that the symbolic manipulation capability of computers is an important asset.

Prime numbers as an example of integer calculation As an example of integer calculations, let us consider a simple way to find a list of prime numbers. A prime number may be defined as an integer that is divisible only by unity and the number itself. For our purpose, we shall describe it in the following way. Let $\{p_i\}$ be a complete list of prime numbers up to some maximum value and be arranged in ascending order according to size. For the convenience of discussion, we shall assume that the list starts with 2 and the largest prime number in the list is $p_{\max}$. It is possible to find out whether an integer $n \leq p_{\max}$ is a prime number by dividing n with each member in the list. What do we mean by the statement that an integer n is divisible by another integer m? Here, we shall take it to be that the ratio n/m is also an integer.

Before we design an algorithm to search for prime numbers, let us examine the question of how integer calculations are carried out on a computer. In ordinary calculations, when an integer n is divided by another integer m, the result is an integer if m is divisible by n; otherwise the result is a real number. For example, $6/2 = 3$, the result is an integer, whereas $5/2 = 2.5$, the result is a real number. Since integers and real numbers are not represented in the same way inside a computer, the rules to handle these two types of calculation are somewhat different. In integer mode, the result remains an integer. Thus $5/2 = 2$, with the fractional part truncated. By the same token, $7/8 = 0$. If one wants the result 2.5 from dividing 5 by 2, it is necessary to carry out the calculation in terms of floating numbers. That is, all the numbers involved are transformed first into their equivalent floating number representations, with the 5 in the numerator changed to 5.0 and the 2 in the denominator to 2.0. The need for these additional steps can be a nuisance at times. On the other hand, we can also turn it into an advantage, as we shall see below.

Because of the way computers carry out integer calculations, we can decide whether an integer is divisible by another integer in the following way. Let us assume that, in integer mode, the result of n divided by m is an integer k,

$$\frac{n}{m} \longrightarrow k.$$

If n is divisible by m, then we have the situation that

$$k * m = n.$$

If n is not divisible by m, a part of the quotient is lost due to truncation and

$$k * m < n.$$

We shall make use of this property to construct a simple algorithm to generate a list of prime numbers.

Let us assume that we have already found the first n_p prime numbers, p_1, p_2, $p_3, \ldots, p_{\max}$, arranged in ascending order according to size. To find the next prime number, we start with an integer $N = p_{\max} + 1$, where $p_{\max}$ is the largest prime number known so far. We can test whether N is prime by dividing it with $p_1, p_2, \ldots$. If N is divisible by any of the existing prime numbers, it is not a prime number. We shall therefore repeat the process with the next integer obtained by increasing again the value of N by 1. This process goes on until we reach a value of N that is not divisible by any of the existing prime numbers. Such a number is therefore a prime

Box 1-1 Program DM_PRIME
Generate the first K prime numbers

Initialization:
- (a) Set K as the total number of prime numbers required.
- (b) Set N_BEG as the number of prime numbers to start with.
- (c) Input the first N_BEG prime numbers explicitly.

1. Store the last prime number in the list as NEW.
2. Construct a new integer by adding 2 to NEW.
3. Test if NEW is prime by dividing it by all the existing prime numbers $\leq$ NEW/2:
   - (a) If divisible, try a new integer by adding 2 to NEW and repeat the test.
   - (b) If not divisible, it is a new prime number.
     - (a) Add the new one to the list.
     - (b) Increase the number of members by one.
4. Repeat steps 2 and 3 to find the next prime number in the list.
5. Stop if the total number is K.

number and may be added to our list as the new member with the largest value. Except for a few refinements to be described next, this is the algorithm outlined in Box 1-1.

A few improvements may be incorporated into the method to make it more efficient. First, we recognize that 2 is a prime number. As a result, all the other even numbers are not prime numbers. For this reason, when we look for the next integer to be tested, we can increase N by 2 each time, rather than 1 as given in the previous paragraph. In this way, the speed of the search is increased by a factor of 2. By the same argument, there is no need to include 2 in the list of prime numbers for our tests, as we have already excluded all the even numbers from our considerations. A second improvement is that we do not need to go through the entire list of existing prime numbers in our test. Obviously, if N is not divisible by any member up to $p_i > N/2$, it will not be divisible by any of the prime numbers in the list that are larger than p_i.

To start off the calculation, we need to have at least one prime number in the list. Since 1 is not suitable and 2 is not needed, our list starts with 3. With such a list, the rest of the prime numbers can be generated.

Our method is not optimal for producing large prime numbers. As the list grows, it takes longer and longer to find the next prime number. In addition, the limitation imposed by the largest integer we can store in a computer word also poses a problem for the method. Our calculation must therefore terminate at some large values of $p_{\max}$, much smaller than the sizes of interest to mathematicians. For such large numbers, methods of testing for prime numbers far superior than the simple one presented here must be used (see, for example, "The Search for Prime Numbers" by Pomerance[42]).

The value of π as an example of floating point calculation As an example of floating number calculations, let us evaluate π. There are several ways to do this. The first is to realize that $\sin\pi/2 = 1$. As a result, we can make use of the inverse relation

$$\sin^{-1} 1 = \frac{\pi}{2}.$$

(The more direct route of using the inverse of $\sin\pi = 0$ does not work if we use the arc sine function in the intrinsic library, as it normally returns values only in the range $[-\frac{\pi}{2}, +\frac{\pi}{2}]$.) In general, the use of trigonometry functions is a good way, for example, to set the value of π in a program to the same precision as the rest of the calculations we wish to carry out on the computer. However, the method is not of interest here, as it depends on the way the arc sine function is evaluated on the computer.

As an illustration also of several important aspects in programming a computer, we shall calculate the value of π using a more elementary approach. Here we encounter the first consideration in solving problems on a computer: selecting a suitable method. For the value of π, we shall make use of series expansion. There are, in fact, quite a few such series from which we may choose. For example, we can use one of the following two expansions:

$$\frac{\pi}{2\sqrt{2}} = 1 + \frac{1}{3} - \frac{1}{5} - \frac{1}{7} + \frac{1}{9} + \frac{1}{11} - \cdots \tag{1-6}$$

$$\frac{\pi}{4} = 1 - \frac{1}{3} + \frac{1}{5} - \frac{1}{7} + \frac{1}{9} - \frac{1}{11} + \cdots = \sum_{n=1}^{\infty} \frac{(-1)^n}{2n+1}. \tag{1-7}$$

We shall reject Eq. (1-6) in favor of (1-7), since the former involves the factor $\sqrt{2}$. As a result, the accuracy one can obtain for the value of π depends also on how well we know $\sqrt{2}$.

It is fairly straightforward to program Eq. (1-7). All we need to do is a simple loop that alternates between adding to and subtracting from the sum of the inverse of the next odd integer. On the other hand, if we examine the number of terms required to achieve a given accuracy in this approach, we will be surprised. For example, if we wish to achieve an accuracy of 10^{-5}, the last term to be included in the summation must be much smaller than 10^{-5}. This means something on the order of 250,000 terms! This is an example that a simple, innocent looking expression can be time consuming to evaluate. It is worthwhile to examine our expression a little closer to see if some analytical work can save us a lot of computer time. For our simple example here, we can see that, if we take the difference between two adjacent terms, the relation given by Eq. (1-7) reduces to the form

$$\frac{\pi}{4} = 1 - 2\Big(\frac{1}{3\times 5} + \frac{1}{7\times 9} + \frac{1}{11\times 13} + \cdots\Big). \tag{1-8}$$

As an analytical expression, this is identical to Eq. (1-7). However, as a way to calculate π on a computer, there is a major difference.

The steps to carry out the calculation using Eq. (1-8) on a computer are shown in the form of a flow chart in Fig. 1-1. The program is only slightly more involved than

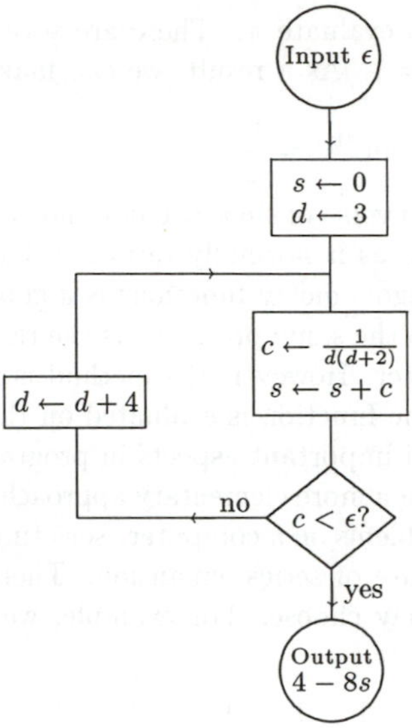

Figure 1-1: A flow chart for calculating the value of π using Eq. (1-8).

the one using Eq. (1-7) directly. However, the number of terms required to achieve the same accuracy of 10^{-5} is now decreased to 62,500, a factor of 4 reduction. Note that the saving is more than the factor of 2 expected from contracting two terms to one in going from Eq. (1-7) to (1-8).

As far as the numerical calculation is concerned, Eq. (1-7) represents an example of a poor approach, as successive terms in the sum alternate in sign. For this reason, the calculation approaches the asymptotic value of $\pi/4$ through a zigzag path, as shown by the dashed curve in Fig. 1-2. In contrast, the summation on the right side of Eq. (1-8) approaches the limit smoothly, as shown by the solid curve. In general, it is desirable to avoid the kind of "fluctuations" in the intermediate results given by Eq. (1-7). For more realistic cases, the improvement in computational speed by a slight change in the algorithm may not be as obvious as in our example, but the savings can often be much greater.

What is the best accuracy we can achieve in calculating the value of π by the method illustrated? In single precision, we have seen earlier that the ultimate limitation is around seven significant figures, because of the fact that only 23 bits are assigned to store the mantissa. In practice, it is difficult to achieve this degree of accuracy in a lengthy calculation due to truncation errors. It is more likely that, in reality, only five significant figures can be obtained. As a result, there is very little value in going much beyond the 62,500 terms we have used earlier for Eq. (1-8). For a more detailed discussion on the calculation of π see, for example, the article on

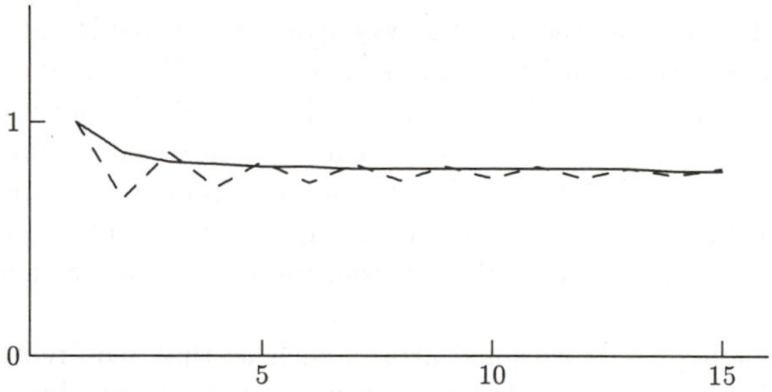

Figure 1-2: Convergence of two different series to the value of $\pi/4$. The dashed curve is obtained with Eq. (1-7) and the solid curve with Eq. (1-8).

"Ramamujan and Pi" in *Scientific American* by Borwein and Borwein.[5]

We have mentioned earlier that computational efficiency is an important consideration; however, simplicity in the algorithm is also an essential feature. Often, these two considerations may be in conflict with each other. As a result, the choice of an algorithm may need some balance between these two requirements. The difference between the two approaches in calculating the value of π is an elementary example of the type of problems often encountered in computational work.

1-5 Examples of unconventional techniques

We shall end this introductory chapter with a few examples of applying techniques not commonly associated with solving numerical problems on computers. This is done in part to illustrate the fact there are many ways to make the computer work for us other than using conventional methods. We shall only attempt three cases here, namely: squaring a matrix with the help of sorting, the use of spread sheet to obtain the trajectory of a projectile, and calculation of binomial coefficients using a table of the logarithm of factorials,.

Square of a matrix with the help of sorting The diagonal elements of the square of a matrix $\boldsymbol{H}$ is given by

$$S_{ii} = \sum_{j=1}^{N} H_{ij} H_{ji} \tag{1-9}$$

where N is the dimension of the matrix and H_{ij} is the matrix element of $\boldsymbol{H}$ in row i and column j. It is a simple thing to program Eq. (1-9) if all $N \times N$ elements of $\boldsymbol{H}$ are stored in the computer memory. However, the calculation can be quite time consuming and the amount of memory required can be extensive for large N.

Let us address first the reason for the slowness in carrying out such a simple calculation. This can be traced to the way arrays are stored in the computer memory. Conceptually one can regard an array as an ordered collection of elements. For a one-dimensional array, say A, of dimension N, it is a list of N quantities, $A(1)$, $A(2)$, $\cdots$, $A(N)$. In the computer memory, this list may be considered as stored in N contiguous locations. Element n is found at the nth location from the starting point of array A. If we wish to carry out a calculation involving A, for example, to find the sum of the N elements, we can take each element from the list, one after another, and add it to the the sum.

For two or higher dimensional arrays, such as that required for $\boldsymbol{H}$ in Eq. (1-9) above, the situation is somewhat more complicated. A two-dimensional array B, in general, may be regarded as a rectangular area of memory locations, each of which is given by two indices i and j. If the dimension of B is $N \times M$, index i goes from 1 to N and index j from 1 to M. Unfortunately, computer memories are usually arranged as linear arrays. That is, the address of each memory location is given in terms of its location relative to some starting point, for example, the first element of our array. As a result, instead of a rectangular area, the actual storage of our array B is a linear array of length $L = N \times M$.

There are two logical ways to convert an array of two or higher dimensions to a linear one, column-major or row-major. In the former case, the first index varies most rapidly, the second index, next most rapidly, and so on for arrays of dimension higher than two. Fortran uses this method. In the case of row-major method, the opposite order is used, with the last index varying most rapidly. The C language is one that adopts this approach. In either case, to find the location of an element in, for example, our two-dimensional array B, requires an (integer) multiplication followed by an (integer) addition. That is, in the case of column-major method, the element B_{ij} is located at

$$\ell = (j - 1) * N + i$$

with respect to the starting point. One may speed up the index calculation somewhat by optimizing techniques, such as using shift operation in the place of multiplication. However, it does not change the basic fact that, for our example of the square of a matrix in Eq. (1-9), the number of the actual operations is dominated by indexing, rather than the multiplication of two floating numbers and the addition of one given explicitly in the equation.

We can avoid the overhead of calculating the location for arrays with two or higher dimensions by changing our algorithm such that we work with one-dimensional arrays and take care of the indexing explicitly in our code in some efficient way. For example, instead of using a $N \times M$ two-dimensional array for $\boldsymbol{B}$, we allocate a one-dimensional array B consisting of $L = N \times M$ storage locations and keeping in mind that the first N ($\ell = 1, 2, \cdots, N$) corresponds to B_{1i} for $i = 1$ to N, the second N ($\ell = N + 1, N + 2, \cdots, 2N$) for B_{2i}, and so on. Any calculation that requires $\boldsymbol{B}$ is arranged, if possible, to start from the beginning of a row and go to the end of the same row before proceeding to the next one.

However, this technique is not quite adequate to solve our problem for the square of a matrix. As we can see from Eq. (1-9), the product required is H_{ij} with H_{ji}. If the matrix $\boldsymbol{H}$ is stored in a row-major manner, as we have done implicitly for B in the previous paragraph, it is good for H_{ij} but very inconvenient for locating the necessary corresponding element H_{ji}, and vice versa. To get around this difficulty, we can store a second one-dimensional array for $\boldsymbol{H}$ except that now we arrange it in column-major order. Conceptually, we can think of Eq. (1-9) as

$$S_{ii} = \sum_{j=1}^{N} H_{ij}\tilde{H}_{ij}$$

where $\tilde{\boldsymbol{H}}$ is the transpose of $\boldsymbol{H}$. That is,

$$\tilde{H}_{ij} = H_{ji}. \tag{1-10}$$

If both $\boldsymbol{H}$ and $\tilde{\boldsymbol{H}}$ are stored as two one-dimensional arrays in column-major order, the calculations for $\boldsymbol{S}$ can proceed by taking the products between the first N pairs of $\boldsymbol{H}$ and $\tilde{\boldsymbol{H}}$ and sum them to form S_{11}, the second N pairs to form S_{22} and so on. In this way, we have avoided calculations to convert two-dimensional indices into one-dimensional ones.

A closer examination will reveal that the new method may not save us any computer time either, as it takes time to construct the array for $\tilde{\boldsymbol{H}}$. Furthermore, the amount of memory required is increased by the fact we need to store two arrays. We shall first find a better way to construct $\tilde{\boldsymbol{H}}$ and then return to address the question of memory requirement.

In general, if we want to carry out the transformation given in Eq. (1-10) it will take just as many indexing operations as to carry out the summation of the products given originally in Eq. (1-10). An alternative is to use sorting techniques. Since sorting is such an important part of data processing, very efficient algorithm and system libraries are available. For the ease of explanation, we shall cast the discussions in terms of external files. In other words, as each element of matrix $\boldsymbol{H}$ is generated, we write to a file the value of the element H_{ij} as well as the two indices i and j. Each record in the file consists of three quantities, i, j, and H_{ij}, two integers and one floating number. At the end of the calculation for the elements of the matrix $\boldsymbol{H}$, we can use a sorting program that usually accompanies the operating system of a computer to put the elements into the order needed by sorting on the indices. For example, if the elements are generated in column-major order, the sorting can produce a new file with elements in row-major order.

The end result is that we have two files, one in column-major order and the other in row-major order, as shown schematically in Fig. 1-3. It is now a simple matter to read one element from each file, take their products and sum the first N of them to form S_{11}, the second N to form S_{22}, and so on. In fact, in this way, we do not even need the array `H` in memory, as no elements of matrix $\boldsymbol{H}$ have to be stored. This is

1	1	H_{11}
1	2	H_{12}
⋮	⋮	⋮
1	N	H_{1N}
2	1	H_{21}
⋮	⋮	⋮
2	N	H_{2N}
⋮	⋮	⋮
N	N	H_{NN}

1	1	H_{11}
2	1	H_{21}
⋮	⋮	⋮
N	1	H_{N1}
1	2	H_{12}
⋮	⋮	⋮
N	2	H_{N2}
⋮	⋮	⋮
N	N	H_{NN}

Figure 1-3: Two files of the elements of matrix $\boldsymbol{H}$, one in column-major order and the other in row-major order.

especially advantageous if we wish to carry out calculations larger than the physical memory of the computer can handle.

In principle, we are trading between central processor unit (cpu) and input-output (i/o) time in the above approach. Where such is not desirable, it is also possible not to write the files. In this case, we can store the elements together with their corresponding indices in arrays and call appropriate system subroutines to carry out the sorting. Depending on the size of the matrix and the computer used, this approach may be more efficient in some cases.

Needless to say, the above approach of using sorting techniques can be extended to calculations other than taking the diagonal elements of the square of a matrix. In fact, the two basic points of this example, namely reduction of the number of dimensions of an array where possible and use of data-processing and other techniques not usually associated with mathematical calculations, can be important in many cases. For modern computers optimized to handle floating point calculations, indexing can often be a bottle neck in getting the best performance from the system. Considerations of the type discussed above can make a difference in many problems.

Projectile motion using a spread sheet program We shall illustrate the use of spread-sheet programs, such as Excel (Microsoft Corporation) and Lotus 1-2-3 (Lotus Development Corp.), to carry out calculations of interest. The example selected is the motion of a projectile leaving the surface of the earth with initial velocity v_0

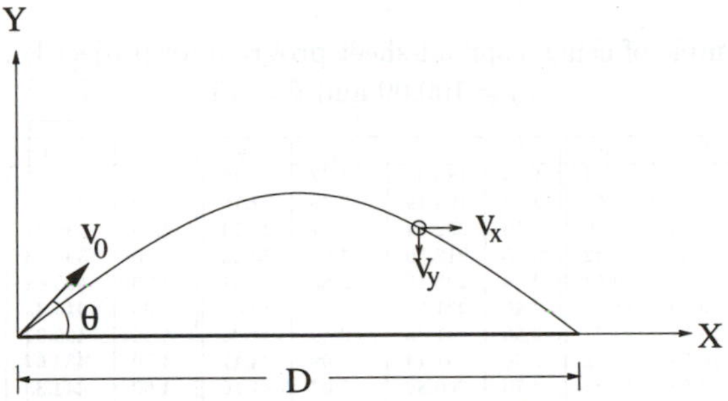

Figure 1-4: Trajectory of a projectile with initial velocity $\boldsymbol{v}_0$ and angle of inclination θ.

and angle of elevation θ. In the absence of forces other than gravity, the horizontal velocity is constant

$$v_x = v_0 \cos\theta$$

and the vertical velocity changes in time because of acceleration due to gravity

$$v_y(t) = v_0 \sin\theta - gt$$

where $g = 9.80$ m/s^2. Among the questions we can ask in this problem, let us focus on the quantity D, the horizontal distance from the origin where the projectile lands. For our purpose, the question may be stated in the following way. Starting from $x = y = 0$, what is the value of x when $y = 0$ again? For the moment, we shall ignore the analytical solution except coming back at the end to use it as a check against our calculations.

Our spread-sheet appears something like that shown in Table 1-3. In the first column we have time t starting from 0 and increasing by Δt as we move down a column. In other words, we divide time into many equal-size slices, each one of duration Δt. The value of time (at the beginning) of interval i is

$$t(i) = t(i-1) + \Delta t$$

starting with $t(1) = 0.0$. The horizontal distance x traveled in Δt is

$$\Delta x = v_x \Delta t.$$

The value of x at the beginning of time slice i is therefore

$$x(i) = x(i-1) + \Delta x$$

and this is given in column 2. In column 3, we list the corresponding value of v_y, as it is varying with time.

Table 1-3: Example of using a spread-sheet program for projectile motion with $v_0 = 100.00$ and $\theta = 15°$.

t	x	$\overline{v}_y$	y	t	x	$\overline{v}_y$	y	t	x	$\overline{v}_y$	y
0.00	0.00	25.39	0.00	1.90	183.53	6.77	31.49	3.80	367.05	-11.85	27.60
0.10	9.66	24.41	2.54	2.00	193.19	5.79	32.16	3.90	376.71	-12.83	26.41
0.20	19.32	23.43	4.98	2.10	202.84	4.81	32.74	4.00	386.37	-13.81	25.13
0.30	28.98	22.45	7.32	2.20	212.50	3.83	33.22	4.10	396.03	-14.79	23.75
0.40	38.64	21.47	9.57	2.30	222.16	2.85	33.61	4.20	405.69	-15.77	22.27
0.50	48.30	20.49	11.72	2.40	231.82	1.87	33.89	4.30	415.35	-16.75	20.69
0.60	57.96	19.51	13.77	2.50	241.48	0.89	34.08	4.40	425.01	-17.73	19.02
0.70	67.61	18.53	15.72	2.60	251.14	-0.09	34.17	4.50	434.67	-18.71	17.15
0.80	77.27	17.55	17.57	2.70	260.80	-1.07	34.16	4.60	444.33	-19.69	15.18
0.90	86.93	16.57	19.32	2.80	270.46	-2.05	34.05	4.70	453.99	-20.67	13.11
1.00	96.59	15.59	20.98	2.90	280.12	-3.03	33.85	4.80	463.64	-21.65	10.95
1.10	106.25	14.61	22.54	3.00	289.78	-4.01	33.55	4.90	473.30	-22.63	8.68
1.20	115.91	13.63	24.00	3.10	299.44	-4.99	33.14	5.00	482.96	-23.61	6.32
1.30	125.57	12.65	25.37	3.20	309.10	-5.97	32.65	5.10	492.62	-24.59	3.86
1.40	135.23	11.67	26.63	3.30	318.76	-6.95	32.05	**5.20**	**502.28**	**-25.57**	**1.31**
1.50	144.89	10.69	27.80	3.40	328.41	-7.93	31.35	**5.30**	**511.94**	**-26.55**	**-1.35**
1.60	154.55	9.71	28.87	3.50	338.07	-8.91	30.56	5.40	521.60	-27.53	-4.10
1.70	164.21	8.73	29.84	3.60	347.73	-9.89	29.67	5.50	531.26	-28.51	-6.95
1.80	173.87	7.75	30.71	3.70	357.39	-10.87	28.68	5.60	540.92	-29.49	-9.90

Check: $t = 2v_{0y}/g = 5.28$ s $\qquad x = v_{0x}t = 510.20$ m

For the convenience of calculating y, our primary interest here, we shall find the average value of v_y in time interval Δt,

$$\overline{v}_y(i) = \frac{1}{\Delta t}\int_{t(i)}^{(i+1)} v_y\,dt = \begin{cases} v_0\sin\theta - \frac{1}{2}g\Delta t & i = 1 \\ v_y(i-1) + gt & i > 1. \end{cases}$$

The vertical distance $y(i)$ traveled in time interval i may be approximated as

$$\Delta y(i) = \overline{v}_y(i)\Delta t$$

and hence the value of $y(i)$ given in column 4 is

$$y(i) = y(i-1) + \Delta y(i).$$

The value of D, the horizontal distance traveled when $y = 0$ again, can be obtained from the table as the value of $x(i)$ in the row where y changes sign. The advantage of using a spread sheet program is that all the "instructions" to the computer can be entered in a simple way and the results obtained without actually doing any programming.

As a check, we can use the fact that the vertical distance traveled is given by

$$y = \int v_y\,dt = v_0\sin\theta t - \frac{1}{2}gt^2.$$

By solving for the value of t when $y = 0$ (for $t > 0$), we obtain the time of travel as

$$t = \frac{2v_0 \sin\theta}{g}.$$

From this we obtain the result

$$D = (v_0 \cos\theta)t = \frac{v_0^2}{g} \sin 2\theta.$$

The numerical value for $v_0 = 100$ m and $\theta = 15°$ is given at the end of Table 1-3.

The accuracy of the method is governed by the size of Δt. For the value of $\Delta t = 0.1$ s used in the example, we can only expect an accuracy of $v_0 \sin\theta\,\Delta t \sim 10$ m and this is borne out of our result of $D = 502$ to 512 m, compared to 510 m in the analytical result. We can reduce the "uncertainty" of 10 m by, for example, changing Δt from 0.1s to 0.05s (and doubling the number of rows).

In general, spread sheet is a convent way to solve the problem where only simple calculations are involved. It is especially useful in dealing with input that are not given by a function, such as those obtained from measurements. It can often save the trouble of writing a computer program and going through repeatedly the procedure of compiling, testing, and correcting it.

The method of using logarithms For many calculations, it is more convenient to use a floating number representation even though the final values are integers. This occurs, for example, when the results are to be used in another part of the calculation that is in floating number representation. In addition, a floating number representation becomes the only choice if the values involved are too large to be stored as integers in the computer. Some loss in accuracy is unavoidable, but this may not be a serious problem if the other parts of the calculation are carried out in terms of floating numbers anyway.

As an example, we shall consider binomial coefficient

$$\binom{n}{m} = \frac{n!}{m!(n-m)!}. \tag{1-11}$$

On the surface, this is a very easy calculation. For example, we can evaluate each one of the three factorials involved using the recursion relation

$$n! = n \times (n-1)!.$$

The process stops at $n = 1$ [and take $0! = 1$, as we shall find later in Eq. (4-108)]. We have seen earlier that this method works only for small values of n and m, as a four-byte integer can only store values of $n!$ for $n \leq 12$. An eight-byte word extends the range somewhat but does not solve the fundamental question of the limited range of values that can be stored.

Box 1-2 Program DM_BNML
Binomial coefficients using a table of $\ln(k!)$

1. Input a table of the logarithms of factorials using Eq. (1-13).
2. Input n and m.
3. Calculate the logarithm of $\binom{n}{m}$ using Eq. (1-14).
4. Take the exponential of the logarithm of $\binom{n}{m}$ and return the result.

One way to get around the problem is to evaluate all the factorials involved as floating numbers in terms of their logarithms. On taking the logarithm of both sides of Eq. (1-11) we obtain the result

$$\ln\binom{n}{m} = \ln(n!) - \ln(m!) - \ln(\{n-m\}!). \tag{1-12}$$

Unfortunately, logarithm is a relatively slow function to evaluate. To speed up the calculation, we can store, once and for all, a table of the logarithms of $k!$ for $k = 0$, 1, 2, ..., N. This can be done fairly efficiently using the recursion relation

$$\ln(k!) = \ln k + \ln(\{k-1\}!) \tag{1-13}$$

starting with

$$\ln(0!) = 0.$$

If we have $\log(k!)$, up to some large value of k, in an array $L_G(k)$, the calculation of Eq. (1-12) reduces to

$$\ln\binom{n}{m} = L_G(n) - L_G(m) - L_G(n-m) \equiv c. \tag{1-14}$$

To find the value of $\binom{n}{m}$, we need to take the inverse of the logarithm,

$$\binom{n}{m} = \exp c.$$

In this approach, the main part of the computer time in calculating a binomial coefficient is to evaluate the exponential function at the end. The savings come from having an existing table of the logarithms of the factorials. Because of this, the only other calculations involved are two subtractions, as can be seen in Eq. (1-14). The algorithm is outlined in Box 1-2. The advantages of such an approach over that of a simple floating number calculation for Eq. (1-11) are left as an exercise (see Problem 1-14).

Problems

1-1 If a is any one of the five integers 3, 4, 5, 6, and 7, and b is any one of the three integers 2, 3, and 4, what are the 15 values of $c = a/b$ if the calculation is carried out in integer mode on a computer?

1-2 If a and b are two real numbers, design an algorithm to calculate $c = a*b$ so that the product c is an integer and has the value equal to the nearest integer to the product $a * b$. The purpose of this function is equivalent to that of NINT($a * b$) in the intrinsic function library of a Fortran compiler.

1-3 Design an algorithm using the function SIN in the Fortran intrinsic function library to calculate $\sin x$ for x from 0° to 90° at every 10°. Note that the SIN function takes arguments in radians.

1-4 Describe a way to calculate the square root of a complex number using a function that can only take the square root of real numbers.

1-5 On some machines, each computer word is made of 36 binary bits. Find the range of integers it can store. Design a way to store floating numbers and state the range of values that can be represented.

1-6 Imagine a simple computer which has 64 storage locations and one register (a special storage area), and can perform only the following 11 operations: input, output, addition, subtraction, multiplication, division, store, if-zero, if-positive, go-to, and stop. Assuming that all the operations are for real numbers, design an "assembly" language (i.e., a language that gives symbolic representations to all the operations so that a program written in the language can be understood relatively easily by a person) for this machine and write a program with it to calculate the values of y for $y = x^2$ for a set of ten equally spaced values of x in the range $x = 0$ to 1.

1-7 Give two different ways to find on a computer the roots of the quadratic equation $ax^2 + bx + c = 0$ for a set of input values of the three coefficients a, b, and c.

1-8 Design an algorithm for carrying out an indefinite integral

$$I = \int P_n(x)\,dx$$

where the integrand is a power series of the form

$$P_n(x) = a_0 + a_1 x + a_2 x^2 + \cdots + a_n x^n$$

Write a computer program to carry out this algebraic calculation.

1-9 Use a computer algebra software package to prove the identity of Eq. (1-5) by induction.

1-10 Calculate the algebraic result of a determinant of order $n = 4$ using minors according to Eq. (5-7).

1-11 Construct a function that returns the binomial coefficient $\binom{n}{m}$ of Eq. (1-11) using a table of $\ell n(k!)$ as outlined in Box 1-2.

1-12 In the ancient Roman system of notation, the letter I is used to stand for numerical value 1, V for 5, X for 10, L for 50, C for 100, D for 500, and M for 1000. If a letter is followed by one of equal or lesser value, the two values are added. On the other hand, if one of the above letters is followed by one of greater value, it is subtracted from the one following. Thus MCMLXXIV means 1974. (*a*) Design an algorithm to convert Roman numerals to integers and vice versa. (*b*) Think of a way to add two integers expressed in Roman numerals without first converting them to their corresponding numerical values.

1-13 Very often we are in need of two real numbers a and b subject to the condition $a^2 + b^2 = 1$. As a result, there is only one degree of freedom among them. If both a and b are positive (and less than 1), there are two ways to take this degree of freedom. The first is to define $a = \sin\theta$ and $b = \cos\theta$ and consider θ as the independent quantity. The second is to define $b = (1 - a^2)^{1/2}$ and consider a as the independent quantity. Use a random number generator to generate a set of values of θ in the first case and a in the second. Check the relative efficiency of the two methods to obtain a and b by comparing the amount of time to calculate, for example, 5000 different sets of values in each case.

1-14 Instead of using logarithm, we can also evaluate the binomial coefficient of Eq. (1-11) for large n and m by calculating the factorials involved in terms of floating numbers. What are the advantages and disadvantages of the present method compared with that of Box 1-2?

Chapter 2

Integration and Differentiation

Numerical integration has been an active field in mathematics even before the introduction of computers. This comes from the simple fact that not all integrations can be carried out analytically, and numerical methods, however tedious without the help of a computer, become the only way to solve the problem. Furthermore, the results of many common integrals, such as those for the error function erf(x) and gamma function $\Gamma(x)$, do not exist as analytical functions and their values must be found numerically. This is quite different from differentiation. For most of the functions encountered in physics and engineering, differentiations can usually be carried out by analytical means. However, there are exceptions. On many occasions, we encounter functions that are given numerically. In this case, the functional form is not available and the derivatives can only be obtained using numerical methods. Furthermore, numerical differentiation is a good way to be introduced to the wide world of finite difference methods for solving a large variety of problems, including differential equations.

2-1 Numerical integration

One of the more common forms of integration may be represented by

$$I_{[a,b]} = \int_a^b f(x)\,dx. \tag{2-1}$$

This is a standard one-dimensional definite integral with both upper and lower limits of the integration specified. For simplicity, we can assume that the integrand $f(x)$ is greater than or equal to zero everywhere in the interval $x = [a, b]$. Under such conditions, the integral $I_{[a,b]}$ may be interpreted as the area bound above by $f(x)$, below by the x-axis, on the left by $x = a$, and on the right by $x = b$, as illustrated schematically in Fig. 2-1 by the dotted area. This form of the integral is known in mathematics as the (definite) Riemann integral and is the only form with which we shall be dealing here. Certain types of indefinite integrals can also be evaluated using computers. The subject belongs to symbolic manipulation, or computer algebra, and we shall not be concerned with it here.

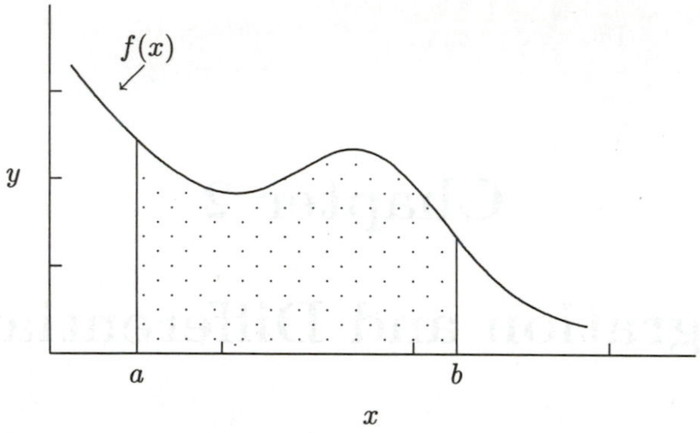

Figure 2-1: Schematic illustration of a definite integral $\int_a^b f(x)\,dx$ as the area between $f(x)$ and the x-axis in the region between $x = a$ and $x = b$.

The use of numerical integration may be illustrated using as an example the distance traveled by a particle moving along the x-direction with velocity v. If v is constant, the distance d covered in the interval between times $t = a$ and b is simply given by

$$d = v \times (b - a).$$

On the other hand, if the velocity varies as a function of time, it is given, instead, by the integral

$$d = \int_a^b v(t)\,dt. \tag{2-2}$$

If $v(t)$ can be expressed in terms of some analytical function, the integral can usually be carried out without resorting to numerical methods. However, there are many occasions where $v(t)$ is available only as a table of data or in a form that cannot be integrated analytically.

As another example, think of a car traveling on a highway. If the only way we can find the velocity is to sample it once a minute, what is the distance traveled in, say, 10 minutes? Let us assume that the sample results are those listed in column 2 of Table 2-1. The concept that the distance traveled is the integral of velocity over time is still correct here except that, since $v(t)$ is not a known function, the integral of Eq. (2-2) cannot be carried out analytically. However, we can get a very good "estimate" of the distance using the following approach. Since we do not know the exact velocity at all times, we shall use the sampled result as the average velocity $\bar{v}_t$ for that minute. The distance traveled in the minute is then $\bar{v}_t \times 60$. This is given in column 3 of the table. By summing all ten results in the column, we obtain the distance traveled in 10 minutes. This is basically the spirit of numerical integration. Instead of a knowledge of the integrand for all values of the variable of integration, we

Table 2-1: Distance traveled in ten minutes.

Time (minutes)	Velocity (m/s)	Distance in 1 minute (m)	Total distance traveled (m)
1	26	1,560	1,560
2	27	1,620	3,180
3	25	1,500	4,680
4	24	1,440	6,120
5	23	1,380	7,500
6	22	1,320	8,820
7	24	1,440	10,260
8	26	1,560	11,820
9	25	1,500	13,320
10	24	1,440	14,760
Total distance traveled in 10 minutes =			14,760 m

divide the range into a number of subintervals. For each subinterval, the value of the integrand is approximated as a constant or by some extremely simple function, such as a linear one. The value for the entire interval is then the sum of the contributions from each part, the same as we have done for our example of the distance traveled by the car.

Evaluation of an integral by numerical methods is often called numerical quadrature for reasons that will become obvious in §2-4. Although most methods follow the basic spirit outlined in the previous paragraph, accuracy and efficiency considerations dictate many modifications.

2-2 Rectangular and trapezoidal rules

As a simple application of the basic idea of numerical integration for Eq. (2-1), we shall divide the interval between $x = a$ and $x = b$ into a number of smaller ones, with the ith subinterval starting at $x = x_{i-1}$ and ending at $x = x_i$. For simplicity, we shall begin by taking all the subintervals to be of equal size,

$$h = x_i - x_{i-1}$$

and there are N such subintervals with

$$Nh = b - a.$$

The area underneath the curve we wish to calculate is now divided into N narrow strips, each of width h. If h is sufficiently small, we may approximate each slice of the area by some simple shape. For the rectangular rule, the shape is taken to be a rectangle with the height given by some reasonable average of $f(x)$ in the subinterval.

For the trapezoidal rule, the shape is a trapezoid with the height $f(x_{i-1})$ at $x = x_{i-1}$ and $f(x_i)$ at $x = x_i$. In the limit $h \to 0$, the two rules become equivalent to each other and the value of the integral is exact.

Rectangular rule To apply the rectangular rule, we need the average of $f(x)$ in each subinterval. This is given by

$$\overline{f}_i = \frac{1}{h}\int_{x_{i-1}}^{x_i} f(x)\,dx.$$

If the exact values of $\overline{f}_i$ are available for all N subintervals, we have the result

$$I_{[a,b]} = \int_a^b f(x)\,dx = h\sum_{i=1}^{N} \overline{f}_i.$$

However, this is impractical. The purpose of the rectangular rule is to approximate $\overline{f}_i$ so that the integration can be carried out efficiently with a minimum amount of calculation.

For a slowly varying function, a good estimate of its average in a small subinterval is given by its value in the middle of the range,

$$\overline{f}_i \approx f(x_{i-1/2})$$

where $x_{i-1/2} \equiv \frac{1}{2}(x_{i-1} + x_i)$. To simplify the notation, we shall use

$$f_i \equiv f(x_i) \tag{2-3}$$

on those occasions where there is no ambiguity. That is, f_i is the value of $f(x)$ at $x = x_i$. The integral of $f(x)$ in the interval $[a, b]$ is then

$$I_{[a,b]} = h\sum_{i=1}^{N} f_{i-1/2}. \tag{2-4}$$

Note that, since we have taken $x_0 = a$, the estimate of $f(x)$ in the first interval $[a, a+h]$ is denoted as $f_{1/2} \equiv f(x_{1/2})$.

The method of numerical integration given by Eq. (2-4) is known as the rectangular rule. It is easy to show that the error in this method due to the finite step size decreases as N, the number of subintervals, is increased. Consider the contribution of the subinterval $[x_{i-1}, x_i]$

$$I_{[x_{i-1},x_i]} = \int_{x_{i-1}}^{x_i} f(x)\,dx \approx hf_{i-1/2}. \tag{2-5}$$

We can check the accuracy of this approximation by expanding the function $f(x)$ in terms of the Taylor series around the middle point $x_{i-1/2}$,

$$\begin{aligned} f(x) = f_{i-1/2} &+ \frac{1}{1!}f'_{i-1/2}(x - x_{i-1/2}) + \frac{1}{2!}f''_{i-1/2}(x - x_{i-1/2})^2 \\ &+ \frac{1}{3!}f^{(3)}_{i-1/2}(x - x_{i-1/2})^3 + \cdots \end{aligned} \tag{2-6}$$

where we have used $f'_{i-1/2}$, $f''_{i-1/2}$, and $f^{(3)}_{i-1/2}$ to denote, respectively, the first, second, and third derivatives of $f(x)$ evaluated at $x = x_{i-1/2}$.

Using Eq. (2-6), the value of the integral in the subinterval may be expressed as

$$\begin{aligned}\int_{x_{i-1}}^{x_i} f(x)\,dx = f_{i-1/2}\int_{x_{i-1}}^{x_i} dx &+ \frac{1}{1!}f'_{i-1/2}\int_{x_{i-1}}^{x_i}(x-x_{i-1/2})\,dx \\ &+ \frac{1}{2!}f''_{i-1/2}\int_{x_{i-1}}^{x_i}(x-x_{i-1/2})^2\,dx \\ &+ \frac{1}{3!}f^{(3)}_{i-1/2}\int_{x_{i-1}}^{x_i}(x-x_{i-1/2})^3\,dx \\ &+ \cdots. \end{aligned} \tag{2-7}$$

The first term is the approximation used in Eq. (2-5). The second term vanishes because of the integral associated with it. The leading order error in the rectangular rule for a single subinterval of width h is therefore given by the third term,

$$\begin{aligned}\Delta I_{[x_{i-1},x_i]} &\equiv \int_{x_{i-1}}^{x_i} f(x)\,dx - hf_{i-1/2} \\ &\approx \frac{1}{2!}f''_{i-1/2}\int_{x_{i-1}}^{x_i}(x-x_{i-1/2})^2\,dx = \frac{h^3}{24}f''_{i-1/2}\,.\end{aligned}$$

The total error in the entire interval $[a,b]$ is obtained by summing the contributions from all N subintervals,

$$\Delta I_{[a,b]} = \int_a^b f(x)\,dx - I_{[a,b]} \approx \frac{b-a}{24}h^2 f''(\xi) = \frac{1}{24}\frac{(b-a)^3}{N^2}f''(\xi)$$

where we have made use of the fact that $Nh = (b-a)$ and have taken $f''(\xi)$ to be the average value of the second derivative of $f(x)$ in $[a,b]$.

Trapezoidal rule As an alternative to the rectangular rule, we can approximate the area of each subinterval by a trapezoid with the four corners at coordinates $(x_{i-1}, 0)$, (x_{i-1}, f_{i-1}), (x_i, f_i), and $(x_i, 0)$. The area is given by

$$I_{[x_{i-1},x_i]} = \frac{h}{2}(f_{i-1}+f_i). \tag{2-8}$$

Schematically, the difference between the two rules is shown in Fig. 2-2.

For the entire interval $[a,b]$, the integral is given by the sum over all the slices,

$$I_{[a,b]} = \frac{1}{2}h\sum_{i=1}^{N}(f_{i-1}+f_i) = h(\frac{1}{2}f_0 + f_1 + f_2 + \cdots + f_{N-1} + \frac{1}{2}f_N). \tag{2-9}$$

This is a very reasonable result which we could have anticipated from the beginning. Compared with the others, the contributions of f_0 and f_N are only half as important, as they are located at the two ends of the interval. The result follows also from the

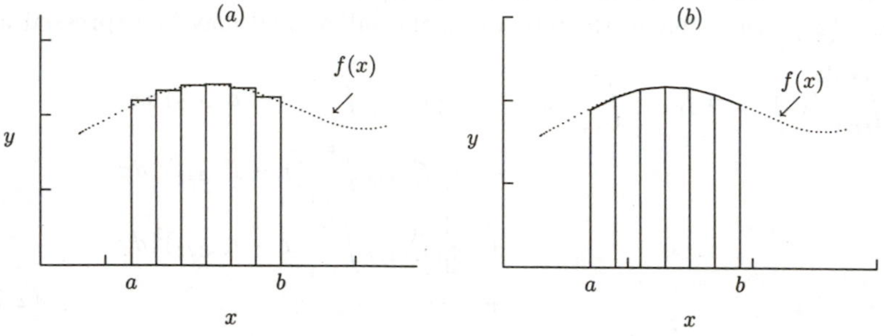

Figure 2-2: Schematic illustrations of (a) rectangular rule and (b) trapezoidal rule. In (a), each subinterval is approximated by a rectangle of width h and height at the middle point of the subinterval. In (b), the approximation is a trapezoid, bounded by $(x_{i-1}, 0)$, (x_{i-1}, f_{i-1}), (x_i, f_i), and $(x_i, 0)$.

Euler-Maclaurin summation formula for integrals (see, for example, Abramowitz and Stegun[1])

$$\begin{aligned}\int_a^b f(x)\,dx &= h\Big\{\frac{1}{2}f_0 + f_1 + f_2 + \cdots + f_{N-1} + \frac{1}{2}f_N\Big\} \\ &\quad -\frac{B_2}{2!}h^2(f'_N - f'_0) - \cdots - \frac{B_{2k}}{(2k)!}h^{2k}(f_N^{(2k-1)} - f_0^{(2k-1)}) - \cdots .\end{aligned} \tag{2-10}$$

The coefficients B_{2k} are the Bernoulli numbers, with

$$B_1 = -\frac{1}{2}$$

and all the other odd order terms vanish. The first few even ones have the values:

$$B_0 = 1 \qquad B_2 = \frac{1}{6} \qquad B_4 = -\frac{1}{30} \qquad B_6 = \frac{1}{42}.$$

More generally, they are given by the generating function,

$$\frac{t}{e^t - 1} = \sum_{k=0}^{\infty} B_k \frac{t^k}{k!}. \tag{2-11}$$

Explicit values of B_k up to B_{60} can be found on page 810 of Abramowitz and Stegun.[1] Note that, among other things, Eq. (2-11) demonstrates that the errors of the trapezoidal rule are given by the even powers of h.

Box 2-1 Subroutine TRAPZ(A,B,KEY,RSLT,H)
Trapezoidal rule integration

Arguments:
- A: Lower integration limit.
- B: Upper integration limit.
- KEY = 0: Initiate a new integration; > 0: subsequent iterations.
- RSLT: Value of the integral.
- H: Integration step size.

Initialization:
Define the integrand as an external function.

1. First call (KEY=0):
   - (a) Calculate the integral using only two subintervals.
     - (i) Define the number of subintervals to be 2.
     - (ii) Let SUM be the average of the the two end points plus the midpoint.
     - (iii) Define the step size H to be half of the distance from A to B.
   - (b) Change the value of KEY to 1.
   - (c) Return the product of H and SUM.
2. Subsequent calls (KEY> 0):
   - (a) Calculate the contributions from points halfway between old ones:
     - (i) Start from the middle of the first subinterval.
     - (ii) Add the value of the integrand at this point to SUM.
     - (iii) Move to the next subinterval and repeat steps (i) and (ii).
   - (b) Double the number of subintervals and halve the step size H.
   - (c) Return H times SUM.

The advantage of the trapezoidal rule is that the values of $f(x)$ are evaluated only at the *grid* or *mesh* points, that is, at $x = x_0, x_1, \ldots, x_N$. As can be seen from Problem 2-1, the errors due to finite step sizes associated with this method are expected to be a factor of 2 larger than those of the rectangular rule. On the other hand, the simplicity of the trapezoidal rule lends itself to further improvements. We shall see later in §3-5 that the method may be used in conjunction with extrapolation techniques to construct a very efficient algorithm for numerical integration. The algorithm for the trapezoidal rule is outlined in Box 2-1.

The algorithm includes also a method to start with a minimum number of subintervals and increase the number in subsequent iterations without repeating any work in evaluating the integrand at the grid points. This, as we shall see later, is useful in applying the method as the core for more advanced techniques, such as extrapolation to find the approximate value of the integral as $h \to 0$.

2-3 Simpson's rule

In the previous section, we saw that the errors associated with both the rectangular and trapezoidal rules are on the order of $1/N^2$. To improve the accuracy without increasing very much the amount of computation involved, we can use Simpson's rule. In the Taylor series expansion of the integral given in Eq. (2-7), we saw that terms involving odd-order derivatives of $f(x)$ vanish if the limits of the integration are taken symmetrically around the point where we do the expansion. To make use of this property, we can take a Taylor series expansion of the area in two adjacent subintervals. If the expansion is done around x_i, the mid-point of the two subintervals combined, we obtain the result

$$\begin{aligned} I_{[x_{i-1},x_{i+1}]} &= \int_{x_{i-1}}^{x_{i+1}} f(x)\,dx \\ &= f_i \int_{x_{i-1}}^{x_i+1} dx + \frac{1}{1!} f_i' \int_{x_{i-1}}^{x_{i+1}} (x - x_i)\,dx + \frac{1}{2!} f_i'' \int_{x_{i-1}}^{x_{i+1}} (x - x_i)^2 dx \\ &\qquad + \frac{1}{3!} f_i^{(3)} \int_{x_{i-1}}^{x_{i+1}} (x - x_i)^3 dx + \cdots \\ &= f_i \times 2h + 0 + \frac{1}{2!} f_i'' \times \frac{2}{3} h^3 + 0 + O(h^5 f_i^{(4)}). \end{aligned} \tag{2-12}$$

We see that the accuracy of integration can be improved by two orders, to $1/N^4$, if we can also include into the numerical integration terms involving the second-order derivative of $f(x)$.

This may be achieved in the following way. Using the central difference introduced later in §2-7, we can approximate the second-order derivative of $f(x)$ as

$$f_i'' \equiv \left. \frac{d^2}{dx^2} f(x) \right|_{x=x_i} \approx \frac{1}{h^2} (f_{i-1} - 2f_i + f_{i-1}). \tag{2-13}$$

On substituting this result into Eq. (2-12), the integral in the subinterval $[x_{i-1}, x_{i+1}]$ may be approximated as

$$\begin{aligned} I_{[x_{i-1},x_{i+1}]} &= 2hf_i + \tfrac{1}{3}h(f_{i-1} - 2f_i + f_{i-1}) + O(h^5 f_i^{(4)}) \\ &= h(\tfrac{1}{3}f_{i-1} + \tfrac{4}{3}f_i + \tfrac{1}{3}f_{i-1}) + O(h^5 f_i^{(4)}). \end{aligned} \tag{2-14}$$

For the complete interval $[a, b]$, the result is

$$I_{[a,b]} = \tfrac{1}{3}h(f_0 + 4f_1 + 2f_2 + 4f_3 + \cdots + 2f_{N-2} + 4f_{N-1} + f_N) \tag{2-15}$$

and we have assumed that the number of intervals N is even. The contributions from the odd grid points are more important than those from even ones by a factor of 2, as illustrated in Fig. 2-3. The difference in the weight comes from our effort to correct the first-order results of more elementary methods by including the contributions from

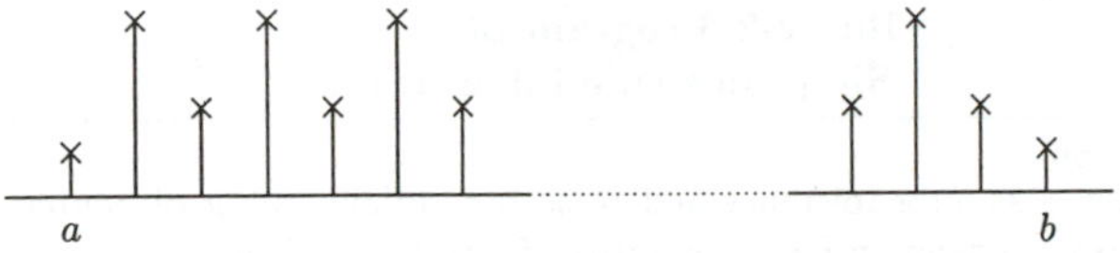

Figure 2-3: Weighting factors in Simpson's rule integration. The odd points are twice as important, and the two end points only half as much as the even ones.

second-order derivatives of $f(x)$. The end points, f_0 and f_N, enter with a weight that is only half that of the even points.

In addition to better accuracy, Simpson's rule given by Eq. (2-15) has also the advantage that it leads naturally to a *strategy* for evaluating an integral to the desired accuracy by an iterative procedure. In principle, we can find the error in the numerical calculation by making an estimate of the size of the contributions from the leading order term ignored in Eq. (2-12). However, this is not always easy to do, as it involves a knowledge of the average value of some high-order derivatives, such as $f^{(4)}(\xi)$ in Eq. (2-12). A simpler, but less certain, way is to compare the results of two consecutive iterations. If the difference in the calculated values of $I_{[a,b]}$ is smaller than some preset tolerance ϵ, it is likely that the required accuracy is reached.

For Simpson's rule, the strategy may be implemented in the following way. Consider the case of integrating a function $f(x)$ between $x = a$ and $x = b$ up to some accuracy ϵ. From Eq. (2-15), we see that the contributions to the integral may be divided into three parts, the end zone, the odd points, and the even points,

$$\begin{aligned} S_{\rm d}(N) &\equiv f_0 + f_N \\ S_{\rm o}(N) &\equiv f_1 + f_3 + f_5 + \cdots + f_{N-3} + f_{N-1} \\ S_{\rm e}(N) &\equiv f_2 + f_4 + f_6 + \cdots + f_{N-4} + f_{N-2} \end{aligned}$$

where $f_k \equiv f(a + kh)$ and $h = (b - a)/N$. In terms of $S_{\rm d}$, $S_{\rm o}$, and $S_{\rm e}$, Eq. (2-15) may be written in the form

$$I_{[a,b]}(N) = \tfrac{1}{3}h\{S_{\rm d}(N) + 2S_{\rm e}(N) + 4S_{\rm o}(N)\}. \tag{2-16}$$

We shall see that, for a new N, we can make use of the values of $S_{\rm d}$, $S_{\rm o}$, and $S_{\rm e}$ we already have for the previous iteration.

To make the calculation into an iterative one, we shall start with $N = N_1$, where N_1 is some small but reasonable number of intervals. For example, $N_1 = 6$. This constitutes the first step in our iterative calculation. For the next iteration we shall double the number of subintervals and halve the step size h by taking $N = N_2$ with $N_2 = 2N_1$. The advantage of changing the number of subintervals in this way is that, for the new N value, the contributions from the two end points, stored as $S_{\rm d}$, are

Box 2-2 Program DM_SIMPS
Simpson's rule integration

Initialization:
- (a) Set ϵ as error tolerance and N as the initial number of points.
- (b) Set a maximum for the number of iterations allowed.

1. Input a as the lower limit of integration and b as the upper limit.
2. Make an initial calculation of the integral with N points:
   - (a) Calculate the contributions from end points.
   - (b) Sum over the contributions from odd points.
   - (c) Sum over the contributions from even points.
   - (d) Calculate the integral from these three terms using Eq. (2-16).
3. Double the number of points and calculate the value of the integral again:
   - (a) Sum the previous even and odd contributions as the new even value.
   - (b) Calculate the contribution from the new odd points.
   - (c) Calculate the integral with Eq. (2-16) using the new values.
4. Compare the new result with the previous one.
   - (a) If the difference is greater than ϵ go back to step 3.
   - (b) If the difference is smaller or equal to ϵ, return the result.

unchanged, and the even grid points are exactly the ones already calculated in the previous iteration. That is, for the new iteration, the value of S_e is the sum of S_o and S_e of the previous iteration. The only new calculations required are the values of the function at the (new) odd grid points. Thus, we have the result

$$S_e(N_2) = S_e(N_1) + S_o(N_1) \qquad\qquad S_o(N_2) = \sum_{i=1}^{N_2/2} f(a + (2i-1)h)$$

where $h = (b-a)/N_2$. We can now apply Eq. (2-16) to obtain the new value of $I_{[a,b]}$ for $N = N_2$. If the new result differs from that obtained in the previous iteration by an amount less than ϵ, we shall assume that the calculation has converged. Otherwise, the number of intervals is doubled again by making $N = N_3$ with $N_3 = 2N_2$. The process is repeated until convergence is achieved or a maximum for the number of iterations is reached. An outline of the algorithm is given in Box 2-2.

A simple example will serve to illustrate the power of Simpson's rule. Consider the integral

$$I = \int_0^{\pi/2} \sin x \, dx.$$

The result is well known: $\cos(0) = 1$. To find the value numerically, we shall start with $N = 6$. In addition to the two end points at 0 and $\pi/2$, we have three odd points at $x_i = \pi/12$, $\pi/4$, and $5\pi/12$ and two even points at $x_i = \pi/6$ and $\pi/3$. In the first iteration, the integrand is evaluated at these seven points. The size of the integration interval h is given in Table 2-2 together with the value of the integral obtained.

Table 2-2: Example of the accuracy of Simpson's rule for different N.

N	$\int_0^{\pi/2} \sin x\, dx$	h	$h^4/60$
6	1.0000262	0.262	8×10^{-5}
12	1.0000015	0.131	5×10^{-6}
24	0.9999999	0.065	3×10^{-7}
48	0.9999999	0.033	2×10^{-8}
exact value	1.0000000	—	—

For this simple example, we can actually use Eq. (2-12) to estimate the error expected for a given N. In a given subinterval the first nonvanishing term beyond the second-order derivatives already included in Simpson's rule is

$$\Delta_i(N) = \frac{1}{4!} f^{(4)}(x_i) \times \frac{2}{5} h^5.$$

For the integral as a whole, we must multiply the result by N and find $f^{(4)}(\xi)$, the average value of the fourth-order derivative of $f(x)$ in $[a, b]$

$$\Delta(N) = \frac{1}{4!} f^{(4)}(\xi) \times \frac{2}{5} h^4 (b-a).$$

Since the even-order derivatives of a sine function are also sine functions, we obtain the condition $|f^{(4)}(x_i)| \leq 1$. For the example in hand, $(b-a) = \pi/2$. We can therefore take $(b-a)f^{(4)}(\xi)$ to be on the order of unity for our error estimate. This gives us the result

$$\Delta(N) \approx \frac{h^4}{60}.$$

The numerical values are listed in the last column of Table 2-2 for different size h used.

The results of the numerical integration are given in the second column. When these are compared with the exact value of unity, we find that our error estimates are of the correct orders of magnitude. If we double the number of subintervals and take $N = 12$, we expect an accuracy of one part in a million. The tabulated values also show that, for a single precision calculation, there is no point to go beyond $N = 24$, as the size of the error associated with the numerical method becomes comparable with the truncation errors of the computer.

There are several variants of Simpson's rule. For example, it is possible to change the relative weight of the contributions from different grid points and achieve an even higher-order accuracy than that of Eq. (2-15). Efficiency can also be improved by using extrapolation techniques, as we shall see later in §3-5 in connection with the Romberg method of integration.

2-4 Gaussian quadrature

The approximations used by trapezoidal and Simpson's rules may be regarded as methods to replace the integrand in each subintervals by, respectively, second- and third-order polynomials. The idea of a polynomial approximation may also be applied to the interval $[a, b]$ as a whole. This leads us to the idea behind Gaussian quadrature, the method of Gauss for numerical integration.

There are several different forms of this method. A typical one is the Gauss-Legendre integration. In general, a function $f(x)$ in a given interval $[a, b]$ may be expanded in terms of a complete set of polynomials $P_\ell(x)$,

$$f(x) = \sum_{\ell=0}^{n} \alpha_\ell P_\ell(x). \tag{2-17}$$

The definition of the expansion coefficients α_ℓ depends somewhat on the polynomial $P_\ell(x)$ used and we shall see examples later. For each polynomial, the subscript ℓ indicates the order, the highest power of the argument x. For $f(x)$ in Eq. (2-17), it is given by n, the upper limit of the summation.

In general, it is more convenient to adopt a set of orthogonal polynomials to expand a function. In principle, any orthogonal polynomial can be used. For our purpose here, we shall take Legendre polynomials, discussed later in §4-2. The orthogonality relation between these polynomials is

$$\int_{-1}^{+1} P_k(x) P_\ell(x)\, dx = \frac{2}{2\ell+1}\delta_{k,\ell}. \tag{2-18}$$

Since Legendre polynomials are defined only in the interval $[-1, +1]$, it is necessary for us to change the interval of integration for an arbitrary integral from $[a, b]$, for finite a and b, to $[-1, +1]$ by making the substitution

$$x \longrightarrow \frac{b-a}{2}x + \frac{b+a}{2}. \tag{2-19}$$

The integral given, for example, by Eq. (2-1) is now written in the form

$$I_{[a,b]} = \int_a^b f(x)\, dx \longrightarrow \frac{b-a}{2}\int_{-1}^{+1} f\Big(\frac{b-a}{2}x + \frac{b+a}{2}\Big)\, dx.$$

For our discussions below, we shall limit ourselves to

$$I_{[-1,+1]} = \int_{-1}^{+1} f(x)\, dx \tag{2-20}$$

and use the integral as our prototype.

Let us consider the case that the integrand may be approximated by a function of polynomial of order $(2n-1)$,

$$f(x) \approx p_{2n-1}(x). \tag{2-21}$$

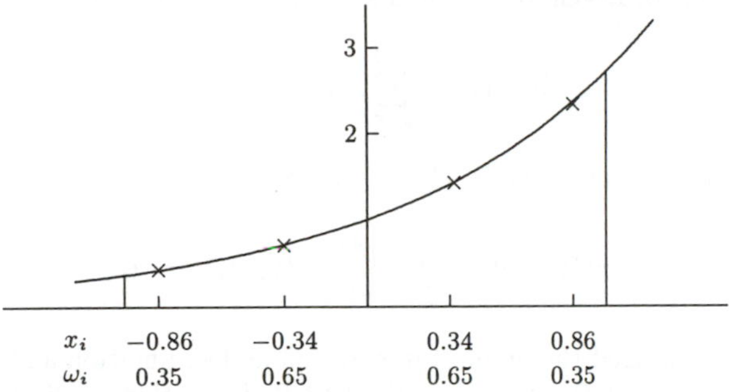

Figure 2-4: Fourth-order Gauss-Legendre integration for $\int_{-1}^{1} e^x\,dx$. Only four values of the integrand at $x_i = \pm 0.3399810$ and ± 0.8611363 are used to achieve an accuracy of seven significant figures.

In other words, if we expand $f(x)$ in terms of Legendre polynomials using the form given by Eq. (2-17), it is sufficient to use $P_\ell(x)$ up to order $\ell = (2n-1)$. If this is a good approximation, it is possible to express the integral of $f(x)$ in term of $2n$ parameters, and the method of Gauss provides us with a way to arrive at the most accurate value. In other words, we wish to write the integral in the form

$$I_{[-1,+1]} = \sum_{i=1}^{n} \omega_i f(x_i) \tag{2-22}$$

where both x_i, the abscissas, and ω_i, the weight factors, are to be determined in such a way as to get the best accuracy possible for the amount of calculations involved. As an example, the integral $\int_{-1}^{+1} e^x dx$, using a fourth-order Gauss-Legendre quadrature is shown in Fig. 2-4.

Abscissas and weight factors To select the abscissas and weight factors required for Eq. (2-22), we shall assume for the time being that the integrand is a polynomial of order $(2n-1)$. That is, Eq. (2-21) is an exact identity rather than an approximation. By polynomial division, we can decompose $p_{2n-1}(x)$ into a sum of two terms,

$$p_{2n-1}(x) = p_{n-1}(x)P_n(x) + q_{n-1}(x) \tag{2-23}$$

where $P_n(x)$ is Legendre polynomial of order n. The other two quantities on the right side, $p_{n-1}(x)$ and $q_{n-1}(x)$, are polynomials of order $(n-1)$.

The first term on the right side of Eq. (2-23) vanishes on integrating from -1 to $+1$. This arises from the orthogonal property of the polynomials in the following way. By construction, $p_{n-1}(x)$ is a polynomial of order $(n-1)$ and, as a result, may be expanded in terms of Legendre polynomial up to order $(n-1)$,

$$p_{n-1}(x) = \sum_{k=0}^{n-1} a_k P_k(x). \tag{2-24}$$

Since the expression does not contain $P_n(x)$, we obtain the result

$$\int_{-1}^{+1} p_{n-1}(x) P_n(x)\, dx = \sum_{k=0}^{n-1} \int_{-1}^{+1} a_k P_k(x) P_n(x)\, dx = 0$$

where we have interchanged the order of integration and summation and made use of the orthogonal property of $P_\ell(x)$ given in Eq. (2-18). From this, we find the equality

$$\int_{-1}^{+1} p_{2n-1}(x)\, dx = \int_{-1}^{+1} q_{n-1}(x)\, dx.$$

Furthermore, as a polynomial of order n, the function $P_n(x)$ has n zeros in the interval $[-1, +1]$. Let these zero points be located at $x = x_1, x_2, \ldots, x_n$. At these points, we have the relation

$$p_{2n-1}(x_i) = q_{n-1}(x_i) \tag{2-25}$$

as can be seen from Eq. (2-23).

We can now expand the polynomial $q_{n-1}(x)$ also in terms of Legendre polynomials in a similar fashion as we have done in Eq. (2-24):

$$q_{n-1}(x) = \sum_{k=0}^{n-1} \omega_k P_k(x)$$

where ω_k is the weight for polynomial $P_k(x)$. Combining this result with Eq. (2-25), we obtain the relation

$$p_{2n-1}(x_i) = \sum_{k=0}^{n-1} \omega_k P_k(x_i)$$

at the roots of $P_n(x)$. This may be written in a matrix notation

$$p_i = \sum_{k=0}^{n-1} P_{ik}\, \omega_k \tag{2-26}$$

where we have used the symbols

$$p_i \equiv p_{2n-1}(x_i) \qquad \text{and} \qquad P_{ik} \equiv P_k(x_i).$$

Note that $P_k(x_i)$ represents the values of Legendre polynomials of order k, up to $k = (n-1)$, evaluated at the zeros of $P_n(x)$. The values of x_i are independent of the polynomial $p_{2n-1}(x)$.

Table 2-3: Abscissas and weight factors for Gauss-Legendre integration.

Order	x_i	ω_i
$n = 4$	±0.339981043584856	0.652145154862546
	±0.861136311594053	0.347854845137454
$n = 5$	0.000000000000000	0.568888888888889
	±0.538469310105683	0.478628670499366
	±0.906179845938664	0.23692688505618
$n = 12$	±0.125233408511469	0.249147045813403
	±0.367831498998180	0.233492536538355
	±0.587317954286617	0.203167426723066
	±0.769902674194305	0.160078328543346
	±0.904117256370475	0.106939325995318
	±0.981560634246719	0.047175336386512

The relation given by Eq. (2-26) may be inverted to find the values of ω_k in terms of p_i:

$$\omega_k = \sum_{i=1}^{n} p_i \{\boldsymbol{P}^{-1}\}_{ik} \tag{2-27}$$

where $\boldsymbol{P}^{-1}$ is the inverse of the matrix $\{P_{ik}\}$. That is $\sum_k \{\boldsymbol{P}^{-1}\}_{ik} P_{kj} = \delta_{i,j}$. This expansion helps us to evaluate the integral in the following way. Since we are assuming that Eq. (2-21) is exact, we have the result

$$I_{[-1,+1]} = \int_{-1}^{+1} f(x)\,dx = \int_{-1}^{+1} p_{2n-1}(x)\,dx = \sum_{k=0}^{n-1} \omega_k \int_{-1}^{+1} P_k(x)\,dx. \tag{2-28}$$

For $k = 0$, we have $P_0(x) = 1$, and this gives the only nonvanishing integral among the n terms in the sum

$$\int_{-1}^{+1} P_k(x)\,dx = \frac{2}{2k+1}\delta_{k,0} = 2\,\delta_{k,0}$$

as can be seen from the orthogonal property of Legendre polynomials. The final result of Eq. (2-28) is then

$$I_{[-1,+1]} = 2\omega_0 = 2\sum_{k=1}^{n} p_k \{\boldsymbol{P}^{-1}\}_{k0}. \tag{2-29}$$

The last equality comes from Eq. (2-27).

For the purpose of numerical integration, Eq. (2-29) may be written in a more familiar form. Recalling that

$$p_i \equiv p_{2n-1}(x_i) \approx f(x_i)$$

Table 2-4: Example of Gauss-Legendre integration for $\int_{-1}^{+1} e^x dx$.

Order	x_i	ω_i	$\exp(x_i)$	$\omega_i \exp(x_i)$
$n=4$	-0.8611363	0.3478549	0.4226815	0.1470318
	-0.3399810	0.6521451	0.7117838	0.4641864
	0.3399810	0.6521451	1.4049209	0.9162124
	0.8611363	0.3478549	2.3658476	0.8229715
			$\sum_i \omega_i \exp(x_i) =$	2.3504021
$n=5$	-0.9061798	0.2369269	0.4040649	0.0957338
	-0.5384693	0.4786287	0.5836409	0.2793473
	0.0000000	0.5688889	1.0000000	0.5688889
	0.5384693	0.4786287	1.7133822	0.8200738
	0.9061798	0.2369269	2.4748502	0.5863585
			$\sum_i \omega_i \exp(x_i) =$	2.3504023
			Exact value =	2.350402387

the integral of $f(x)$ reduces to the form of Eq. (2-22)

$$I_{[-1,+1]} = \sum_{i=1}^{n} \omega_i f(x_i)$$

where ω_i is the weight of $f(x_i)$ at the zeros of $P_n(x)$ and is given by

$$\omega_i = 2\{\boldsymbol{P}^{-1}\}_{i0}.$$

We shall not be concerned here with the method to invert the polynomial matrix $\boldsymbol{P} = \{P_k(x_i)\}$, as the general approach will be discussed later in §5-2. The result is known and may be put in the form

$$\omega_i = \frac{2}{(1-x_i^2)[P_n'(x_i)]^2} \tag{2-30}$$

where

$$P_n'(x_i) = \frac{d}{dx} P_n(x)\Big|_{x=x_i}.$$

The values of $P_n'(x_i)$ may be calculated using the methods given in §4-2. Extensive tables of both x_i and ω_i for Gaussian quadrature of different orders may be found in Abramowitz and Stegun,[1] and examples for orders n = 4, 5, and 12 are given in Table 2-3 for illustration.

Example of Gaussian-Legendre integration As an example, let us calculate the integral

$$I = \int_{-1}^{+1} e^x \, dx. \tag{2-31}$$

Box 2-3 Program DM_QUAD
Gauss-Legendre quadrature for $\int_{-a}^{+a} e^x dx$

Initialization:
- (a) Convert the integral to the integration interval $[-1, +1]$.
- (b) Input the order, abscissas, and weights.

1. Input the value of a.
2. Calculate the contribution to the integral from each point:
   - (a) Find the value of the integrand at each point.
   - (b) Taking the product with the weight factor.
   - (c) Multiply the result by a as required in Eq. (2-32).
   - (d) Sum the contributions.
3. Output the sum.

The values of the various quantities that enter into the calculation are summarized in Table 2-4. The abscissas and weight factors are taken from Table 2-3, and the values of the integrand at each of the points x_i are given in columns 2 and 3 of Table 2-4. Compared with the exact value of

$$\int_{-1}^{+1} e^x\,dx = e^1 - e^{-1} = 2.350402387$$

we find that, even for an $n = 4$ approximation, the result obtained using a Gauss-Legendre integration is accurate up to seven significant figures.

In principle, the exponential function is an infinite series in terms of its argument x,

$$e^x = 1 + \frac{x}{1!} + \frac{x^2}{2!} + \frac{x^3}{3!} + \cdots.$$

When we use $n = 4$ for the Gauss-Legendre integration, we are approximating the function by a series up to order $(2n - 1) = 7$. In this limit, terms involving powers higher than x^8 are ignored. For this reason, the accuracy may be improved by using instead order $n = 5$, whereby terms up to order x^9 are also included. As we can see from the second part of Table 2-4, the calculated value is now 2.3504023, essentially the same as the best value that can be achieved in a single precision calculation.

To carry out the same integral for limits other than ± 1, we can make the following change of variable:

$$\int_{-a}^{+a} e^x dx = a \int_{-1}^{+1} e^{ay} dy. \tag{2-32}$$

The integral now takes on the form of Eq. (2-31) and may therefore be carried out in the same way. The algorithm is outlined in Box 2-3. As expected, the accuracy of the calculation deteriorates as a increases beyond 1.

An interesting exercise in the accuracy of Gaussian quadrature is to carry out the integral $\int_{-1}^{+1} x^8 dx$. With $n = 4$, the result is 0.21061227, a 5% difference compared with the exact value of 2/9. The reason for the relatively large error is quite obvious — the method approximates the integrand as a polynomial of order $(2n-1) = 7$. On the other hand, if we use $n = 5$, a result of 0.22222221 is obtained, essentially identical to the exact value in single precision. This illustrates an important limitation of Gaussian quadrature: the order of the polynomial required to achieve a given accuracy depends very much on the nature of the integrand. In general, it is not easy to know what is the order of polynomial approximation required for a given function and accuracy. Furthermore, the method itself does not provide us with a natural way to estimate the error. As a result, it is not possible to know the reliability of the value obtained without carrying out some independent checks.

2-5 Monte Carlo integration

All the methods of numerical integration we have discussed so far require the integrand $f(x)$ evaluated at points spaced at regular intervals along the abscissa. For rectangular, trapezoidal, and Simpson's rules, $f(x)$ are calculated at $(N+1)$ evenly spaced points, $x_0, x_1, \ldots, x_N$. In the case of Gauss-Legendre integration, the points x_i are chosen to be the roots of Legendre polynomials of a given order in the interval $[-1,+1]$. In this section, we shall take a different philosophy and use sampling techniques to choose the points. The advantage of this approach is the convenience it offers for multidimensional integrals and certain types of improper integrals to which we shall return in the next section. For one-dimensional integrals with well-behaved integrands, which we shall continue to use as the example in this section, the method is not as efficient as those we have seen so far.

For the discussions in this section, it is convenient to take the integration interval to be $[0,1]$. The choice is connected with the fact that most generators of random numbers with a uniform distribution are designed for this range. For integrations with lower limit a different from 0 or upper limit b different from 1, it is a simple matter to make the transformation

$$x \longrightarrow \frac{1}{b-a}(x-a)$$

so that the actual numerical integration is carried out in the interval $[0,1]$. This is similar to that given earlier in Eq. (2-19), except there the transformation ends up with the range of integration on $[-1,+1]$.

The Monte Carlo integration method is based on the idea that, instead of selecting the mesh points ahead of time, we can take a random sample. If our sampling is truly random, the points $x_1, x_2, \ldots, x_N$ cover the interval $[0,1]$ with equal probability. For N such points, the average distance between two adjacent ones is

$$h = 1/(N-1) \xrightarrow[\text{large } N]{} 1/N. \tag{2-33}$$

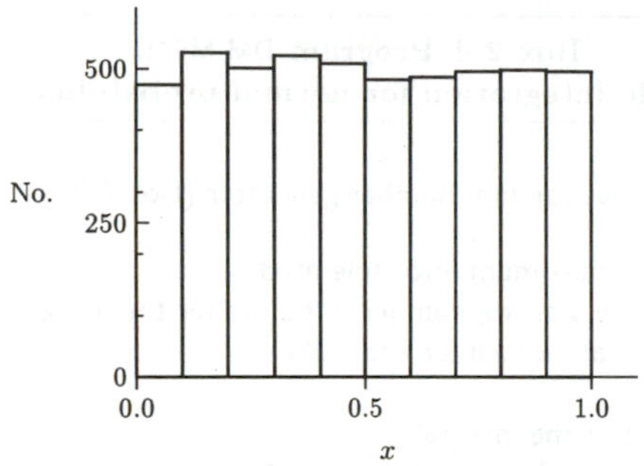

Figure 2-5: Histogram showing the distribution of 5,000 uniform random numbers in the interval [0, 1] generated using the subtractive method of Box 7-2.

The value of the integral is then

$$I_{[0,1]} = \int_0^1 f(x)\,dx \xrightarrow[N\to\infty]{} \frac{1}{N}\sum_{i=1}^{N} f(x_i). \tag{2-34}$$

To select the points along the x-axis, we can use a generator for evenly distributed random numbers, such as the ones discussed in §7-1.

If our sampling is truly random, the number of points selected in a given region $[x-\frac{1}{2}\Delta x, x+\frac{1}{2}\Delta x]$ is independent of the value of x. In other words, if we divide the interval [0, 1] into an arbitrary number of equal-size subintervals, the number of points in each subinterval is the same within statistical fluctuations. Elementary notions of statistics tell us that this can only be true if we take a large sample, such as the one of 5,000 random points shown in Fig. 2-5.

An iterative approach One difficulty inherent in Monte Carlo methods is to know when N is large enough for Eq. (2-34) to be valid. For our purpose here, we can take a practical approach and use an iterative strategy, similar to what we did earlier for Simpson's rule. We start by taking a sample of N_1 points and evaluate the integral using Eq. (2-34). The sampling is repeated for another N_1 different points and the integral is evaluated for these N_1 points as well. If the two values are the same within some tolerance ϵ, it is likely that both samples are large enough. To improve the accuracy further, we can combine the two samples and use the average of the two as the final result. On the other hand, if the difference is too large, we can group both sets together as a single one of $N_2 = 2N_1$ points and take another sample of N_2 different points. The results are compared with each other again. The process is repeated until sufficient accuracy is achieved or the maximum number of iterations we impose on the calculation is exceeded.

In the example for Simpson's rule, we have started with a small number of points ($N = 2 \sim 6$). Since our method here depends on a random sample of evenly distributed points, it is not possible to attain such a distribution with a small N. A

Box 2-4 Program DM_MC1D
Monte Carlo integration for normal probability function

Subprogram used:
RSUB: Subtractive random number generator (Box 7-2)
Initialization:
(a) Set ϵ as the maximum error tolerated.
(b) Let N be the starting number of points for the integration.
(c) Zero the iteration counter and SUM.
1. Input:
(a) Upper limit of the integral.
(b) Seed for the random number generator.
2. Calculate the integral for N random points:
(a) Generate a uniform random number t in $[0, 1]$.
(b) Calculate the value of the integrand at t and add it to SUM.
3. Convert SUM to the value of the integral using Eqs. (2-34) and (2-35).
4. Repeat steps 2 and 3 with a different set of N random points.
5. Compare the two values of the integral obtained:
(a) If the difference is larger than ϵ, then:
(i) Take the average of the two values.
(ii) Double the value of N.
(iii) If less than the maximum number of iterations, go back to step 4.
(iv) Otherwise return with error message.
(b) If the difference is less than or equal to ϵ,
Take the average of the two values and output the result.

cautious person may want to start with a much larger sample, for example, $N_1 \sim 10^2$. To see the rate of convergence in a Monte Carlo integration as we increase the sample size, we shall use as an example the normal probability function

$$A(x) = \frac{1}{\sqrt{2\pi}} \int_{-x}^{+x} e^{-t^2/2} dt \tag{2-35}$$

discussed later in §6-1. For simplicity, we shall consider first the case of $x = 1$. Furthermore, since the integrand is symmetric with respect to $x = 0$, we can change the integral into the form

$$A(x=1) = \sqrt{\frac{2}{\pi}} \int_0^1 e^{-t^2/2} dt.$$

This makes it possible to use a uniformly distributed random number generator in the interval $[0, 1]$ without any conversion. The value is equal to the area underneath a normal distribution curve within one standard deviation on either side of the mean and is given in tables of statistics to be

$$A(x=1) = 2\Big(\frac{1}{\sqrt{2\pi}} \int_{-\infty}^{1} e^{-t^2/2} dt - \frac{1}{2}\Big) = 0.6826895.$$

Table 2-5: Convergence of a Monte Carlo integration for $A(x) = \frac{2}{\sqrt{\pi}} \int_0^1 e^{-t^2/2} dt$

Iteration	No. of points	$A(x=1)$
1	40	0.6931776
2	80	0.6801362
3	160	0.6781273
4	320	0.6887547
5	640	0.6889585
6	1,280	0.6852926
7	2,560	0.6829402
8	5,120	0.6838834
9	10,240	0.6835121
10	20,480	0.6828257
11	40,960	0.6829104
12	81,920	0.6824411
13	163, 840	0.6826339
14	327, 680	0.6826127
15	655, 360	0.6827372
Exact value		0.6826895

The results of numerical integration using the Monte Carlo method for $A(x = 1)$ are given in Table 2-5 for different numbers of points sampled. For purposes of illustration, we shall start with a small value of $N = 40$ points. In each successive iteration, the number of points is doubled. At the fourteenth iteration, a total of 655,360 points is used to reach a difference of one part in 10^5. The algorithm is outlined in Box 2-4.

For values of $A(x)$ other than $x = 1$, we can make the following substitution, similar to what we did earlier in Eq. (2-32):

$$A(a) = \sqrt{\frac{2}{\pi}} \int_0^a e^{-t^2/2} dt = a\sqrt{\frac{2}{\pi}} \int_0^1 e^{-a^2y^2/2} dy \tag{2-36}$$

where $t = ay$. In general, more points are needed to reach the same level of accuracy if a is much larger than 1.

For most Monte Carlo calculations, one of the time-consuming part is in generating the random numbers. In fact, for the one-dimensional example above, most of the time is spent in producing the random numbers. As a result, the method is not efficient compared with those based on a fixed grid of points. For this reason, Monte Carlo integration is favored only in cases where other methods, such as those described in earlier sections, are unsuitable. We shall see examples of such cases in the next section.

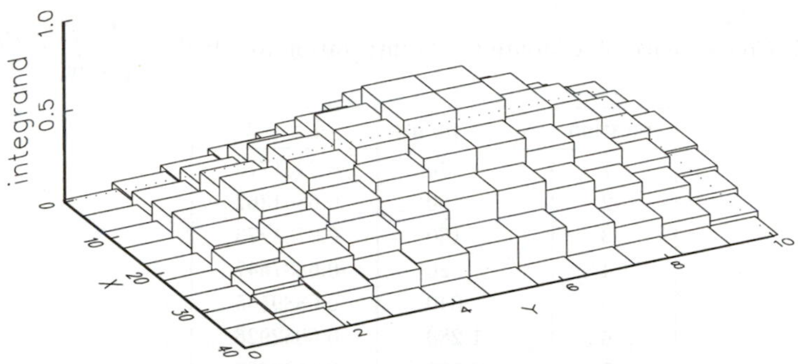

Figure 2-6: Approximation of an integral of two variables by $(M \times N)$ small columns, each with length h, width k, and height $f(x_i, y_j)$ given by the average value of the integrand $f(x, y)$ in the rectangular area $(h \times k)$.

2-6 Multidimensional integrals and improper integrals

In the previous sections, we have used mainly one-dimensional integrals to illustrate the principles behind numerical integration. For the sake of comparison, most of the examples employed were those where the integral can also be carried out analytically. In practice, our interest is primarily in cases where analytical solutions are too difficult or impossible to obtain. This happens frequently in multidimensional integrals and improper integrals. Because of the large variety of such integrals, it is not possible to discuss each type in detail. Instead, we shall give a general discussion to extend the basic methods to situations commonly encountered.

Multidimensional integrals All the approaches we have used so far can be generalized, at least in principle, to integrations over several variables. Let us illustrate this point by considering a two-dimensional integral of the form

$$I = \int_c^d \int_a^b f(x, y)\, dx dy. \tag{2-37}$$

One way to carry out the calculation is to divide the interval $[a, b]$ along the x-axis into M small subintervals of width h each. Similarly, the interval $[c, d]$ along the y-axis may be divided into N subdivisions of width k each. Instead of summing over a number of small areas, as in the one-dimensional case, we are now summing over a number of small volumes. This is shown schematically in Fig. 2-6. In general, we have $h \neq k$.

The xy-plane is now divided into $(M \times N)$ rectangular grids. If the area of each rectangle $(h \times k)$ is sufficiently small, we may approximate the value of $f(x, y)$ within it to be a constant equal to $f(x_i, y_j)$, the value at the center of the grid. The integral

I is then given by the sum of the products of $f(x_i, y_j)$ and the area of each grid,

$$I = \sum_{i=1}^{M} \sum_{j=1}^{N} f(x_i, y_j)\, hk. \tag{2-38}$$

This is nothing more than an extension of the rectangular rule of §2-2 to the case of two independent variables.

We can improve the accuracy of Eq. (2-38) by expanding $f(x, y)$ within a rectangular grid in terms of a Taylor series in both x and y. This is similar to what we did earlier in Eq. (2-6). If we further incorporate the first-order partial derivatives in both x and y as corrections, in the same way we did in Eq. (2-8) for x alone, we obtain the equivalent of the trapezoidal rule for two-dimensional integrals. Similarly, we can also make improvements using second-order derivatives, and an extension of the Simpson's rule is obtained.

To apply the Gauss-Legendre integration to the case of two integration variables, we need to take a somewhat different approach from the one-dimensional case. One possible way is to carry out the integral in Eq. (2-37) in two stages. First, we integrate $f(x, y)$ over x for a given value of y.

$$g(y) = \int_a^b f(x, y)\, dx. \tag{2-39}$$

Once we have the values of $g(y)$ for all y required in a Gaussian quadrature calculation, we can go on to the second stage and carry out the integration over y

$$I = \int_c^d g(y)\, dy. \tag{2-40}$$

Since, in the numerical integration of Eq. (2-40), we need only the values of $g(y)$ at a finite, discrete set of points along the y-axis, there is no difficulty in applying the Gauss-Legendre integration of Eq. (2-22) to the integral in Eq. (2-39). However, the calculation must be carried out for each value of y required in Eq. (2-40). Once this is done for all the grid points along the y-axis, we can evaluate I by integrating $g(y)$ over y. Again, the integral is over a single variable and Eq. (2-22) may be used. For example, if we decide to use a $(2n-1)$-degree polynomial approximation for $g(y)$, we need only the values of $g(y)$ evaluated at y_1, $y_2, \ldots, y_n$, the zeros of nth-order Legendre polynomial. For each $g(y_i)$, we need to evaluate once the integral on the right side of Eq. (2-39), which, in turn, may be approximated by a polynomial of degree $(2m-1)$. More generally, one can apply techniques specially developed for multidimensional quadratures, such as that of Keister.[29]

Monte Carlo methods For any of the methods described above, the amount of computation required in a two-dimensional integral is proportional to the number of mesh points used. For Eq. (2-38), this number is $(M \times N)$, where M is the number of subintervals along the x-axis and N the corresponding number along the y-axis. To simplify the discussion, let us take $M = N$. In this case, the amount of computation for a two-dimensional integral is proportional to N^2. More generally, for integrals with n independent variables, the amount is proportional to N^n, a number that grows very rapidly with n. Regardless of minor improvements one can make, methods based on a fixed mesh become impractical very quickly if the number of integration variables is large.

For multidimensional integrals, Monte Carlo methods may be more economical. Consider as an example the case of a three-dimensional integral:

$$I = \int_e^f \int_c^d \int_a^b f(x, y, z)\, dx\, dy\, dz. \tag{2-41}$$

Again, we can replace the integral by a sum over the products of the values of the integrand in each cell and the volume of the cell:

$$I \longrightarrow \sum_{i=1}^{N} f(x_i, y_i, z_i)\, \Delta v \tag{2-42}$$

where N is the number of points in the three-dimensional space in which the integrand is evaluated and

$$\Delta v = \frac{(b-a)(d-c)(f-e)}{N}$$

is the average volume of each cell in the three-dimensional space.

Instead of evaluating the integrand in the three-dimensional space at a fixed set of mesh points, we can take a random sample in the $[a, b] \times [c, d] \times [e, f]$ cube. As a practical matter, the location of each point in the cube is represented by three random numbers, x_i, y_i, and z_i. As we saw in the previous section, if the total number of points sampled is large enough, and our random numbers are not biased in any way, the collection of N random points (x_i, y_i, z_i) will eventually fill the cube uniformly. As a result, we can expect that the approximation of an integral by the sum in Eq. (2-42) to be a good one. Except for the number of random numbers required to determine the location of a point, such an approach is essentially *independent of the number of integration variables involved.*

The number of random points required in a Monte Carlo integration depends on the accuracy required and the nature of the integrand. For smooth functions, sufficient accuracy of an integral can often be achieved with a relatively small sample. Furthermore, the same iterative strategy as used in the previous section may also be applied to multidimensional integrals. That is, we make a comparison of the results of two different samples of N points each. If the difference is smaller than the tolerance, the value is accepted. Otherwise, the sample size is doubled. The process is repeated until the desired accuracy is reached. In this way our algorithm naturally includes an estimate of the accuracy of the result as well.

Let us consider as an example the problem of calculating the total amount of charge Q in a cubic volume of length 2 on each side in some arbitrary units. For the charge distribution inside this volume, we shall take

$$\rho(x, y, z) = \exp(-r)/8\pi$$

Box 2-5 Program MC_3D
Three dimensional Monte Carlo integration

Subprogram used:
 RSUB: Subtractive random number generator (Box 7-2)
Initialization:
 (a) Set ϵ as the maximum error to be tolerated.
 (b) Let N be the starting number of points for the integration.
 (c) Zero the iteration counter and SUM.
1. Input:
 (a) Value of a, the scale factor for the charge distribution.
 (b) Seed for the random number generator.
2. Calculate the integral with N random points:
 (a) Generate three uniform random numbers in $[0, 1]$ as (x, y, z).
 (b) Calculate the integrand at (x, y, z) and add the value to SUM.
3. Convert SUM to the value of the integral using Eq. (2-42).
4. Repeat steps 2 and 3 with a different set of N random points.
5. Compare the two values of the integral obtained.
 (a) If the difference is larger than ϵ, then:
  (i) Take the average of the two values.
  (ii) Double the value of N.
  (iii) Go back to step 4 if the number of iterations is less than the maximum.
 (b) If the difference is less or equal to ϵ, then:
  Take the average of the two values and output the result.

where $r = \sqrt{x^2 + y^2 + z^2}$. The form is very close to a Yukawa or screened charge distribution used in a variety of applications. The peak of the distribution is at the center of the cube where $r = 0$. The total amount of charge inside this volume is given by the integral:

$$Q = \int_{-1}^{+1} \int_{-1}^{+1} \int_{-1}^{+1} \frac{1}{8\pi} e^{-r} \, dx \, dy \, dz.$$

We shall make use of the symmetry of the problem and evaluate only one-eighth of the space by changing the integral into the form

$$Q = \frac{1}{\pi} \int_{0}^{+1} \int_{0}^{+1} \int_{0}^{+1} e^{-r} \, dx \, dy \, dz.$$

This is also convenient for making use of evenly distributed random numbers in the interval $[0, 1]$. The result, $Q \approx 0.4/\pi$, is reached with $N = 8,000$ points to an accuracy of 0.0002. Obviously, the efficiency of the method depends to a great deal on the amount of time required to generate the random numbers. In fact, since the integrand is a simple function, a relatively larger fraction of the computer time is expended in generating the random numbers than for the rest of the calculations. The algorithm used is outlined in Box 2-5.

As a side interest, we can use the following calculation to check if the computer program is giving the correct result. For this purpose, we can make use of the fact that the method may be extended to arbitrary integration limits by applying a transformation similar to that used in Eq. (2-36). The number of samplings required to reach the same degree of accuracy is, however, proportional to the volume involved in the integration. On the other hand, the charge density $\rho(x, y, z)$ drops off very quickly with distance from the center of the cube. As a result, very few contributions to the integral come from regions where the absolute values of x, y, and z are large. For this reason, we shall take the value 10 as the "infinity" in this problem. In other words, it is sufficient for us to carry out the integration only in the interval $[-10, +10]$ for each variable. The result may be compared with the exact value, obtained by the following calculation:

$$\begin{aligned}\frac{1}{8\pi}\int_{-\infty}^{+\infty}\int_{-\infty}^{+\infty}\int_{-\infty}^{+\infty} e^{-r}\,dxdydz &= \frac{1}{8\pi}\int_{0}^{+\infty}\int_{0}^{\pi}\int_{0}^{2\pi} e^{-r}r^2\sin\theta\, d\phi d\theta dr\\ &= \frac{1}{2}\int_{0}^{+\infty} e^{-r}r^2 dr\\ &= 1.\end{aligned}$$

To reach the same accuracy of 10^{-3} as we have done earlier, the number of samplings must be increased by a factor of 10^3!

Improper integrals In calculus, an *improper integral* is defined as one with either the range of integration being infinite or the integrand containing infinities within the integration range. Obviously, the singularities must be of a form that they do not make contributions such that the integral itself becomes infinite. For example, the integrand of the integral

$$I = \int_0^1 \frac{dx}{\sqrt{1-x^2}} \tag{2-43}$$

has a singularity at the upper limit. However, the integral itself exists and has the value $I = \pi/2$. On the other hand, the integral

$$I(y) = \int_0^y \frac{dx}{x} \tag{2-44}$$

does not exist (meaning, it does not have a finite value), since

$$\int \frac{dx}{x} = \ln(x).$$

At the lower limit of 0, we have $\ln(0) = -\infty$ and $I(y)$ is undefined. We shall call the type of singularity at $x = 1$ in the integrand of Eq. (2-43) an *integrable singularity* and the corresponding one at $x = 0$ in Eq. (2-44) a *nonintegrable singularity.*

The use of numerical methods to evaluate improper integrals has certain obvious advantages over analytical methods. However, if an integral does not have a finite value, numerical techniques must also fail. We shall not go into the conditions under which an integral exists. Our interest here is only in the problem of how to find the values of integrals that do not have nonintegrable singularities.

Special problems of numerical integration Before discussing numerical methods to evaluate improper integrals, it is worthwhile noting some of the special problems in numerical calculations involving infinities and the related question of zeros in the denominator. For example, it is well known that

$$\frac{\sin x}{x} \xrightarrow[x\to 0]{} 1.$$

If we wish to find the ratio of $\sin x$ over x for a range of x values, the normal procedure is to calculate $\sin(x)$ and x separately and then take the quotient. This method fails at $x = 0$, as the denominator vanishes. Even though the numerator vanishes as well, the computer does not know how to handle the situation of zero divided by zero and gives us an overflow condition as a result. The problem is a nuisance but not a fundamental difficulty. All that is required is to check whether the denominator is zero before making the division. In fact, it is good programming practice to anticipate for the possibilities of overflows and install the equivalent of L'Hospital rule in the program where needed. For our example, we can avoid any overflow by putting in the condition that if $|x| \leq \epsilon$ let $\sin(x)/x = 1$. The choice for the value of some small, positive number ϵ as the practical criterion for "zero" depends on the computer used and accuracy required in the calculation.

A second difficulty in numerical integration occurs in *infinite integrals* where one or both integration limits are at infinity. In mathematics, infinity is a quantity that is larger than any finite value. We have seen in Chapter 1 that the range of values that can be stored in a computer word is limited. As a result, it may be tempting to define infinity as the absolute value of the largest number that can be represented by a computer word. This is not very useful for our purpose here. A more practical way is to make use of the fact that we can only calculate an integral to some finite accuracy ϵ. An infinity in the integration limits may therefore be taken as the value beyond which the total contribution to the integral is less than ϵ.

Consider the following integral:

$$I(b) = \int_0^b e^{-x^2} dx.$$

If the upper limit b is at infinity, it is an infinite integral. The integral $I(b = \infty)$ exists and has the value $\sqrt{\pi}/2 = 0.886227$. On the other hand, we find that at $b = 3$ the value of the integral is already 0.886207. Thus the value 3 may be taken as infinity for this integral if we are willing to accept an error of $\epsilon = 10^{-4}$. On the other hand, for the integral

$$I = \int_0^\infty \frac{\sin x}{x} dx$$

the value of the integrand fluctuates rapidly between positive and negative values for large x. As a result, we need a value of the upper limit on the order $x \gg 1/\epsilon$ before we can obtain a result by numerical integration that is close to the exact value of $\pi/2$. In addition to the large value for the upper limit, rapid fluctuations in the integrand also make it difficult to achieve good accuracy for integrals of this type.

A third difficulty with infinite integrals is that some methods, such as the Gauss-Legendre method, may not work. The root of the problem lies with the fact that the polynomials are defined only in the interval $[-1, +1]$. We can, in principle, transform the integration variable such that the domain coincides with $[-1, +1]$, as we have done in Eq. (2-19). However, the accuracy may turn out to be rather poor. The proper alternative is to use polynomials defined in the appropriate intervals. For example, for the interval $[0, \infty]$, we can use the Laguerre integration method in which the integral is approximated by the relation

$$\int_0^\infty f(x)\,dx = \sum_{i=1}^{n} \omega_i\, e^{x_i} g(x_i) \tag{2-45}$$

in a similar way as in Eq. (2-22) for the Gauss-Legendre method. Here, instead of Legendre polynomials, x_i are the zeros of Laguerre polynomials (see §4-4). It is also possible to make use of the Hermite integration method

$$\int_{-\infty}^\infty f(x)\,dx = \sum_{i=1}^{n} \omega_i\, e^{x_i^2} g(x_i) \tag{2-46}$$

where x_i are the zeros of Hermite polynomials (see §4-1). The values of the abscissas and the weight factors for different orders of both types of quadrature are given in Abramowitz and Stegun.[1]

Numerical methods for improper integrals Let us now turn our attention to integrals containing integrable singularities inside the domain. Most of the numerical methods we have discussed so far may be used also for this type of integral, provided some minor modifications are made to avoid the singular points. Since singularities cause overflows in the calculation, it is essential that they not coincide with any of the mesh points. For most of the methods, it is necessary to know ahead of time the locations of the singular points and avoid them in defining the mesh. The exception is the Monte Carlo integration method. The advantage comes from the fact that, in a random sample of the integration domain, the probability of encountering one of the singularities is small. Furthermore, in the event that such a singular point does occur in the sampling, we can sacrifice a little accuracy by discarding the point. In this way, it is possible to essentially ignore the existence of such singularities in the calculation. For this reason the Monte Carlo method is often the more convenient one to use for improper integrals.

As an example, let us consider again the integral

$$I(b) = \int_0^b \frac{dx}{\sqrt{1-x^2}}\,dx. \tag{2-47}$$

The integrable singularity occurs at $x = 1$. To carry out a Monte Carlo calculation for this integral, we shall assume that our random numbers are distributed uniformly in the interval $[0, b]$. Unless we happen to obtain a random number $x_i = 1$ (that is, x differs from 1 by some small amount ϵ), the singularity does not give us any difficulty in practice. If the value of one of the randomly selected x_i is essentially 1, we can avoid the problem by making a small compromise. The random number is discarded and another one is generated to take its place.

In some cases, an algorithm based on a fixed mesh of points may be preferred. For our discussion here, let us consider the trapezoidal rule and use the integral of Eq. (2-47) for $b = 1$ as our prototype. We cannot use Eq. (2-9) here because f_N, the value of the integrand at the upper limit of the integral, is infinite. Since the integral itself exists, the fault must be with our approach. It is possible to find an alternative algorithm such that the value of the integrand at the upper limit of the integration is not needed. This leads us to the second Euler-Maclaurin summation formula,

$$\begin{aligned}\int_a^b f(x)\,dx &= h\Big\{f_{1/2} + f_{3/2} + f_{5/2} + \cdots + f_{N-3/2} + f_{N-1/2}\Big\} \\ &\quad + \frac{B_2}{4}h^2\{f'_N - f'_0\} + \cdots \\ &\quad + \frac{B_{2k}}{(2k)!}h^{2k}\{1 - 2^{-2k+1}\}\{f_N^{(2k-1)} - f_0^{(2k-1)}\} - \cdots. \qquad (2\text{-}48)\end{aligned}$$

This result may be obtained from Eq. (2-10) by subtracting from it twice that of a similar expression except that the step size is $h/2$. In contrast with the trapezoidal rule given by Eq. (2-9), we have here a method to approximate an integral by the sum of the values of the integrand in the middle of each subinterval. One small point of caution must be observed in applying Eq. (2-48). Since the formula involves middle points (half-integer grid points), we cannot extend the method to an iterative one by simply doubling the number of intervals as we did earlier. In each iteration, the values calculated in the previous iteration are those for the new integer grid points, rather than at half-integer points required by the method. It is easy to get around this point — instead of doubling the number of grid points in each successive iteration, we can triple the number or take some other odd multiples.

In general, it is not advisable to use any of the Gaussian quadrature formulas for integrals with singular values in the integrand. One of the premises of Gauss's approach is that the function can be approximated by a polynomial of finite order, and this, in turn, implies that the function is fairly smooth in the interval $[a, b]$. Functions with singularities, except perhaps those occurring at the ends of the interval, are usually not smooth, and it may not be appropriate to approximate them using finite polynomial series.

2-7 Numerical differentiation

The derivative of a function $f(x)$ at a point x is usually defined in terms of limiting values. Let x and $(x+h)$ be two nearby points separated by some small distance h and the values of the function at these two points be represented, respectively, as $f(x)$ and $f(x+h)$. The first-order derivative of $f(x)$ at x is defined as

$$f'(x) = \lim_{h\to 0} \frac{f(x+h) - f(x)}{h}. \tag{2-49}$$

Graphically, we can think of $f'(x)$ as the slope, or tangent, of $f(x)$ at x. In general, we do not have difficulties in differentiating a function. A concise account of some of the necessary cautions may be found in an article by Press and Teukolsky.[43]

Numerical differentiation becomes important as a result of increasing reliance on computers to solve problems. For example, if a function exists only as tabulated values at discrete intervals along the abscissa, such as data taken by instruments for different input values, the derivatives of the function must be calculated numerically. A more important role of derivatives in numerical calculations is in solving differential equations, to which we shall come to later.

Since derivatives of a function exist only as limiting values, such as that given by Eq. (2-49), it is usually impossible in numerical work to deal with them directly. Instead, we use *finite differences*, meaning small differences in the function at nearby points. A straightforward generalization of Eq. (2-49) is to make use of the difference

$$\Delta f(x) \equiv f(x+h) - f(x) \tag{2-50}$$

and equate $f'(x)$ with the quotient $\Delta f(x)/h$ for finite h. The quantity $\Delta f(x)$ is known as the *forward difference* of $f(x)$ at x, as it is the difference of $f(x)$ at x and a point ahead. This is not the only way to define the difference of a function at two adjacent points. Two others can also be used,

$$\nabla f(x) = f(x) - f(x-h) \tag{2-51}$$

$$\delta f(x) = f(x+h/2) - f(x-h/2). \tag{2-52}$$

The former, $\nabla f(x)$, is known as *backward difference*, and the latter, $\delta f(x)$, *central difference*. Graphically, the three types of differences are compared in Fig. 2-7. In many discussions, it is useful to consider Δ, ∇, and δ as operators that, when acting on a function $f(x)$, generate, respectively, the forward, backward, and central differences given in Eqs. (2-50) through (2-52).

The most natural way to compute the first-order derivative for $f(x)$ at $x = x_i$ is to use the central difference. In the notation of Eq. (2-3), this may be expressed as

$$f'_k = \frac{f_{k+1/2} - f_{k-1/2}}{h}. \tag{2-53}$$

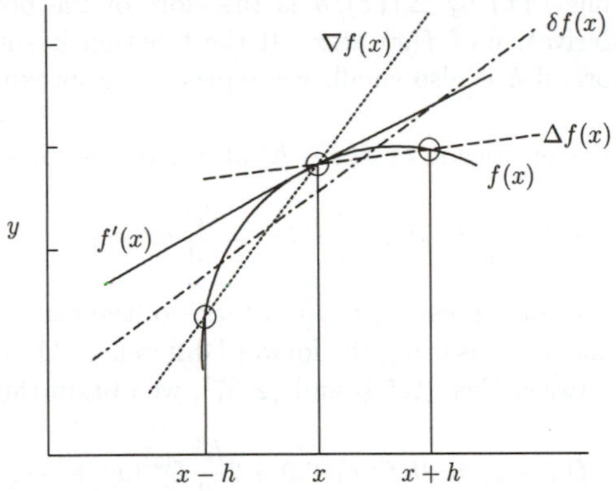

Figure 2-7: Schematic illustration of the various finite difference approximations to $f'(x)$: dashed line, forward difference $\Delta f(x)$; dotted line, backward difference $\nabla f(x)$; dash-dot line, central difference $\delta f(x)$; and exact value; solid line.

Mathematically, the relation is exact only in the limit $h \to 0$, as we have seen in Eq. (2-44). Here, we shall assume that h is sufficiently small that the relation holds for the numerical accuracy required. As usual, the expression implies that the abscissa is divided into grids of equal width h.

Often it may be undesirable to use half-intervals. In such cases, we can take the central difference over two adjacent subintervals

$$f'_k = \frac{f_{k+1} - f_{k-1}}{2h}. \tag{2-54}$$

A generalization of the idea behind Eq. (2-53) gives us the central difference form of the second-order derivative of $f(x)$:

$$f''_k = \frac{f'_{k+1/2} - f'_{k-1/2}}{h} = \frac{f_{k+1} - 2f_k + f_{k-1}}{h^2} \tag{2-55}$$

a form used earlier in Eq. (2-13).

For many applications, the central difference is preferred over forward and backward differences. This may be seen using a Taylor series expansion of $f(x+h)$ at x,

$$f(x+h) = f(x) + \frac{h}{1!}f'(x) + \frac{h^2}{2!}f''(x) + \frac{h^3}{3!}f^{(3)}(x) + \cdots. \tag{2-56}$$

To obtain the forward difference approximation for $f'(x)$, we can subtract $f(x)$ from both sides and rearrange terms. This gives us the result

$$f'(x) = \frac{1}{h}\{f(x+h) - f(x)\} - \frac{h}{2}f''(x) + \cdots.$$

The error in approximating $f'(x)$ by $\Delta f(x)/h$ is therefore of the order of h times $f''(x)$, the second-order derivative of $f(x)$ at x. If the function is smooth, $f''(x)$ is small in value. Furthermore, if h is also small, we expect the approximation to be a good one.

Similarly, a Taylor series expansion of $f(x-h)$ at x gives us the expression

$$f(x-h) = f(x) - \frac{h}{1!}f'(x) + \frac{h^2}{2!}f''(x) - \frac{h^3}{3!}f^{(3)}(x) + \cdots. \tag{2-57}$$

From this we obtain the error of expressing $f'(x)$ in terms of the backward difference. As expected, it is of the same order as using the forward difference. On the other hand, by taking the difference between Eqs. (2-56) and (2-57), we obtain the expression

$$f(x+h) - f(x-h) = 2hf'(x) + 0 + 2\frac{h^3}{3!}f^{(3)}(x) + \cdots.$$

All the terms involving odd powers of h vanish on the right side. From this we obtain the first-order derivative of $f(x)$ in the form

$$f'(x) = \frac{1}{2h}\{f(x+h) - f(x-h)\} - \frac{h^2}{3!}f'''(x) + \cdots. \tag{2-58}$$

The first term on the right side may be identified with the central difference given by Eq. (2-54). The error in approximating $f'(x)$ by the central difference here is smaller by one order in h than forward and backward differences.

In practice, it is difficult to achieve good accuracy in numerical differentiation as compared, for example, with numerical integration. The main reason comes from the fact that we are taking the ratio of two differences, for example, $\{f(x+h/2) - f(x-h/2)\}$ and $\{(x+h/2) - (x-h/2)\}$, as we saw earlier in Eq. (2-53). Each of these two differences is between two quantities of very similar values. Underlying our discussions, there is the implicit assumption that the function, whose derivative we are seeking, is a smooth one. If this is true, the values of the function at two nearby points must be very close to each other. The value of the derivative is then related to the small difference between two numbers that are large in comparison. Consider the case of a function that exists only as tabulated values obtained by measuring a physical quantity y at different values x_i. If the measurements are made with n significant figures, and the variation of y as a function of x is a smooth one, the difference between two adjacent values y_{i+1} and y_i is expected to be much smaller than either y_{i+1} and y_i. As a result, the number of significant figures in $\Delta y_i \equiv (y_{i+1} - y_i)$ is less than n. The same argument applies to the difference between x_{i+1} and x_i. The net result is that the number of significant figures we can have for any finite differences is smaller than n.

In addition to practical difficulties, there are also intrinsic limitations in replacing derivatives by finite differences. This is best illustrated by an example. Consider the exponential function

$$f(x) = e^{i\omega x}.$$

Here, the symbol i stands for the imaginary number, $i^2 = -1$, and ω represents the frequency of oscillation. Analytically, the first-order derivative of the function is

$$f'(x) = i\omega e^{i\omega x}. \tag{2-59}$$

To find the finite difference for $f(x)$, let us divide the abscissa x into equally spaced intervals of width h each and, as usual, use the symbol

$$x_k = x_0 + kh$$

to represent the location of the kth grid points from x_0 along the x-axis. The value of $f(x)$ at x_k is then

$$f_k = e^{i\omega x_0} e^{i\omega kh}.$$

The derivative calculated using a forward difference approximation is

$$f'(x_k) \approx \frac{1}{h}\Delta f_k = \frac{f_{k+1} - f_k}{h} = e^{i\omega x_0} e^{i\omega kh} \frac{e^{i\omega h} - 1}{h}.$$

Similarly, the backward and central difference results are, respectively,

$$\begin{aligned} f'(x_k) &\approx \frac{1}{h}\nabla f_k = \frac{f_k - f_{k-1}}{h} = e^{i\omega x_0} e^{i\omega kh} \frac{1 - e^{-i\omega h}}{h} \\ f'(x_k) &\approx \frac{1}{2h}\delta f_k = \frac{f_{k+1} - f_{k-1}}{2h} = e^{i\omega x_0} e^{i\omega kh} \frac{e^{i\omega h} - e^{-i\omega h}}{2h}. \end{aligned}$$

To check the accuracy of various finite difference approximations, let us take the ratio of each with the exact value given by Eq. (2-59). The results for the three differences are, respectively,

$$e^{i\omega h/2} \frac{\sin \frac{1}{2}\omega h}{\frac{1}{2}\omega h} \qquad e^{-i\omega h/2} \frac{\sin \frac{1}{2}\omega h}{\frac{1}{2}\omega h} \qquad e^{i\omega h} \frac{\sin \omega h}{\omega h}.$$

For any finite size h, the absolute values of these three ratios are always less than unity, unless $\omega \to 0$. In other words, the numerical results constantly underestimate the derivative except for $\omega = 0$. The reason is easy to understand. In all three cases, we have essentially used a linear approximation to the tangent of $f(x)$ at x, whereas $f(x)$ is a function that oscillates with a frequency related to ω.

It is not difficult to make some improvement on the accuracy for derivatives in numerical calculations. In the case of forward and backward difference approximations, we have been using essentially only two points of $f(x)$ to calculate the derivative. For the central difference approximation, we have gone one step further and made a three-point approximation by using f_{k-1}, f_k, and f_{k+1}, even though f_k itself does not appear explicitly in the final result. We can also adopt a five-point approximation by making use of a Taylor series expansion for $f(x \pm 2h)$, in a way analogous to what we did for $f(x \pm h)$ in Eqs. (2-57) and (2-58). The result is

$$f'_k = \frac{1}{12h}(f_{k-2} - 8f_{k-1} + 8f_{k+1} - f_{k+2}) + \frac{h^4}{30} f_k^{(5)} + \cdots. \tag{2-60}$$

The error is now in the order of h^4, an improvement of two orders in h over that of Eq. (2-58). Similarly, the central difference result for the second-order derivative becomes

$$f_k'' = \frac{1}{12h^2}(-f_{k-2} + 16f_{k-1} - 30f_k + 16f_{k+1} - f_{k+2}) + \frac{h^4}{90}f^{(6)}(x) + \cdots . \qquad (2\text{-}61)$$

The improvement is again two orders in h over that given by Eq. (2-55).

Problems

2-1 Show that the error associated with the trapezoidal rule of integration for the definite integral

$$I_{[a,b]} = \int_a^b f(x)\,dx$$

is $-(b-a)h^2 f''(\xi)/12$, where $f''(\xi)$ is the average value of the second-order derivative of $f(x)$ in the interval $x = [a, b]$.

2-2 Compare the efficiencies of calculating the normal probability function integral

$$A(x) = \frac{1}{\sqrt{2\pi}}\int_{-x}^{+x} e^{-t^2}\,dt$$

using the trapezoidal rule, Simpson's rule, and a Monte Carlo method.

2-3 Use a Gauss-Legendre quadrature of order $n = 4$ to evaluate the integral

$$I(a) = \int_{-a}^{+a} x^9\,dx$$

for an arbitrarily large value of a. The exact value for this integral is zero. Use the same method to calculate the integral

$$I(b) = \int_{-a}^{+a} x^8\,dx.$$

The exact value is $2a^9/9$. Compare the accuracies obtained for the two integrals and explain why there is a large difference.

2-4 For $0 < n < 1$, the following integral has the analytical solution

$$\int_0^\infty \frac{x^{n-1}}{1+x}\,dx = \frac{\pi}{\sin n\pi}$$

Use the trapezoidal or Simpson's rule to evaluate the integral. Compare the efficiency and accuracy of carrying out the calculation by splitting the integral into two parts

$$\int_0^\infty \frac{x^{n-1}}{1+x}\,dx = \int_0^a \frac{x^{n-1}}{1+x}\,dx + \int_a^\infty \frac{x^{n-1}}{1+x}\,dx.$$

For sufficiently large values of a, the second integral on the right side may be approximated by the result

$$\int_a^\infty \frac{x^{n-1}}{1+x}\,dx \approx \int_a^\infty x^{n-2}\,dx = \frac{a^{n-1}}{1-n}.$$

When does such a substitution improve the efficiency of the integration without sacrificing any accuracy?

2-5 Laguerre and Hermite polynomials are defined, respectively, in the ranges $[0, \infty]$ and $[-\infty, \infty]$ by Eqs. (4-81) and (4-13). Derive the numerical quadrature formulas analogous to Eq. (2-22) using Laguerre and Hermite polynomials instead of Legendre polynomials. Find the roots and weights for order $n = 4$ and check the results with those given in Abramowitz and Stegun.[1]

2-6 Use a Taylor series expansion to derive the central difference result given in Eq. (2-55) for the second-order derivative of $f(x)$ at $x = x_k$,

$$f''(x_k) = \frac{f_{k+1} - 2f_k + f_{k-1}}{h^2}.$$

Calculate the leading order correction term in powers of h.

2-7 Use the same method as for Problem 2-6 to derive the five-point approximate result

$$f''(x_k) = \frac{1}{12h^2}(-f_{k-2} + 16f_{k-1} - 30f_k + 16f_{k+1} - f_{k+2}) + \frac{h^4}{90}f^{(6)}(x) + \cdots$$

given earlier in Eq. (2-61).

2-8 The values of a function $f(x)$ are given only at a number of discrete points x_0, x_1, x_2, $\cdots$, x_N. If the spacings between the neighboring points are not the same, derive the equivalent forms for the three-point approximation for f''_k given in Eq. (2-55) and the five-point approximation for f'_k given in Eq. (2-60).

2-9 Use a Monte Carlo method to carry out the integral

$$I = \int_0^1 \frac{dx}{1+x^2}.$$

The result may be tested with the help of the relation

$$\int_0^1 \frac{dx}{1+x^2} = \tan^{-1} x\Big|_0^1 = \frac{\pi}{4}.$$

Is this a good way to obtain the value of π or just a way to test the random number generator used in the calculation?

2-10 Use a Monte Carlo approach to evaluate the integral

$$I(q) = \int_0^\infty \frac{e^{-qx}}{\sqrt{x}} dx$$

for $q = 0.5$, 1, and 2. Compare the results with the exact value of $\sqrt{\pi/q}$. It is also possible to use Eq. (2-48) for the numerical integration. Compare the accuracies with the Monte Carlo method for the same number of points.

Chapter 3

Interpolation and Extrapolation

A function is a mathematical expression that describes the relation between variables. For simplicity, let us consider a single-valued one with only one independent variable x. If y is a function of x, we can associate a value of y for every one of x. Often the dependence of y on x is given by some known analytical form but this may not always be the case. For example, we can measure the distances d_0, $d_1, \ldots$, d_N traversed by a car from some fixed reference point at times t_0, $t_1, \ldots$, t_N, respectively. In this case, the relation between d and the independent variable t is given by N pairs of numbers (t_0, d_0), $(t_1, d_1), \ldots$, (t_N, d_N). Unless we are provided with some additional information, the relation between distance d and time t is not available to us in terms of an analytical function. What happens if we wish to find out the value of d at some time t that is not one of those measured? If t is within the range of time $[t_0, t_N]$ where measurements have been taken, the value of d may be approximated using interpolation techniques. If t is outside the range, the corresponding technique is called extrapolation. As in any other approximation schemes, it is important that the method we apply to do calculations also supply us with an estimate of the error associated with the result. Interpolation and extrapolation techniques are also useful in cases where the functional relations are known but are too time consuming to evaluate. We shall see several examples of this kind.

3-1 Polynomial interpolation

In interpolation, we are interested in finding the most likely value of y for a given x using as input a table of the values of y at a finite number of points x_0, $x_1, \ldots$, x_N. This is illustrated by Fig. 3-1. In general, we do not have any knowledge of the relationship between y and x other than the $(N+1)$ pairs of values. Furthermore, there is usually no control over the values of x_i where those of y are given and, in general, we have no reason to assume that they are equally spaced.

In general, a continuous function $f(x)$ in a finite interval $x = [a, b]$ can always be fitted by a polynomial $P(x)$. This is known as the Weierstrass convergence theorem, a proof of which can be found on pages 273-74 in volume II of Courant and Hilbert.[12]

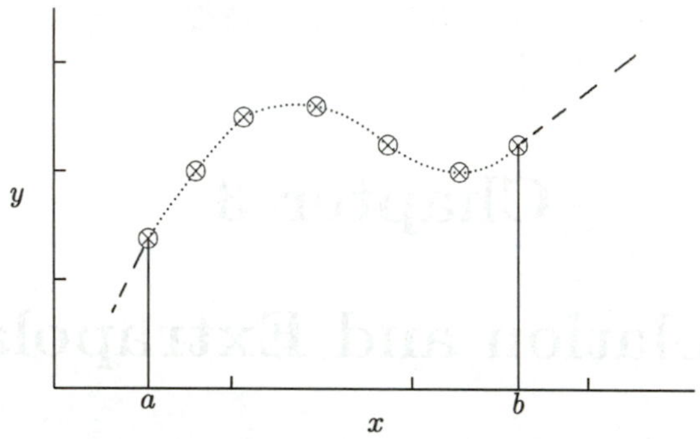

Figure 3-1: Interpolation and extrapolation of y as a function of x from input values indicated as $\otimes$. Interpolation gives other values of y for $a < x < b$ (dotted curve) and extrapolation, for x outside $[a, b]$ (dashed curve).

Our interest here is to find a polynomial approximation for $f(x)$, given to us in the form of $(N+1)$ pairs of numbers $\{x_i, f(x_i)\}$ for $i = 0, 1, \ldots, N$. This task may be achieved by constructing a polynomial of the form

$$P(x) = \sum_{k=0}^{N} p_k(x) f(x_k).$$

It is clear that if

$$p_k(x) = \frac{(x-x_0)(x-x_1)\cdots(x-x_{k-1})(x-x_{k+1})\cdots(x-x_N)}{(x_k-x_0)(x_k-x_1)\cdots(x_k-x_{k-1})(x_k-x_{k+1})\cdots(x_k-x_N)}$$

we have the result that $P(x_i) = f(x_i)$ for $i = 0, 1, \ldots, N$. Such a scheme is known as the Lagrange polynomial interpolation of $f(x)$. The polynomial $p_k(x)$ is of degree N, since the numerator is a product of N terms each of which has the form $(x - x_i)$, for $i = 0$ to N except for $i = k$. We shall assume that no two values of x_i are equal to each other, otherwise there will be singularities in $p_k(x)$.

Neville's algorithm A polynomial of degree m may also be built in the form of a power series with $(m+1)$ coefficients, $a_0, a_1, \ldots, a_m$:

$$P_m(x) = \sum_{k=0}^{m} a_k x^k. \tag{3-1}$$

If our function $f(x)$ has values given at $(N+1)$ points, it is possible to construct a polynomial of this form, with degree N, in such a way that $P_N(x)$ goes through each of the points where the values of the function are known. However, for the purpose of computer calculations, the Lagrange method given by Eq. (3-1) is not most suitable. Furthermore, it is not easy in such an approach to make an estimate of the errors associated with the calculated result.

In interpolation, our interest is purely local. For this reason, the polynomial needs only to fit the known values of the function in the small region where the results are needed. This is different from curve fitting where, as we shall see in Chapter 6, it is desirable to have a good overall representation in the entire interval. For this reason, there is often no point of making use of too many points in an interpolation. It complicates the calculations without necessarily improving the reliability of the final result.

Consider the following example of finding the approximate value of a function at x using the known values at five nearby points. Let the abscissas of these five points be x_1, x_2, x_3, x_4, and x_5. The corresponding values of y at these five points are, respectively, f_1, f_2, f_3, f_4, and f_5, and we have adopted the notation

$$f_i \equiv f(x_i) \tag{3-2}$$

as we have done in the previous chapter. Although the method outlined below is more general, we shall assume, for the sake of simplifying the explanation, that x_3 is the nearest point to x and x lies between x_3 and x_4.

The most naive estimate for the value y at x is given by the known value of $f(x)$ that is closest to x. In our case, we have assumed that this is the point x_3. The zeroth-order approximation to y is then f_3. For a better estimate, we note that since x lies between x_3 and x_4, we can make a linear interpolation between the known values of $f(x)$ at these two points. Let us represent this new approximate value as

$$f_{34} = \frac{1}{x_3 - x_4}\{(x - x_4)f_3 - (x - x_3)f_4\}. \tag{3-3}$$

The notation here is that f_{ij} is the linearly interpolated value between f_i and f_j. It is easy to see that f_{34} is a better approximation to the value of y at x than f_3, as it makes use of the information provided at two nearby points.

A further improvement may be achieved by making a quadratic interpolation. In addition to f_3 and f_4, we shall also include f_2 as a part of the input. There are several ways to do this. The most convenient one for the purpose of programming a computer is to adopt *Neville's algorithm* and make a linear interpolation between the values of f_{34} and f_{23}. This gives us the result

$$f_{234} = \frac{1}{x_2 - x_4}\{(x - x_4)f_{23} - (x - x_2)f_{34}\}. \tag{3-4}$$

This is very similar in spirit to what we have done in Eq. (3-3). The advantage in this approach comes from the fact that Eqs. (3-3) and (3-4) involve the same types of operation and, as a result, may be put into a recursive form, as we shall see later in Eq. (3-6).

Table 3-1: Neville's method of interpolation from five known values.

x_1	f_1				
		f_{12}			
x_2	f_2		f_{123}		
		f_{23}		f_{1234}	
x_3	f_3		f_{234}		f_{12345}
		f_{34}		f_{2345}	
x_4	f_4		f_{345}		
		f_{45}			
x_5	f_5				

To go even further, it is necessary to make a cubic approximation. This may be done by using f_{2345}, a linear interpolation between f_{234} and f_{345}. Similarly, by a linear interpolation between f_{123} and f_{234}, we obtain f_{1234}. Using these two values, we can carry out the next and final step for our interpolation with five pairs of input values. The result is

$$f_{12345} = \frac{1}{x_1 - x_5}\{(x - x_5)f_{1234} - (x - x_1)f_{2345}\}. \tag{3-5}$$

Eqs. (3-3) to (3-5) may be put in the form of a recursion relation,

$$f_{ij\ldots k\ell} = \frac{1}{x_i - x_\ell}\{(x - x_\ell)f_{ij\ldots k} - (x - x_i)f_{j\ldots k\ell}\}. \tag{3-6}$$

That is, the value of $f_{ij\ldots k\ell}$ is the result of a linear interpolation of two similar quantities that are nearest it but with one subscript less. The quantities $f_{ij\ldots k\ell}$, with m subscripts, are polynomials of x with leading order $(m-1)$. The starting point of this set of relations is the f_i given in Eq. (3-2). Note that, in calculating the values of $f_{ij\ldots m}$ with m subscripts, we need x, x_i, x_m, together with the values of $f_{i,j,\ldots}$ and $f_{jk\ldots}$ having one less subscript, as can be seen from examining Eq. (3-6).

The successive approximations to $f(x)$ in terms of $f_{ij\ldots k\ell}$ may be represented by the diagram shown in Table 3-1. In general, for an interpolation with N input values, the highest polynomial order we can use is $(N-1)$. This, in turn, means that the maximum number of subscripts for $f_{ij\ldots k\ell}$ is N. The essence of the method is to make small adjustments in each step to the calculated value of y by incorporating one more piece of input information. In the process, the polynomial degree is also increased by 1. The uncertainty in the value obtained may be estimated from the difference between two successful steps, given essentially by the difference between two adjacent columns in Table 3-1. Thus, the difference between f_{2345} (or f_{1234}) and f_{12345} provides us with an error estimate of the final result.

A simple numerical illustration will give us a feeling of the way the algorithm works in practice. Let us try to find the value of the error function

$$\mathrm{erf}(x) = \frac{2}{\sqrt{\pi}}\int_0^x e^{-t^2}\,dt \tag{3-7}$$

Table 3-2: Interpolation of erf(x) for $x = 0.52$ from five values given in $x = [0.3, 0.7]$.

x_i	f_i	f_{ij}	f_{ijk}	$f_{ijk\ell}$	$f_{ijk\ell m}$
0.3	0.3286268				
		0.5481111			
0.4	0.4283924		0.5380024		
		0.5389214		0.5379062	
0.5	0.5204999		0.5378712		0.5378986
		0.5371711		0.5378923	
0.6	0.6038561		0.5379240		
		0.5447000			
0.7	0.6778012				

near the point $x = 0.5$ by interpolation. As input, we shall use the five values of erf(x) at $x = 0.3$, 0.4, 0.5, 0.6, and 0.7. These are given in the second column of Table 3-2. If we wish to find the value of erf(x) at $x = 0.52$, we can calculate the four linear terms f_{ij} associated with this value of x using Eq. (3-6), and these are listed in the third column. The fourth column gives the three values of f_{ijk} in the second-order approximation, the fifth column gives the two values in the third-order, and so on. The final value for a fourth-order approximation is $f_{12345} = 0.5378986$ given in the last column. The error of the interpolation process is estimated to be around 7×10^{-6}, as can be seen from the differences between the values in the fifth and sixth columns. For smooth functions, the actual accuracy is usually better than that indicated by the error estimate. For this reason, it should not be a surprise for us to find that, for our error function example, the calculated result happens to agree with the exact value up to the seven significant figures listed.

The example also serves as a good illustration of the applications of interpolation techniques. Since the error function is defined only in terms of an integral, the values are cumbersome to calculate and are usually found by looking up a table. However, the entries in a mathematical table are limited and interpolation is a convenient way to obtain accurate results for values not listed.

Improvements on the algorithm The recursion relation for $f_{ij\ldots k\ell}$ in Eq. (3-6) is not the ideal one for computation. The reason is that we are making small corrections by taking weighted averages between two relatively large numbers. Better accuracies may be obtained if, at each stage, we deal directly with the small corrections themselves. For this purpose, we shall define two related types of difference. The first is between two values of $f_{ij\ldots k\ell}$ with the last subscript differing by unity,

$$D_{i,\ell} \equiv f_{i,i+1,\ldots,i+\ell} - f_{i,i+1,\ldots,i+\ell-1}. \tag{3-8}$$

For the lowest order of $\ell = 1$, we take

$$D_{i,1} = f_i.$$

Box 3-1 Subroutine NVLLE(X_IN,F_IN,N,X,FX,DF)
Neville's algorithm for interpolation

Argument list:
- X_IN(I): Input abscissa.
- F_IN(I): Input values of $f(x)$ at x =X_IN(I).
- N: Number of input points.
- X: Value of x to be calculated.
- FX: Returned value of $f(x)$.
- DF: Estimated error for FX.

Initialization: (in the calling program)
- (a) Store a table of the values of $\{x_i, f(x_i)\}$ for $i = 1$ to N.
- (b) Input the value of x.

1. Define U and D of Eq. (3-10) and let IDX= i to indicate the x_i nearest to x.
2. Make an initial estimate using only the table entry nearest to x. Determine if x is above x_{IDX} (UP = .TRUE.) or below (UP = .FALSE.).
3. Improving the estimate by higher-order interpolations:
   - (a) Let the order $L = 1$.
   - (b) If UP is .TRUE.:
     - (i) Decrease IDX by 1.
     - (ii) If IDX≥ 1, increase the estimate using U_{IDX}.
     - (iii) Otherwise, set IDX=1 and use D_{IDX} instead.
     - (iv) Change UP to .FALSE. and go to step (d).
   - (c) If UP is .FALSE.:
     - (i) Increase the estimate using D_{IDX} if IDX $> (N - L)$.
     - (ii) Otherwise, set IDX equal to $(N - L)$ and use D_{IDX}.
     - (iii) Change UP to .TRUE..
   - (d) Increase L by 1 and go back to step (b) until L reaches the maximum.
4. Return the interpolated value and the last improvement as the estimate of error.

In terms of the quantities given in Table 3-1, we are moving downward along the diagonal, as $f_{i,i+1,\ldots,i+\ell-1}$ is one row above and one column to the left of $f_{i,i+1,\ldots,i+\ell}$.

In the absence of a better name, we shall call $D_{i,\ell}$ the "downward difference." There is no point in indicating the intervening subscripts of f from which $D_{i\ell}$ are constructed, as they are always increasing in value by unity each time we move one index to the right. Similarly, we can also define an "upward difference" by taking the difference between two values of $f_{i,i+1,\ldots,i+m}$, with the first subscript differing by unity:

$$U_{i,\ell} \equiv f_{i,i+1,\ldots,i+\ell} - f_{i+1,i+2,\ldots,i+\ell}. \tag{3-9}$$

The starting point for this set of differences is $U_{i,1} = f_i$. We can work out the recursion relations between $D_{i,\ell}$ and $U_{i,\ell}$ by substituting $f_{ij\ldots k\ell}$ given in Eq. (3-6) into Eqs. (3-8)

Table 3-3: Interpolated values of erf(x) using the same five inputs as Table 3-2.

x	Tabulated value	Interpolated value	Estimated error
0.46	0.4846554	0.4846556	9.7×10^{-6}
0.47	0.4937451	0.4937451	8.4×10^{-6}
0.48	0.5027497	0.5027497	6.2×10^{-6}
0.49	0.5116683	0.5116683	3.4×10^{-6}
0.51	0.5292436	0.5292436	3.4×10^{-6}
0.52	0.5378986	0.5378986	6.2×10^{-6}
0.53	0.5464641	0.5464640	8.4×10^{-6}
0.54	0.5549393	0.5549392	9.7×10^{-6}

and (3-9). The results are

$$D_{i,\ell+1} = \frac{x_i - x}{x_i - x_{i+\ell+1}}\{D_{i+1,\ell} - U_{i,\ell}\} \qquad U_{i,\ell+1} = \frac{x_{i+\ell+1} - x}{x_i - x_{i+\ell+1}}\{D_{i+1,\ell} - U_{i,\ell}\}. \tag{3-10}$$

The derivation is left as an exercise (Problem 3-3).

Since both $D_{i,\ell}$ and $U_{i,\ell}$ are "corrections" to the previous order of approximation, we must make a choice as to which of these two quantities to adopt at each order. This can be done by following the way we make use of the successive orders of $f_{ij\ldots k\ell}$. Earlier, we have assumed that the value of x is near the value of x_i in the middle of the table of known values. For the convenience of discussion, let us give the name i_x to this value of the subscript. At the starting point of the interpolation, we have $i_x = i$. In the zeroth-order approximation, we take $f(x) = f_i$.

If $x > x_i$, the next-order approximation comes from $f_{i-1,i}$, following the way the subscripts are defined here. Since the difference between $f_{i-1,i}$ and f_i is given by $U_{i-1,2}$, we shall use $U_{i-1,2}$ to generate the next-order correction in the present scheme. The difference between $f_{i-1,i}$ and $f_{i-1,i,i+1}$ is given by $D_{i-1,3}$, and we shall make use of $D_{i-1,3}$ next. Continuing in this way with the help of Table 3-1, we come quickly to the conclusion that, for $x > x_i$, we start from $f_i = D_{i,1}$ and the successive corrections are $U_{i-1,2}$, $D_{i-1,3}$, $U_{i-2,4}$, $D_{i-2,5}$, and so on. A general scheme to obtain successive corrections to the interpolated value of $f(x)$ for $x > x_i$ may now be constructed in the following way. We start with f_i and alternate between using U and D for each higher order. Every time we use U, we decrease the value of the first subscript i_x by 1 and increase the value of the second subscript by an equal amount.

A similar approach can also be obtained for $x < x_i$. In this case we start with f_i and alternate between using D and U, with the value of i_x decreased by 1 every time we encounter an U. A summary of Neville's algorithm in terms of the upward and downward differences is given in Box 3-1. As an example, we shall calculate the values of erf(x) for $x = 0.46$ to 0.54 in steps of 0.01, using the same set of five input pairs as in Table 3-2 and the results are compared with exact values in Table 3-3.

3-2 Interpolation using rational functions

In general, the use of polynomials in approximations works best for smooth functions without any singularities either within the range of interest or near one of the end points. Unfortunately, these conditions are not met by a large class of functions, such as $\tan x$ near $x = \pi/2$ and $\ln x$ for small values of x. In these cases, it is more advantageous to make use of rational functions consisting of ratios of two polynomials. In fact, for the purpose of interpolation, rational functions are often superior even when the function is well behaved.

Just as in the case of polynomial approximations, there are many different ways to construct an approximation based on rational functions. The more popular approach is to use the Padé approximation. However, for computational purposes, a slightly different approach is suggested by Stoer and Bulirsch.[54] The idea is very similar to Neville's algorithm for polynomial interpolation. Among other advantages, the method provides with us an error estimate for the results as well as a natural way to extend the approximation to higher orders when needed.

The method of Stoer and Bulirsch Let us again assume that the function $f(x)$ in the interval $x = [a, b]$ is given to us in terms of $(N + 1)$ pairs of values (x_0, f_0), $(x_1, f_1), \ldots, (x_N, f_N)$. For convenience, we shall arrange the values in ascending order according to x, that is, $x_0 < x_1 < \cdots < x_N$. For such a table, we can write the rational function for interpolation in the following way:

$$R_{\mu\nu s}(x) = \frac{P_{\mu\nu s}(x)}{Q_{\mu\nu s}(x)} = \frac{p_{0;\mu\nu s} + p_{1;\mu\nu s}x + p_{2;\mu\nu s}x^2 + \cdots + p_{\mu;\mu\nu s}x^\mu}{1 + q_{1;\mu\nu s}x + q_{2;\mu\nu s}x^2 + \cdots + q_{\nu;\mu\nu s}x^\nu} \tag{3-11}$$

where μ is the degree of the polynomial in the numerator and ν, that of the polynomial in the denominator. The third subscript s is needed to distinguish the different independent rational functions that can be constructed. Some further explanation for this label is required.

Since the constant term in the denominator is set to unity, the function $R_{\mu\nu s}(x)$ has only $(\mu + \nu + 1)$ parameters. It is therefore possible to make $R_{\mu\nu s}(x)$ equal to $f(x)$ at $(\mu + \nu + 1)$ points. We shall choose $R_{\mu\nu s}(x)$ in such a way that

$$R_{\mu\nu s}(x_i) = f_i \qquad \text{at} \qquad x = x_s, x_{s+1}, \ldots, x_{s+\mu+\nu}.$$

In other words, the function $\{R_{\mu\nu s}(x) - f(x)\}$ has $(\mu + \nu + 1)$ nodes and they occur at $x = x_s$, $x_{s+1}, \ldots, x_{s+\mu+\nu}$. As a result, rational functions having the same degrees μ and ν but differing in the location of their first node may be distinguished by the label s. Furthermore, the numerator polynomial $P_{\mu\nu s}(x)$ and denominator polynomial $Q_{\mu\nu s}(x)$ for a given μ, ν, and s are related to each other, since, in order to have roots at the given locations, it is impossible to change one of the two in a nontrivial way without having to modify the other. For this reason, coefficients $p_{k;\mu\nu s}$ of the numerator polynomial $P_{\mu\nu s}(x)$, and $q_{k;\mu\nu s}$ of the denominator polynomial $Q_{\mu\nu s}(x)$, are labeled by four indices, k, the power of x it is associated with, as well as μ, ν, and s.

One strength of the Stoer-Bulirsch algorithm is that rational functions $R_{\mu\nu s}(x)$ may be constructed in a recursive manner. This is especially true if we restrict ourselves to those with $\nu = \mu$ or $\nu = (\mu+1)$. In these cases, it is possible to simplify the notation somewhat by using a single index,

$$m = \mu + \nu$$

to label the polynomial degree. It is obvious that m is even for $\nu = \mu$ and odd for $\nu = \mu+1$. Starting with $R_{0s} = f_s$ and $R_{ms}(x) = 0$ for $m < 0$, Stoer and Bulirsch give

$$R_{ms}(x) = R_{(m-1)(s+1)}(x) + \frac{R_{(m-1)(s+1)}(x) - R_{(m-1)s}(x)}{\frac{\alpha_s}{\alpha_{m+s}}\left\{1 - \frac{R_{(m-1)(s+1)}(x) - R_{(m-1)s}(x)}{R_{(m-1)(s+1)}(x) - R_{(m-2)(s+1)}(x)}\right\} - 1}. \tag{3-12}$$

The derivation of this relation is fairly straightforward but somewhat too tedious to be repeated here.

Recursion relation between differences The recursion relation for rational functions given by Eq. (3-12) is analogous to that given by Eq. (3-6) for polynomials. As we have seen in the previous section, it is advantageous for computational purposes to use instead recursion relations for the differences between rational functions of different degrees, as we have done in Eq. (3-10) for Neville's algorithm. For this purpose, we shall again define two difference functions:

$$\Delta_{ms} \equiv R_{ms}(x) - R_{(m-1)s}(x) \tag{3-13}$$

$$\Theta_{ms} \equiv R_{ms}(x) - R_{(m-1)(s+1)}(x). \tag{3-14}$$

To find the recursion relations between Δ_{ms} and Θ_{ms}, we can start with Eq. (3-12) and obtain the result

$$\Theta_{ms} = \frac{R_{(m-1)(s+1)}(x) - R_{(m-1)s}(x)}{\frac{\alpha_s}{\alpha_{m+s}}\left\{1 - \frac{R_{(m-1)(s+1)}(x) - R_{(m-1)s}(x)}{R_{(m-1)(s+1)}(x) - R_{(m-2)(s+1)}(x)}\right\} - 1}. \tag{3-15}$$

Note that, by taking the difference between Eqs. (3-13) and (3-14) for $(m+1)$ instead of m, we have the identity

$$\Delta_{(m+1)s} - \Theta_{(m+1)s} = R_{m(s+1)}(x) - R_{ms}(x). \tag{3-16}$$

Similarly, if we change s to $(s+1)$ in Eq. (3-13) before taking the difference, we have

$$\Delta_{m(s+1)} - \Theta_{ms} = R_{m(s+1)}(x) - R_{ms}(x). \tag{3-17}$$

On substituting these two equations into the right side of Eq. (3-15), we obtain one of the two recursion relations for Δ_{ms} and Θ_{ms},

$$\Theta_{(m+1)s} = \frac{(\Delta_{m(s+1)} - \Theta_{ms})\Delta_{m(s+1)}}{\frac{\alpha_s}{\alpha_{m+s+1}}\Theta_{ms} - \Delta_{m(s+1)}}. \tag{3-18}$$

Box 3-2 Subroutine RATNV(X_IN,F_IN,N,X,FX,DF)
Rational function interpolation using Neville's algorithm

Argument list:
- X_IN: Input abscissa array
- F_IN: Input array of $f(x)$ corresponding to X_IN
- N: Number of input points
- X: Value of x to be interpolated
- FX: Output value of f(x)
- DF: Estimated error for FX

1. Lowest-order approximation.
   - (a) Set up index IDX to indicate the table entry nearest to x.
   - (b) Define Δ_{0s} and Θ_{0s} to be equal to $f(x_s)$.
   - (c) Let $f(x)$ equal to the nearest input value.
   - (d) Put UP as (.TRUE.) for $x < x_i$ and (FALSE.) otherwise.
2. Improve the estimate using higher-order approximations.
   For $L = 1$ to maximum order allowed, repeat the following steps:
   - (a) Calculate Δ and Θ according to Eqs. (3-18) and (3-19).
   - (b) If UP is .TRUE.:
     - (i) Decrease IDX by 1.
     - (ii) If IDX≥ 1, increase the estimate of $f(x)$ using Θ.
     - (iii) Otherwise, set IDX=1 and use Δ.
     - (iv) Change UP to .FALSE. and go to the next L.
   - (c) If UP is .FALSE.:
     - (i) Increase the estimate by Δ if IDX> $(N - L)$.
     - (ii) Otherwise, set IDX= $(N - L)$ and use Θ.
     - (iii) Change UP to .TRUE. and go to the next L.
3. Return the interpolated value and the last improvement as the error estimate.

The other may be found by noting the fact that, from Eqs. (3-16) and (3-17), we have the expression

$$\Delta_{(m+1)s} - \Theta_{(m+1)s} = \Delta_{m(s+1)} - \Theta_{ms}.$$

Using this equivalence, we can write

$$\begin{aligned}
\Delta_{(m+1)s} &= \Theta_{(m+1)s} + \Delta_{m(s+1)} - \Theta_{ms} \\
&= \frac{(\Delta_{m(s+1)} - \Theta_{ms})\Delta_{m(s+1)}}{\frac{\alpha_s}{\alpha_{m+s+1}}\Theta_{ms} - \Delta_{m(s+1)}} + (\Delta_{m(s+1)} - \Theta_{ms}) \\
&= \frac{\frac{\alpha_s}{\alpha_{m+s+1}}\Theta_{ms}(\Delta_{m(s+1)} - \Theta_{ms})}{\frac{\alpha_s}{\alpha_{m+s+1}}\Theta_{ms} - \Delta_{m(s+1)}}.
\end{aligned} \tag{3-19}$$

These two relations are analogous to Eq. (3-10) for polynomials. By starting with

$$\Delta_{0s} = \Theta_{0s} = f_s$$

Table 3-4: Rational function interpolation for $\tan x$ near $x = \pi/2$. Values in parentheses are error estimates from interpolation.

Input		Interpolation			
x	$\tan x$	x	Exact	Rational	Polynomial
1.1	1.9647597	1.15	2.2344969	2.2344921 (2×10^{-4})	2.0980797 (2×10^{-2})
1.2	2.5721516	1.25	3.0095697	3.0095730 (2×10^{-4})	3.0862589 (1×10^{-1})
1.3	3.6021024	1.35	4.4552218	4.4552164 (6×10^{-4})	4.3438010 (1×10^{-1})
1.4	5.7978837	1.45	8.2380928	8.2381449 (3×10^{-3})	8.7133064 (2×10^{-1})
1.5	14.1014200				

we have now a method for rational function interpolation that is very similar to Neville's algorithm for polynomial interpolation. This is true also for the strategy for making a choice between Δ_{ms} and Θ_{ms} to improve the interpolated result at each stage. The algorithm is summarized in Box 3-2.

As an application, we shall use the method of rational function interpolation to calculate the values of $\tan x$ for x near the singular point at $x = \pi/2$. The input values are chosen to be at x = 1.1, 1.2, 1.3, 1.4, and 1.5. Since tangent rises very fast as $x \to \pi/2$, the spacing of 0.1 between two adjacent input x values is rather large. This can be seen from the values given in the second column of Table 3-4. We have made this choice on purpose so as to demonstrate the power of rational function interpolation. For the purpose of comparison, we have included also the results obtained with polynomial interpolation, as well as the exact values. We see from the table that the results obtained by using rational functions are always superior to those with polynomials. Later in §3-5, we shall once again see the power of using rational polynomials in extrapolations with singularities nearby.

3-3 Continued fraction

A natural extension of rational polynomial approximation for a function is continued fractions. Here, the value f of a function at x is given in the form

$$f = b_0 + \frac{a_1}{b_1 + \frac{a_2}{b_2 + \frac{a_3}{b_3 + \cdots}}}. \tag{3-20}$$

The coefficients a_1, $a_2, \ldots$, are the partial numerators and b_0, $b_1, \ldots$, the partial denominators. In many mathematical tables, Eq. (3-20) is often written in a more compact form as

$$f = b_0 + \frac{a_1}{b_1 +}\frac{a_2}{b_2 +}\frac{a_3}{b_3 +}\cdots. \tag{3-21}$$

In addition to its role in interpolation, a continued fraction is also a convenient way to evaluate certain functions. For example, on page 81 of Abramowitz and Stegun,[1] the inverse of tangent is given as

$$\tan^{-1} x = \frac{x}{1+}\frac{x^2}{3+}\frac{4x^2}{5+}\frac{9x^2}{7+}\frac{16x^2}{9+}.$$

Similarly, coefficients for many other commonly used functions are available from a variety of sources. Derivations of these results can be found in, for example, Hildebrand.[26] Our interest here is to construct a method to evaluate the continued fraction.

Method of evaluation The form of a continued fraction, as it stands in the way given by Eq. (3-21), must be evaluated from right to left. This is different from the usual way a series is evaluated — from left to right, one term at a time until convergence. As a result, different techniques are needed to arrive at an efficient algorithm.

The following theorems, given on page 19 of Abramowitz and Stegun,[1] are useful for calculating the values of a continued fraction:

(I) If a_i and b_i are positive, then

$$f_{2n} < f_{2n+2} \qquad \text{and} \qquad f_{2n-1} > f_{2n+1} \tag{3-22}$$

where

$$f_n = \frac{A_n}{B_n} = b_0 + \frac{a_1}{b_1+}\frac{a_2}{b_2+}\cdots\frac{a_n}{b_n}. \tag{3-23}$$

(II) The quantities A_n and B_n obey the recursion relations

$$A_n = b_n A_{n-1} + a_n A_{n-2} \qquad B_n = b_n B_{n-1} + a_n B_{n-2} \tag{3-24}$$

with

$$B_{-1} = 0 \qquad A_{-1} = B_0 = 1 \qquad A_0 = b_0. \tag{3-25}$$

The proof of these two theorems may be carried out by induction but we shall not do it here. Our interest lies in the fact that A_n and B_n may be constructed using Eq. (3-24), with coefficients a_i and b_i given to us, for example, from a mathematical table. From the ratio of these two quantities, we can calculate f_n using Eq. (3-23). Once this is done, these two steps are repeated for $(n+1)$, and so on.

The accuracy of a calculation may be estimated from the differences in the results for two successive values of n. However, because of item (I) above, it is better if we compare f_n with f_{n+2}, rather than with f_{n+1}. If the difference is smaller than the accuracy required, we can say that convergence is achieved.

As an application, we shall calculate the incomplete gamma function described later in §4-5 and used in §6-5 to evaluate $Q(\chi^2|\nu)$, the probability integral for a χ^2 distribution. The continued fraction form of the function is given on page 263 of Abramowitz and Stegun:[1]

$$\Gamma(a,x) = \int_x^\infty e^{-t}t^{a-1}dt = e^{-x}x^a\left\{\frac{1}{x+}\frac{1-a}{1+}\frac{1}{x+}\frac{2-a}{1+}\frac{2}{x+}\cdots\right\}. \tag{3-26}$$

Box 3-3 Function GMMAI(a, x)
Continued fraction approximation to incomplete gamma function

Initialization:
- (a) Set the maximum number of steps to be 100 and accuracy to be 5×10^{-7}.
- (b) Let $A_0 = 0$, $B_1 = A_0 = 1$, $B_1 = x$, and $f = 1$ according to Eq. (3-25).

For $n = 1$ to the maximum of steps, carry out the following:
- (a) Calculate A_{2n} and B_{2n} using Eq. (3-24) and divide the results by f.
- (b) Divide A_{2n-1} and B_{2n-1} by f and find A_{2n+1} and B_{2n+1} using Eq. (3-24).
- (c) If B_{2n+1} does not vanish,
  - (i) Set $f = B_{2n+1}$.
  - (ii) Calculate $f_{2n+1} = A_{2n+1}/B_{2n+1}$.
- (d) Go to the next n if $|f_{2n+1} - f_{2n-1}| >$ tolerance for error.
  Otherwise return $\Gamma(a, x) = e^{-x}x^a f_{2n+1}$.

As a practical matter, Eq. (3-26) is usually used only for $x < (a+1)$ where convergence is fast. For other values of x, a series form given later in §4-5 is preferred.

Let us start by constructing a table of the partial numerators a_i and denominators b_i from the continued fraction of Eq. (3-26):

i	0	1	2	3	4	5	6	7	
a_i		1	$1-a$	1	$2-a$	2	$3-a$	3	$\cdots$
b_i	0	x	1	x	1	x	1	x	$\cdots$

Using Eq. (3-24), we find that

$$A_1 = b_1 A_0 + a_1 A_{-1} = 1 \qquad B_1 = b_1 B_0 + a_1 B_{-1} = x. \tag{3-27}$$

The table gives that, for $i > 1$,

$$a_i = \begin{cases} i/2 - a & \text{for } i = \text{even} \\ (i-1)/2 & \text{for } i = \text{odd} \end{cases} \qquad b_i = \begin{cases} 1 & \text{for } i = \text{even} \\ x & \text{for } i = \text{odd}. \end{cases} \tag{3-28}$$

This allows us to simplify the recursion relations Eq. (3-24) to the form

$$\begin{aligned} A_{2n} &= A_{2n-1} + (n-a)A_{2n-2} & B_{2n} &= B_{2n-1} + (n-a)B_{2n-2} \\ A_{2n+1} &= xA_{2n} + nA_{2n-1} & B_{2n+1} &= xB_{2n} + nB_{2n-1}. \end{aligned} \tag{3-29}$$

The only other refinement required here is to prevent the values of A_n and B_n from becoming too large. Since we are only interested in their ratio, the magnitudes of the individual coefficients are, in principle, of no direct concern to us. However, in computation, large values of either quantities can cause floating number overflows, even though the ratios between them remain finite. To prevent such problems without

Table 3-5: Examples of incomplete gamma function $\Gamma(a,x)$ by continued fraction.

a	x	$\Gamma(a,x)$	$\Gamma(a)$	ν	χ^2	$Q(\chi^2\|\nu)$
0.5	0.5	0.56241876	$\sqrt{\pi}$	1	1	0.31731
1.0	0.5	0.60653061	1	2	1	0.60653
1.5	0.5	0.71009004	$\frac{1}{2}\sqrt{\pi}$	3	1	0.80125
2.0	0.5	0.90979600	1	4	1	0.90980
2.5	0.5	1.2795792	$\frac{3}{4}\sqrt{\pi}$	5	1	0.96257

changing their ratio, we can divide both A_{2n} and B_{2n} by a constant. For convenience, we can use the value of B_{2n-1} for this purpose.

The calculation is carried out in the following way. Starting from $n = 1$, we evaluate in each step both n and $n+1$, first A_{2n} and B_{2n} and then A_{2n+1} and B_{2n+1}. At the end of each step, the value of f_{2n+1}, calculated using Eq. (3-23), is compared with that of f_{2n-1} to see if convergence is achieved. The reason for taking this approach comes from the fact that convergence is guaranteed among f_n only for n differing by 2, as stated in Eq. (3-22) above. The method is outlined in Box 3-3.

For our example, let us check the value of $\Gamma(a,x)$ obtained in terms of the probability integral for χ^2 distribution

$$Q(\chi^2|\nu) = \frac{1}{2^{\nu/2}\Gamma(\nu/2)} \int_{\chi^2}^{\infty} e^{-t/2} t^{\nu/2-1} dt = \frac{\Gamma(a,x)}{\Gamma(a)} \tag{3-30}$$

where $a = \frac{1}{2}\nu$, $x = \frac{1}{2}\chi^2$, and $\Gamma(a)$ is the ordinary gamma function described later in §4-5. The values of the probability integral are given in standard mathematical tables. For the calculated results shown in Table 3-5, the values of $Q(\chi^2|\nu)$ obtained are identical to those tabulated up to the number of significant figures shown.

3-4 Fourier transform

A periodic function is one that repeats itself after a fixed interval. For example, if

$$f(x) = f(2L + x) \tag{3-31}$$

the function $f(x)$ has a period of $2L$. One important property of such a function is that it can be expressed in terms of a Fourier series,

$$f(x) = \frac{1}{2}a_0 + \sum_{m=1}^{\infty}\Big(a_m \cos\frac{m\pi}{L}x + b_m \sin\frac{m\pi}{L}x\Big). \tag{3-32}$$

Using the orthogonality relations between trigonometry functions, the coefficients a_m and b_m are given by

$$a_m = \frac{1}{L}\int_{-L}^{L} f(x)\cos\frac{m\pi}{L}x\,dx \qquad b_m = \frac{1}{L}\int_{-L}^{L} f(x)\sin\frac{m\pi}{L}x\,dx. \tag{3-33}$$

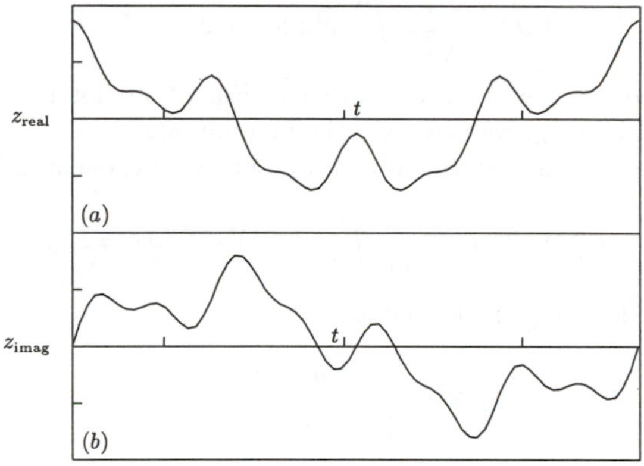

Figure 3-2: Real and imaginary parts of $z = \sum_{\ell=1}^{3} a_\ell \exp\{i\omega_\ell t\}$ for $(a_\ell, \omega_\ell) = (1.0, 1.0)$, $(0.5, 2.0)$, and $(0.25, 8.0)$. Using Fourier transform, the values of the three amplitudes and frequencies may be obtained from those of z.

Alternatively, we can make use of exponential functions,

$$f(x) = \sum_{\ell=-\infty}^{\infty} g_\ell e^{i\frac{\ell\pi}{L}x} \tag{3-34}$$

with

$$g_\ell = \frac{1}{2L}\int_{-L}^{L} f(x)e^{-i\frac{\ell\pi}{L}x}dx. \tag{3-35}$$

The factor $e^{i\ell x}$ is a complex number with absolute value equal to unity. Two such factors, e^{imx} and e^{inx} with $m \neq n$, differ from each other by only a phase angle. For this reason, the right side of Eq. (3-34) is often called a phase polynomial. The relation between a function $f(x)$ and its Fourier coefficients is illustrated by Fig. 3-2.

Since the relation given by either Eq. (3-32) or (3-34) is exact, a complete set of Fourier coefficients $\{a_m, b_m\}$, for $m = 0, 1, \ldots, \infty$ (with $b_0 = 0$), or $\{g_\ell\}$ for $\ell = -\infty$ to $+\infty$, provides an alternative way to specify the function. On the other hand, if the functional form of $f(x)$ is unknown and the relation between x and $f(x)$ is given to us in the form of $(N+1)$ pairs of values (x_i, f_i) for $i = 0, 1, \ldots, N$, we can determine the values of at most $(N+1)$ coefficients. Furthermore, since the function is periodic, all the coefficients may be determined from the information on the function within a single cycle.

Fourier transform If we take the limit $L \to \infty$, the summation index m in Eq. (3-32) becomes a continuous variable. For later convenience, we shall write it as ω. Similarly, the summation over ℓ becomes an integration over ω and Eq. (3-34) takes on the form

$$f(x) = \frac{1}{\sqrt{2\pi}} \int_{-\infty}^{\infty} g(\omega) e^{i\omega x} d\omega. \tag{3-36}$$

A normalization factor of $1/\sqrt{2\pi}$ is introduced in Eq. (3-36) for later convenience and, in place of coefficients g_ℓ, we now have the function $g(\omega)$.

For m and n integers, the orthogonal relation between exponential functions is

$$\frac{1}{2\pi} \int_{-\pi}^{\pi} (e^{imx})^* e^{inx} dx = \frac{1}{2L} \int_{-L}^{L} (e^{i\frac{m\pi}{L}x})^* e^{i\frac{n\pi}{L}x} dx = \delta_{m,n} \tag{3-37}$$

where the Kronecker delta $\delta_{m,n}$ is defined as

$$\delta_{m,n} \equiv \begin{cases} 1 & \text{if } m = n \\ 0 & \text{if } m \neq n. \end{cases} \tag{3-38}$$

For real numbers s and r,

$$\frac{1}{2\pi} \int_{-\infty}^{\infty} (e^{irx})^* e^{isx} dx = \frac{1}{2\pi} \int_{-\infty}^{\infty} e^{i(s-r)x} dx = \delta(s - r) \tag{3-39}$$

we have the Dirac delta function $\delta(s - r)$ instead. Using the property,

$$\int_{-\infty}^{\infty} f(x)\delta(x - x_0)\, dx = f(x_0)$$

we find that

$$g(\omega) = \frac{1}{\sqrt{2\pi}} \int_{-\infty}^{\infty} f(x) e^{-i\omega x} dx \tag{3-40}$$

in parallel with Eq. (3-35). The function $g(\omega)$ is known as the inverse Fourier transform of $f(x)$. By inserting the factor $1/\sqrt{2\pi}$ in Eq. (3-36), the two functions $f(x)$ and $g(\omega)$ are symmetrical with respect to each other, except for a change of sign in the argument of the exponential function in the integrands. Instead of exponential functions, the relation between $f(x)$ and $g(\omega)$ may also be defined in terms of sines and cosines. This is left as an exercise.

The use of Fourier transformation is usually introduced in conjunction with the study of waveforms. Given two waves of the same frequency, the difference between their shapes may be stated in terms of the differences in their Fourier coefficients $\{a_i, b_i\}$ or $\{g_\ell\}$. This is the reason why the same musical note sounds quite differently on different instruments. Since a musical note is made of a linear combination of the fundamental frequency ν and its harmonics,

$$\psi(x) = \sum_{\ell=1}^{N} a_\ell \sin(\ell 2\pi\nu x)$$

differences in the magnitudes of a_ℓ — related to the intensities of the different components — distinguish one instrument from another. Because of this, it is possible, in principle, to synthesize the musical notes from any instrument by providing the proper mixture of different harmonics. This is very similar in spirit to the inverse of the mathematical operation of a Fourier transformation.

In quantum mechanics, Fourier transform is used, for example, to change a wave function in coordinate representation to momentum representation. In other words, if $f(x)$ represents the wave function as a function of x, then $g(\omega)$ of Eq. (3-40) is the same state vector as a function of momentum ω.

Perhaps the widest application of Fourier transform these days is in "treating" data. For example, if a sample of data contains noise because of imperfections in the measuring instrument, we can use Fourier transform techniques to filter out some of the noise. For this reason Fourier transform is used in image processing (see, for example, Cannon and Hunt[9]). Because of its broad applications, Fourier transform has become one of the more important calculations carried out on computers. (For a historical review as well as a description of some of the modern application of Fourier transform, see Bracewell.[7]) The advantage of such applications is also greatly enhanced by the development of fast Fourier transformation (FFT) algorithms. In fact, the speed of performing FFT is a measure of the power of a computer.

Discrete Fourier transform Our interest here is to calculate the Fourier coefficients $\{a_m\}$ and $\{b_m\}$ of Eq. (3-32) or $\{g_\ell\}$ of (3-34) for a function $f(x)$ given to us in terms of $(N+1)$ values $f_0, f_1, \ldots, f_N$, at respectively $x = x_0, x_1, \ldots, x_N$. Our problem is similar to that of the first two sections of this chapter. Instead of polynomials and rational functions, we are, in essence, using trigonometry functions and the equivalent exponential function to carry out the "interpolations" here. For this reason, Fourier transform for a discrete set of points is sometimes known as trigonometry interpolation.

Although we do not need any information outside the region $x = [x_0, x_N]$ in carrying out the calculations, we shall nevertheless assume that $f(x)$ is a periodic function described by Eq. (3-31). In terms of a set of evenly spaced points along the x-axis, the same condition may be stated as

$$f(x_i) = f(x_{i+N+1}).$$

We shall take x_0 as the starting point of the cycle with which we are concerned. The corresponding point of the next cycle is x_{N+1}. The indexing scheme used here is slightly different from that adopted elsewhere in this volume. For example, we find that

$$2L = x_{N+1} - x_0 = (N+1)h$$

rather than $2L = (x_N - x_0)$. This is done so that the number of subintervals of width h is $(N+1)$ rather than N. We shall soon see that such a system is more convenient for the method of calculation we wish to adopt.

For simplicity, we shall take $x_0 = 0$ and the points are evenly spaced along the x-axis. As a result, we have the relation $x_k = kh$ and Eq. (3-34) for $x = x_k$ may now be written in the form

$$f_k \equiv f(x_k) = \sum_{\ell=0}^{N} g_\ell e^{i\ell\pi x_k/L} = \sum_{\ell=0}^{N} g_\ell \alpha^k \tag{3-41}$$

where

$$\alpha \equiv e^{i\pi h/L} = e^{i2\pi/(N+1)}. \tag{3-42}$$

The summation goes from 0 to N, arising from the fact that we have only $(N+1)$ pieces of input information, $f_0, f_1, \ldots, f_N$, and, consequently, we can determine at most $(N+1)$ coefficients $g_0, g_1, \cdots, g_N$.

For the most part, we shall use the exponential function form of Eq. (3-34). In certain applications, sine and cosine functions may be more useful and the arguments below apply equally well to them. Using Eq. (3-34), we can construct a set of equations relating the unknown Fourier coefficients $\{g_\ell\}$ with the input quantities $\{f_i\}$. For each of the $(N+1)$ values of $f(x)$ given to us, we have an equation of the form of Eq. (3-41). In terms of α, these $(N+1)$ equations may be written as

$$\begin{array}{l}
g_0 + \quad g_1 + \quad g_2 + \quad g_3 + \cdots + \quad g_N = f_0 \\
g_0 + \alpha\, g_1 + \alpha^2 g_2 + \alpha^3 g_3 + \cdots + \alpha^N g_N = f_1 \\
g_0 + \alpha^2 g_1 + \alpha^4 g_2 + \alpha^6 g_3 + \cdots + \alpha^{2N} g_N = f_2 \\
g_0 + \alpha^3 g_1 + \alpha^6 g_2 + \alpha^9 g_3 + \cdots + \alpha^{3N} g_N = f_3 \\
\quad \vdots \qquad\quad \vdots \qquad\quad \vdots \qquad\qquad \vdots \\
g_0 + \alpha^N g_1 + \alpha^{2N} g_2 + \alpha^{3N} g_3 + \cdots + \alpha^{NN} g_N = f_N.
\end{array}$$

In matrix notation, they may be expressed as

$$\boldsymbol{Ag} = \boldsymbol{f} \tag{3-43}$$

where

$$\boldsymbol{A} = \begin{pmatrix} 1 & 1 & 1 & 1 & \cdots & 1 \\ 1 & \alpha & \alpha^2 & \alpha^3 & \cdots & \alpha^N \\ 1 & \alpha^2 & \alpha^4 & \alpha^6 & \cdots & \alpha^{2N} \\ \vdots & \vdots & \vdots & \vdots & \ddots & \vdots \\ 1 & \alpha^N & \alpha^{2N} & \alpha^{3N} & \cdots & \alpha^{NN} \end{pmatrix} \qquad \boldsymbol{g} = \begin{pmatrix} g_0 \\ g_1 \\ g_2 \\ \vdots \\ g_N \end{pmatrix} \qquad \boldsymbol{f} = \begin{pmatrix} f_0 \\ f_1 \\ f_2 \\ \vdots \\ f_N \end{pmatrix}.$$

Our aim here is to solve this equation and obtain the values of g_i.

Formally, the solution of Eq. (3-43) may be written as

$$\boldsymbol{g} = \boldsymbol{Bf} \tag{3-44}$$

where the matrix $\boldsymbol{B} = \boldsymbol{A}^{-1}$. The fact that $\boldsymbol{B}$ is the inverse of $\boldsymbol{A}$ may be expressed in terms of their matrix elements in the following way:

$$(\boldsymbol{AB})_{rs} = \sum_{t=0}^{N} A_{rt} B_{ts} = \delta_{r,s} \tag{3-45}$$

where A_{rt} is the matrix element of $\boldsymbol{A}$ in row r and column t, and B_{ts} is the matrix element of $\boldsymbol{B}$ in row t and column s. For later convenience, we shall number the $(N+1)$ rows and columns of both $\boldsymbol{A}$ and $\boldsymbol{B}$ starting from 0 and ending with N.

Note that the matrix $\boldsymbol{A}$ has a special feature in that its elements are in the form

$$A_{rs} = (\alpha)^{rs} \qquad r = 0, 1, 2, \ldots, N \qquad s = 0, 1, 2, \ldots, N. \tag{3-46}$$

That is, all the matrix elements are integer powers of the factor α defined in Eq. (3-42), with the power given by the product of row and column numbers in the particular way we number the rows and columns. Because of this feature, elements of the inverse matrix $\boldsymbol{B}$ have the value

$$B_{rs} = \frac{1}{N+1}\alpha^{-rs}. \tag{3-47}$$

That is, B_{rs} is proportional to the inverse of A_{rs}. It is worthwhile emphasizing again that the convenient forms of Eqs. (3-46) and (3-47) are, in part, the result of the particular labeling scheme we have adopted.

With the elements of the inverse matrix $\boldsymbol{B}$ given by Eq. (3-47), we can calculate the values of the Fourier coefficients g_ℓ using Eq. (3-44),

$$g_\ell = \sum_{s=0}^{N} B_{\ell s} f_s = \frac{1}{N+1}(f_0 + \alpha^{-\ell} f_1 + \alpha^{-2\ell} f_2 + \alpha^{-3\ell} f_3 + \cdots + \alpha^{-N\ell} f_N). \tag{3-48}$$

This is very similar in form to Eq. (3-41). In fact, we can also express f_k in terms of g_ℓ in an analogous way:

$$f_\ell = g_0 + \alpha^\ell g_1 + \alpha^{2\ell} g_2 + 2 \cdots + \alpha^{N\ell} g_N.$$

The similarity is not a coincidence. It is possible to write the pair of relations Eqs. (3-43) and (3-44) in the form

$$\boldsymbol{f} = \boldsymbol{A}\boldsymbol{g} \qquad \boldsymbol{g} = \boldsymbol{A}^{-1}\boldsymbol{f}. \tag{3-49}$$

They are essentially the same relations as given by Eqs. (3-36) and (3-40) except, here, the Fourier transformation is for a set of discrete points rather than for a continuous function. Furthermore, since the matrix elements of $\boldsymbol{A}$ and its inverse differ only by the sign in the powers of α, it is possible to use a single computer program to carry out both types of transformation, Fourier transforming from $\boldsymbol{g}$ to $\boldsymbol{f}$ using the matrix $\boldsymbol{A}$ (positive powers of α) and inverse Fourier transforming using the matrix $\boldsymbol{B} = \boldsymbol{A}^{-1}$ (negative powers of α).

In principle, we have completed our primary goal of carrying out a discrete Fourier transform by expressing the value of the coefficients g_ℓ in terms of the input quantities f_i. The only trouble is that there are, in general, $(N+1)$ Fourier coefficients to be calculated, and each requires somewhere between one to two times $(N+1)$ basic operations – additions, subtractions, multiplications, divisions, and so on. As a result, the total amount of computation required to obtain all the coefficients is proportional to $(N+1)^2$. In any realistic applications, $(N+1)$ can be a very large number, perhaps on the order of 10^6, and the calculations involved become quite prohibitive. For this reason, more efficient algorithms for Fourier transform were developed in the 1960s. An example of such an approach is given below.

Fast Fourier transform Algorithms for fast fourier transform (FFT) usually work best when the number of available points $(N+1)$ is equal to some integer powers of 2. For the convenience of discussion, let us consider the case of $(N+1) = 2^\eta$ evenly spaced input points. For illustration, we shall use $\eta = 3$ (and $N+1 = 8$), even though the method itself is designed for much larger numbers. From Eq. (3-42), we see that

$$\alpha^{N+1} = e^{i2\pi} = 1.$$

Similarly, we have

$$\alpha^{(N+1)/2} = -1 \qquad \alpha^{(N+1)/4} = i \qquad \alpha^{(N+1)/2+\ell} = -\alpha^{\ell}.$$

As a result, the matrix $\boldsymbol{A}$ in Eq. (3-43) for $\eta = 3$ takes on a particularly simple form

$$\boldsymbol{A} = \begin{pmatrix} 1 & 1 & 1 & 1 & 1 & 1 & 1 & 1 \\ 1 & \alpha & i & i\alpha & -1 & -\alpha & -i & -i\alpha \\ 1 & i & -1 & -i & 1 & i & -1 & -i \\ 1 & i\alpha & -i & \alpha & -1 & -i\alpha & i & -\alpha \\ 1 & -1 & 1 & -1 & 1 & -1 & 1 & -1 \\ 1 & -\alpha & i & -i\alpha & -1 & \alpha & -i & i\alpha \\ 1 & -i & -1 & i & 1 & -i & -1 & i \\ 1 & -i\alpha & -i & -\alpha & -1 & i\alpha & i & \alpha \end{pmatrix}.$$

There are many "symmetries" in this matrix. For example, the elements in the first four rows are similar to those in the next four rows, except that the odd elements (start the counting of the elements in each row from zero) have the opposite signs.

These symmetries carry over to $\boldsymbol{B}$, the inverse of $\boldsymbol{A}$. For example,

$$\boldsymbol{B} = \frac{1}{8} \begin{pmatrix} 1 & 1 & 1 & 1 & 1 & 1 & 1 & 1 \\ 1 & -i\alpha & -i & -\alpha & -1 & i\alpha & i & \alpha \\ 1 & -i & -1 & i & 1 & -i & -1 & i \\ 1 & -\alpha & i & -i\alpha & -1 & \alpha & -i & i\alpha \\ 1 & -1 & 1 & -1 & 1 & -1 & 1 & -1 \\ 1 & i\alpha & -i & \alpha & -1 & -i\alpha & i & -\alpha \\ 1 & i & -1 & -i & 1 & i & -1 & -i \\ 1 & \alpha & i & i\alpha & -1 & -\alpha & -i & -i\alpha \end{pmatrix}$$

where we have made use of the fact that $\alpha^{-\ell} = \alpha^{N+1-\ell}$. Using the symmetries in $\boldsymbol{B}$, it is possible to reduce by a large extent the number of operations required to obtain

Table 3-6: Reduction of g_ℓ to sums of two terms.

$8g_0 = f_{10} + \ f_{11}$ $8g_4 = f_{10} - \ f_{11}$	$f_{10} = f_0 + \ f_2 + f_4 + \ f_6$ $f_{11} = f_1 + \ f_3 + f_5 + \ f_7$
$8g_1 = f_{12} - i\alpha f_{12}$ $8g_3 = f_{12} + i\alpha f_{13}$	$f_{12} = f_0 - if_2 - f_4 + if_6$ $f_{13} = f_1 - if_3 - f_5 + if_7$
$8g_2 = f_{14} - i\ f_{14}$ $8g_6 = f_{14} + i\ f_{15}$	$f_{14} = f_0 - \ f_2 + f_4 - \ f_6$ $f_{15} = f_1 - \ f_3 + f_5 - \ f_7$
$8g_3 = f_{16} - \ \alpha f_{17}$ $8g_7 = f_{16} + \ \alpha f_{17}$	$f_{16} = f_0 + if_2 - f_4 - if_6$ $f_{17} = f_1 + if_3 - f_5 - if_7$

all $(N+1)$ Fourier coefficients, usually from $(N+1)^2$ to $(N+1)\log_2(N+1)$, a very significant factor when $(N+1)$ is large.

Before designing an algorithm for the general case, it is instructive to work out the $(N+1) = 8$ case explicitly. Because of the symmetries in $\boldsymbol{B}$, we shall carry out the calculations in pairs, with the two members in each pair separated by $(N+1)/2$ in the values of their indices. For the $(N+1) = 8$ example we are working on here, we shall work on g_0 together with g_4, then g_1 with g_5, g_2 with g_6, and finally g_3 with g_5. For the first pair, we find from Eq. (3-48) that

$$\begin{aligned} 8g_0 &= f_0 + f_1 + f_2 + f_3 + f_4 + f_5 + f_6 + f_7 \\ 8g_4 &= f_0 - f_1 + f_2 - f_3 + f_4 - f_5 + f_6 - f_7. \end{aligned}$$

This can be put in a more symmetric form by defining

$$f_{10} = f_0 + f_2 + f_4 + f_6 \qquad\qquad f_{11} = f_1 + f_3 + f_5 + f_7.$$

Using these intermediate quantities, we have the results

$$8g_0 = f_{10} + f_{11} \qquad\qquad 8g_4 = f_{10} - f_{11}.$$

Similar grouping can also be constructed for the other three pairs, (1,5), (2,6), and (3,7). The complete list is given in Table 3-6.

The intermediate quantities f_{1k}, for $k = 0, 1, \ldots, 7$, in the table again display a symmetry between pairs of elements (0,4), (1,5), (2,6), and (3,7). In fact, a similar table, Table 3-7, can be built for f_{1k}. In an actual calculation, we start with the construction of the eight $f_{2\ell}$ from pairs of input f_k. Next we calculate the eight $f_{1\ell}$ from pairs of f_{2k} just obtained. The final step involves calculating the eight g_ℓ from pairs of f_{1k}. The total number of operations is therefore $(N+1)\log_2(N+1) = 8\times 3 = 24$, rather than $(N+1)^2 = 64$. The reduction is not very significant here, since we are only working with a small example. However, the calculation does provide us with a demonstration of the basic principle of FFT.

Table 3-7: Linear combinations of f_k for calculating g_ℓ.

$f_{10} = f_{20} + f_{21}$ $f_{14} = f_{20} - f_{21}$	$f_{20} = f_0 + f_4$ $f_{21} = f_2 + f_6$
$f_{11} = f_{22} + f_{23}$ $f_{15} = f_{22} - f_{23}$	$f_{22} = f_1 + f_5$ $f_{23} = f_3 + f_7$
$f_{12} = f_{24} - if_{25}$ $f_{16} = f_{24} + if_{25}$	$f_{24} = f_0 - f_4$ $f_{25} = f_2 - f_6$
$f_{13} = f_{26} - if_{27}$ $f_{17} = f_{26} + if_{27}$	$f_{26} = f_1 - f_5$ $f_{27} = f_3 - f_7$

FFT algorithm To apply the method outlined above as a practical algorithm, we need a way to find the "phase factor" between each pair of elements so that it may be programmed for arbitrary 2^η number of elements. For this purpose, let us rewrite the two sets of equations in Eq. (3-49), one for Fourier transform and the other for inverse Fourier transform, as

$$\psi_{\eta,\ell} = \sum_{s=0}^{N} \beta^{\ell s}\phi_s \tag{3-50}$$

where, according to the definitions laid down in Eqs. (3-43) and (3-44), we have

	Fourier transform	Inverse Fourier transform
$\psi_{\eta,\ell} =$	$(N+1)g_\ell$	f_ℓ
$\phi_s =$	f_s	g_s
$\beta =$	$\alpha^{-1} = e^{-i2\pi/(N+1)}$	$\alpha = e^{i2\pi/(N+1)}$

As we shall see later, the addition of a second subscript to $\psi_{\eta,\ell}$ is necessary if we want to use the same symbol to represent also the intermediate results. The new index labels which of the η steps the results belong to. For the convenience of later discussions, we shall number this index starting from η and decreasing by unity for each step.

For the phase factor $\beta^{\ell s}$, it is useful to recall that, since $\beta = \alpha^{\pm 1}$, we have

$$\beta^{N+1} = +1 \qquad \beta^{(N+1)/2} = -1 \qquad \beta^{\pm(N+1)/4} = \pm i.$$

The terms on the right side of Eq. (3-50) may be separated into an even group involving ϕ_{2s} and an odd group involving ϕ_{2s+1}:

$$\begin{aligned}\psi_{\eta,\ell} &= \sum_{s=0}^{[N/2]} \beta^{\ell(2s)}\phi_{2s} + \sum_{s=0}^{[N/2]} \beta^{\ell(2s+1)}\phi_{2s+1} \\ &= \sum_{s=0}^{[N/2]} \beta^{\ell(2s)}\phi_{2s} + \beta^\ell \sum_{s=0}^{[N/2]} \beta^{\ell(2s)}\phi_{2s+1}\end{aligned} \tag{3-51}$$

where for simplicity we have used $[N/2]$ to represent the integer part of the result of N divided by 2. Since $(N+1) = 2^\eta$, we have $[N/2] = (N-1)/2$, and there are $(N+1)/2$ or $2^{\eta-1}$ number of terms in each of the sums above.

Similarly, for $\ell < (N+1)/2$, we have

$$\begin{aligned}
\psi_{\eta,\ell+\frac{N+1}{2}} &= \sum_{s=0}^{[N/2]} \beta^{(\ell+\frac{N+1}{2})2s}\phi_{2s} + \sum_{s=0}^{[N/2]} \beta^{(\ell+\frac{N+1}{2})(2s+1)}\phi_{2s+1} \\
&= \sum_{s=0}^{[N/2]} (\beta^{N+1})^s \beta^{\ell(2s)}\phi_{2s} + \beta^{\frac{N+1}{2}}\beta^{\ell} \sum_{s=0}^{[N/2]} (\beta^{N+1})^s \beta^{\ell(2s)}\phi_{2s+1} \\
&= \sum_{s=0}^{[N/2]} \beta^{\ell(2s)}\phi_{2s} - \beta^{\ell} \sum_{s=0}^{[N/2]} \beta^{\ell(2s)}\phi_{2s+1}.
\end{aligned} \tag{3-52}$$

This result is the same as that for $\psi_{\eta,\ell}$ in Eq. (3-51) except for a different sign in the second term of the final form. Each pair of terms, $\psi_{\eta,\ell}$ and $\psi_{\eta,\ell+\frac{n+1}{2}}$, for $\ell = 0, 1, \ldots, [N/2]$, is formed of the even and odd sums of two other terms,

$$\begin{aligned}
\psi_{\eta,\ell} &= \psi_{\eta-1,s} + P_{\eta-1,\ell}\psi_{\eta-1,s+\frac{N+1}{2}} \\
\psi_{\eta,\ell+\frac{n+1}{2}} &= \psi_{\eta-1,s} - P_{\eta-1,\ell}\psi_{\eta-1,s+\frac{N+1}{2}}
\end{aligned}$$

where

$$\psi_{\eta-1,\ell} \equiv \sum_{s=0}^{[N/2]} \beta^{\ell(2s)}\phi_{2s} \qquad\qquad \psi_{\eta-1,s+\frac{N+1}{2}} \equiv \sum_{s=0}^{[N/2]} \beta^{\ell(2s)}\phi_{2s+1}$$

and the phase factor

$$P_{\eta-1,\ell} \equiv \beta^{\ell}.$$

This is exactly what we have shown earlier in Table 3-6 for the $(N+1) = 8$ example. For later convenience, we shall adopt the order of putting the $2^{\eta-1}$ even sums ahead of the odd sums, instead of the chronological order given in the table.

The relations given in Eqs. (3-51) and (3-52) are recursive. For example, it is possible also to express $\psi_{\eta-1,\ell}$ as a sum of two terms:

$$\begin{aligned}
\psi_{\eta-1,\ell} &= \sum_{s=0}^{[N/2]} \beta^{2\ell s}\phi_{2s} \\
&= \sum_{s=0}^{[N/4]} \beta^{2\ell 2s}\phi_{4s} + \sum_{s=0}^{[N/4]} \beta^{2\ell(2s+1)}\phi_{2(2s+1)} \\
&= \sum_{s=0}^{[N/4]} \beta^{4\ell s}\phi_{4s} + \beta^{2\ell}\sum_{s=0}^{[N/4]} \beta^{4\ell s}\phi_{4s+2} \equiv \psi_{\eta-2,\ell} + P_{\eta-2,\ell}\psi_{\eta-2,\ell+\frac{N+1}{4}}.
\end{aligned}$$

It is obvious that

$$\psi_{\eta-1,\ell+\frac{N+1}{4}} = \psi_{\eta-2,\ell} - P_{\eta-2,\ell}\,\psi_{\eta-2,\ell+\frac{N+1}{4}} \tag{3-53}$$

since

$$\psi_{\eta-1,\ell+\frac{N+1}{4}} = \sum_{s=0}^{[N/4]} \beta^{4s\frac{N+1}{4}} \beta^{2\ell 2s} \phi_{4s} + \sum_{s=0}^{[N/4]} \beta^{(4s+2)\frac{N+1}{4}} \beta^{2\ell(2s+1)} \phi_{2(2s+1)}$$

and

$$P_{\eta-2,\ell} = (\beta^2)^{\ell}.$$

There are altogether four different groups of $\psi_{\eta-2,\ell}$ here, corresponding to sums of terms of the forms ϕ_{4s}, ϕ_{4s+2}, ϕ_{4s+1}, and ϕ_{4s+3}, for $s = 0, 1, \ldots, [N/4]$. In each group there are $2^{\eta-2}$ elements, distinguished from each other by the label $\ell = 0, 1, \ldots, [N/4]$.

In the next step, the number of groups is doubled and the number of elements in each group is reduced to half of that in each group for the previous step. The process continues until step η, where there are 2^{η} groups of a single element in each, the individual ϕ_ℓ in the input. That is,

$$\psi_{0,0}^{k} = \phi_\ell$$

where we have adopted temporarily a superscript k on the left side of the equation indicating to which of the 2^{η} groups the element belongs.

In principle, this completes the set of recursion relations for carrying out FFT. However, we still do not have an easy way in each step to program the linear combination of two $\psi_{r,s}$ to form the two new $\psi_{r+1,t}$ for the next step. To achieve this, we need to define a system to order the elements in such a way that it is convenient to form the groups.

For this purpose, we shall adopt the *bit-reversed order* given in §A-2, related to the degree of "evenness" of the index. The advantage of arranging the elements in bit-reversed order may be seen by working out the step before the last one in our FFT algorithm. The number of groups at this stage is $(N+1)/2$ and there are two elements in each group. The elements of the most even group have the form

$$\psi_{1,\ell} = \sum_{s=0}^{1} \beta^{2^{\eta-1}\ell s} \phi_{2^{\eta-1}s} = \phi_0 \pm \phi_{2^{\eta-1}}$$

where the $\pm$ signs correspond to even or odd ℓ's, respectively. Since there are two members in this group, we shall label them $\ell = 0$ and 1. The next group of two elements is formed from linear combinations of $\phi_{2^{\eta-2}}$ and $\phi_{2^{\eta-2}+2^{\eta-1}}$:

$$\psi_{1,\ell} = \sum_{s=0}^{1} \beta^{2^{\eta-1}(\ell-2)s} \phi_{2^{\eta-2}+2^{\eta-1}s} = \phi_{2^{\eta-2}} \pm \phi_{2^{\eta-2}+2^{\eta-1}}.$$

We shall label these two $\ell = 2$ and 3. It is easy to check that the phase factor between the two elements on the right side of each one of the two equations is either $+1$ or -1 and that the values of the subscripts the two elements are separated by is $2^{\eta-1}$. Other linear combinations of two elements may be taken up in this way for all the $2^{\eta-1}$ groups in this step. The resulting order of the elements is shown in Table 3-8.

Table 3-8: Order of input for FFT.

r	Subscript		Binary		Bit-reversed	
r	ℓ	$\ell+2^{\eta-1}$	ℓ	$\ell+2^{\eta-1}$	ℓ	$\ell+2^{\eta-1}$
	0	$2^{\eta-1}$	000···000	100···000	000···000	000···001
2	$2^{\eta-2}$	$2^{\eta-2}+2^{\eta-1}$	010···000	110···000	000···010	000···011
3	$2^{\eta-3}$	$2^{\eta-3}+2^{\eta-1}$	001···000	101···000	000···100	000···101
	$2^{\eta-3}+2^{\eta-2}$	$2^{\eta-3}+2^{\eta-2}+2^{\eta-1}$	011···000	111···000	000···110	000···111
⋮	⋮	⋮	⋮	⋮	⋮	⋮
r	$2^{\eta-r}$	$2^{\eta-r}+2^{\eta-1}$	000···10··· ·00···000	100···10··· ·00···000	000···00··· ·01···000	000···00··· ·01···001
	$2^{\eta-r}+2^{\eta-2}$	$2^{\eta-r}+2^{\eta-2}+2^{\eta-1}$	010···10··· ·00···000	110···10··· ·00···000	000···00··· ·01···010	000···00··· ·01···011
	⋮	⋮	⋮	⋮	⋮	⋮
	$2^{\eta-r}+2^{\eta-r+1}$ $+\cdots+2^{\eta-2}$	$2^{\eta-r}+2^{\eta-r+1}$ $+\cdots+2^{\eta-2}+2^{\eta-1}$	011···10··· ·00···000	111···10··· ·00···000	000···00··· ·01···110	000···00··· ·01···111
⋮	⋮	⋮	⋮	⋮	⋮	⋮
	$2^{\eta-1}-1$	$2^{\eta}-1$	011···111	111···111	111···110	111···111

Before describing an algorithm to take advantage of the bit-reversed order for FFT, let us find out why the arrangement is useful. In a system consisting of 2^{η} elements in the bit-reversed order, the first element is the zeroth element and the second is $\ell = 2^{\eta-1}$. Since these are the two most even elements, the sum constructed from these two elements is the most even of any two elements. The pair of elements following these two are the next most even and they form the next most even sum of two elements. By going through the list of input elements arranged in bit-reversed order in this way, we can form all the sums of two elements in order of decreasing evenness. This gives us half the sums we need in the first of η steps of FFT. The other half of the sums comes from taking the differences between pairs of input elements. Again, the bit-reversed order gives us these sums ordered according to the order we have defined.

Storage considerations There is one more improvement we can make to the method before implementing it as an algorithm. We have implicitly assumed that the ordered list we generate in each step is stored in a new array of $(N+1)$ elements. To carry out the complete transformation in this way, we will need a total of η arrays, each of length $(N+1)$ to store the results of each step. Since we wish to handle cases with large numbers of elements, the method proves to be uneconomical in terms of storage locations. We can solve this problem by defining two arrays each of length $(N+1)$, one for the input list and one for the output list. In each of the η steps, we move the output list of the previous step into the input list at the start and store the calculated values in the output list. The problem with this method is that the number of operations is increased by almost $(N+1)\log_2(N+1)$, a significant fraction

Table 3-9: Order to carry out FFT for eight elements.

s	$\psi_{3,s}$	$\psi_{2,s}$	$\psi_{1,s}$
0	$0+4$	$(0+4)+(2+6)$	$\{(0+4)+(2+6)\}+\{(1+5)+(3+7)\}=g_0$
4	$0-4$	$(0-4)+(2-6)$	$\{(0-4)+(2-6)\}+\{(1-5)+(3-7)\}=g_1$
2	$2+6$	$(0+4)-(2+6)$	$\{(0+4)-(2+6)\}+\{(1+5)-(3+7)\}=g_2$
6	$2-6$	$(0-4)-(2-6)$	$\{(0-4)-(2-6)\}+\{(1-5)-(3-7)\}=g_3$
1	$1+5$	$(1+5)+(3+7)$	$\{(0+4)+(2+6)\}-\{(1+5)+(3+7)\}=g_4$
5	$1-5$	$(1-5)+(3-7)$	$\{(0-4)+(2-6)\}-\{(1-5)+(3-7)\}=g_5$
3	$3+7$	$(1+5)-(3+7)$	$\{(0+4)-(2+6)\}-\{(1+5)-(3+7)\}=g_6$
7	$3-7$	$(1-5)-(3-7)$	$\{(0-4)-(2-6)\}-\{(1-5)-(3-7)\}=g_7$

compared with the total number of operations required to carry out the entire FFT. On the other hand, the storage requirement is now reduced to $2(N+1)$ locations.

It is possible to do even better in terms of storage locations. Since each pair of elements in the input list are used only to form the sum and difference of a pair of elements (together with a phase factor), they are not needed in any of the subsequent calculations. As a result, we can store the two output quantities at the end of each group of calculations into the two storage locations no longer needed. For the convenience of the next step, we shall put the result obtained by summing in the array ahead of that obtained by taking the difference. In this way, a single array is adequate for both input and output. The only trouble with this method is that the output list is only partially ordered. This can be seen from Table 3-9, where we have displayed the output from each of the three steps in the eight-element example used earlier. In step 0, the list of f_k is arranged in the bit-reversed order on input and is given in the first column. At the end of step 1, the linear combinations of two f_k's are in the order as shown in the second column. If we number the most "symmetric" pair as 0 and arrange the rest of the elements according to the symmetry we have adopted, the order of the entries in the second column is 0, 2, 1, 3, 4, 6, 5, and 7. It is essentially in ascending order except that, within each group of four elements, there is a reversal between the second and third members. However, the partial order is sufficiently simple and, as we shall see later, it is not difficult to spot and compensate for it in a computer program.

The same type of departure from an ordered list happens again in the output of the third step. The order is now 0, 4, 2, 6, 1, 5, 3, and 7. For our eight-element example, this is the bit-reversed order from the input list. Since this is the last step in this example, it is not surprising that the order is exactly the "opposite" to the input one. As the input to step 1 is arranged according to bit-reversed order, the output at the last step is in the ascending order for the Fourier coefficients g_ℓ we want.

The calculations involved may now be summarized in the following way. We as-

Box 3-4 Program DM_FFT
Fast Fourier Transform

1. Input the list of complex amplitudes.
2. Arrange the input amplitudes into bit-reversed order:
   (a) Generate a bit-reversed order list using the method of Box A-2.
   (b) Store the results temporarily in the array for the output.
   (c) Replace each element of the array by the input element to which it is pointing.
3. Initialization for taking sums and differences of two elements:
   (a) Define the number of elements in each group to be 1.
   (b) Set $\alpha = \exp\{-2\pi/(2N+1)\}$.
   (c) Set the base of phase factor equal $n_b = (2N+1)$.
4. Carry out the η steps of taking sums and differences of two elements:
   (a) Set the spacing between elements in the same group.
   (b) Double the number of elements in each group from the previous step.
   (c) Reduce n_b to half and calculate the basic phase factor α^{n_b}.
   (d) Go through all the groups:
      (i) Form sums and differences in each group.
      (ii) Store the sum and difference.
5. Output the Fourier coefficients.

sume that the input for the calculation is arranged in the bit-reversed order. From this list, we can take each pair of input elements in turn and form sums and differences. These are stored back into the two locations from which the input pair were taken. When this operation is carried to the end of the list, the first of the η steps is completed. Because of the partial order in the results, the next step involves the formation of sums and differences of a pair of members separated by one element. Here, we must also incorporate the phase factor

$$P_{k,\ell} = (\beta^m)^\ell \qquad \text{with} \qquad m = \eta - k \tag{3-54}$$

where $\ell = 0, 1, 2, \ldots, 2^{k+1}$ labels the pair and $k = 0, 1, 2, \ldots, (\eta - 1)$ labels the step, as can be seen by generalizing Eq. (3-53) to step k. Again, these results are stored in the locations left vacant by the pair used as the input for this step. The process continues until step η, where a complete list of the output in the usual ascending order is produced. This is shown by the $\eta = 3$ number of steps in our $(N+1) = 8$ example earlier. In each step, the pair of input elements with which we wish to work is separated by a distance of $(2^{k-1} - 1)$ elements, where $k = 1, 2, \ldots, \eta$ is the step number. That is, in step $k = 1$, the input list is in the bit-reversed order and all the pairs are located adjacent to each other. In step $k = 2$, they are separated by one element, step $k = 3$ by three elements, step $k = 4$ by seven elements, and so on. In the final step, the separation is $(2^{\eta-1} - 1)$ and each element is combined with another one halfway down the list. This is shown explicitly by the $\eta = 3$ example given in Table 3-9. For simplicity, only the signs, but not the phase factors, of the terms are given. Note also that, since the phase factor depends only on the order of the members in a group, it is more efficient to carry out the transformation for the same pair of members in all the groups first and then proceed to the next pair of members.

The algorithm outlined in Box 3-4 assumes that one wishes to preserve the input list and a separate array is used for the output. To this end, we can use the output array for the dual purpose of generating the list of bit-reversed indices as well as storing intermediate results in the calculation. This is not always desirable. Since FFT is designed to handle cases where the number of elements is large, one may not be able to afford a second array of $(N+1)$ elements. It is possible to modify the algorithm to use only a single array. However, this will require a slightly modified method to order the input. If we can interchange the elements in the input array in a clever way so that it can be reordered into the bit-reversed pattern, all the subsequent calculations can be performed within the same array, with the output delivered to the array used originally for the input. This, as well as other more specialized FFT calculations, may be found in Press and coauthers,[44] as well as many of the standard subroutine libraries like NAG.[41]

3-5 Extrapolation

Extrapolation is the process of deducing the approximate values of a function $f(x)$ outside the interval $x = [x_0, x_N]$ where the known values are. It is inherently a very delicate operation, since the available information says very little about the behavior of $f(x)$ outside $[x_0, x_N]$. For example, a common practice to describe a minimum in a potential is to use a parabola, as shown schematically in Fig. 3-3. For our purpose here, we can approximate the function around some point $x = x_0$ near the minimum in the form

$$f(x) \approx a_0 + a_1(x - x_0) + a_2(x - x_0)^2.$$

With three or more values of $f(x)$ available around x_0, we can have a fairly good description of the function in the neighborhood. On the other hand, we have no way of predicting that, for the example shown in Fig. 3-3, there is a maximum beyond $x = x_5$ and the way the function increases in value for $x < x_1$. The example amplifies the fact that extrapolation can only be carried out reliably in regions close to the interval where the known values are.

Since extrapolation is also based on an approximate form of the unknown function $f(x)$, the basic philosophy of any method of extrapolation is essentially the same as that for interpolation. Consequently, it is possible to make some minor adjustments to the algorithms for interpolation and adapt the same techniques for extrapolations. For this reason, we have taken care of the possibility of x being outside the range of the input values in the algorithm for Neville's interpolation described in Box 3-1. Had we restricted the value of x to the central region, a somewhat simpler procedure

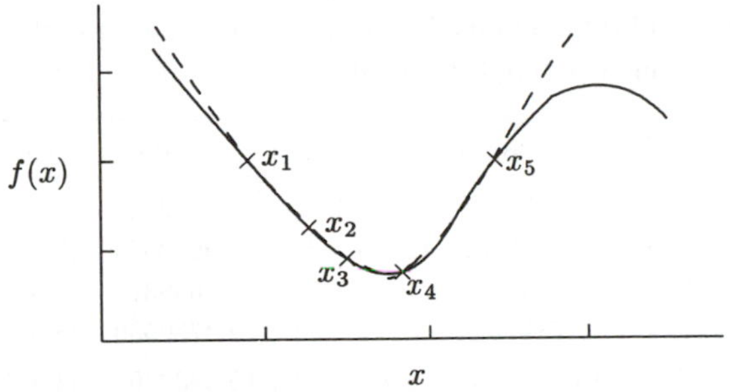

Figure 3-3: Example illustrating the danger of extrapolation. A second-degree polynomial fit to the vicinity of a minimum (dashed curve) fails to predict the maximum beyond $x = x_5$ and gives a poor representation for $x \leq x_1$.

could have been constructed.

As an example, we shall consider again the function $\tan x$, used earlier in Table 3-4 to demonstrate interpolation. The input values, at $x = 1.1$, 1.2, 1.3, 1.4, and 1.5, are the same as before. However, instead of interpolating for values within the range $[1.1, 1.5]$, our interest here is to find those for $x > 1.5$ and $x < 1.1$. Because of the singularity at $x = \pi/2$, it presents somewhat of a challenge to extrapolation techniques for x near $\pi/2$. The results obtained with polynomial and rational function approximations are compared with the exact values in Table 3-10. We see that, even at $x = 1.57$, the rational function approximation is able to produce a result with two significant figures, as well as an error estimate that give the correct information on the accuracy of the calculation.

Richardson extrapolation An important application of extrapolation techniques is found in the method of *Richardson's deferred approach to the limit.* We shall use numerical integration as an illustration of the method. Later, in Chapter 8, we shall see that the same approach is useful also in solving differential equations.

In §2-2, we saw that the errors in numerical integration are proportional to powers in the step size h and one way to improve the accuracy is to reduce h. Here, we shall make use of extrapolation techniques to achieve the same goal. For this purpose, it is convenient to express the integral in terms of the Euler-Maclaurin summation formula given earlier in Eq. (2-10):

$$\begin{aligned} I &\equiv \int_a^b f(x)\,dx \\ &= h\Big\{\frac{1}{2}f_0 + f_1 + f_2 + \cdots + f_{N-1} + \frac{1}{2}f_N\Big\} - \frac{B_2}{2!}h^2\big\{f'(b) - f'(a)\big\} - \cdots \\ &\quad -\frac{B_{2k}}{(2k)!}h^{2k}\big\{f^{(2k-1)}(b) - f^{(2k-1)}(a)\big\} - \cdots \end{aligned}$$

Table 3-10: Comparison of rational function and polynomial extrapolation for $\tan x$ using the same five input values in $x = [1.1, 1.5]$ as Table 3-4.

x	Exact	Rational		Polynomial	
1.00	1.5774077	1.5574998	(4×10^{-3})	5.2353148	(4×10^{0})
1.02	1.6281304	1.6281912	(2×10^{-3})	3.9640360	(3×10^{0})
1.04	1.7036146	1.7036510	(1×10^{-3})	3.0659318	(2×10^{0})
1.06	1.7844248	1.7844439	(8×10^{-4})	2.4730530	(8×10^{-1})
1.51	16.4280917	16.427917	(8×10^{-3})	15.583240	(1×10^{-1})
1.53	24.4984104	24.496742	(7×10^{-2})	19.040157	(5×10^{-1})
1.55	48.0784825	48.063660	(6×10^{-1})	23.236004	(1×10^{0})
1.57	1255.7655915	1237.2043	(4×10^{2})	28.274174	(2×10^{0})

$$= h\Big\{\frac{1}{2}f_0 + f_1 + f_2 + \cdots + f_{N-1} + \frac{1}{2}f_N\Big\}$$
$$-\frac{h^2}{12}\Big\{f'(b) - f'(a)\Big\} + \frac{h^4}{720}\Big\{f''(b) - f''(a)\Big\} + \cdots \tag{3-55}$$

where we have divided the range $[a, b]$ into N equal subintervals, each of size h. The first term in the final form is the result given by the trapezoidal rule of integration

$$I(h) \equiv h\Big\{\frac{1}{2}f_0 + f_1 + f_2 + \cdots + f_{N-1} + \frac{1}{2}f_N\Big\}. \tag{3-56}$$

In the limit of zero step size, all the other terms vanish and we have the equality

$$I = \lim_{h\to 0} I(h). \tag{3-57}$$

In a numerical calculation, it is not possible to evaluate $I(h)$ for $h = 0$. However, there is nothing preventing us to extrapolate its value to the limit $h \to 0$.

The Richardson technique is similar to what we did earlier in §3-1 for polynomial interpolation. To start with, we need to construct a difference table in $I(h)$ for different step sizes. Up to terms involving h^2, Eq. (3-55) may be written in the form

$$I_N = I(h) - \frac{1}{12}\Big\{f'(b) - f'(a)\Big\}h^2 \tag{3-58}$$

where a subscript is added to I to indicate the number of subintervals used. This expression is not useful as it stands since we do not know the values of the first-order derivative $f'(x)$ at the two integration limits. On the other hand, the corresponding value of the integral for step size $h/2$ is given by

$$I_{2N} = I(h/2) - \frac{1}{12}\Big\{f'(b) - f'(a)\Big\}\Big(\frac{h}{2}\Big)^2. \tag{3-59}$$

Box 3-5 Program DM_RCHRD
Integration by extrapolation using Richardson's deferred approach to limit

Subprograms used:
  TRAPZ: Trapezoidal rule integration (Box 2-1).
Initialization:
  (a) Set the tolerance for error to be $\epsilon = 10^{-5}$.
  (b) Set the number of points for the extrapolation to be $m = 5$.
1. Input the integration limits.
2. Initialize the trapezoidal integration.
3. Set up the input to the extrapolation step:
  (a) Calculate the integral for m different step sizes using TRAPZ.
  (b) Store the value of the integral and the square of the step size.
4. Extrapolate to zero step size.
5. Output the result.

Between Eqs. (3-58) and (3-59), we can eliminate the unknown factor $\{f'(b) - f'(a)\}$ and obtain the result

$$I_{N,2N} = \frac{4}{3}I(h/2) - \frac{1}{3}I(h) = I(h/2) + \frac{1}{3}\big\{I(h/2) - I(h)\big\}. \tag{3-60}$$

This is a better approximation than that given by $I(h)$ in Eq. (3-58), as errors of the order of h^2 have been eliminated.

We can also achieve the same result of eliminating terms of order h^2 using a linear combination I_{2N} and I_{4N}, approximate values of the integral obtained, respectively, with step sizes $h/2$ and $h/4$. Analogous to Eq. (3-60), we have

$$I_{2N,4N} = I(h/4) + \frac{1}{3}\big\{I(h/4) - I(h/2)\big\}. \tag{3-61}$$

The error in both Eqs. (3-60) and (3-61) is of the order of h^4. This can be seen by going back to Eq. (3-55) and derive these two expressions with terms up to h^4 included:

$$\begin{aligned} I &\approx I_{N,2N} + \frac{1}{720}\big\{f''(b) - f''(a)\big\}\frac{1}{4}h^4 \\ I &\approx I_{2N,4N} + \frac{1}{720}\big\{f''(b) - f''(a)\big\}\frac{1}{4}\frac{h^4}{2^4}. \end{aligned}$$

By eliminating terms that depend on the unknown factor $\{f''(b) - f''(a)\}$ from these two equations, we obtain an approximation that is accurate to order h^4:

$$I_{N,2N,4N} = I_{2N,4N} + \frac{1}{2^4 - 1}\{I_{2N,4N} - I_{N,2N}\}.$$

Table 3-11: erf(x) for $x = 0.5$ using Richardson's deferred approach to the limit.

N	I_i	$I_{i,i+2}$	$I_{i,i+2,i+4}$	$I_{i,i+2,i+4,i+6}$	$I_{i,i+2,i+4,i+6,i+8}$
2	0.5158988				
		0.5205060			
4	0.5193542		0.5204999		
		0.5205003		0.5204999	
8	0.5202138		0.5204999		0.5204999
		0.5204999		0.5204999	
16	0.5204284		0.5204999		
		0.5204999			
32	0.5204821				
		Exact value:	erf($x = 0.5$)	=	0.5204999

The result is similar to Eq. (3-60), except for a further improvement in the accuracy by order h^2. It is obvious that we can repeat the process and achieve even better accuracies. For example, by making use of $I(h/2)$, $I(h/4)$, and $I(h/6)$, we obtain $I_{2N,4N,6N}$. This, together with $I_{N,2N,4N}$, enables us to eliminate errors of order h^6. In each such step, we need the approximate value of the integral given by Eq. (3-56) with a step size that is half that of the last one used in the previous step.

The calculations we have carried out in the previous paragraphs may be put into the form of a recursion relation, similar to that in Eq. (3-6):

$$I_{i,i+2,\ldots,i+m-2,i+m} = I_{i+2,\ldots,i+m-2,i+m} + \frac{1}{2^m - 1}\{I_{i+2,\ldots,i+m-2,i+m} - I_{i,i+2,\ldots,i+m-2}\} \tag{3-62}$$

where, to simplify the notation, we have adopted the symbol

$$I_{i,i+2,\ldots,i+m-2,i+m} \equiv I_{2^iN,2^{i+2}N,\ldots,2^{i+m-2}N,2^{i+m}N}.$$

Here m indicates that terms up to order h^m have been eliminated. Only even values of i appear in the expression, as there are no odd powers of h beyond the first term in the Euler-Maclaurin summation formula. The relation between the various order approximations can also be represented by a difference table similar to Table 3-1, but we shall not repeat it here. The algorithm is outlined in Box 3-5.

As an example, we shall apply the technique to error function defined in Eq. (3-7). To obtain the value of the integral for a given step size, we can use the trapezoidal rule given in Box 2-1. To apply the Richardson's deferred approach to the limit, we need a number of such values, each calculated with a different step size. We can start with only a few values of $I(h)$ obtained with minimum numbers of subintervals, such as 2, 4, 8, 16, and 32. From these five "input" values, we can apply the recursion relation Eq. (3-62) to find the four values of I_{02}, I_{24}, I_{46}, and I_{68} and eliminate errors of order h^2. Next, we calculate I_{024}, I_{246}, and I_{468} to eliminate errors of order h^4. Any

Box 3-6 Program DM_RMBRG
Romberg integration using Neville's algorithm

Subprograms used:
NVLLE: Neville's algorithm for interpolation (Box 3-1).
TRAPZ: Trapezoidal rule integration (Box 2-1).
Initialization:
(a) Set $\epsilon = 10^{-5}$ as the error tolerance.
(b) Set $m = 5$ as the number of points for extrapolation.
(c) Set $I_x = 10$ as the maximum number of iterations.
1. Input the integration limits.
2. Initialize the trapezoidal integration.
3. Set up the input to the extrapolation step:
   (a) Calculate the integral for m different step sizes using TRAPZ (Box 2-1).
   (b) Store the m values of the integral and squares of the step sizes.
4. Use NVLLE to extrapolate to $h^2 = 0$.
5. Check the accuracy using the error estimates provided by NVLLE:
   (a) If the accuracy is not enough,
      (i) Calculate the integral for a smaller step size.
      (ii) Store the results and repeat the extrapolation with the last m values.
   (b) Output the result if the required accuracy is reached.

dependence on h^6 is taken out when we obtain the values for I_{0246} and I_{2468}. Our final result for a five-point extrapolation is I_{02468}, which is, in principle, accurate to order h^{10}, as all terms up to h^8 have been eliminated. The values at each stage of the calculation for erf(x) at $x = 0.5$ are given in Table 3-11. We see that it is possible to obtain a good agreement to the exact value for a single-precision calculation with no more than five input points in the extrapolation.

Romberg integration Another way of using extrapolation techniques for numerical integration is the method of Romberg. The difference from Richardson's deferred approach to the limit may be seen by expressing $I(h)$ as a power series in h^2:

$$I(h) = a_0 + a_1 h^2 + a_2 h^4 + \cdots + a_k h^{2k} + \cdots. \tag{3-63}$$

To identify the coefficients a_0, $a_1, \ldots$ in the expression, we shall rewrite the Euler-Maclaurin formula of Eq. (3-55) in the following form with the help of Eq. (3-56):

$$I(h) = I + \frac{h^2}{12}\{f'(b) - f'(a)\} - \frac{h^4}{720}\{f''(b) - f''(a)\} + \cdots + \cdots \tag{3-64}$$

where I is the exact value defined by Eq. (3-55) and $I(h)$ is the value obtained through numerical integration using Eq. (3-56) with step size h.

A comparison of Eqs. (3-63) and (3-64) gives a_0 as the exact value of the integral,

$$a_0 = I = \int_a^b f(x)\,dx.$$

Table 3-12: Values of erf(x) obtained using Romberg integration with error estimates in parentheses.

x	3-Point extrapolation	4-Point extrapolation	5-Point extrapolation	Exact value
0.3	$0.32862678(3\times10^{-8})$	$0.32862675(3\times10^{-10})$	$0.32862678(1\times10^{-10})$	0.32862676
0.4	$0.42839241(1\times10^{-7})$	$0.42839241(2\times10^{-10})$	$0.42839241(8\times10^{-11})$	0.42839236
0.5	$0.52049989(3\times10^{-7})$	$0.52049994(5\times10^{-10})$	$0.52049994(1\times10^{-10})$	0.52049988
0.6	$0.60385603(6\times10^{-7})$	$0.60385615(2\times10^{-9})$	$0.60385621(4\times10^{-11})$	0.60385609
0.7	$0.67780131(1\times10^{-6})$	$0.67780131(1\times10^{-10})$	$0.67780131(1\times10^{-10})$	0.67780119

The other coefficients $a_1, a_2, \ldots$ are related to the derivatives of the integrand

$$a_1 = \frac{1}{12}\{f'(b) - f'(a)\} \qquad \cdots \qquad a_k = \frac{B_{2k}}{(2k)!}\{f^{(2k-1)}(b) - f^{(2k-1)}(a)\}.$$

In the limit $h \to 0$, all the terms on the right side of Eq. (3-63) vanish except the first one, and we recover the result given by Eq. (3-57), stating that $I(h)$ is equal to the exact value I as $h \to 0$.

Our interest is in a_0. To extrapolate for this value we need again a number of $I(h)$ calculated with different step sizes:

$$h_1 = \frac{b-a}{n_1} \qquad h_2 = \frac{b-a}{n_2} \qquad h_3 = \frac{b-a}{n_3} \qquad \cdots$$

for $n_1 < n_2 < n_3 < \cdots$. This is the essence of the Romberg method of integration. Conceptually, it is simpler than the Richardson's deferred approach to the limit and the method allows more freedom in choosing the step sizes.

The algorithm is outlined in Box 3-6. At the start, the integration interval $[a, b]$ is divided into n_1 subintervals, where n_1 is some reasonably small number like 6. We can make use of the trapezoidal rule to calculate the value of $I(h)$ for this step size and the result is stored as $I(h_1)$. Next, the same integral is evaluated with $n_2 = 2n_1$ subintervals and this gives us $I(h_2)$. The process is repeated for a total of m times with $n_3 = 3n_1$, $n_4 = 4n_1$, …, $n_m = mn_1$. These m pairs of values $\{I(h_1), h_1^2\}$, $\{I(h_2), h_2^2\}$, …, $\{I(h_m), h_m^2\}$ provide us with the input to Neville's algorithm of Box 3-1 to extrapolate for $I(h)$ at $h^2 = 0$.

An estimate of the accuracy of the integration is provided by the uncertainty $\mathcal{E}$ of the extrapolation process. If $\mathcal{E}$ is larger than the error that can be tolerated, the numerical integration is carried out again with $(m+1)n_1$ subintervals, and the extrapolation process is repeated with $\{I(h_{m+1}), h_{m+1}^2\}$ as a part of the input. To keep the calculation simple, we can discard $\{I(h_1), h_1^2\}$ at this stage and use only the last m pairs of values of $I(h)$ and h^2 in the calculation. The process of going to smaller and smaller step sizes is repeated until the error is small enough or a maximum number of steps is reached.

As an example, we shall evaluate the error function of Eq. (3-7) for $x = 0.3$ to 0.7. Three different sets of results are shown in Table 3-12. The first one, given in column 1, consists of the extrapolated values obtained with three pairs of input calculated with the integration range divided into 6, 12, and 18 subintervals ($m = 3$ and $n_1 = 6$). The second set in column 2 is obtained with one additional input of 24 subintervals ($m = 4$) and the last set in column 5, with the addition of one more input consisting of 30 subintervals ($m = 5$). As we can see from the table, the final results do not differ much from the corresponding ones in the previous column. In each case, the error estimates provided by the extrapolation algorithm are given in parentheses.

3-6 Inverse interpolation

In interpolation, we are interested to find the value of y at a point x within the interval $[a, b]$, using as input a table of the values of y at a finite number of points $x = x_0, x_1, \ldots, x_N$. On many occasions, we may be interested, instead, in the opposite question of finding the value of x for a given value of y from the same set of input. For example, we may wish to locate the point along x at which $f(x)$ takes on a particular value C. This is the problem of inverse interpolation and is, in general, different from that of interpolation, as it is not always possible to simply interchange the roles of x and y in the table of input.

There are two main reasons for the difference. First, in interpolation, the value of y as a function of x is assumed to be single valued; otherwise the problem is not well defined. However, for the same function, the value of x as a function of y may not be single valued. This happens, for example, around an extremum, as illustrated schematically in Fig. 3-4. Second, most of the interpolation techniques approximate the function $y = f(x)$ by a polynomial or a rational function in x. The fact that such an approximation can be made does not guarantee that x may be represented by a polynomial or a rational function in terms of y. For this reason, it is not possible to expect that we can use interpolation techniques to carry out inverse interpolation. Only in simple cases, where $f(x)$ is a smooth, monotonically increasing or decreasing function of x (that is, $df/dx \neq 0$ in the interval), we can be assured of achieving any accuracy in using standard interpolation procedures to carry out an inverse interpolation.

The problem of inverse interpolation may be posed in the following way. Given $f(x)$, we wish to find the value of x for which

$$f(x) - C = 0 \tag{3-65}$$

where C is the value of $f(x)$ we wish to locate. To have a unique solution, it is necessary that $f(x)$ is not a constant equal to C in any finite size region of x. For this reason, we also need to add the stipulation that $f(x + \delta) - C$ and $f(x + \delta) - C$ are different in sign, at least for some small δ. In this way, the problem is made equivalent to one of finding the root of Eq. (3-65). There are several ways to obtain

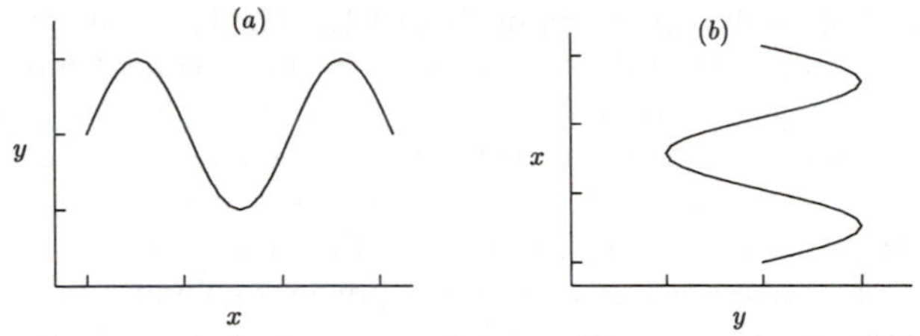

Figure 3-4: Relation between y as a function of x and its inverse. If there is a local minimum or maximum in y, as shown in (a), there are more than one value of x corresponding to a given y, as shown in (b).

a solution. One possibility, mentioned in the previous paragraph, is to think in terms of a function $g(y) = x$. If the function $y = f(x)$ is monotonic, standard interpolation algorithms may be used. In general, a rational function interpolation is preferred here to ensure better results. However, the usefulness of such an approach is limited.

The bisection method A simple technique that works for a large class of functions is to use the method of bisection to find the root for Eq. (3-65). Let us define a new function

$$\phi(x) = f(x) - C. \tag{3-66}$$

The problem of inverse interpolation for the value of x for which $f(x) = C$ is now transformed into one of searching for the zero of $\phi(x)$ in the range $x = [a, b]$. For the convenience of discussion, we shall assume that there is only one zero in the interval and that $\phi(a) < 0$ and $\phi(b) > 0$.

To start with, let us find the value of $\phi(x)$ at the midpoint of the interval

$$x_m = \frac{1}{2}(a + b).$$

If $\phi(x_m) < 0$, the zero of $\phi(x)$ must be in the interval $[x_m, b]$. On the other hand, if $\phi(x_m) > 0$, the zero must be in the interval $[a, x_m]$. In either case, the range of our search is reduced to half of the original one, either $[x_m, b]$ or $[a, x_m]$. The procedure is now repeated for the new, smaller interval. This gives us an iterative procedure to carry out the search with the interval reduced in each step by a factor of 2 from its previous value. When the size of the interval is smaller than the accuracy of x required, we can stop and adopt the midpoint of the final interval as the root. The algorithm is summarized in Box 3-7.

The major difficulty with this procedure is that, for functions $\phi(x)$ with more than one zero in the interval, we need to know the approximate locations of all the roots. One easy way to solve this particular problem is to make a rough plot of the function using, if necessary, the method of cubic spline described in the next section.

Box 3-7 Inverse interpolation using bisection.
Find the only root of $\phi(x)$ in an interval

Initialization:
  (a) Let ϵ be the required accuracy.
  (b) Input a as the lower limit of the range and b the upper limit.

1. Check whether $\phi(a) < 0$ and $\phi(b) > 0$ (increasing function) or $\phi(a) > 0$ and $\phi(b) < 0$ (decreasing function).
2. Find the midpoint $x_m = \frac{1}{2}(a + b)$.
   (a) If $\phi(x_m) > 0$:
      (i) replace b by x_m for an increasing function, or
      (ii) replace a by x_m for a decreasing function.
   (b) If $\phi(x_m) < 0$:
      (i) replace a by x_m for an increasing function, or
      (ii) replace b by x_m for a decreasing function.
   (c) If $\phi(x_m) = 0$, exit. The root of $\phi(x)$ is at x_m.
3. Repeat step 2 if $(b - a) > \epsilon$. Otherwise, return $x = \frac{1}{2}(a + b)$.

The plot or, alternatively, a list of the values from which a plot is made gives us the approximate locations of all the roots with sufficient accuracy to be used as the starting points for the bisection method.

A second, but less serious, difficulty with the bisection method is that the convergence rate is not slow — it takes η bisections to reduce the interval by a factor of 2^η. This may be quite adequate if the need for inverse interpolation is only occasional. For more general applications, it is better to use one of the methods described below.

Newton's method of interpolation It is useful, partly for historical interest, to derive here the interpolation method of Newton using finite differences. For this purpose, we shall examine only polynomial interpolation.

Consider the case where we are given $(N+1)$ values of $f(x)$ at $x = x_0, x_1, x_2, \ldots, x_N$. It is possible to construct a polynomial expression of degree N,

$$p(x) = c_0 + \sum_{j=1}^{N} c_j(x - x_0)(x - x_1)\cdots(x - x_{j-1}) \tag{3-67}$$

such that it goes through all the $(N + 1)$ points given. That is, we can find a set of values for the $(N + 1)$ coefficient $c_0, c_1, \ldots, c_N$, such that $f(x) - p(x) = 0$ at $x = x_0, x_1, x_2, \ldots, x_N$. At other values of x, the difference between $p(x)$ and $f(x)$ is given by

$$f(x) - p(x) = \frac{f^{(N+1)}(\xi)}{(N+1)!}(x - x_0)(x - x_1)\cdots(x - x_N) \tag{3-68}$$

where $f^{(m)}(x)$ is the mth derivative of $f(x)$ and ξ is a point in the interval $[x_0, x_N]$. The result follows from the fact that, by definition, the $(N + 1)$th derivative of $p(x)$

Table 3-13: Divided differences in Newton's interpolation.

x_0	$f(x_0)$					
		f_{x_0,x_1}				
x_1	$f(x_1)$		f_{x_0,x_1,x_2}			
		f_{x_1,x_2}		f_{x_0,x_1,x_2,x_3}		
x_2	$f(x_2)$		f_{x_1,x_2,x_3}		f_{x_0,x_1,x_2,x_3,x_4}	
		f_{x_2,x_3}		f_{x_1,x_2,x_3,x_4}		$f_{x_0,x_1,x_2,x_3,x_4,x_5}$
x_3	$f(x_3)$		f_{x_2,x_3,x_4}		f_{x_1,x_2,x_3,x_4,x_5}	
		f_{x_3,x_4}		f_{x_2,x_3,x_4,x_5}		
x_4	$f(x_4)$		f_{x_3,x_4,x_5}			
		f_{x_4,x_5}				
x_5	$f(x_5)$					

vanishes and that of the product $(x - x_0)(x - x_1)\cdots(x - x_N)$ on the right side of Eq. (3-68) is equal to $(N + 1)!$.

The coefficients c_j of $p(x)$ in Eq. (3-67) may be found from the *divided differences* of the input values $f(x_0)$, $f(x_1), \ldots,$ $f(x_N)$. The lowest-order difference between x and x_0 is defined as

$$f_{x_0,x} \equiv \frac{f(x) - f(x_0)}{x - x_0}.$$

Similarly, second-order divided difference may be written as

$$f_{x_0,x_1,x} \equiv \frac{f_{x_1,x} - f_{x_0,x_1}}{x - x_0}.$$

Higher-order divided differences are defined using the recursion relation

$$f_{x_i,x_{i+1},x_{i+2},\ldots,c_{k-1},x_k,x} \equiv \frac{f_{x_{i+1},x_{i+2},\ldots,c_{k-1},x_k,x} - f_{x_i,x_{i+1},x_{i+2},\ldots,x_{k-1},x_k}}{x - x_i}. \tag{3-69}$$

It is not difficult to identify that

$$c_{k+1} = f_{x_0,x_1,x_2,\ldots,x_{k-1},x_k,x_{k+1}} \tag{3-70}$$

with $c_0 = f(x_0)$, $c_1 = f_{x_0,x_1}$, $c_2 = f_{x_0,x_1,x_2}$, and so on. This is known as Newton's interpolation formula. An illustration using $(N + 1) = 6$ is given in Table 3-13.

For input involving equally spaced values of x, with $x_k - x_0 = kh$ for h a constant, the divided differences of Eq. (3-69) may be expressed in terms of the forward difference $\Delta f(x)$ defined in Eq. (2-50). For example,

$$\begin{aligned} c_1 &= f_{x_0,x_1} = \frac{1}{1!h}\Delta f(x_0) \\ c_2 &= f_{x_0,x_1,x_2} = \frac{1}{2!h^2}\Delta^2 f(x_0) \end{aligned} \tag{3-71}$$

where we have used the notation $\Delta^2 f(x_0) \equiv \Delta f(x_1) - \Delta f(x_0)$. The general case is

$$c_k = f_{x_0,x_1,x_2,\ldots,x_{k-1},x_k} = \frac{1}{k!h^k}\left\{\Delta^{k-1}f(x_1) - \Delta^{k-1}f(x_0)\right\} \equiv \frac{1}{k!h^k}\Delta^k f(x_0). \qquad (3\text{-}72)$$

In terms of these forward differences, Newton's formula takes on a simpler form. For $x = (x_0 + ph)$, Eq. (3-67) may be written as

$$\begin{aligned} f(x_0 + ph) &= f_0 + \sum_{j=1}^{N} \Delta^j f(x_0) \frac{p(p-1)(p-2)\cdots(p-j+1)}{j!} \\ &\quad + f^{(N+1)}(\xi) h^{N+1} \frac{p(p-1)(p-2)\cdots(p-N)}{(N+1)!} \end{aligned} \qquad (3\text{-}73)$$

where, in arriving at the result, we have made use of the relation

$$\begin{aligned} (x-x_0)(x-x_1)(x-x_2)\cdots(x-x_{k-1}) &= ph\,(p-1)h\,(p-2)h\cdots(p-k+1)h \\ &= h^k p(p-1)(p-2)\cdots(p-k+1) \end{aligned}$$

obtained from the fact that $x_k = (x_0 + kh)$.

Bessel's formula for finite difference A more convenient method for inverse interpolation is named after Bessel, and it may be obtained from that of Newton in the following way. Instead of $\Delta^k f(x_0)$ in Eq. (3-72), we can make use of other forward differences. For this purpose, it is worth recalling that the general result of Eq. (3-70) does not depend on any particular order of arranging the points $x_0, x_1, \ldots, x_N$. Instead of having x_0 at one end of the interval, we can put it in the middle. The $(N+1)$ input points are now labeled instead as $x_{-N/2}, x_{-N/2+1}, \ldots, x_{-1}, x_0, x_1, \ldots, x_{N/2}$, with $x_{-k} = (x_0 - kh)$ and $x_k = (x_0 + kh)$. In the discussions above, we have always made use of the input values $f(x_0), f(x_1), \ldots$, in ascending order according to the subscript, as illustrated in Table 3-13. In the new labeling scheme, the corresponding order is $f(x_{-N/2}), f(x_{-N/2+1}), \ldots$. There is no compelling reason to follow this order. In fact, we shall see that it is more convenient to form divided differences by taking $f(x)$ in the order $f(x_0), f(x_1), f(x_{-1}), f(x_2), f(x_{-2}), f(x_3), f(x_{-3}), \ldots$.

In this new scheme, the coefficient c_1 remains unchanged in form from that given by Eq. (3-71). However, the actual input points that enter into the calculation are now taken from the middle of the interval rather than from one end. The expressions for the other coefficients are somewhat different from those in Newton's scheme. For example, in the place of Eq. (3-71), we have

$$c_2 = f_{x_{-1},x_0,x_1} = \frac{1}{2!h^2}\Delta^2 f(x_{-1}).$$

Similarly,

$$\begin{aligned} c_3 &= f_{x_{-1},x_0,x_1,x_2} = \frac{1}{3!h^3}\Delta^3 f(x_{-1}) \\ c_4 &= f_{x_{-2},x_{-1},x_0,x_1,x_2} = \frac{1}{4!h^4}\Delta^4 f(x_{-2}). \end{aligned}$$

The function $f(x)$ may now be expressed in the form

$$\begin{aligned} f(x_0+ph) &= f(x_0)+\frac{p}{1!}\Delta f(x_0)+\frac{p(p-1)}{2!}\Delta^2 f(x_{-1})+\frac{(p+1)p(p-1)}{3!}\Delta^3 f(x_{-1}) \\ &\quad +\frac{(p+1)p(p-1)(p-2)}{4!}\Delta^4 f(x_{-2})+\cdots \end{aligned} \tag{3-74}$$

instead of Newton's formula Eq. (3-72).

On the other hand, Problem 3-8 shows that, by making use of the input values of $f(x)$ in the order $f(x_0)$, $f(x_1)$, $f(x_2)$, $f(x_{-1})$, $f(x_3)$, $f(x_{-2})$, ..., and expanding $f(x)$ around $x = x_1$, a slightly different approximation for $f(x)$ is obtained:

$$\begin{aligned} f(x_0+ph) &= f(x_1)+\frac{p-1}{1!}\Delta f(x_0)+\frac{p(p-1)}{2!}\Delta^2 f(x_0)+\frac{p(p-1)(p-2)}{3!}\Delta^3 f(x_{-1}) \\ &\quad +\frac{(p+1)p(p-1)(p-2)}{4!}\Delta^4 f(x_{-1})+\cdots. \end{aligned} \tag{3-75}$$

The average of Eqs. (3-74) and (3-75) gives us the Bessel's formula for interpolation

$$\begin{aligned} f(x_0+ph) &= \frac{1}{2}\{f(x_0)+f(x_1)\}+(p-\frac{1}{2})\Delta f(x_0)) \\ &\quad +\frac{p(p-1)}{4}\{\Delta^2 f(x_{-1})+\Delta^2 f(x_0)\}+\frac{p(p-1)(2p-1)}{12}\Delta^3 f(x_{-1}) \\ &\quad +\frac{(p+1)p(p-1)(p-2)}{48}\{\Delta^4 f(x_{-2})+\Delta^4 f(x_{-1})\} \\ &\quad +\cdots. \end{aligned} \tag{3-76}$$

This is not the only way to express $f(x+ph)$ in term of forward differences. In fact, several other interpolation relations can also be derived starting from Newton's formula given by Eq. (3-72), but we shall not attempt them here.

The advantage of Bessel's formula is that the coefficients are small in general and vary only slowly with p. Let us see why this is useful in inverse interpolation. For this purpose, it is convenient to rewrite Eq. (3-76) in terms of the central difference operator $\hat{\delta}_i$, defined earlier in Eq. (2-52) as

$$\hat{\delta}_i f \equiv \hat{\delta}_i^1 f = f(x_{i+1/2}) - f(x_{i-1/2})$$

and

$$\hat{\delta}_i^n \equiv \hat{\delta}_{i+1/2}^{n-1} - \hat{\delta}_{i-1/2}^{n-1}.$$

Similarly, we can define an averaging operator $\hat{\mu}$ as

$$\hat{\mu}\phi(x) \equiv \frac{1}{2}\{\phi(x-\frac{1}{2}h)+\phi(x+\frac{1}{2}h)\}.$$

Table 3-14: Input to a fourth-order Bessel's interpolation formula.

$f(x_{-2})$				
	$\hat{\delta}_{-3/2}f$			
$f(x_{-1})$		$\hat{\delta}^2_{-1}f$		
	$\hat{\delta}_{-1/2}f$		$\hat{\delta}^3_{-1/2}f$	
$\underline{f(x_0)}$		$\underline{\hat{\delta}^2_0 f}$		$\underline{\hat{\delta}^4_0 f}$
	$\underline{\hat{\delta}_{1/2}f}$		$\underline{\hat{\delta}^3_{1/2}f}$	
$\underline{f(x_1)}$		$\underline{\hat{\delta}^2_1 f}$		$\underline{\hat{\delta}^4_1 f}$
	$\hat{\delta}_{3/2}f$		$\hat{\delta}^3_{3/2}f$	
$f(x_2)$		$\hat{\delta}^2_2 f$		
	$\hat{\delta}_{5/2}f$			
$f(x_3)$				

In terms of these two operators, Eq. (3-76) may be expressed as

$$\begin{aligned} f(x_0+ph) &= \hat{\mu}f(x_{1/2}) + \frac{p-\frac{1}{2}}{1!}\hat{\delta}_{1/2}f + \frac{p(p-1)}{2!}\,\hat{\mu}\hat{\delta}^2_{1/2}f + \frac{p(p-1)(p-\frac{1}{2})}{3!}\,\hat{\delta}^3_{1/2}f \\ &\quad + \frac{(p+1)p(p-1)(p-2)}{4!}\,\hat{\mu}\hat{\delta}^4_{1/2}f + \cdots. \end{aligned} \tag{3-77}$$

The general form of each term in the series (beyond the first two) is determined by whether it involves even or odd order central differences; that is, whether it involves $\hat{\delta}^k_j f$ for $k = 2m$ or $(2m+1)$:

$$\begin{cases} \dfrac{p(p^2-1^2)(p^2-2^2)\cdots\{p^2-(m-1)^2\}(p-m)}{(2m)!}\,\hat{\mu}\hat{\delta}^{2m}_{1/2}f & \text{for } k = 2m \\ \dfrac{p(p^2-1^2)(p^2-2^2)\cdots\{p^2-(m-1)^2\}(p-m)(p-\frac{1}{2})}{(2m+1)!}\,\hat{\delta}^{2m+1}_{1/2}f & \text{for } k = 2m+1. \end{cases}$$

The derivation of this relation is given in Hildebrand[26] and we shall not repeat it here.

A method for inverse interpolation The inverse interpolation problem we are interested in is now reduced to one of finding the value of p in Eq. (3-77) for which $f(x_0+ph) = C$. If we use Bessel's formula involving central differences up to $\hat{\delta}^4_i f$, the five input quantities are

$$\begin{aligned} \hat{\mu}f_{1/2} = \frac{1}{2}(f_1+f_0) \qquad \hat{\delta}_{1/2}f &= f(x_1) - f(x_0) \qquad \hat{\mu}\hat{\delta}^2_{1/2}f = \frac{1}{2}(\hat{\delta}^2_1 f + \hat{\delta}^2_0 f) \\ \hat{\delta}^3_{1/2}f = \hat{\delta}^2_1 f - \hat{\delta}^2_0 f & \qquad\qquad \hat{\mu}\hat{\delta}^4_{1/2}f = \frac{1}{2}(\hat{\delta}^4_1 f + \hat{\delta}^4_0 f). \end{aligned} \tag{3-78}$$

These are underlined in Table 3-14. The six equidistant values of $f(x_i)$, required to calculate the central differences involved, are given in the first column.

Table 3-15: A different indexing scheme for central differences (cf. Table 3-14).

$f(x_1)$				
	$\hat{\delta}_1 f$			
$f(x_2)$		$\hat{\delta}_1^2 f$		
	$\hat{\delta}_2 f$		$\hat{\delta}_1^3 f$	
$f(x_3)$		$\underline{\hat{\delta}_2^2 f}$		$\underline{\hat{\delta}_1^4 f}$
	$\underline{\hat{\delta}_3 f}$		$\underline{\hat{\delta}_2^3 f}$	
$f(x_4)$		$\underline{\hat{\delta}_3^2 f}$		$\underline{\hat{\delta}_2^4 f}$
	$\hat{\delta}_4 f$		$\hat{\delta}_3^3 f$	
$f(x_5)$		$\hat{\delta}_4^2 \mathrm{f}$		
	$\hat{\delta}_5 f$			
$f(x_6)$				

Once the five quantities in Eq. (3-78) are available, Eq. (3-77) may be put in the form of finding the value of p satisfying the equation

$$\begin{aligned} C &\approx \hat{\mu} f_{1/2} + (p - \frac{1}{2})\hat{\delta}_{1/2} f + \frac{p(p-1)}{2}\,\hat{\mu}\hat{\delta}_{1/2}^2 f + \frac{p(p-1)(2p-1)}{12}\,\hat{\delta}_{1/2}^3 f \\ &\quad + \frac{(p+1)p(p-1)(p-2)}{24}\,\hat{\mu}\hat{\delta}_{1/2}^4 f \\ &= f_0 + p\hat{\delta}_{1/2} f + \frac{p(p-1)}{2}\,\hat{\mu}\hat{\delta}_{1/2}^2 f + \frac{p(p-1)(2p-1)}{12}\,\hat{\delta}_{1/2}^3 f \\ &\quad + \frac{(p+1)p(p-1)(p-2)}{24}\,\hat{\mu}\hat{\delta}_{1/2}^4 f. \end{aligned}$$

Since the coefficients of all the terms vary only slowly with p, we can rewrite this relation in the following form:

$$\begin{aligned} p &\approx \frac{1}{\hat{\delta}_{1/2} f}\Big\{C - f(x_0) - \frac{p(p-1)}{2}\,\hat{\mu}\hat{\delta}_{1/2}^2 - \frac{p(p-1)(2p-1)}{12}\,\hat{\delta}_{1/2}^3 f \\ &\qquad - \frac{(p+1)p(p-1)(p-2)}{24}\,\hat{\mu}\hat{\delta}_{1/2}^4 f\Big\}. \end{aligned} \tag{3-79}$$

This is not a solution for p, as it appears on both sides of the equation. However, it may be used in an iterative procedure to find its value to the desired accuracy.

The lowest order approximation for p from Eq. (3-79) is

$$p_0 \approx \frac{1}{\hat{\delta}_{1/2} f}\Big\{C - f(x_0)\Big\}.$$

If we express the right side of Eq. (3-79) as a function in the following way:

$$\psi(p) \equiv \frac{1}{\hat{\delta}_{1/2} f}\Big\{C - f(x_0) - \frac{p(p-1)}{2}\,\hat{\mu}\hat{\delta}_{1/2}^2 f - \frac{p(p-1)(2p-1)}{12}\,\hat{\delta}_{1/2}^3 f$$

Table 3-16: Input for inverse interpolation of $\frac{1}{\sqrt{2\pi}}\int_{-\infty}^{x} e^{-t^2/2}dt$.

x_i	$f(x_i)$	$\hat{\delta}_i f$	$\hat{\delta}_i^2 f$	$\hat{\delta}_i^3 f$	$\hat{\delta}_i^4 f$
0.0	0.50000000				
		0.07925971			
0.2	0.57925971		-0.00309768		
		0.07616203		-0.00273921	
0.4	0.65542174		-0.00583689		0.00064868
		0.07032514		-0.00209051	
0.6	0.72574688		-0.00792742		0.00082037
		0.06239772		-0.00127016	
0.8	0.78814460		-0.00919758		0.00085319
		0.05320014		0.00041697	
1.0	0.84134474		-0.00961455		0.00075894
		0.04358559		0.00034197	
1.2	0.88493033		-0.00927258		0.00057497
		0.03431301		0.00091694	
1.4	0.91924334		-0.00835564		0.00035030
		0.02595737		0.00126724	
1.6	0.94520071		-0.00708840		0.00013238
		0.01886897		0.00139962	
1.8	0.96406968		-0.00568878		
		0.01318019			
2.0	0.97724987				

$$- \frac{(p+1)p(p-1)(p-2)}{24}\hat{\mu}\hat{\delta}_{1/2}^4 f\Big\} \tag{3-80}$$

successive higher order approximations of p may be put into the form of a recursive relation

$$p_{k+1} = \psi(p_k). \tag{3-81}$$

Note that the approach is based on the fact that the value of $\psi(p)$ varies only slowly with p.

To carry out an inverse interpolation in practice, we need to construct a difference table of the form of Table 3-14. The five quantities that must be calculated in Eq. (3-78) to provide the necessary input to Eq. (3-81) come from $\hat{\delta}_{1/2}f$, $\hat{\delta}_1 f$, $\hat{\delta}_0 f$, $\hat{\delta}_{1/2}^3 f$, $\hat{\delta}_1^4 f$, and $\hat{\delta}_0^4 f$. Such an approach may be adequate for hand calculations — to construct a computer program to carry out the work, it is essential that we have a more automated method.

To start with, we need a method to obtain the value of any central difference $\hat{\delta}_m^n f$ from a list of $f(x_i)$. From Table 3-13, it is easy to see that $(n+1)$ values of $f(x_i)$ are required to calculate central differences up to order n. If we use the standard method for numbering the subscripts, such as that given in Table 3-14, the first element of this list is $(m - n/2)$, the next one $(m - n/2 + 1)$, and so on. A simpler approach, as far as programming a computer is concerned, is to use the scheme that each element — the input $f(x_i)$ values as well as finite differences of a given order — is numbered according to the order in which it appears. This is given in Table 3-15. The correspondence with the scheme used in Table 3-14 may be found by comparing

Box 3-8 Program DM_BSNTP
Inverse interpolation using Bessel's formula
Assuming only one root in the interval

Subprogram used:
  CENT_DIFF: Calculate the central difference of $f(x)$ for a given order.
Initialization:
  (a) Set the accuracy required to be ϵ.
  (b) Define the interpolation formula Eq. (3-80) for p as a function.
1. Input the list of $\{x_i, f(x_i)\}$ values.
2. Find the size of spacing in x.
3. Input y for which the value of x is to be interpolated.
4. Start the interpolation:
   (a) Locate the nearest point in the table.
   (b) Use CENT_DFF to obtain the five input quantities of Eq. (3-78):
      (i) Define the $(n+1)$ zeroth-order differences $\{d_i\}$ to be equal to $\{f_i\}$.
      (ii) Calculate the next order by storing $(d_{i+1} - d_i)$ in the location for d_i. For order k, there are $(n+1-k)$ such differences.
      (iii) Repeat step (ii) until $k = n$
      (iv) Return the value of d_1 as the order-n central difference.
5. Calculate the value of p using the recursion relation of Eq. (3-81).
6. Output the result of each iteration.

the elements at the same locations in the two tables. In this way, the mth finite difference of order n may be calculated from input $f(x_i)$ values starting from the mth element in the list. The algorithm for constructing a function that returns the finite difference of an arbitrary order is given as a part of Box 3-8.

With a general routine to calculate the central differences, it is now relatively easy to implement Bessel's formula for inverse interpolation. As an example, we shall work out the value of x for which the normal probability integral has the value 0.75. That is, we wish to find the value of x satisfying the relation

$$\frac{1}{\sqrt{2\pi}} \int_{-\infty}^{x} e^{-t^2/2} dt = 0.75.$$

The input pairs of quantities, taken from a standard mathematical table, are listed in the first two columns of Table 3-16, and the maximum error we can tolerate in the inverse interpolation is set to be $\epsilon = 10^{-3}$.

From the table, we see that the value of x we wish to find must occur between 0.6 and 0.8. Since $f(x = 0.6) = 0.72574688$ and $f(x = 0.8) = 0.78814459$, we choose $x_0 = 0.6$. It then follows that

$$f(x_0) = 0.72574688 \qquad \hat{\delta}_{1/2} f = 0.06239772$$
$$\hat{\delta}_0^2 f = -0.00792742 \qquad \hat{\delta}_1^2 f = -0.00919759 \qquad \hat{\delta}_{1/2}^3 f = -0.00127017$$
$$\hat{\delta}_0^4 f = 0.00082034 \qquad \hat{\delta}_1^4 f = 0.00085324.$$

From this set of values we obtain

$$p_0 = 0.38868618 \qquad p_1 = 0.37217590 \qquad p_2 = 0.37246385.$$

The difference between p_2 and p_1 is less than ϵ. As a result, no further iterations of Eq. (3-81) are needed. The final result, $p = 0.372$, corresponds to

$$x = x_0 + ph = 0.6 + 0.372 * 0.2 = 0.674$$

and this is our calculated result for $f(x) = 0.75$.

Coefficients of the interpolation polynomial There are occasions where the values of the coefficients in the interpolation polynomial are of interest. For simplicity, we shall again examine only the case of polynomial interpolation here. Using Eq. (3-1), the order-N polynomial approximation to $f(x)$ has the form

$$f(x) \approx a_0 + a_1 x + a_2 x^2 + \cdots + a_N x^N. \tag{3-82}$$

Our problem here is that, given a table of $(N+1)$ pairs of $\{x_i, f(x_i)\}$, we wish to find the values of the coefficients a_0, $a_1, \ldots, a_N$.

Since we have $(N+1)$ pairs of values, Eq. (3-82) gives us a set of $(N+1)$ equations:

$$\begin{array}{ccccccccccc} f(x_0) & = & a_0 & + & a_1 x_0 & + & a_2 x_0^2 & + & \cdots & + & a_N x_0^N \\ f(x_1) & = & a_0 & + & a_1 x_1 & + & a_2 x_1^2 & + & \cdots & + & a_N x_1^N \\ \vdots & & \vdots & & \vdots & & \vdots & & \vdots & & \vdots \\ f(x_{N-1}) & = & a_0 & + & a_1 x_{N-1} & + & a_2 x_{N-1}^2 & + & \cdots & + & a_N x_{N-1}^N \\ f(x_N) & = & a_0 & + & a_1 x_N & + & a_2 x_N^2 & + & \cdots & + & a_N x_N^N. \end{array} \tag{3-83}$$

We can use interpolation (or extrapolation) to find the value of $f(x)$ at $x = 0$. This gives the value of a_0, as can be seen from Eq. (3-82).

Once a_0 is known, we can take it out from $f(x_0)$, $f(x_1), \ldots, f(x_N)$. On dividing the resulting differences in each case by the corresponding value of x, we obtain a new set of polynomial expressions that is one order lower than those given in Eq. (3-83):

$$\begin{array}{l} \frac{f(x_0)-a_0}{x_0} = a_1 + a_2 x_0 + \cdots + a_N x_0^{N-1} \\ \frac{f(x_1)-a_0}{x_1} = a_1 + a_2 x_1 + \cdots + a_N x_1^{N-1} \\ \quad \vdots \qquad \vdots \qquad \vdots \qquad \vdots \qquad \vdots \\ \frac{f(x_{N-1})-a_0}{x_{N-1}} = a_1 + a_2 x_{N-1} + \cdots + a_N x_{N-1}^{N-1} \\ \frac{f(x_N)-a_0}{x_N} = a_1 + a_2 x_N + \cdots + a_N x_N^{N-1}. \end{array}$$

Any N of these $(N+1)$ relations may now be used to find the value of a_1 by interpolating (or extrapolating) for $x = 0$ and this gives us a_1 in the same way as we did earlier for the value of a_0. This process is repeated N times until all $(N+1)$ coefficients are found.

In matrix notation, Eq. (3-83) may be written as

$$\begin{vmatrix} 1 & x_0 & x_0^2 & \cdots & x_0^N \\ 1 & x_1 & x_1^2 & \cdots & x_1^N \\ \vdots & \vdots & \vdots & \ddots & \vdots \\ 1 & x_{N-1} & x_{N-1}^2 & \cdots & x_{N-1}^N \\ 1 & x_N & x_N^2 & \cdots & x_N^N \end{vmatrix} \begin{vmatrix} a_0 \\ a_1 \\ \vdots \\ a_{N-1} \\ a_N \end{vmatrix} = \begin{vmatrix} f(x_0) \\ f(x_1) \\ \vdots \\ f(x_{N-1}) \\ f(x_N) \end{vmatrix}. \tag{3-84}$$

We shall see later in §5-1 that the unknown coefficients a_0, $a_1, \ldots, a_N$ may be found by inverting the determinant on the right side. The particular form of the determinant we have here is known as a Vandermonde determinant and it may be inverted more easily than the general case. However, we shall not go into this topic.

3-7 Cubic spline

A spline is a piece of flexible wood or plastic that can be bent into arbitrary smooth shapes. In the days before computers, it was commonly used in tracing a smooth curve between points on a sheet of graph paper. For this reason, the word is now associated with numerical techniques that perform the same function. The basic idea behind most such methods is to use a simple function to approximate the relation between the dependent and independent variables, similar to what we did in interpolation. For most cases, it is adequate to use a third-degree polynomial of the form

$$f(x) \approx a_0 + a_1 x + a_2 x^2 + a_3 x^3 \tag{3-85}$$

and hence the name *cubic spline.*

Our interest is mainly in a function given to us in terms of a set of $(N+1)$ values f_0, $f_1, \ldots, f_N$ taken, respectively, at $x = x_0$, $x_1, \ldots, x_N$. In general, we cannot assume that the points $\{x_i\}$ are evenly spaced. The goal in a cubic spline calculation is to find a smooth third-degree polynomial approximation to the unknown function $f(x)$ that makes the best attempt to go through all the given points.

The interpolation polynomials discussed in the first two sections of this chapter are not suitable for our purpose here, as our aim is limited to finding the best approximate values for the function in small local regions. As an alternative, we can use an interpolation function made of a sum of terms, each one of which is a third-degree polynomial within a subinterval. For two adjacent subintervals, the coefficients for the two interpolation polynomials can be different from each other and, as a result, the interpolation function may not be a smooth one. However, this is not a problem, as we are not interested in the global behavior.

Mathematically, a smooth function is one that has derivatives of all orders and they are continuous everywhere in the region. This is not possible if we use different finite degree polynomials in different subintervals. For the third-degree polynomial approximation we wish to use here, the best we can do is to have derivatives up to the second order continuous across the boundary. To construct such a polynomial, let us consider first an arbitrary subinterval $[x_i, x_{i+1}]$. For the approximation to go through f_i and f_{i+1} at the two ends of this subinterval, it is sufficient to use a first-order polynomial of the form

$$f(x) = f_i\,\lambda_i(x) + f_{i+1}\,\omega_i(x) \tag{3-86}$$

where

$$\lambda_i(x) = \frac{x_{i+1} - x}{x_{i+1} - x_i} \qquad \omega_i(x) = 1 - \lambda(x) = \frac{x - x_i}{x_{i+1} - x_i}. \tag{3-87}$$

To extend the functional form to a third-degree polynomial, we need information in addition to f_i and f_{i+1}. Since all the known values of the function within the subinterval are used up already, the new ones have to be found from the condition that both the first- and second-order derivatives must be continuous across the boundaries to the two subintervals on either side.

For later convenience, we shall build the third-order polynomial by starting from Eq. (3-86) and considering it as the first term of a series. To this, we shall add terms involving x^2 and x^3. Furthermore, since the first term already has the correct values at the two ends of the subinterval, contributions from the additional terms must vanish at these points. One possible way to satisfy these requirements is to use

$$\begin{aligned} f(x)_{[x_i,x_{i+1}]} &= f_i\lambda_i(x) + f_{i+1}\omega_i(x) + f_i''\frac{(x_{i+1}-x_i)^2}{6}\{\lambda_i^3(x) - \lambda_i(x)\} \\ &\quad + f_{i+1}''\frac{(x_{i+1}-x_i)^2}{6}\{\omega_i^3(x) - \omega_i(x)\} \end{aligned} \tag{3-88}$$

where the two coefficients f_i'' and f_{i+1}'' will be identified later as the values of the second-order derivatives of $f(x)$ at, respectively, x_i and x_{i+1}. The constant factor $(x_{i+1} - x_i)^2/6$ is included for later convenience.

We shall now examine the continuity of the polynomial $f(x)$ and its derivatives across the boundaries to the subinterval on the left at $x = x_i$ and to that on the right at $x = x_{i+1}$. It is obvious that we have $f_{[x_i,x_{i+1}]}(x_i) = f_i$ and $f_{[x_i,x_{i+1}]}(x_{i+1}) = f_{i+1}$, since both $\{\lambda^3(x) - \lambda(x)\}$ and $\{\omega^3(x) - \omega(x)\}$ vanish at $x = x_i$ and x_{i+1}. Using Eq. (3-88), we find that the first-order derivative of $f_{[x_i,x_{i+1}]}(x)$ is

$$\begin{aligned} \left.\frac{df}{dx}\right|_{[x_i,x_{i+1}]} &= \frac{f_{i+1} - f_i}{x_{i+1} - x_i} - f_i''\frac{x_{i+1} - x_i}{6}\{3\lambda_i^2(x) - 1\} \\ &\quad + f_{i+1}''\frac{x_{i+1} - x_i}{6}\{3\omega_i^2(x) - 1\}. \end{aligned} \tag{3-89}$$

For it to be continuous across the boundary, its value at $x = x_i$ must be equal to that for the subinterval $[x_{i-1}, x_i]$ at the same point. Similarly, its value at $x = x_{i+1}$ must be the same as that for $[x_{i+1}, x_{i+2}]$. As we shall see later, these requirements provide us with the two conditions to determine coefficients f_i'' and f_{i+1}''.

The second-order derivative of $f_{[x_i,x_{i+1}]}(x)$ is

$$\left.\frac{d^2 f}{dx^2}\right|_{[x_i,x_{i+1}]} = f_i''\lambda_i(x) + f_{i+1}''\omega_i(x).$$

Since $\lambda(x) = 1$ and $\omega(x) = 0$ at $x = x_i$, the right side is equal to f_i''. As a result, d^2f/dx^2 is continuous across $x = x_i$ if f_i'' is the value of the second-order derivative of polynomial $f_{[x_{i-1},x_i]}(x)$ from the subinterval $[x_{i-1}, x_i]$ to the left. Similarly, $\lambda(x) = 0$ and $\omega(x) = 1$ at $x = x_{i+1}$, and d^2f/dx^2 is continuous across $x = x_{i+1}$ if f_{i+1}'' is the value of the second-order derivative of polynomial $f_{[x_{i+1},x_{i+2}]}(x)$ from the subinterval $[x_{i+1}, x_{i+2}]$ to the right. From this, we see that the second-order derivatives can be continuous across all the boundaries between subintervals by imposing the condition that the value of f_i'' at $x = x_i$ is the same for the two subintervals on either side.

The only calculation remaining now is to ensure that the first-order derivatives from two adjacent subintervals are also equal at to each other at the boundary. Consider now two adjacent subintervals $[x_{i-1}, x_i]$ and $[x_i, x_{i+1}]$. At their boundary $x = x_i$, we have from Eq. (3-89) the results

$$\begin{aligned}
\left.\frac{df}{dx}\right|_{[x_i,x_{i+1}]}(x = x_i) &= \frac{f_{i+1} - f_i}{x_{i+1} - x_i} - \frac{x_{i+1} - x_i}{3} f_i'' - \frac{x_{i+1} - x_i}{6} f_{i+1}'' \\
\left.\frac{df}{dx}\right|_{[x_{i-1},x_i]}(x = x_i) &= \frac{f_i - f_{i-1}}{x_i - x_{i-1}} + \frac{x_i - x_{i-1}}{6} f_{i-1}'' + \frac{x_i - x_{i-1}}{3} f_i''.
\end{aligned}$$

For these two quantities to be equal to each other, the following relation must be satisfied:

$$(x_i - x_{i-1})f_{i-1}'' + 2(x_{i+1} - x_{i-1})f_i'' + (x_{i+1} - x_i)f_{i+1}'' = 6\left(\frac{f_{i+1} - f_i}{x_{i+1} - x_i} - \frac{f_i - f_{i-1}}{x_i - x_{i-1}}\right). \tag{3-90}$$

There are three unknown quantities in this equation, f_{i-1}'', f_i'', and f_{i+1}''. Two of the three quantities appear also in a similar equation at $x = x_{i+1}$, obtained from the boundary between subintervals $[x_i, x_{i+1}]$ and $[x_{i+1}, x_{i+2}]$. When we can apply the same consideration to all the boundaries between subintervals in our domain, we end up with $(N-1)$ equations, one for each boundary point between two subintervals. There are $(N+1)$ unknown f_i'', one for each point in the region, including the two end points. To obtain a solution, two additional pieces of information must be supplied. These are usually taken as external conditions we must impose on the second derivative of $f(x)$.

It is helpful to write out Eq. (3-90) explicitly at the two points of $i = 1$ and $(N-1)$, just inside the two end points:

$$(x_1 - x_0)f_0'' + 2(x_2 - x_0)f_1'' + (x_2 - x_1)f_2'' = 6\left(\frac{f_2 - f_1}{x_2 - x_1} - \frac{f_1 - f_0}{x_1 - x_0}\right) \tag{3-91}$$

Box 3-9 Program DM_CBSPL
Cubic spline with Gaussian elimination

Subprograms used:
CUBIC_SPLINE: Evaluates the second-order derivatives for cubic spline.
SMOOTH: Generates evenly spaced interpolated values.
Initialization:
(a) Set the limits for the maximum number of input and output points.
(b) Define the two second-order derivatives at the boundaries to be zero.
1. Input:
(a) $\{x_i, f_i\}$ for $i = 0, 1, 2, \ldots, N$.
(b) m, the number of evenly spaced output points required.
2. Use CUBIC_SPLINE for second-order derivatives at the internal mesh points.:
(a) Define the elements of the tridiagonal matrix using Eq. (3-94).
(b) Reduce the matrix to an upper bidiagonal one using Eq. (A-7).
(c) Back substitution to obtain y_i'' using Eqs. (A-7) and (A-8).
3. Use derivatives from SMOOTH to interpolate for a set of evenly spaced points:
(a) Find the step size.
(b) Set the first output point to be the same as the first input point.
(c) Interpolate the value of $f(x)$ for all the points:
(i) Find the subinterval in the input to which x belongs.
(ii) Calculate the value of $f(x)$ using Eq. (3-88).
(d) Return x and the interpolated values of $f(x)$.
4. Output the values of $f(x)$ at the m evenly spaced points between x_0 and x_N.

$$(x_{N-1} - x_{N-2})f''_{N-2} + 2(x_N - x_{N-2})f''_{N-1} + (x_N - x_{N-1})f''_n = 6\left(\frac{f_N - f_{N-1}}{x_N - x_{N-1}} - \frac{f_{N-1} - f_{N-2}}{x_{N-1} - x_{N-2}}\right). \qquad (3\text{-}92)$$

If the two additional input quantities are given to us as the values of f''_0 and f''_1, the only unknown remaining in Eq. (3-91) is f''_2. This gives us the value of f''_2 in terms of the input f''_0 and f''_1. Once we have f''_2, we can use it together with f''_1 to solve for f''_3 using Eq. (3-90) by putting $i = 2$. This process may be continued until we come to the end, where $i = (N-1)$. Similarly, if we have the values f''_N and f''_{N-1} instead of f''_0 and f''_1, we can use Eq. (3-92) and solve the problem backward.

Since the function $f(x)$ is unknown, it is difficult in general for us to have an idea on what the second-order derivatives should be anywhere in the region of interest. A more realistic approach is for us to make some estimates of the values of f''_0 and f''_N at the two ends of the interval. Once this is done, we have enough information to solve the set of $(N-1)$ equations given by Eq. (3-90). The solution is now slightly more complicated because of the different boundary conditions. Let us first rewrite Eqs. (3-91) and (3-92) into a form with the unknown second-order derivatives of $f(x)$

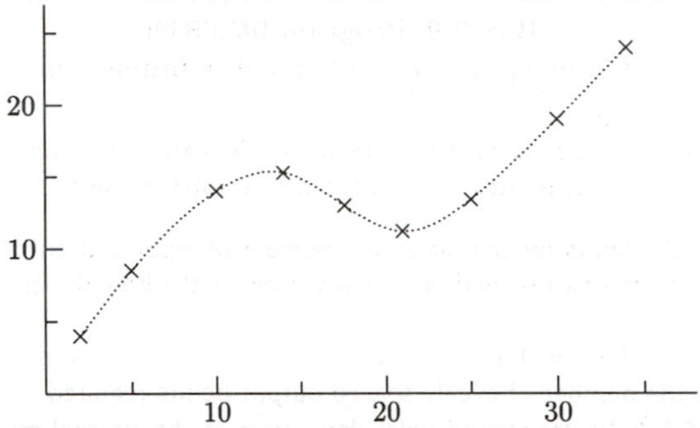

Figure 3-5: Smoothing a function by cubic spline. The crosses are the nine values of $f(x)$ supplied and the dots are those generated using a cubic spline routine.

only on the left side:

$$2(x_2 - x_0)f_1'' + (x_2 - x_1)f_2'' = 6\Big(\frac{f_2 - f_1}{x_2 - x_1} - \frac{f_1 - f_0}{x_1 - x_0}\Big) - (x_1 - x_0)f_0''$$

$$(x_{N-1} - x_{N-2})f_{N-2}'' + 2(x_N - x_{N-2})f_{N-1}'' = 6\Big(\frac{f_N - f_{N-1}}{x_N - x_{N-1}} - \frac{f_{N-1} - f_{N-2}}{x_{N-1} - x_{N-2}}\Big) - (x_N - x_{N-1})f_N''.$$

Together with Eq. (3-90) for $i = 2, 3, \ldots, (N-2)$, we have a set of $(N-1)$ equations, and these may be written in the form of a matrix product:

$$\begin{pmatrix} b_1 & c_1 & 0 & 0 & \ldots & 0 \\ a_2 & b_2 & c_2 & 0 & \ldots & 0 \\ 0 & a_3 & b_3 & c_3 & \ldots & 0 \\ \vdots & \vdots & \vdots & \ddots & \vdots & \vdots \\ 0 & \ldots & 0 & a_{N-2} & b_{N-2} & c_{N-2} \\ 0 & \ldots & 0 & 0 & a_{N-1} & b_{N-1} \end{pmatrix} \begin{pmatrix} f_1'' \\ f_2'' \\ f_3'' \\ \ldots \\ f_{N-2}'' \\ f_{N-1}'' \end{pmatrix} = \begin{pmatrix} d_1 \\ d_2 \\ d_3 \\ \ldots \\ d_{N-2} \\ d_{N-1} \end{pmatrix} \tag{3-93}$$

where

$$a_i = \begin{cases} 0 & \text{for } i = 1 \\ x_i - x_{i-1} & \text{for } i = 2, 3, \ldots, (N-1) \end{cases}$$

$$b_i = 2(x_{i+1} - x_{i-1})$$

$$c_i = \begin{cases} x_{i+1} - x_i & \text{for } i = 1, 2, 3, \ldots, (N-2) \\ 0 & \text{for } i = (N-1) \end{cases}$$

$$d_i = \begin{cases} 6\Big(\dfrac{f_2 - f_1}{x_2 - x_1} - \dfrac{f_1 - f_0}{x_1 - x_0}\Big) - (x_1 - x_0)f_0'' & \text{for } i = 1 \\ 6\Big(\dfrac{f_{i+1} - f_i}{x_{i+1} - x_i} - \dfrac{f_i - f_{i-1}}{x_i - x_{i-1}}\Big) & \text{for } i = 2, 3, \ldots, (N-2) \\ 6\Big(\dfrac{f_N - f_{N-1}}{x_N - x_{N-1}} - \dfrac{f_{N-1} - f_{N-2}}{x_{N-1} - x_{N-2}}\Big) - (x_N - x_{N-1})f_N'' & \text{for } i = (N-1). \end{cases} \tag{3-94}$$

Since the square matrix involved has a tridiagonal form, the solution can be obtained using the method of Gaussian elimination given in §A-3.

Once all the values of f_i'' are known, we have a complete set of continuous, third-order polynomial approximations to $f(x)$. Each polynomial has the form given by Eq. (3-88). Using such an expression, it is possible for us to calculate the approximate value of $f(x)$ for any x within $[x_0, x_N]$. The algorithm is summarized in Box 3-9. As an example, an arbitrary function, defined by the values at nine input points in the interval $x = [2, 34]$, is shown by the crosses in Fig. 3-5. To draw a smooth curve through these nine points, we interpolate the values of $f(x)$ for a number of intermediate points using a cubic spline routine based on the algorithm outlined above and these are shown as dots. As can be seen from the figure, the interpolated values do form a smooth curve joining the input points.

Problems

3-1 If a function $f(x)$ is a polynomial of degree 2, having the form

$$f(x) = a_0 + a_1 x + a_2 x^2$$

show that the second-order difference of Neville's algorithm given by Eq. (3-4) is exact.

3-2 In *Computing Methods for Scientists and Engineers* by Fox and Mayers,[22] the following table of the values of a function $f(x)$ is given for $x = -0.2$ to 1.2 in steps of 0.2:

$$-0.7328 \quad -0.7071 \quad -0.6528 \quad -0.3981 \quad 0.5721 \quad 3.1165 \quad 8.4372 \quad 18.0797.$$

Find the value of x for $f(x) = 0.0$. Compare the results using Neville's method and Bessel's formula.

3-3 Prove the recursion relation of Eq. (3-10) by working out the difference between $D_{i+1,\ell}$ and $U_{i,\ell}$. Express the result in terms of $f_{i,i+1,\ldots,i+\ell}$ and $f_{i,i+1,\ldots,i+\ell+1}$.

3-4 The value of $\sin x$ for $x = 2.5$ is positive, and for $x = 3.5$, the value is negative. Use the bisection method of Box 3-7 on a calculator to find the value of x in the interval $[2.617994, 3.665191]$ for which $\sin x = 0$. Deduce the number of steps required to get an agreement with $x = \pi$ to five significant figures.

3-5 Apply the Fourier transform of Eq. (3-32) to a square wave

$$f(x) = \begin{cases} 0 & \text{for } 0 \le x < 1,\ 2 \le x < 3,\ \cdots \\ 1 & \text{for } 1 \le x < 2,\ 3 \le x < 4,\ \cdots \end{cases}$$

and find the values of the first nine coefficients a_0, a_1, b_1, a_2, b_2, a_3, b_4, a_4, and a_5 using Eq. (3-33). Reconstruct the waveform from the Fourier coefficients obtained and compare the results with the original. The differences come from the higher-order terms ignored.

3-6 Repeat Problem 3-5 with a sawtooth wave form

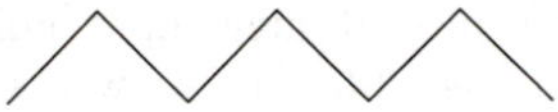

$$f(x) = \begin{cases} x & \text{for } 0 \le x \le 1 \\ 2 - x & \text{for } 1 \le x \le 2 \\ x - 2 & \text{for } 2 \le x \le 3 \\ 4 - x & \text{for } 3 \le x \le 4 \\ \cdots \end{cases}$$

3-7 Derive Eq. (3-75) using a polynomial approximation of the form

$$\begin{aligned} f(x) &= c_0 + c_1(x - x_0) + c_2(x - x_0)(x - x_1) \\ &\quad + c_3(x - x_0)(x - x_1)(x - x_2) \\ &\quad + c_4(x - x_0)(x - x_1)(x - x_2)(x - x_{-1}) + \cdots \end{aligned}$$

where $x_k = (x_0 + kh)$ for both positive and negative integer values of k and $ph = (x - x_0)$.

3-8 Use the method of bisection of Box 3-7 to find the two values of x in the interval $x = [0, 15]$ for which the function

$$f(x) = 2.5 - 3.0x + 0.2x^2 + 0.015x^3$$

vanishes.

3-9 Apply a cubic spline to the periodic function given by the following nine pairs of values:

$x =$	0	$\frac{1}{2}\pi$	π	$\frac{3}{2}\pi$	2π	$\frac{5}{2}\pi$	3π	$\frac{7}{2}\pi$	4π
$f(x) =$	0	1	0	-1	0	1	0	-1	0

Use $f''(x = 0) = f''(x = 4\pi) = 0$ as the two additional conditions required. These sets of points can also be fitted by a sine function. What are the differences between the results of a cubic spline and $\sin x$?

3-10 Show that a matrix with elements given by Eq. (3-47) is the inverse of $\boldsymbol{A}$.

Chapter 4

Special Functions

A large number of mathematical functions have been developed over the years because of the needs in physics and engineering. In this chapter, we shall examine some of these functions, partly for their intrinsic usefulness and partly as the vehicle to introduce certain computational techniques. Although each function is discussed in relation to a particular type of problem, the interest is usually much broader. However, we shall not try to make any attempt to cover the full range of possible applications for any of the special functions.

4-1 Hermite polynomials and harmonic oscillator

A one-dimensional harmonic oscillator is often used to illustrate certain fundamental aspects in both classical and quantum mechanics. The reason for its popularity comes from the fact that the potential is simple in form

$$V_{\text{h.o.}}(x) = \frac{1}{2}\mu\omega^2 x^2 \tag{4-1}$$

where μ is the mass and ω is the angular frequency of the oscillator. It is an idealization of the potential experienced by an object trapped near a minimum, as shown schematically in Fig. 4-1. In the vicinity of such a point at x_0, we can expand the potential $V(x)$ in terms of a series,

$$V(x) = c_0 + c_1(x - x_0) + c_2(x - x_0)^2 + c_3(x - x_0)^3 + \cdots.$$

The first term c_0 is a constant, which may be removed by redefining the zero point of the potential, as we shall do soon. At a minimum, the first-order derivative of $V(x)$ vanishes and, as a result, $c_1 = 0$. The quadratic term becomes the dominant one and Eq. (4-1) emerges if we make the replacements $x \longrightarrow (x - x_0)$ and $V(x) \longrightarrow V(x) - c_0$.

One-dimensional harmonic oscillator in quantum mechanics The particular interest we have here is the wave function of a one-dimensional harmonic oscillator in quantum mechanics. Many different ways are available in standard texts to find the

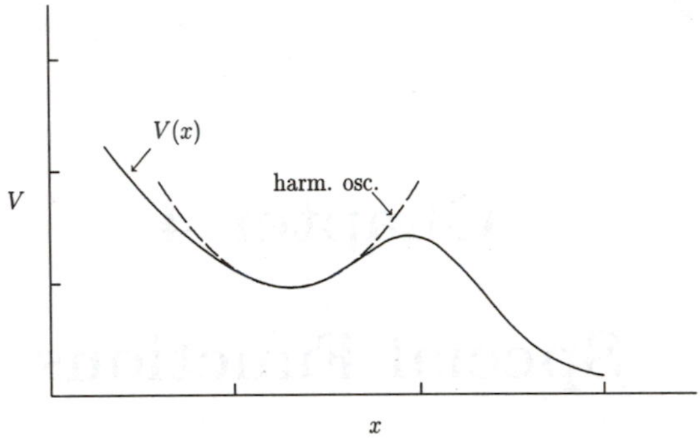

Figure 4-1: Approximation of the shape of a potential $V(x)$ near a minimum by harmonic oscillator $V_{\text{h.o.}}(x) = \frac{1}{2}\mu\omega^2 x^2$.

solution. Here, we shall concentrate on the one that makes use of Hermite polynomials. In addition to the one-dimensional harmonic oscillator, Hermite polynomials are also useful, for example, in a Gram-Charlier expansion of a distribution that is almost a normal one (see, for example, Cramer.[13])

The Hamiltonian of a particle of mass μ moving in an harmonic oscillator potential well of Eq. (4-1) is given by

$$H = \frac{1}{2\mu}p_x^2 + \frac{1}{2}\mu\omega^2 x^2 \tag{4-2}$$

where p_x is the linear momentum of the particle. To obtain a quantum mechanical solution of the problem, we must first express H in the form of an operator:

$$\hat{H} = -\frac{\hbar^2}{2\mu}\frac{d^2}{dx^2} + \frac{1}{2}\mu\omega^2 x^2 \tag{4-3}$$

where we have made use of the substitution $p_x \to (\hbar/i)(d/dx)$. Using $\hat{H}$, we can construct the time-independent Schrödinger equation ,

$$\hat{H}\psi(x) = E\psi(x) \tag{4-4}$$

from which we obtain the eigenvalues and eigenfunctions that describe the system.

Mathematically, the problem is equivalent to one of solving a second-order ordinary differential equation of the form

$$-\frac{\hbar^2}{2\mu^2}\frac{d^2\psi}{dx^2} + \Big(\frac{1}{2}\mu\omega^2 x^2 - E\Big)\psi(x) = 0. \tag{4-5}$$

It is convenient to make a change of the variables here. If we measure energies in units of $\frac{1}{2}\hbar\omega$,

$$E \longrightarrow \lambda = \frac{E}{\frac{1}{2}\hbar\omega} \tag{4-6}$$

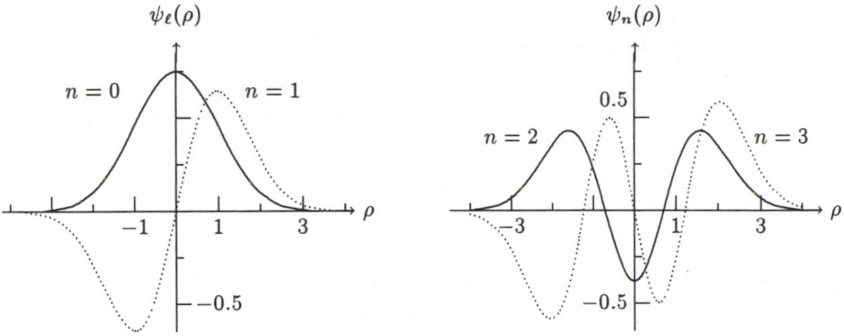

Figure 4-2: One-dimensional harmonic oscillator wave function $\psi_n(\rho)$ of Eq. (4-9) for $n = 0$, 1, 2, 3, and 4.

and express length in terms of the dimensionless quantity

$$\rho = \sqrt{\frac{\mu\omega}{\hbar}}x \tag{4-7}$$

the differential equation takes on the form

$$\frac{d^2\psi}{d\rho^2} + (\lambda - \rho^2)\psi(\rho) = 0. \tag{4-8}$$

To keep the expression simple, we have dropped an overall factor of $\frac{1}{2}\hbar\omega$.

Harmonic oscillator wave function and Hermite polynomial Eq. (4-5) is an eigenvalue equation, as the solution exists only for certain values of E. It is well known from quantum mechanics that the allowed values are

$$E_n = (2n+1)\frac{1}{2}\hbar\omega \qquad\qquad n = 0, 1, \cdots$$

or λ of Eq. (4-6) odd integers. The solution for a given n is known as an eigenfunction and may be written in the form

$$\psi_n(\rho) = \frac{1}{\sqrt{2^n n!\sqrt{\pi}}}e^{-\rho^2/2}H_n(\rho) \tag{4-9}$$

where $H_n(\rho)$ is Hermite polynomial of degree n. The shapes of $\phi_0(\rho)$, $\phi_1(\rho)$, $\phi_2(\rho)$, and $\phi_3(\rho)$ are given in Fig. 4-2 as illustration.

By substituting Eq. (4-9) into (4-8), it is easy to see that Hermite polynomials are the solution of

$$\frac{d^2}{d\rho^2}H_n(\rho) - 2\rho\frac{d}{d\rho}H_n(\rho) + (\lambda - 1)H_n(\rho) = 0. \tag{4-10}$$

It may be expressed as a power series consisting of either even or odd powers of ρ

$$H_n(\rho) = \sum_{k=0,1}^{n} a_{n,k}\rho^k \tag{4-11}$$

where

$$a_{n,k+2} = \frac{2k - \lambda + 1}{(k+1)(k+2)}a_{n,k}. \tag{4-12}$$

For $\lambda = 2n + 1$, the series terminates at $k = n$. Note that the equation actually represents two sets of solution, one for k even and the other for k odd. This is expected from the fact that Eq. (4-10) is a second-order differential equation. The two sets may be distinguished by having one set starting with

$$a_0 \neq 0 \qquad\qquad a_1 = 0$$

and the other set with

$$a_0 = 0 \qquad\qquad a_1 \neq 0.$$

We shall not show the proof here that these two sets of solution are linearly independent of each other.

The orthogonality relation for $H_n(\rho)$ is given by

$$\int_{-\infty}^{+\infty} e^{-\rho^2} H_n(\rho)H_m(\rho)\,d\rho = 2^n n!\sqrt{\pi}\delta_{n,m}. \tag{4-13}$$

Explicitly, the lowest few orders are

$$\begin{aligned} &H_0(\rho) = 1 && H_1(\rho) = 2\rho \\ &H_2(\rho) = 4\rho^2 - 2 && H_3(\rho) = 8\rho^3 - 12\rho \\ &H_4(\rho) = 16\rho^4 - 48\rho^2 + 12 && H_5(\rho) = 32\rho^5 - 160\rho^3 + 120\rho \end{aligned} \tag{4-14}$$

where we have taken

$$a_{n,n} = 2^n$$

to be the coefficient of the leading term for each order so as to satisfy the normalization given in Eq. (4-13).

A method to generate Hermite polynomials Once we have the forms of a few low-order polynomials, we can generate the higher order ones through the following recursion relation:

$$H_{n+1}(\rho) = 2\rho H_n(\rho) - 2nH_{n-1}(\rho). \tag{4-15}$$

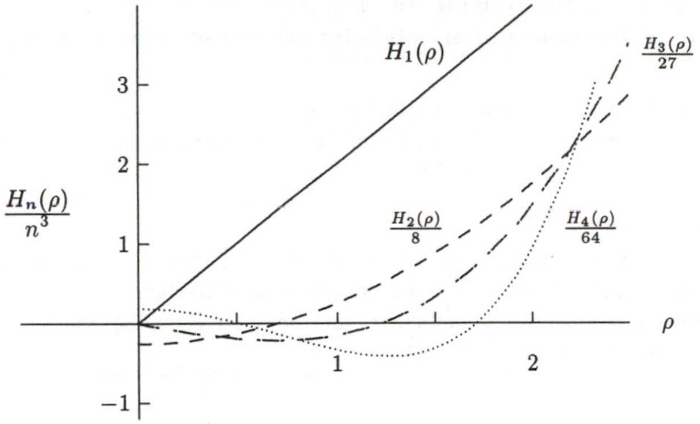

Figure 4-3: Hermite polynomials $H_n(\rho)$ for $n = 1$ to 4. The plotted values are $H_n(r)/n^3$ so as to bring all four curves to roughly the same range.

In terms of the coefficients, this may be expressed as

$$a_{n+1,k} = 2a_{n,k-1} - 2na_{n-1,k}. \tag{4-16}$$

That is, to find the coefficient for order $(n+1)$ for a given power k, we need as input those for orders n and $(n-1)$ of the same k. Using the fact that $H_0(\rho) = 1$ and $H_1(\rho) = 2\rho$, we have the starting values

$$a_{0,0} = 1 \qquad a_{1,0} = 0 \qquad a_{1,1} = 2$$

(and $a_{n,k} = 0$ for $k > n$). From these, it is possible to find the coefficients $a_{2,0}$ and $a_{2,2}$ for $H_2(\rho)$ ($a_{2,1} = a_{2,k} = 0$ for $k > 2$). Once the coefficients for $H_2(\rho)$ are known, we can use them, together with those of $H_1(\rho)$, to produce those for $H_3(\rho)$, and so on. Since all the coefficients are integers, the method is efficient, limited only by the largest integer that can be stored in the computer. The steps in the calculation are outlined in Box 4-1.

We can also use Eq. (4-15) to calculate the numerical value of $H_n(\rho)$ for a given ρ. Again we start with $H_0(\rho) = 1$ and $H_1(\rho) = 2\rho$. The value of $H_2(\rho)$ is obtained from the expression

$$H_2(\rho) = 2\rho H_1(\rho) - 2H_0(\rho)$$

using the numerical values for the quantities on the right hand side. Next we can apply Eq. (4-15) to produce $H_3(\rho)$ using the value of $H_2(\rho)$ for the ρ just obtained and the value of $H_1(\rho)$ we already have. The process is repeated for the next order, and so on, until the desired order is reached. In this way, the values of Hermite

Box 4-1 Subroutine HRMTE(K_ORDER,NARY,NDMN) **Coefficients of Hermite polynomials by recursion relation Eq. (4-16)**
Argument list: K_ORDER: Order of the Hermite polynomial. NARY: Two-dimensional array for all the coefficients up to order K_ORDER. NDMN: Dimension of array NARY. Initialization: (a) Zero the array NARY. (b) Define coefficients $a_{0,0} = 1$ for $H_0(\rho)$, and $a_{1,0} = 0$ and $a_{1,1} = 2$ for $H_1(\rho)$. Start the propagation from order 2 and continue till K_ORDER: (a) Find the nonvanishing coefficients for the order using Eq. (4-16). (b) Return the results in array NARY.

polynomials for a given ρ are propagated starting from those of $H_0(\rho)$ and $H_1(\rho)$, one order at a time, up to any arbitrary high orders. The method is simple, but the numerical accuracy is likely to be poor if n is large. A better alternative is to use the method in the previous paragraph to calculate the algebraic form of $H_n(\rho)$ first. The numerical value of $H_n(\rho)$ is then produced by substituting the value of ρ into the algebraic expression. As an illustration, the results for the lowest few orders of $H_n(\rho)$ for $\rho < 5$ are plotted in Fig. 4-3.

4-2 Legendre polynomials and spherical harmonics

In many problems, it is possible to make use of spherical symmetry to simplify the calculations. One such example is the Helmholtz equation,

$$\nabla^2\psi(\boldsymbol{r}) + k^2\psi(\boldsymbol{r}) = 0. \tag{4-17}$$

It is a second-order partial differential equation that appears in a variety of problems in quantum mechanics, electromagnetism, and wave phenomena, among others. If the coefficient k^2 is a constant or only a function of the radial coordinate r, the solution may be expressed as a product of radial and angular parts. To carry out the separation of variables, we shall first rewrite the Laplacian operator in terms of its components in spherical polar coordinates:

$$\nabla^2 = \frac{1}{r^2}\frac{\partial}{\partial r}\left(r^2\frac{\partial}{\partial r}\right) + \frac{1}{r^2}\left[\frac{1}{\sin\theta}\frac{\partial}{\partial\theta}\left(\sin\theta\frac{\partial}{\partial\theta}\right) + \frac{1}{\sin^2\theta}\frac{\partial^2}{\partial\phi^2}\right] \tag{4-18}$$

where θ is the polar angle and ϕ, the azimuthal angle, as shown in Fig. 4-4. In quantum mechanics, the angular part of the expression,

$$\frac{1}{\hbar^2}\hat{\boldsymbol{L}}^2 = -\left[\frac{1}{\sin\theta}\frac{\partial}{\partial\theta}\left(\sin\theta\frac{\partial}{\partial\theta}\right) + \frac{1}{\sin^2\theta}\frac{\partial^2}{\partial\phi^2}\right] \tag{4-19}$$

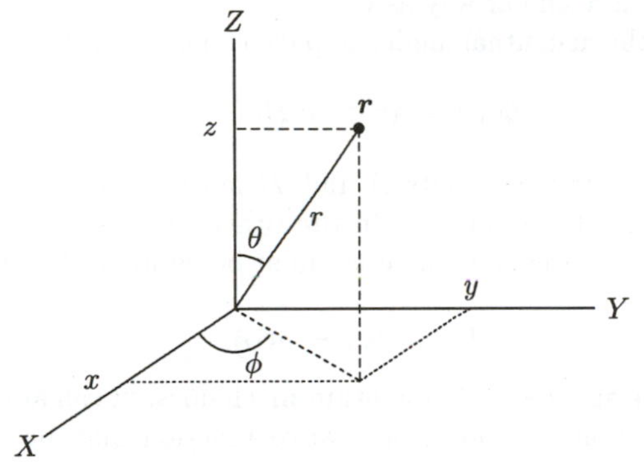

Figure 4-4: Spherical polar coordinate system. The vector $\boldsymbol{r}$ is specified by giving its length r, its polar angle θ, and azimuthal angle ϕ.

is identified as the negative of the square of the orbital angular momentum operator $\hat{\boldsymbol{L}}^2$ in units of $\hbar^2$. Here $\hbar$ is Planck's constant divided by 2π.

Separation of variables For problems with spherical symmetry, we can write the function $\psi(\boldsymbol{r})$ in Eq. (4-17) as

$$\psi(\boldsymbol{r}) = R(r)\,Y(\theta,\phi). \tag{4-20}$$

If k^2 is independent of θ and ϕ, the Helmholtz equation may be separated into two, one dealing with radial dependence and the other with angular dependence:

$$\begin{aligned} \frac{d}{dr}\left(r^2\frac{d}{dr}\right)R(r) + (kr)^2R(r) &= CR(r) \\ \left[\frac{1}{\sin\theta}\frac{\partial}{\partial\theta}\left(\sin\theta\frac{\partial}{\partial\theta}\right) + \frac{1}{\sin^2\theta}\frac{\partial^2}{\partial\phi^2}\right]Y(\theta,\phi) &= -CY(\theta,\phi). \end{aligned} \tag{4-21}$$

Our concern in this section is with the angular part, and we shall delay further discussions of the radial part to the next two sections.

A second separation of variables may be carried out for Eq. (4-21). By writing $Y(\theta,\phi)$ as a product of θ- and ϕ-dependent parts,

$$Y(\theta,\phi) = P(\theta)\Phi(\phi)$$

we obtain the following two equations:

$$\begin{aligned} \frac{1}{P(\theta)\sin\theta}\frac{d}{d\theta}\left(\sin\theta\frac{d}{d\theta}\right)P(\theta) + C + \frac{1}{\sin^2\theta}D &= 0 \\ \frac{1}{\Phi(\phi)}\frac{d^2}{d\phi^2}\Phi(\phi) &= D \end{aligned} \tag{4-22}$$

where D is constant, in a similar way as C.

The solution for the azimuthal angle, ϕ, part is quite simple,

$$\Phi(\phi) = Ae^{im\phi} + Be^{-im\phi} \tag{4-23}$$

where $m^2 = -D$. The two constants A and B must be determined by boundary conditions supplied with the problem. On the other hand, since ϕ can only take on values in the range $[0, 2\pi]$, the function $\Phi(\phi)$ must be periodic. For a classical system, it has the property

$$\Phi(\phi + 2\pi) = \Phi(\phi).$$

As a result, m must be an integer. In quantum mechanics, m can also be half-integers, arising from the fact that the *parity* of a wave function may be either positive or negative

$$\Phi(\phi + 2\pi) = \pm\Phi(\phi).$$

However, we shall ignore such a possibility for the moment.

Polar angle dependence and Legendre equation Let us return to the polar angle dependence of the system given by Eq. (4-22),

$$\frac{1}{\sin\theta}\frac{d}{d\theta}\left(\sin\theta\frac{d}{d\theta}\right)P(\theta) + \left(C - \frac{m^2}{\sin^2\theta}\right)P(\theta) = 0$$

where we have made use of the fact that $D = -m^2$. This equation may be simplified by writing

$$\eta = \cos\theta.$$

The form in Eq. (4-22) is now reduced to

$$\frac{d}{d\eta}\left[(1-\eta^2)\frac{dP}{d\eta}\right] + \left(C - \frac{m^2}{1-\eta^2}\right)P(\eta) = 0 \tag{4-24}$$

where $P(\eta)$ is a function of η, with η restricted to the range $[-1, +1]$.

Let us start by considering the simple case that the system has no ϕ dependence. As a result, $m = 0$. The equation satisfied by $P(\eta)$ is

$$\frac{d}{d\eta}\left[(1-\eta^2)\frac{dP}{d\eta}\right] + CP(\eta) = 0. \tag{4-25}$$

This is known as the Legendre equation. The more general case, with $m \neq 0$, is the associated Legendre equation and we shall return to it later.

Since Eq. (4-24) is an eigenvalue equation for the angular momentum operator $\boldsymbol{L}^2$ of Eq. (4-19), only values

$$C = \ell(\ell+1) \qquad \ell = 0, 1, 2, \cdots$$

Table 4-1: Legendre polynomials of low orders.

$P_0(\eta) = 1$	$P_1(\eta) = \eta$
$P_2(\eta) = \frac{1}{2}(3\eta^2 - 1)$	$P_3(\eta) = \frac{1}{2}(5\eta^3 - 3\eta)$
$P_4(\eta) = \frac{1}{8}(35\eta^4 - 30\eta^2 + 3)$	$P_5(\eta) = \frac{1}{8}(63\eta^5 - 70\eta^3 + 15\eta)$

are allowed. The solutions of Eq. (4-25) are known as Legendre polynomials and they can be written as a power series in η

$$P_\ell(\eta) = \sum_{k=0}^{\infty} a_{\ell,k}\eta^k$$

with

$$a_{\ell,k+2} = \frac{k(k+1) - \ell(\ell+1)}{(k+1)(k+2)} a_{\ell,k}. \tag{4-26}$$

Similar to Hermite polynomials, Legendre polynomials $P_\ell(\eta)$ of even order involves only even powers of η and those with odd ℓ, only odd powers of η.

Legendre polynomials of different ℓ are orthogonal to each other. The normalization is chosen such that

$$\int_{-1}^{+1} P_m(\eta)P_n(\eta)\,d\eta = \frac{2}{2n+1}\delta_{m,n}. \tag{4-27}$$

Since $P_0(\eta)$ is, by definition, a polynomial of degree zero, it cannot have any dependence on η. With the normalization condition above, it is necessary that

$$P_0(\eta) = 1.$$

Similarly, we obtain

$$P_1(\eta) = \eta$$

as it must be a polynomial of degree 1 in η and orthogonal to $P_0(\eta)$.

With the values of $P_0(\eta)$ and $P_1(\eta)$ given above as the starting point, we can make use of the recursion relation

$$(2k+1)\eta P_k(\eta) = (k+1)P_{k+1}(\eta) + kP_{k-1}(\eta) \tag{4-28}$$

to generate the higher-order polynomials. For this purpose, it is convenient to rewrite the equation in the form

$$P_\ell(\eta) = \frac{1}{\ell}\{(2\ell-1)\eta P_{\ell-1}(\eta) - (\ell-1)P_{\ell-2}(\eta)\}. \tag{4-29}$$

This can also be used, for example, to find the value of $P_\ell(\eta)$ for a given η from those of $P_{\ell-1}(\eta)$ and $P_{\ell-2}(\eta)$. However, unlike Hermite polynomials, it is not the preferred method to calculate the coefficients of Legendre polynomials. The difference comes from the fact that the normalization we have chosen for Legendre polynomials produces coefficients that are ratios of polynomials, as can be seen from the explicit forms of the lowest few listed in Table 4-1. As a result, we cannot easily make use of an integer representation in the calculations, as we have done earlier for the Hermite polynomials in Box 4-1.

Legendre polynomials are often given by the Rodrigues formula

$$P_\ell(\eta) = \frac{1}{2^\ell \ell!}\frac{d^\ell}{d\eta^\ell}\left(\eta^2 - 1\right)^\ell. \tag{4-30}$$

The more practical way to generate them on a computer is to use the alternate representation

$$P_\ell(\eta) = \sum_{k=0}^{[\ell/2]} a_{\ell,k}\eta^k = \sum_{k=0}^{[\ell/2]}(-1)^k \frac{(2\ell-2k)!}{2^\ell k!(\ell-k)!(\ell-2k)!}\eta^{\ell-2k} \tag{4-31}$$

where the symbol for the upper limit has the meaning

$$\left[\frac{\ell}{2}\right] = \begin{cases} j & \text{if } \ell = 2j \\ j-1 & \text{if } \ell = 2j-1 \end{cases}$$

for integer j. In terms of the the coefficient for η^k, we have

$$a_{\ell,k} = (-1)^{(\ell-k)/2}\frac{(\ell+k)(\ell+k-1)\cdots(k+2)(k+1)}{2^\ell \ell!}\binom{\ell}{\frac{\ell+k}{2}}. \tag{4-32}$$

Note that, since k and ℓ must be both odd or both even, the factors $(\ell \pm k)/2$ appearing in the phase and the binomial coefficient are integers.

Associated Legendre polynomials Let us now return to the more general case of a system that is a function of both the polar angle θ and azimuthal angle ϕ. As far as the θ-dependent part is concerned, we no longer have the simplification offered by $m = 0$ in Eq. (4-25). Starting from the Rodrigues formula Eq. (4-30), it is not difficult to see that the function

$$P_{\ell,m}(\eta) = (1-\eta^2)^{m/2}\frac{d^m}{d\eta^m}P_\ell(\eta) = \frac{1}{2^\ell \ell!}(1-\eta^2)^{m/2}\frac{d^{\ell+m}}{d\eta^{\ell+m}}(\eta^2-1)^\ell \tag{4-33}$$

for $0 \leq |m| \leq \ell$, satisfies the associated Legendre equation

$$\frac{d}{d\eta}\Big[(1-\eta^2)\frac{d}{d\eta}P_{\ell,m}\Big] + \Big[\ell(\ell+1) - \frac{m^2}{1-\eta^2}\Big]P_{\ell,m} = 0$$

obtained from Eq. (4-24) by replacing C by $\ell(\ell+1)$. The proof can be found, for example, in Arfken[3]. The quantity $P_{\ell,m}(\eta)$ is known as the associated Legendre polynomial.

Table 4-2: Associated Legendre polynomials $P_{\ell,m}(\eta)$ for $m > 0$ and $\ell \leq 5$

$P_{1,1}(\eta) = (1-\eta^2)^{1/2}$	$P_{2,1}(\eta) = 3\eta(1-\eta^2)^{1/2}$
$P_{2,2}(\eta) = 3(1-\eta^2)$	$P_{3,1}(\eta) = \frac{3}{2}(5\eta^2-1)(1-\eta)^{1/2}$
$P_{3,2}(\eta) = 15\eta(1-\eta^2)$	$P_{3,3}(\eta) = 15(1-\eta^2)^{3/2}$
$P_{4,1}(\eta) = \frac{5}{2}\eta(7\eta^2-3)(1-\eta^2)^{1/2}$	$P_{4,2}(\eta) = \frac{15}{2}(7\eta^2-1)(1-\eta^2)$
$P_{4,3}(\eta) = 105\eta(1-\eta^2)^{3/2}$	$P_{4,4}(\eta) = 105(1-\eta^2)^2$

The orthonormal condition of $P_{\ell,m}(\eta)$ for different ℓ but the same m is

$$\int_{-1}^{+1} P_{\ell,m}(\eta)P_{\ell',m}(\eta)\,d\eta = \frac{2}{2\ell+1}\frac{(\ell+m)!}{(\ell-m)!}\,\delta_{\ell,\ell'}. \tag{4-34}$$

However, the orthogonality relation for associated Legendre polynomials of the same ℓ but different m requires an additional weighting factor,

$$\int_{-1}^{+1} P_{\ell,m}(\eta)P_{\ell,m'}(\eta)\frac{1}{1-\eta^2}d\eta = \frac{(\ell+m)!}{m(\ell-m)!}\delta_{m,m'}. \tag{4-35}$$

We shall see later that it is more convenient to express the same two conditions in terms of spherical harmonics. A few of the low-order associated Legendre polynomials for $m > 0$ are given in Table 4-2 as examples. Later, we shall construct a simple algorithm to generate the algebraic form of $P_{\ell,m}(\eta)$ to arbitrary orders starting from Eq. (4-32).

It is also possible to show that, instead of Eq. (4-33), the expression for $P_{\ell,m}(\eta)$ may also be written as

$$P_{\ell,m}(\eta) = (-1)^m\frac{1}{2^\ell \ell!}\frac{(\ell+m)!}{(\ell-m)!}(1-\eta^2)^{-m/2}\frac{d^{\ell-m}}{d\eta^{\ell-m}}(\eta^2-1)^\ell.$$

By comparing with that given in Eq. (4-33), we find that two associated Legendre polynomials, having the same ℓ but with m value opposite in sign, are related to each other in the following way:

$$P_{\ell,-m}(\eta) = (-1)^m\frac{(\ell-m)!}{(\ell+m)!}P_{\ell,m}(\eta). \tag{4-36}$$

The same relation can also be obtained from Eq. (4-33) through the use of Leibnitz's formula for the derivatives of a product of two functions:

$$\frac{d^n}{dx^n}\big\{A(x)B(x)\big\} = \sum_{k=0}^{n}\binom{n}{k}\Big\{\frac{d^{n-k}}{dx^{n-k}}A(x)\Big\}\Big\{\frac{d^k}{dx^k}B(x)\Big\}.$$

By taking $A(x) = (\eta-1)^\ell$ and $B(x) = (\eta+1)^\ell$, we obtain Eq. (4-36) after rearranging terms.

Spherical harmonics Since the square of the angular momentum operator $\hat{\boldsymbol{L}}^2$ given in Eq. (4-19) is a function of both the polar and azimuthal angles, its eigenfunctions are products of $e^{\pm im\phi}$ of Eq. (4-23) and associated Legendre polynomials $P_{\ell,m}(\cos\theta)$,

$$\begin{aligned} Y_{\ell,m}(\theta,\phi) &= (-1)^m \sqrt{\frac{(2\ell+1)}{4\pi}\frac{(\ell-m)!}{(\ell+m)!}}\, e^{im\phi} P_{\ell,m}(\cos\theta) \\ &= \frac{(-1)^m}{2^\ell \ell!}\sqrt{\frac{(2\ell+1)}{4\pi}\frac{(\ell-m)!}{(\ell+m)!}}\, e^{im\phi}(1-\eta^2)^{m/2}\left(\frac{d}{d\eta}\right)^{\ell+m}(\eta^2-1)^\ell. \end{aligned} \tag{4-37}$$

These functions are known as the spherical harmonics and the explicit forms of the lowest few are:

$$\begin{aligned} &Y_{0,0}(\theta,\phi) = \sqrt{\tfrac{1}{4\pi}} && Y_{1,0}(\theta,\phi) = \sqrt{\tfrac{3}{4\pi}}\cos\theta \\ &Y_{1,\pm1}(\theta,\phi) = \mp\sqrt{\tfrac{3}{8\pi}}e^{\pm i\phi}\sin\theta && Y_{2,0}(\theta,\phi) = \sqrt{\tfrac{5}{16\pi}}(3\cos^2\theta - 1) \\ &Y_{2,\pm1}(\theta,\phi) = \mp\sqrt{\tfrac{15}{8\pi}}e^{\pm i\phi}\cos\theta\,\sin\theta && Y_{2,\pm2}(\theta,\phi) = \sqrt{\tfrac{15}{32\pi}}e^{\pm 2i\phi}\sin^2\theta \\ &Y_{3,0}(\theta,\phi) = \sqrt{\tfrac{7}{16\pi}}(5\cos^3\theta - 3\cos\theta) && Y_{3,\pm1}(\theta,\phi) = \mp\sqrt{\tfrac{21}{64\pi}}e^{\pm i\phi}(5\cos^2\theta - 1)\sin\theta \\ &Y_{3,\pm2}(\theta,\phi) = \sqrt{\tfrac{105}{32\pi}}e^{\pm 2i\phi}\cos\theta\,\sin^2\theta && Y_{3,\pm3}(\theta,\phi) = \mp\sqrt{\tfrac{35}{64\pi}}e^{\pm 3i\phi}\sin^3\theta\ . \end{aligned} \tag{4-38}$$

The multiplicative constants for these functions are selected in such a way that $Y_{\ell m}(\theta,\phi)$ are normalized to unity. The orthogonality relation between those with different values of ℓ and m is given by the following integral:

$$\int_0^{2\pi}\int_0^{\pi} Y^*_{\ell m}(\theta,\phi) Y_{\ell' m'}(\theta,\phi)\,\sin\theta\,d\theta\,d\phi = \delta_{\ell\ell'}\delta_{mm'}. \tag{4-39}$$

The complex conjugate of $Y_{\ell m}(\theta,\phi)$ is

$$Y^*_{\ell m}(\theta,\phi) = (-1)^m Y_{\ell,-m}(\theta,\phi) \tag{4-40}$$

as can be seen from Eqs. (4-36) and (4-37).

Being the product of $e^{\pm im\phi}$ and associated Legendre polynomials, spherical harmonics are also the eigenfunctions of the z-component of the operator $\hat{\boldsymbol{L}}$ for angular momentum

$$\hat{\boldsymbol{L}}_z Y_{\ell,m}(\theta,\phi) = -i\hbar\frac{\partial}{\partial\phi}Y_{\ell,m}(\theta,\phi) = m\hbar Y_{\ell,m}(\theta,\phi).$$

The other two components of the operator $\hat{\boldsymbol{L}}$ may be written in the form of the angular momentum raising operator $\hat{\boldsymbol{L}}_+$ and lowering operator $\hat{\boldsymbol{L}}_-$, and they have

Box 4-2 Subroutine LGNDR(L_SIGN,LIST,L_RANK,K_RANK,MD)
Coefficients of Legendre polynomials using Eq. (4-44)

Argument list:
- L_SIGN: Sign of $a_{\ell,k}$.
- LIST: Array of prime number decomposition of $a_{\ell,k}$.
- L_RANK: Rank ℓ of $a_{\ell,k}$.
- K_RANK: Rank k of $a_{\ell,k}$.
- MD: Dimension of LIST.

Subprogram used:
- PRMDP: Prime number decomposition of Box A-1.

1. Initialize prime number decomposition of integers and factorials using PRMDP.
2. Zero the array LIST.
3. Check for special cases:
   (a) Return zero if $(\ell + k)$ odd.
   (b) Return 1 if both ℓ and k are zero.
4. Calculate the nonvanishing coefficients using Eq. (4-44):
   (a) Form the product $(l+k)(l+k-1)\cdots(k+2)(k+1)$.
   (b) Include contributions from the binomial coefficient and the denominator.
   (c) Include the phase factor $(-1)^{(\ell-k)/2}$.
5. Return the prime number decomposition of $a_{\ell,k}$.

the properties:

$$\begin{aligned}
\hat{\boldsymbol{L}}_{\pm}Y_{\ell,m}(\theta,\phi) &= \big(\hat{\boldsymbol{L}}_x \pm i\hat{\boldsymbol{L}}_y\big)Y_{\ell,m}(\theta,\phi) \\
&= \pm\hbar e^{\pm i\phi}\Big(\frac{\partial}{\partial\theta} \pm i\cot\theta\frac{\partial}{\partial\phi}\Big)Y_{\ell,m}(\theta,\phi) \\
&= \hbar\sqrt{(\ell\mp m)(\ell\pm m+1)}\; Y_{\ell m\pm 1}(\theta,\phi).
\end{aligned} \tag{4-41}$$

For $m = 0$, the spherical harmonics may be expressed in term of Legendre polynomials,

$$Y_{\ell 0}(\theta) = \sqrt{\frac{2\ell+1}{4\pi}}\, P_\ell(\cos\theta). \tag{4-42}$$

The same relation can also be obtained from Eq. (4-37) by putting $m = 0$. We can use it together with Eq. (4-41) to generate spherical harmonics of other m values.

In terms of $Y_{\ell,m}(\theta,\phi)$, the separation of $\psi(\boldsymbol{r})$, for a given ℓ and m, as a product of radial and angular parts given earlier in Eq. (4-20) takes on the form

$$\psi_{\ell m}(\boldsymbol{r}) = R(r)Y_{\ell,m}(\theta,\phi). \tag{4-43}$$

For this reason, spherical harmonics occur frequently in problems having spherical symmetry.

Coefficients of Legendre polynomials and spherical harmonics We can now construct a practical method to find the coefficients of the Legendre polynomials $P_\ell(\eta)$, associated Legendre polynomials $P_{\ell,m}(\eta)$, and spherical harmonics $Y_{\ell,m}(\theta,\phi)$. For $P_\ell(\eta)$, it is possible to use the recursion relation Eq. (4-28) to produce higher order ones from the starting point of $P_0(\eta) = 1$ and $P_1(\eta) = \eta$. However, a more direct method is to use the explicit expression for the coefficients $a_{\ell,k}$ of $P_{\ell,m}(\eta)$ given by Eq. (4-32). That is,

$$P_\ell(\eta) = \sum_{k=0}^{\ell} a_{\ell,k}\eta^k$$

with

$$a_{\ell,k} = \begin{cases} 0 & \text{for } (\ell+k) \text{ odd} \\ (-1)^{(\ell-k)/2}\dfrac{(\ell+k)(\ell+k-1)\cdots(k+2)(k+1)}{2^\ell \ell!}\dbinom{\ell}{\frac{\ell+k}{2}} & \text{otherwise.} \end{cases} \tag{4-44}$$

The algorithm, given in Box 4-2, is particularly simple if we use the method of decomposition into powers of prime numbers given in §A-3.

For associated Legendre polynomials, we have two indices, ℓ and m. As a result, the recursion relations are more complicated. Furthermore, the coefficients are ratios of integers because of the normalization condition of Eq. (4-35). For these reasons, it is more convenient to make use of the first form of Eq. (4-33) for $P_{\ell,m}(\eta)$ with $m \geq 0$. This gives us

$$P_{\ell,m}(\eta) = (1-\eta^2)^{m/2}\frac{d^m}{d\eta^m}P_\ell(\eta) = (1-\eta^2)^{m/2}\sum_{k=0}^{\ell-m} b_{\ell,m;k}\,\eta^k$$

where the coefficients $b_{\ell,m;k}$ may be obtained using Eq. (4-44):

$$\begin{aligned} b_{\ell,m;k} = a_{\ell,m+k}\frac{d^m}{d\eta^m}\eta^{k+m} &= a_{\ell,m+k}(k+m)(k+m-1)\cdots(k+2)(k+1) \\ &= \frac{(k+m)!}{k!}a_{\ell,m+k}. \end{aligned} \tag{4-45}$$

For $m < 0$, we can use Eq. (4-36) to relate the coefficients to those with the opposite sign of m. The result is

$$b_{\ell,-m;k} = (-1)^m\frac{(\ell-m)!}{(\ell+m)!}\frac{(k+m)!}{k!}a_{\ell,m+k} \tag{4-46}$$

where m is a positive integer. The algorithm to calculate $b_{\ell,m;k}$, using $a_{\ell,k}$ as the starting point, is included as a part of Box 4-3.

To find the spherical harmonics, we can take the product of the $\exp\{im\phi\}$ factor for the azimuthal angle dependence with the appropriate associated Legendre polynomials. The latter may be obtained using the relation given by the first form of

Box 4-3 Subroutine YLM(KIND,L_RANK,M_RANK)
Associated Legendre Polynomial and spherical harmonics $Y_{\ell,m}$

Argument list:
KIND = 2 associated Legendre polynomial, 3 spherical harmonics.
L_RANK: Rank ℓ for $b_{\ell,m;k}$.
M_RANK: Rank m for $b_{\ell,m;k}$.
Subprogram used:
LGNDR: Coefficients of Legendre polynomials of Box 4-2.
1. Obtain the coefficient $b_{\ell,m;k}$ of $P_{\ell,m}$:
   (a) Use LGNDR to generate $a_{\ell,k}$.
   (b) Convert $a_{\ell,k}$ to $b_{\ell,m;k}$:
      (i) Use Eq. (4-45) for $m > 0$.
      (ii) Use Eq. (4-46) for $m < 0$.
2. Convert $b_{\ell,m;k}$ to $c^2_{\ell,m;k}$ using Eq. (4-47).
3. Return $c^2_{\ell,m;k}$ for KIND=2 and $b_{\ell,m;k}$ for KIND=3.

Eq. (4-37). In terms of a power series in η, spherical harmonics $Y_{\ell,m}(\theta,\phi)$ may be written in the form

$$Y_{\ell,m}(\theta,\phi) = (-1)^m\sqrt{\frac{1}{4\pi}}\, e^{im\phi}\sin^m\theta\sum_{k=0}^{\ell-m} c_{\ell,m;k}\,\eta^k$$

where we have made use of the fact that

$$(1-\eta^2)^{m/2} = (1-\cos^2\theta)^{m/2} = \sin^m\theta.$$

The coefficient $c_{\ell,m;k}$ may be calculated from $b_{\ell,m;k}$ of $P_{\ell,m;k}(\eta)$ given in Eqs. (4-45) and (4-46):

$$c_{\ell,m;k} = \sqrt{\frac{(2\ell+1)(\ell-m)!}{(\ell+m)!}}\, b_{\ell,m;k}.$$

As a practical matter, it is easier to find the squares of the coefficients:

$$c^2_{\ell,m;k} = \frac{(2\ell+1)(\ell-m)!}{(\ell+m)!} b^2_{\ell,m;k}. \tag{4-47}$$

In this way, $c^2_{\ell,m;k}$ may be expressed in terms of a product of integer powers of prime numbers and the sign of $c_{\ell,m;k}$ following that of $b_{\ell,m;k}$. The algorithm for such a calculation is given in Box 4-3.

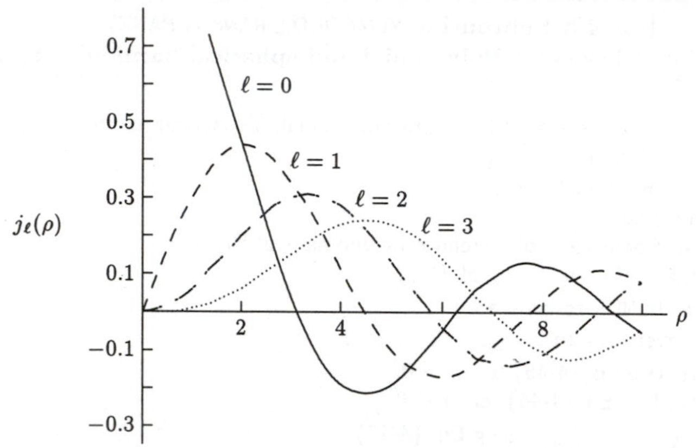

Figure 4-5: Spherical Bessel functions $j_\ell(\rho)$ for $\ell = 0$, 1, 2, and 3.

4-3 Spherical Bessel functions

Let us return to the radial part of the Helmholtz equation given by Eq. (4-17). If k^2 is a constant in the region of interest, the radial equation Eq. (4-20) reduces to

$$\frac{d}{dr}\left(r^2\frac{dR_\ell}{dr}\right) + \{k^2r^2 - \ell(\ell+1)\}R_\ell(r) = 0 \tag{4-48}$$

where we have used $R_\ell(r)$ to represent the radial part of $\psi(r)$ for a given angular momentum ℓ. Since k here has the dimension of inverse length, it is convenient to use, in place of r, the dimensionless variable

$$\rho = kr.$$

The differential equation Eq. (4-48) may now be expressed in the form

$$\rho^2\frac{d^2R_\ell}{d\rho^2} + 2\rho\frac{dR_\ell}{d\rho} + \{\rho^2 - \ell(\ell+1)\}R_\ell(r) = 0. \tag{4-49}$$

It is easy to show that if we make the substitutions

$$W(\rho) = \rho^{1/2}R_\ell(\rho) \qquad \text{and} \qquad \lambda = \ell + \frac{1}{2}.$$

Eq. (4-49) may be put into the standard form of a Bessel equation for $W(\rho)$

$$\rho^2\frac{d^2W}{d\rho^2} + \rho\frac{dW}{d\rho} + (\rho^2 - \lambda^2)W(\rho) = 0. \tag{4-50}$$

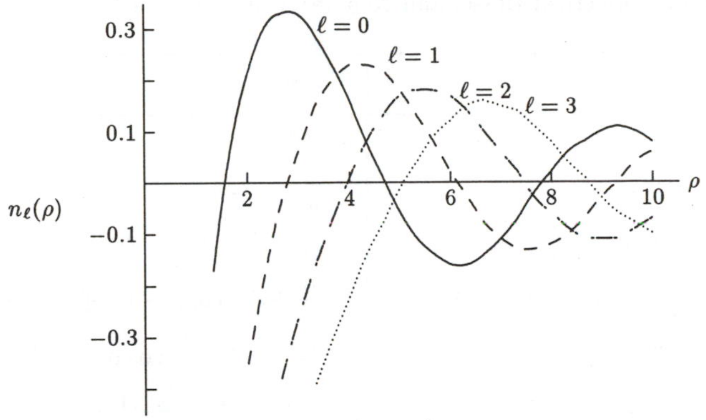

Figure 4-6: Spherical Neumann functions $n_\ell(\rho)$ for $\ell = 0$, 1, 2, and 3.

Since λ is a half-integer here, the solutions are related to Bessel functions of half integer ranks. Our main interest here is in Eq. (4-49), and we shall only return briefly to Eq. (4-50) at the end of this section.

For $\ell = 0$, Eq. (4-49) has a particularly simple form:

$$\rho^2 \frac{d^2 R_0}{d\rho^2} + 2\rho \frac{dR_0}{d\rho} + \rho^2 R_0(r) = 0. \tag{4-51}$$

There are two linearly independent solutions. The one that is regular at the origin is the spherical Bessel function of rank zero,

$$j_0(\rho) = \frac{\sin \rho}{\rho}. \tag{4-52}$$

The other,

$$n_0(\rho) = -\frac{\cos \rho}{\rho} \tag{4-53}$$

is the spherical Neumann function of rank zero. Particular linear combinations of them,

$$h_0^{(1)}(\rho) = j_0(\rho) + i n_0(\rho) \qquad\qquad h_0^{(2)}(\rho) = j_0(\rho) - i n_0(\rho)$$

are known as the spherical Hankel functions of, respectively, the first and second kind.

For $\ell = 1$, we can obtain from Eq. (4-49)

$$j_1(\rho) = \frac{\sin \rho}{\rho^2} - \frac{\cos \rho}{\rho} \qquad\qquad n_1(\rho) = -\frac{\cos \rho}{\rho^2} - \frac{\sin \rho}{\rho}. \tag{4-54}$$

Table 4-3: Spherical Bessel and Neumann functions of $\ell = 0$ to 5.

ℓ	$j_\ell(\rho)$	$n_\ell(\rho)$
0	$\frac{\sin\rho}{\rho}$	$-\frac{\cos\rho}{\rho}$
1	$\frac{\sin\rho}{\rho^2} - \frac{\cos\rho}{\rho}$	$-\frac{\cos\rho}{\rho^2} - \frac{\sin\rho}{\rho}$
2	$(\frac{3}{\rho^3} - \frac{1}{\rho})\sin\rho - \frac{3}{\rho^2}\cos\rho$	$(-\frac{3}{\rho^3} + \frac{1}{\rho})\cos\rho - \frac{3}{\rho^2}\sin\rho$
3	$(\frac{15}{\rho^4} - \frac{6}{\rho^2})\sin\rho + (-\frac{15}{\rho^3} + \frac{1}{\rho})\cos\rho$	$(-\frac{15}{\rho^4} + \frac{6}{\rho^2})\cos\rho + (-\frac{15}{\rho^3} + \frac{1}{\rho})\sin\rho$
4	$(\frac{105}{\rho^5} - \frac{45}{\rho^3} + \frac{1}{\rho})\sin\rho$ $+(-\frac{105}{\rho^4} + \frac{10}{\rho^2})\cos\rho$	$(-\frac{105}{\rho^5} + \frac{45}{\rho^3} - \frac{1}{\rho})\cos\rho$ $+(-\frac{105}{\rho^4} + \frac{10}{\rho^2})\sin\rho$
5	$(\frac{945}{\rho^6} - \frac{420}{\rho^4} + \frac{15}{\rho^2})\sin\rho$ $+(-\frac{945}{\rho^5} + \frac{105}{\rho^3} - \frac{1}{\rho})\cos\rho$	$(-\frac{945}{\rho^6} + \frac{420}{\rho^4} - \frac{15}{\rho^2})\cos\rho$ $+(-\frac{945}{\rho^5} + \frac{105}{\rho^3} - \frac{1}{\rho})\sin\rho$

The forms of the higher orders are usually given by the Rayleigh formulas

$$\begin{aligned} j_n(\rho) &= \rho^n\Big(-\frac{1}{\rho}\frac{d}{d\rho}\Big)^n\frac{\sin\rho}{\rho} \\ n_n(\rho) &= -\rho^n\Big(-\frac{1}{\rho}\frac{d}{d\rho}\Big)^n\frac{\cos\rho}{\rho}. \end{aligned} \tag{4-55}$$

The recursion relations for both functions are identical to each other and may be expressed in the following generic form:

$$(2n+1)f_n(\rho) = \rho\{f_{n+1}(\rho) + f_{n-1}(\rho)\}. \tag{4-56}$$

Here $f_n(\rho)$ can stand for either $j_n(\rho)$ or $n_n(\rho)$. The forms of $j_n(\rho)$ for $\ell \leq 5$ are listed in Table 4-3 as examples. Their variations as functions of ρ for $\ell \leq 3$ are given in Fig. 4-5 for spherical Bessel functions and in Fig. 4-6 for spherical Neumann functions.

Calculation of spherical Bessel and Neumann functions It is obvious from the explicit forms of spherical Bessel functions given in Table 4-3 that $j_n(\rho)$ for an arbitrary n consists of a linear combination of $\sin\rho$ and $\cos\rho$. To make use of Eq. (4-56) for generating $j_n(\rho)$, we shall start from the expression

$$j_n(\rho) = \sin\rho\sum_{k=0}^{n}\frac{a_{n,k}}{\rho^{k+1}} + \cos\rho\sum_{k=0}^{n}\frac{b_{n,k}}{\rho^{k+1}} \tag{4-57}$$

with $a_{n,k} = 0$ for $(n+k)$ an odd integer and $b_{n,k} = 0$ for $(n+k)$ an even integer.

Box 4-4 Program DM_SBES
Coefficients for spherical Bessel function

Initialization:
- (a) Set arrays for the coefficients $a_{n-1,k}$, $a_{n,k}$, and $a_{n+1,k}$ of $\sin\rho$ to zero.
- (b) Set arrays for the coefficients b_{n-1}, b_n, and b_{n+1} of $\cos\rho$ to zero.
- (c) Define the starting values using Eq. (4-59).
- (d) Set $n = 1$.

1. Input the order required.
2. Propagate to $(n+1)$ using Eq. (4-58):
   - (a) For $k = 0$, let $a_{n+1,0} = -a_{n-1,0}$ and $b_{n+1,0} = -b_{n-1,0}$.
   - (b) For $k > 0$, obtain $a_{n+1,k}$ from $a_{n,k}$ and $(a_{n-1,k}, b_{n+1,k})$ from $(b_{n,k}, b_{n-1,k})$.
3. If $(n+1)$ is less than the required rank,
   - (a) Store $a_{n,k}$ and $b_{n,k}$ into the locations of $a_{n-1,k}$ and $b_{n-1,k}$, respectively.
   - (b) Store $a_{n+1,k}$ and $b_{n+1,k}$ into the locations of $a_{n,k}$ and $b_{n,k}$, respectively.
   - (c) Increase the value of n by 1 and go back to step 2.
4. If $(n+1)$ equals the required rank, output the coefficients for $(n+1)$.

The recursion relation of Eq. (4-56) implies that the nonvanishing coefficients of Eq. (4-57) are related in the following way:

$$\begin{aligned} a_{n+1,k} &= (2n+1)a_{n,k-1} - a_{n-1,k} \\ b_{n+1,k} &= (2n+1)b_{n,k-1} - b_{n-1,k} \end{aligned} \tag{4-58}$$

with the understanding that

$$a_{n,k} = b_{n,k} = 0 \qquad \text{for} \qquad k < 0 \qquad \text{and} \qquad k > n.$$

These may be used as the recursion relations to generate $a_{n,\ell}$ and $b_{n,\ell}$ of arbitrary n and k. As starting values, we can use $j_0(\rho)$ and $j_1(\rho)$ given in Table 4-3 to obtain the nonvanishing coefficients of the two lowest-orders as

$$a_{0,0} = 1 \qquad a_{1,1} = 1 \qquad b_{1,0} = -1. \tag{4-59}$$

Since both $a_{n,k}$ and $b_{n,k}$ are integers, we can apply a straightforward propagation procedure, similar to the one used earlier for Hermite polynomials in §4-1. However, for technical interests, we shall take a slightly different approach here and keep only the coefficients $a_{n,k}$ and $b_{n,k}$ of $j_n(\rho)$ and $a_{n-1,k}$ and $b_{n-1,k}$ of $j_{n-1}(\rho)$ at the end of each step of the propagation. This is implemented in the algorithm given in Box 4-4. In principle, it is a more flexible method than the one used for the Hermite polynomials, where coefficients of all orders are kept in a two-dimensional array. However, for the low ranks of interest to us in most practical applications, there are very little differences in the choice between the two methods. If the interest is in the

Box 4-5 Program DM_BESP
Propagation of the values of spherical Bessel functions

1. Input the rank required.
2. Input the value of ρ.
3. Calculate the starting values:
   (a) Evaluate $j_0(\rho)$ and $j_1(\rho)$ using Eqs. (4-52) and (4-54).
   (b) Assign $n = 1$.
4. Calculate the value for order $(n+1)$ using Eq. (4-56).
5. Increment the value of n by 1.
6. Go back to step 4 until n reaches the order required.
7. Output the calculated value.

numerical values, we can start from the algebraic form of Eq. (4-57) and make use of the coefficients obtained using Box 4-4.

For the spherical Neumann functions $n_n(\rho)$, an identical procedure may be used to calculate the coefficients. The only difference is in the starting values. Instead of those given in Eq. (4-59), the nonvanishing coefficients are

$$a_{0,0} = -1 \qquad\qquad a_{1,1} = b_{1,0} = -1$$

as can be seen from Table 4-3.

We can also make use of Eq. (4-56) to propagate the numerical value of $j_n(\rho)$. The starting values are those for $j_0(\rho)$ and $j_1(\rho)$ which may be calculated using Eqs. (4-52) and (4-54). The steps are given in Box 4-5. The same method can also be used to find the value of $n_\ell(\rho)$ by starting from those of $n_0(\rho)$ using Eq. (4-53) and $n_1(\rho)$ using Eq. (4-54). As usual, it is also possible to use the power series forms of $j_n(\rho)$ and $n_n(\rho)$. However, because of the sign, contributions from successive terms have a tendency to cancel each other. This can be seen by examining the examples given in Table 4-3. For this reason, the numerical accuracies are, in general, only comparable with those obtained by propagation and may not be adequate for many purposes. It is also possible to develop special methods for a particular rank of spherical Bessel or Neumann function that are better in accuracy, especially for small values of ρ. These methods are available in references that specialize on the subject.

Bessel functions of integer ranks For completeness, we shall end this section with a brief discussion of Bessel functions of integer ranks, the solution to Eq. (4-50). The regular solution, analogous to $j_n(\rho)$, is

$$J_\lambda(\rho) = \sum_{k=0}^{\infty} \frac{(-1)^k}{k!(\lambda+k)!}\left(\frac{\rho}{2}\right)^{\lambda+2k}. \tag{4-60}$$

Alternatively, we can also use an integral form,

$$J_\lambda(\rho) = \frac{1}{\pi}\int_0^\pi \cos(\lambda\theta - \rho\sin\theta)\,d\theta$$

to represent the function. The irregular solution is the Neumann function $N_\lambda(\rho)$ and it may be defined as a sum of Bessel functions of positive and negative orders

$$N_\lambda(\rho) = \frac{J_\lambda(\rho)\cos(\lambda\pi) - J_{-\lambda}(\rho)}{\sin(\lambda\pi)}. \tag{4-61}$$

They are related to their half-integer rank counterparts in the following way:

$$j_n(\rho) = \sqrt{\frac{\pi}{2\rho}} J_{n+1/2}(\rho) \qquad\qquad n_n(\rho) = \sqrt{\frac{\pi}{2\rho}} N_{n+1/2}(\rho). \tag{4-62}$$

Several polynomial approximations to $J_0(\rho)$, $J_1(\rho)$, $N_0(\rho)$, and $N_1(\rho)$ are given on pages 369-70 of Abramowitz and Stegun,[1] and may be used to obtain the numerical values of these functions. From the values of $\lambda = 0$ and 1, it is also possible to use the recursion relation

$$2\lambda J_\lambda(\rho) = \rho\Big\{J_{\lambda+1}(\rho) + J_{\lambda-1}(\rho)\Big\} \tag{4-63}$$

to produce those for higher ranks. For fast and accurate calculations, one should make use of routines available in standard subroutine libraries, such as NAG.[41]

4-4 Laguerre polynomials

We return now to the more general case for the radial part of Eq. (4-20). Unlike the previous section, we shall be concerned with the situation in which the factor k^2 is a function of the radial coordinate instead of a constant. This occurs, for example, in quantum mechanical problems where the potential of the system is a central one. The two most common examples in this category are the isotropic three-dimensional harmonic oscillator potential and the Coulomb potential.

We have already seen the one-dimensional case of a harmonic oscillator potential in §4-1. Here, we have a more general form,

$$V(r) = \frac{1}{2}\mu\omega^2 r^2 \tag{4-64}$$

where, as we saw earlier, μ is the mass and ω is the angular frequency of the oscillator. It is called an *isotropic* harmonic oscillator potential, as there is no angular dependence. Physically, the force increases in strength linearly with the distance r from the center of the well. In contrast, the strength for a Coulomb potential, similar to a gravity, is inversely proportional to the square of the distance. For example, if an electron is outside a nucleus with Z units of positive charges, the potential in cgs units has the form

$$V(r) = -\frac{Ze^2}{r} \tag{4-65}$$

where e is the charge of a proton (equal in magnitude but opposite in sign as the charge on an electron) and the negative sign indicates it is an attractive potential. If we wish to use SI units instead, we shall regard Z as the number of protons divided by $4\pi\epsilon_0$. (Here ϵ_0 is the permittivity of free space.)

In both cases, the radial part of the Helmholtz equation of Eq. (4-20) reduces to a Laguerre differential equation

$$t\frac{d^2v}{dt^2} + (\alpha + 1 - t)\frac{dv}{dt} + kv = 0. \tag{4-66}$$

Before attempting to solve the equation, we shall see how this reduction can be carried out for the two potentials of interest to us here.

Isotropic harmonic oscillator potential If we generalize the Schrödinger equation for the harmonic oscillator given in Eq. (4-4) to three spatial dimensions, we obtain the following second-order differential equation:

$$-\frac{\hbar^2}{2\mu}\nabla^2\psi(\boldsymbol{r}) + \left(\frac{1}{2}\mu\omega^2 r^2 - E\right)\psi(\boldsymbol{r}) = 0.$$

Since the potential has only radial dependence, the angular part of the wave function is given by spherical harmonics, as we saw earlier in Eq. (4-43). For this reason, we can write the wave function $\psi(\boldsymbol{r})$ as a product of $R_\ell(r)$, the radial wave function for angular momentum ℓ, and spherical harmonics $Y_{\ell,m}(\theta, \phi)$:

$$\psi_{\ell m}(\boldsymbol{r}) = R_\ell(r)Y_{\ell,m}(\theta, \phi).$$

The equation for the radial dependence is given by

$$\frac{1}{r^2}\frac{d}{dr}\left(r^2\frac{dR_\ell}{dr}\right) + \left\{\frac{2\mu}{\hbar^2}\left(E - \frac{1}{2}\mu\omega^2 r^2\right) - \frac{\ell(\ell+1)}{r^2}\right\}R_\ell(r) = 0. \tag{4-67}$$

We shall rewrite this equation in the form of a Laguerre differential equation given by Eq. (4-66).

Similar to the one-dimensional case, it is convenient to express both the energy E and the radial coordinate r in terms of dimensionless quantities. For this purpose, we shall define λ and ρ, in the same way as we did in Eqs. (4-6) and (4-7), by the following relations:

$$E = \lambda\left(\frac{1}{2}\hbar\omega\right) \qquad\qquad r = \rho\sqrt{\frac{\hbar}{\mu\omega}}. \tag{4-68}$$

The factor $\sqrt{\hbar/(\mu\omega)}$ has the dimension of length and is often referred to as the harmonic oscillator length parameter. In terms of λ and ρ, the radial equation Eq. (4-67) becomes

$$\frac{1}{\rho^2}\frac{d}{d\rho}\left(\rho^2\frac{dR_\ell}{d\rho}\right) + \left\{\lambda - \rho^2 - \frac{\ell(\ell+1)}{\rho^2}\right\}R_\ell(\rho) = 0$$

or

$$\frac{d^2R_\ell}{d\rho^2} + \frac{2}{\rho}\frac{dR_\ell}{d\rho} + \left\{\lambda - \rho^2 - \frac{\ell(\ell+1)}{\rho^2}\right\}R_\ell(\rho) = 0. \tag{4-69}$$

We shall be using these two equivalent forms in our discussions, depending on which is more convenient for the occasion.

We can transform Eq. (4-69) into a Laguerre differential equation by letting

$$R_\ell(\rho) = \rho^\ell e^{-\rho^2/2}\phi(\rho). \tag{4-70}$$

On substituting this form into Eq. (4-69), we obtain the equation that must be satisfied by $\phi(\rho)$:

$$t\frac{d^2\phi}{dt^2} + \left(\ell + \frac{3}{2} - t\right)\frac{d\phi}{dt} + k\phi(t) = 0 \tag{4-71}$$

where

$$t = \rho^2 \qquad\qquad k = \frac{1}{4}\left(\lambda - 2\ell - 3\right). \tag{4-72}$$

We see that Eq. (4-71) is the same as (4-66) if we let $\alpha = (\ell + \frac{1}{2})$.

Coulomb potential and the hydrogenlike atom For an atomic electron, the electrostatic attraction provided by the nucleus is given by the Coulomb potential of Eq. (4-65). The simplest case is that of a hydrogen atom, as the nucleus has only a single unit of positive charge ($Z = 1$). For other atoms, $Z > 1$, and the neutral atom has more than one electron outside the nucleus. The net effect of the other electrons is to screen the force acting on an individual electron from the nucleus and this complicates the situation. For this reason, the potential given in Eq. (4-65) applies only to the idealized situation of a hydrogenlike atom in which all the atomic electrons except one are stripped away. The result is similar to a hydrogen atom except that the number of positive charges in the nucleus is Z instead of one. We shall use this simplified situation as an example for Laguerre polynomials. Analogous to the case of a particle in an isotropic harmonic oscillator potential in Eq. (4-67), the radial equation in this case takes on the form

$$\frac{1}{r^2}\frac{d}{dr}\left(r^2\frac{dR_\ell}{dr}\right) + \left\{\frac{2\mu}{\hbar^2}\left(E + \frac{Ze^2}{r}\right) - \frac{\ell(\ell+1)}{r^2}\right\}R_\ell(r) = 0. \tag{4-73}$$

We shall be interested only in the bound states for the electron. For such cases, the energy E is a negative quantity.

It is convenient here to define a dimensionless quantity

$$\rho = \kappa r. \tag{4-74}$$

For $E < 0$, the quantity κ is a real, positive number:

$$\kappa = 2\sqrt{-\frac{2\mu E}{\hbar^2}}.$$

In terms of ρ, the differential equation has the form

$$\frac{1}{\rho^2}\frac{d}{d\rho}\left(\rho^2\frac{dR_\ell}{d\rho}\right) + \left\{\frac{\eta}{\rho} - \frac{1}{4} - \frac{\ell(\ell+1)}{\rho^2}\right\}R(\rho) = 0 \tag{4-75}$$

Table 4-4: Laguerre polynomials $L_k(t)$ up to order $k = 6$.

$$L_0(t) = 1$$
$$L_1(t) = 1 - t$$
$$L_2(t) = \tfrac{1}{2!}(2 - 4t + t^2)$$
$$L_3(t) = \tfrac{1}{3!}(6 - 18t + 9t^2 - t^3)$$
$$L_4(t) = \tfrac{1}{4!}(24 - 96t + 72t^2 - 16t^3 + t^4)$$
$$L_5(t) = \tfrac{1}{5!}(120 - 600t + 600t^2 - 200t^3 + 25t^4 - t^5)$$
$$L_6(t) = \tfrac{1}{6!}(720 - 4320t + 5400t^2 - 2400t^3 + 450t^4 - 36t^5 + t^6)$$

where

$$\eta = \frac{2\mu Z e^2}{\kappa \hbar^2} = \sqrt{-\frac{\mu Z^2 e^4}{2E\hbar^2}} \tag{4-76}$$

is a dimensionless quantity related to the square root of the energy of the system. By substituting

$$R_\ell(\rho) = \rho^\ell e^{-\rho/2}\phi(\rho) \tag{4-77}$$

into Eq. (4-75), we obtain the equation that must be satisfied by $\phi(\rho)$,

$$\rho\frac{d^2\phi}{d\rho^2} + (2\ell + 2 - \rho)\frac{d\phi}{d\rho} + (\eta - \ell - 1)\phi(\rho) = 0. \tag{4-78}$$

This is the same equation as Eq. (4-66) if we let $\alpha = (2\ell + 1)$ and $k = (\eta - \ell - 1)$.

Solution of the Laguerre equation From Eqs. (4-71) and (4-78), we see that Laguerre equation is important in the radial solution of central-potential problems represented by the isotropic harmonic oscillator and the hydrogenlike atom. As a start, we shall consider first the simpler case with $\alpha = 0$ in Eq. (4-66). The differential equation reduces to the form

$$t\frac{d^2\phi}{dt^2} + (1 - t)\frac{d\phi}{dt} + k\phi(t) = 0. \tag{4-79}$$

For integer values of k, the solution is a polynomial in t

$$L_k(t) = \sum_{j=0}^{k}(-1)^j \frac{k!}{(k-j)!(j!)^2} t^j. \tag{4-80}$$

This is known as a Laguerre polynomial of degree t. The explicit forms for the lowest few orders are given in Table 4-4 as illustration.

The particular form shown in Eq. (4-80) has the orthogonality relation

$$\int_0^\infty e^{-t} L_m(t) L_k(t)\, dt = \delta_{m,k}. \tag{4-81}$$

Different normalizations from that given in Eq. (4-80) are sometimes used. For example, many authors prefer an additional factor of $k!$, as this leads to the result that all the coefficients are integers. Such a choice changes the appearance of, for example, the orthogonality condition and the recursion relations but does not make any difference substantial to our discussions.

For the form adopted in Eq. (4-80), the recursion relations for Laguerre polynomials are

$$(k+1)L_{k+1}(t) = (2k+1-t)L_k(t) - kL_{k-1}(t) \tag{4-82}$$

$$t\frac{dL_k}{dt} = kL_k(t) - kL_{k-1}(t). \tag{4-83}$$

It is also possible to express Laguerre polynomials in a differential form

$$L_k(t) = \frac{1}{k!}e^t\frac{d^k}{dt^k}\{t^k e^{-t}\} \tag{4-84}$$

that are useful for some our later discussions.

Associated Laguerre polynomials If we differentiate each term in the Laguerre equation Eq. (4-79) once with respect to t, we obtain

$$\frac{d^2L_k}{dt^2} + t\frac{d^3L_k}{dt^3} - \frac{dL_k}{dt} + (1-t)\frac{d^2L_k}{dt^2} + k\frac{dL_k}{dt} = 0.$$

On rearranging terms, this equation may be written as

$$t\frac{d^2}{dt^2}\left(\frac{dL_k}{dt}\right) + \{1+(1-t)\}\frac{d}{dt}\left(\frac{dL_k}{dt}\right) + (k-1)\frac{dL_k}{dt} = 0.$$

Comparing with Eq. (4-66), we find that dL_k/dt is the solution for $\alpha = 1$ and k replaced by $(k-1)$. If we carry out the same type of differentiation p times, we obtain the solution of Laguerre equation for degree k and $\alpha = p$. In other words,

$$L_k^p(t) = (-1)^p\frac{d^p}{dt^p}L_{k+p}(t). \tag{4-85}$$

The function $L_k^p(t)$ is known as the associated Laguerre polynomial and the forms of the lowest few orders are given in Table 4-5 as an illustration.

Using the series form of $L_k(t)$ given by Eq. (4-80) and the definition of $L_k^p(t)$ in Eq. (4-85), we obtain a power series expansion for the associated Laguerre polynomials:

$$\begin{aligned} L_k^p(t) &= \sum_{j=0}^{k}(-1)^j\frac{(k+p)!}{(k-j)!(p+j)!j!}t^j \\ &= \sum_{j=0}^{k}\binom{k+p}{k-j}\frac{(-t)^j}{j!} \qquad \text{for} \qquad p > -1. \end{aligned} \tag{4-86}$$

Table 4-5: Associated Laguerre polynomials $L^{\ell+1/2}_{(n-\ell)/2}(t)$.

n	ℓ	$L^{\ell+1/2}_{(n-\ell)/2}(t)$	ℓ	$L^{\ell+1/2}_{(n-\ell)/2}(t)$	ℓ	$L^{\ell+1/2}_{(n-\ell)/2}(t)$
0	0	$L_0^{1/2}(t) = 1$				
1	0	$L_0^{3/2}(t) = 1$				
2	0	$L_1^{1/2}(t) = \frac{3}{2} - t$	2	$L_0^{3/2} = 1$		
3	1	$L_1^{3/2}(t) = \frac{5}{2} - t$	3	$L_0^{5/2} = 1$		
4	0	$L_2^{1/2}(t) = \frac{15}{8} - \frac{5}{2}t + \frac{1}{2}t^2$	2	$L_1^{5/2} = \frac{7}{2} - t$	4	$L_0^{9/2}(t) = 1$
5	1	$L_2^{3/2}(t) = \frac{35}{8} - \frac{7}{2}t + \frac{1}{2}t^2$	3	$L_1^{7/2} = \frac{9}{2} - t$	5	$L_0^{11/2}(t) = 1$

Similarly, by starting from Eq. (4-84), we obtain a differential form for the same function:

$$L_k^p(t) = \frac{1}{k!}t^{-p}e^t\frac{d^k}{dt^k}\left\{t^{k+p}e^{-t}\right\}.$$

The orthogonality relation

$$\int_0^\infty t^p e^{-t} L_m^p(t) L_k^p(t)\, dt = \frac{(k+p)!}{k!}\delta_{m,k} \tag{4-87}$$

differs slightly from that for Laguerre polynomials in Eq. (4-81) and requires $t^p e^{-t}$ as the weighting factor.

The recursion relations for $L_k^p(t)$ are similar in form to those for Laguerre polynomials given by Eqs. (4-82) and (4-83)

$$(k+1)L_{k+1}^p(t) = (2k+p+1-t)L_k^p(t) - (k+p)L_{k-1}^p(t) \tag{4-88}$$

$$t\frac{dL_k^p}{dt} = kL_k^p(t) - (k+p)L_{k-1}^p(t). \tag{4-89}$$

By subtracting Eq. (4-89) from (4-88), we obtain one further recursion relation

$$(k+1)L_{k+1}^p(t) = t\frac{dL_k^p}{dt} + (k+p+1-t)L_k^p(t). \tag{4-90}$$

This is useful in propagating associated Laguerre polynomials.

Eigenvalues and radial wave functions of hydrogenlike atoms We are now in a position to make use of associated Laguerre polynomials as a part of the solution for the radial wave function of hydrogenlike atoms. Let us start from Eq. (4-78) for $\phi(\rho)$:

$$\rho\frac{d^2\phi}{d\rho^2} + (2\ell + 2 - \rho)\frac{d\phi}{d\rho} + (\eta - \ell - 1)\phi = 0.$$

This may be compared with the standard form of an associated Laguerre differential equation given by Eq. (4-66) by rewriting it as

$$t\frac{d^2v}{dt^2} + (p+1-t)\frac{dv}{dt} + kv = 0.$$

The solution to this differential equation is $L_k^p(t)$, as we saw in Eq. (4-85). By making the identifications that $\rho = t$, $k = (\eta - \ell - 1)$, and $p = (2\ell + 1)$, we obtain

$$\phi(\rho) = \frac{1}{\mathcal{N}} L_{\eta-\ell-1}^{2\ell+1}(\rho). \tag{4-91}$$

We shall return later to the factor $\mathcal{N}$ by treating it as part of the normalization for the radial wave function we are interested in here.

Since $k = (\eta - \ell - 1)$ and must be an integer, only certain values of η are allowed. This, in turn, means that the energies of a hydrogenlike atom are restricted to a discrete set of values, the well-known success of quantum mechanics in the early days. From the definition of η given in Eq. (4-76), we obtain

$$E_n = -\frac{\mu Z^2 e^4}{2n^2\hbar^2} = -\frac{Z^2 e^2}{2a_0 n^2}$$

where the principal quantum number

$$n = \eta = k + \ell + 1$$

is an integer greater than or equal to 1, and $a_0 = \hbar^2/\mu e^2$ is the Bohr radius for the hydrogenlike atom with μ as the reduced mass of an electron. In terms of n and a_0, the relation between r and ρ given in Eq. (4-74) simplifies to

$$\rho = \frac{2Z}{na_0} r.$$

This is the form of ρ commonly found in standard textbooks.

Using Eq. (4-91), the normalized radial wave function of Eq. (4-77) may now be written in the form

$$R_{n\ell}(r) = \left(\frac{2Z}{na_0}\right)^{3/2} \sqrt{\frac{(n-\ell-1)!}{2n(n+\ell)!}}\,(\kappa r)^\ell e^{-\kappa r/2} L_{n-\ell-1}^{2\ell+1}(\kappa r). \tag{4-92}$$

Since $\kappa = 2Z/na_0$, the values of $\rho = \kappa r$ for the same r are different for states with different principal quantum number n. It is therefore more convenient for many purposes to use a different dimensionless quantity,

$$t = \frac{Z}{a_0} r$$

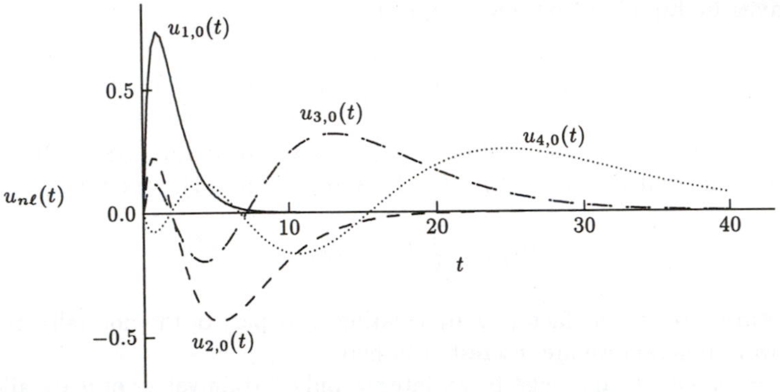

Figure 4-7: Examples of the modified radial wave functions of hydrogenlike atoms, $u_{n\ell}(t) \equiv tR_{n\ell}(t)$, for $\ell = 0$ and $n = 1$ to 4.

instead of ρ. Physically, we may interpret t as the length measured in units of the Bohr radius for a hydrogenlike atom consisting of Z units of charge in the nucleus. In terms of t, the radial wave function takes on the form

$$R_{n\ell}(t) = \left(\frac{2}{n}\right)^{3/2} \sqrt{\frac{(n-\ell-1)!}{2n(n+\ell)!}} (2t/n)^{\ell} e^{-t/n} L_{n-\ell-1}^{2\ell+1}(2t/n). \tag{4-93}$$

Since both n and ℓ are integers, $L_{n-\ell-1}^{2\ell+1}(2t/n)$ are given by Eq. (4-86). Using these, the radial wave functions $R_{n\ell}(t)$ for low values of n and ℓ may be written down explicitly. The forms of those with $\ell = 0$ and $n \leq 3$ are given here as examples:

$$\begin{aligned} R_{10}(t) &= 2e^{-t} \\ R_{20}(t) &= \frac{1}{2\sqrt{2}}(2-t)e^{-t/2} \\ R_{30}(t) &= \frac{2}{81\sqrt{3}}(27 - 18t + 2t^2)e^{-t/3}. \end{aligned}$$

Their variations as functions of ρ are displayed in Fig. 4-7 in terms of the modified radial wave function $u_{n\ell}(t) = tR_{n\ell}(t)$.

Harmonic oscillator radial wave functions Associated Laguerre polynomials appear also as a part of the eigenfunctions for a particle in an isotropic harmonic oscillator potential well. This may be carried out in a way similar to what we have done above for the hydrogenlike atom. From the fact that k of Eq. (4-72) has to be an integer, we obtain the condition that must be satisfied by the parameter λ of Eq. (4-68)

$$\lambda = 4m + 2\ell + 3$$

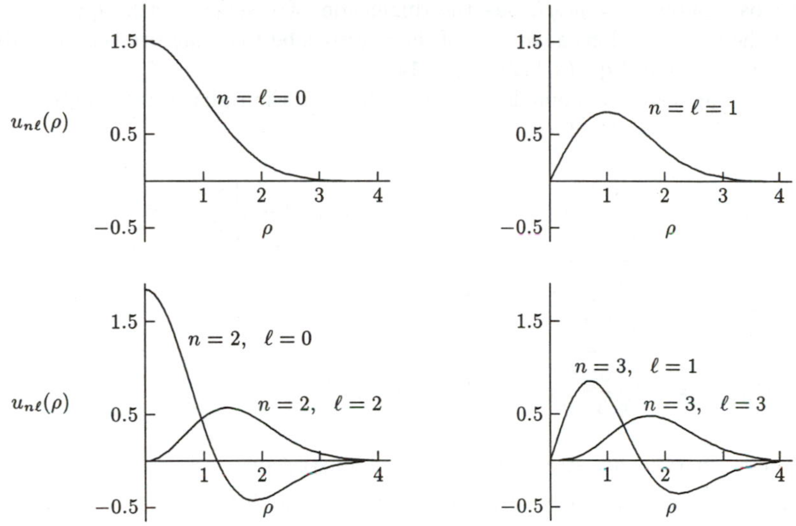

Figure 4-8: Modified radial wave functions $u_{n\ell}(\rho) \equiv \rho R_{n\ell}(\rho)$ for a particle in an isotropic three-dimensional harmonic potential well.

where $m = 0, 1, 2, \ldots$. Since λ is the energy of the particle in units of $\frac{1}{2}\hbar\omega$, we obtain the familiar result

$$E_n = \left(n + \frac{3}{2}\right)\hbar\omega.$$

Here $n = (2m + \ell)$ is the principal quantum number, measuring the number of harmonic oscillator quanta above the zero-point energy of $\frac{3}{2}\hbar\omega$.

Except a normalization factor, the solution to the differential equation for $\phi(t)$ of Eq. (4-71) may be obtained by comparing it with the standard form of a Laguerre equation given by Eq. (4-66). This gives us the result $\alpha = (\ell + 1/2)$ and $k = (\lambda - 2\ell - 3)/4 = (n - \ell)/2$, or

$$\phi(t) = L_k^{\alpha}(t) = L_{(\lambda-2\ell-3)/4}^{\ell+1/2}(t) = L_{(n-\ell)/2}^{\ell+1/2}(\rho^2).$$

Since k is an integer, we find the condition that both n and ℓ must be either even or odd at the same time. For odd ℓ values, α is a half-integer and we shall see later how to obtain associated Laguerre polynomials for noninteger p values.

The complete radial wave function for a given n and ℓ is then

$$R_{n\ell}(r) = \sqrt{\frac{2^{\ell+2}\nu^{\ell+3/2}(n-\ell)!!}{(n+\ell+1)!!\sqrt{\pi}}}\, r^{\ell} e^{-\frac{\nu r^2}{2}} L_{(n-\ell)/2}^{\ell+1/2}(\nu r^2) \tag{4-94}$$

where the parameter $\nu = \mu\omega/\hbar$ has the dimension of inverse length squared. In arriving at the result, we have made use of the relations between factorials and double factorials given later in Eqs. (4-112) to (4-114).

Using the Laguerre polynomials given in Table 4-5, the radial wave functions for $n \leq 3$ and $\ell \leq 1$ take on the forms

$$\begin{aligned} R_{00}(r) &= \sqrt{\frac{4\nu^{3/2}}{\sqrt{\pi}}}e^{-\nu r^2/2} \qquad\qquad R_{11}(r) = \sqrt{\frac{8\nu^{5/2}}{3\sqrt{\pi}}}re^{-\nu r^2/2} \\ R_{20}(r) &= \sqrt{\frac{6\nu^{3/2}}{\sqrt{\pi}}}\Big(1-\frac{2}{3}\nu r^2\Big)e^{-\nu r^2/2} \\ R_{31}(r) &= 2\sqrt{\frac{5\nu^{5/2}}{3\sqrt{\pi}}}re^{-\nu r^2/2}\Big(1-\frac{2}{5}\nu r^2\Big). \end{aligned}$$

Plots of these functions are shown in Fig. 4-8 as illustrations.

Evaluation of Laguerre polynomials It is fairly straightforward to evaluate Laguerre polynomials to any order. The general form may be written as a power series in the following way:

$$L_k(t) = \sum_{j=0}^{k} a_{k,j}t^j. \tag{4-95}$$

Using Eq. (4-80), we find that the coefficient for t^j is

$$a_{k,j} = (-1)^j \frac{k!}{(k-j)!(j!)^2}.$$

Similarly, for the associated Laguerre polynomial of integer order (k,p), we have

$$L_k^p(t) = \sum_{j=0}^{k} b_{k,p;j}t^j \tag{4-96}$$

with coefficients

$$b_{k,p;j} = (-1)^j \frac{(k+p)!}{(k-j)!(p+j)!j!}$$

as can be see from Eq. (4-86).

The relation given by Eq. (4-85) and, consequently, Eq. (4-86) applies only to integer values of p. As a result, the form for $L_k^p(t)$ in Eq. (4-96) cannot be used for p that are half-integers. In its place, we shall make use of the recursion relations of Eqs. (4-88) and (4-89). Again we shall write the polynomials as a power series:

$$L_k^p(t) = \sum_{j=0}^{k} c_{k,p;j}t^j. \tag{4-97}$$

Box 4-6 Program DM_ALAGR

Coefficients of associated Laguerre polynomials $L_n^p(t)$ for integer n

Initialization:

(a) Zero the array for storing the coefficients.

(b) Set the starting value according to Eq. (4-99).

1. Input n and p.
2. Propagate the coefficients of L_k^p for k from 0 to n:
   For each $L_k^p(t)$, calculate all the coefficient for $j = 0$ to k using Eq. (4-98).
3. Output the coefficients $c_{n,p,j}$ for $j = 0$ to n.

To distinguish from Eq. (4-96), the coefficients are represented as $c_{n,p;j}$.

We can use Eq. (4-90) to derive the following recursion relation for $c_{n,p;j}$:

$$(k+1)\sum_{j=0}^{k+1} c_{k+1,p;j}\, t^j = t\sum_{j=0}^{k} c_{k,p;j}\, jt^{j-1} + (k+p+1-t)\sum_{j=0}^{k} c_{k,p;j}\, t^j .$$

On rearranging terms, the same equation may be put into the form

$$\sum_{j=0}^{k+1}\Big\{(k+1)c_{k+1,p;j} - jc_{k,p;j} - (k+p+1)c_{k,p;j} + c_{k,p,j-1}\Big\}t^j = 0$$

where it is understood that $c_{k,p,j} = 0$ for $j > k$. Since the equality holds for arbitrary values of t, the coefficient of t^j must vanish separately for different j. From this we obtain a relation among $c_{k,p,j}$ for orders $(k+1)$ and k:

$$c_{k+1,p,j} = \frac{1}{k+1}\Big\{(k+p+j+1)c_{k,p;j} - c_{k,p;j-1}\Big\}. \tag{4-98}$$

This allows us to calculate, for any given value of p, all the coefficients of order $(k+1)$ from those of order k.

The starting point of the recursion relations is the values of the coefficient for $k = 0$. Using the fact that

$$L_0^p(t) = 1$$

for arbitrary values of $p \geq 0$, we have

$$c_{0,p;j} = \begin{cases} 1 & \text{for } j = 0 \\ 0 & \text{otherwise.} \end{cases} \tag{4-99}$$

The algorithm outlined in Box 4-6 makes use of this to obtain the values of $c_{k,p;j}$ with Eq. (4-98).

As a bonus, we find that Eq. (4-98) applies to integer values of p as well. Because of this, we can use the same procedure to generate the coefficient $b_{k,p;j}$. Furthermore, since

$$L_k(t) = L_k^0(t)$$

the same method may also be used to produce the coefficients $a_{k,j}$ of Eq. (4-97) by letting $p = 0$. In this way, a single computer program, based on the algorithm of Box 4-6, may be used to produce all the Laguerre polynomials of interest.

For the numerical values of $L_k^p(t)$, we can anticipate that some caution is required in order to obtain good accuracies at high orders. The source of the trouble lies in the fact that the coefficients alternate in sign between successive terms. As a result, cancelations take place between neighboring terms and the numerical accuracy is worse than what we can otherwise achieve. The same problem occurs also if we apply the recursion relations directly on the numerical values, as we can see from the forms of Eqs. (4-82) and (4-88).

4-5 Error integrals and gamma functions

Integrals of exponential function appear in a wide variety of problems. In this section, we shall give a short account of two of them, the error function $\text{erf}(x)$ and the gamma function $\Gamma(z)$. A number of other functions are related to these integrals and we shall examine some of them as well.

Error integrals The error function, or error integral, is defined by the relation

$$\text{erf}(x) = \frac{2}{\sqrt{\pi}} \int_0^x e^{-t^2} dt. \tag{4-100}$$

The normalization constant is taken in such a way that

$$\text{erf}(x = \infty) = 1.$$

From this condition we obtain a relation with the complementary error function

$$\text{erfc}(x) = \frac{2}{\sqrt{\pi}} \int_x^\infty e^{-t^2} dt = 1 - \text{erf}(x) \tag{4-101}$$

and the normal probability function

$$P(x) = \frac{1}{\sqrt{2\pi}} \int_{-\infty}^x e^{-t^2/2} dt = \frac{1}{2}\Big\{1 + \text{erf}\Big(\frac{x}{\sqrt{2}}\Big)\Big\}. \tag{4-102}$$

A function related to $P(x)$ was used earlier in §2-5 as an example for numerical integration. As we shall see in Chapter 6, error integrals and the related probability integrals are needed in many applications involving probability and statistics.

Rational polynomial approximations In general, it is not possible to use analytical methods to evaluate error function and functions directly related to it. In Chapter

3, we saw that numerical integration is one possible way to carry out the calculation. An alternative is to use polynomial approximations.

In terms of the inverse of x defined by

$$t = \frac{1}{1+px} \qquad \text{where} \qquad p = 0.47047$$

the error function may be approximated with an uncertainty of $|\epsilon| \leq 2.5 \times 10^{-5}$ by the expression

$$\text{erf}(x) = 1 - (a_1 t + a_2 t^2 + a^3 t^3)e^{-x^2} \tag{4-103}$$

where the three coefficients have the values

$$a_1 = 0.3480242 \qquad a_2 = -0.0958798 \qquad a_3 = 0.7478556.$$

A more accurate form, with uncertainty $|\epsilon| \leq 1.5 \times 10^{-7}$ is given by the expression

$$\text{erf}(x) = 1 - (b_1 t + b_2 t^2 + b_3 t^3 + b_4 t^4 + b_5 t^5)e^{-x^2}$$

with

$$t = \frac{1}{1 + 0.3275911x} \qquad b_1 = 0.254829592 \qquad b_2 = -0.284496736$$

$$b_3 = 1.421413741 \qquad b_4 = -1.453152027 \qquad b_5 = 1.061405429.$$

Two other methods of approximations are also given on page 299 of Abramowitz and Stegun.[1] More elaborate approximations, with even better accuracies, are found in standard subroutine libraries.

Gamma function There are several ways to define the gamma function. In terms of an integral, it may be written as

$$\Gamma(z) = \int_0^\infty e^{-t} t^{z-1} dt. \tag{4-104}$$

Integrating by parts,

$$\Gamma(z+1) = \int_0^\infty e^{-t} t^z dt = \left[-e^{-t} t^z\right]_0^\infty + z \int_0^\infty e^{-t} t^{z-1} dt$$

we see that it has the recursion relation

$$\Gamma(z+1) = z\Gamma(z). \tag{4-105}$$

The variable z is a complex number in general, but we shall be mainly concerned with cases in which it is a purely real number. In particular, when z is a positive integer n, the function is related to the factorial of the argument

$$\Gamma(n) = (n-1)(n-2)\cdots 1 = (n-1)! \tag{4-106}$$

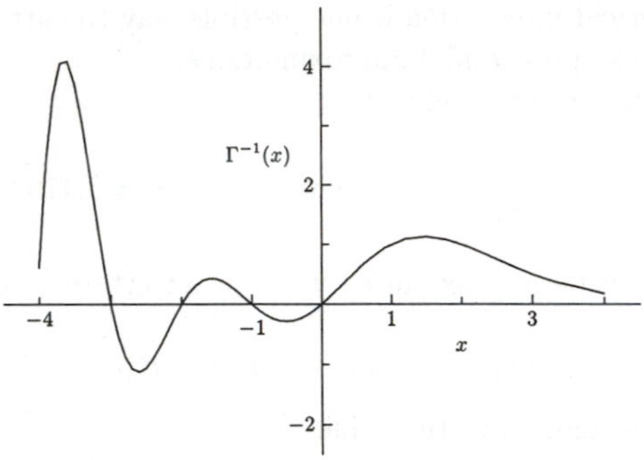

Figure 4-9: Inverse gamma function, $1/\Gamma(x)$, for real arguments $x = [-4, 4]$.

Because of this property, $\Gamma(z)$ is also known as the factorial function. The behavior of the gamma function is shown in terms of its inverse in Fig. 4-9.

Polynomial approximations Several polynomial approximations of $\Gamma(z)$ are available for z not necessarily an integer. Since gamma functions for different arguments are connected by the recursion relation Eq. (4-105), all we need is an expression for $0 < z < 1$. A five-term form for $z = x$, where x is a real number, is given by

$$\Gamma(1+x) = 1 + \sum_{k=1}^{5} a_k x^k.$$

The values of the coefficients are

$$a_1 = -0.5748646 \qquad a_2 = 0.9512363 \qquad a_3 = -0.6998588$$
$$a_4 = 0.4245549 \qquad a_5 = -0.1010678.$$

The results calculated with this approximate expression are accurate to $|\epsilon| \leq 5 \times 10^{-5}$. An eight-term form,

$$\Gamma(1+x) = 1 + \sum_{k=1}^{8} b_k x^k \tag{4-107}$$

with

$$b_1 = -0.577191652 \qquad b_2 = 0.988205891 \qquad b_3 = -0.897056937$$
$$b_4 = 0.918206857 \qquad b_5 = -0.756704078 \qquad b_6 = 0.482199394$$
$$b_7 = -0.193527818 \qquad b_8 = 0.035868343$$

has an accuracy of $|\epsilon| \leq 3 \times 10^{-7}$. Both formulas work also in the limit of $x = 0$ and 1, as can be seen by inspection. A series expansion of $1/\Gamma(z)$ for complex argument, consisting of 26 terms, is also given on page 256 of Abramowitz and Stegun.[1] Using

Box 4-7 Function GMMA8(x)
Gamma function for real arguments
Using an eight-term approximation of $\Gamma(1+x)$ for $0 \leq x < 1$

Argument list:
  x: A real, nonnegative number.
Initialization:
  Store the values of coefficients $b_1, b_2, \ldots, b_8$ of Eq. (4-107).

1. Reject $x \leq 0$.
2. Reduce the argument to the range $[0, 1]$:
   (a) Let $z = (x - 1)$.
   (b) If $0 \leq z < 1$, let $f = 1$ and go to step 3.
   (c) If $z < 0$, use $\Gamma(x) = x^{-1}\Gamma(1+x)$:
      (i) Let $f = x^{-1}$ and $z = (1 + x)$.
      (ii) Go to step 3 to calculate the value of $\Gamma(z)$.
   (d) If $z > 1$, use $\Gamma(1+x) = x\Gamma(x)$ to reduce the argument to ≤ 1:
      (i) Start with $f = 1$.
      (ii) Let $f = fz$ and $z = (z - 1)$.
      (iii) Repeat step (ii) until $z < 1$.
3. Calculate the value of $\Gamma(1+z)$ for $0 \leq z < 1$ using Eq. (4-107).
4. Return $\Gamma(x) = f\Gamma(1+z)$.

a combination of the recursion relation $\Gamma(1+z) = z\Gamma(z)$ given in Eq. (4-105) and the polynomial approximation of Eq. (4-107), a method to calculate $\Gamma(x)$ for $x > 0$ is implemented in the algorithm outlined in Box 4-7.

An efficient way to evaluate gamma functions for large values of the argument is to use Stirling's formula,

$$\Gamma(z) \approx e^{-z} z^{z-\frac{1}{2}} \sqrt{2\pi}\Big\{1 + \frac{1}{12z} + \frac{1}{288z^2} - \frac{139}{51,840z^3} - \frac{571}{2,488,320z^4} + \cdots\Big\}.$$

This result may be derived from applying the Euler-Maclaurin summation formula given in Eq. (2-25) for the integral form of $\Gamma(z)$ in Eq. (4-104).

Special values The values of gamma function for a few special cases are of interest and they may be obtained directly from the properties of $\Gamma(z)$. From the factorial form given in Eq. (4-106) and by making use of the recursion relation Eq. (4-105), we have

$$(z-1)! = \frac{z!}{z}.$$

On letting $z = 1$, we obtain the result

$$0! = 1 \tag{4-108}$$

used earlier in §1-4 for the values of binomial coefficients. Similarly, if $z = 0$, we obtain the result

$$(-1)! = \infty.$$

Putting this together with the recursion relation of Eq. (4-105), we come to the conclusion that the factorial of a negative integer is either $+\infty$ or $-\infty$.

From the fact that

$$\Gamma(z)\Gamma(1-z) = \frac{\pi}{\sin(\pi z)} \tag{4-109}$$

we obtain the value of $\Gamma(z)$ for $z = \frac{1}{2}$ as

$$\Gamma\left(z = \frac{1}{2}\right)^2 = \frac{\pi}{\sin(\frac{1}{2}\pi)} = \pi$$

or

$$\Gamma\left(\frac{1}{2}\right) = \sqrt{\pi}. \tag{4-110}$$

Using similar techniques, it is also possible to show that

$$\Gamma\left(-\frac{1}{2}\right) = -2\sqrt{\pi}. \tag{4-111}$$

Other values of $\Gamma(z)$ for z equal to other ratios of (small) integers can also be obtained in this way. Some of these are listed on page 3 of Abramowitz and Stegun.[1]

Double factorial Using the recursion relation Eq. (4-105), we can show that the gamma function for any positive half-integer has the form

$$\begin{aligned}\Gamma\left(n+\frac{1}{2}\right) &= \frac{2n-1}{2}\Gamma\left(n-\frac{1}{2}\right)\\ &= \frac{(2n-1)(2n-3)(2n-5)\cdots 3\cdot 1}{2^n}\Gamma\left(\frac{1}{2}\right)\\ &= \frac{1}{2^n}(2n-1)!!\sqrt{\pi}\end{aligned} \tag{4-112}$$

where we have used the double factorial symbol in the following sense:

$$k!! \equiv \begin{cases} k(k-2)(k-4)\cdots 3\cdot 1 & \text{for odd } k \\ k(k-2)(k-4)\cdots 4\cdot 2 & \text{for even } k. \end{cases} \tag{4-113}$$

Note that, for Eq. (4-112) to be true for $n = 0$, it is necessary to define $(-1)!! = 1$. It is also possible to express the factorial of an integer in terms of a double factorial,

$$k! = \left(\frac{2k}{2}\right)! = \frac{(2k)(2k-2)(2k-4)\cdots 4\cdot 2}{2^k} = \frac{1}{2^k}(2k)!! \tag{4-114}$$

a result that is useful on many occasions.

Box 4-8 Function S_GMMA(a, x)
Series expansion of the incomplete gamma function $\gamma(a, x)$ for small x

Argument list:
 a: First argument of $\gamma(a, x)$.
 x: Second argument of $\gamma(a, x)$.
Initialization:
 (a) Set maximum number of terms to be 100.
 (b) Let the accuracy be 5×10^{-7}.
 (c) Let $d = a$.
1. Reject $x < 0$.
2. Return 0 if $x = 0$.
3. For $x > 0$, use Eq. (4-118):
 (a) Let the numerator equal 1 and contribution from the first term equal $1/a$.
 (b) Initialize the sum to equal the first term.
 (c) For $n = 1$ to a maximum of terms, carry out the following steps:
  (i) Increase d by 1 and set $t = t * x/d$ as the contribution of this term.
  (ii) Add t to the sum.
  (iii) Go to step 4 if t is less than the accuracy required.
  (iv) Reject the result if the calculation does not converge.
4. Multiply the sum by the other factors in Eq. (4-118) and return the product.

Incomplete gamma function If, instead of ∞, we replace the upper limit of the integral in the definition of the gamma function given in Eq. (4-104) by a variable, we obtain the incomplete gamma functions. There are two related forms of this function,

$$\begin{aligned} \gamma(a, x) &= \int_0^x e^{-t} t^{a-1} dt \\ \Gamma(a, x) &= \Gamma(a) - \gamma(a, x) = \int_x^\infty e^{-t} t^{a-1} dt. \end{aligned} \tag{4-115}$$

A number of functions involving integrals of exponential functions may be written in terms of these integrals. For example, if we change the integration variable t in Eq. (4-115) to $y = t^{1/2}$, the result is proportional to the error function of Eq. (4-100):

$$\operatorname{erf}(x) = \frac{1}{\sqrt{\pi}} \gamma\Big(\frac{1}{2}, x^2\Big) \qquad \operatorname{erfc}(x) = \frac{1}{\sqrt{\pi}} \Gamma\Big(\frac{1}{2}, x^2\Big). \tag{4-116}$$

We shall see later in §6-1 that these functions are also related to the normal probability integral. An example was also given earlier in Eq. (4-102).

We have used $\Gamma(a, x)$ in §3-3 as an application of the continued fraction form of a function. A series expansion form for $\gamma(a, x)$ can also be found on page 262 of Abramowitz and Stegun:[1]

$$\gamma(a, x) = e^{-x} x^a \Gamma(a) \sum_{n=0}^{\infty} \frac{x^n}{\Gamma(a + n + 1)}. \tag{4-117}$$

From the recursion relation Eq. (4-105), we have

$$\begin{aligned}\Gamma(a+n+1) &= (a+n)\Gamma(a+n) = (a+n)(a+n-1)\Gamma(a+n-1)\\ &= (a+n)(a+n-1)\cdots(a)\Gamma(a).\end{aligned}$$

As a result, the series in Eq. (4-117) may also be written in a way that does not involve any gamma functions

$$\gamma(a,x) = e^{-x}x^a \sum_{n=0}^{\infty} \frac{x^n}{(n+a)(n+a-1)\cdots(a)}. \tag{4-118}$$

For computational purposes, this form is easier to handle, as gamma functions are, in general, more time consuming to evaluate. Explicitly, the various terms in the summation are

$$\begin{aligned}\sum_{n=0}^{\infty} \frac{x^n}{(n+a)(n+a-1)\cdots(a)} = \frac{1}{a} &+ \frac{1}{a}\frac{x}{a+1} + \frac{1}{a}\frac{x}{a+1}\frac{x}{a+2} + \cdots\\ &+ \frac{1}{a}\frac{x}{a+1}\frac{x}{a+2}\cdots\frac{x}{a+n} + \cdots.\end{aligned}$$

The convergence of the series becomes much faster once we are beyond the term where the additional factor $(a+n)$ in the denominator is larger than x in the numerator. Since the continued fraction of Eq. (3-26) is rapidly converging only for $x > (a+1)$, the present form is more useful for $x < (a+1)$. The algorithm to evaluate $\gamma(a,x)$ according to Eq. (4-118) is given in Box 4-8.

Problems

4-1 Show that the orthogonal property of Hermite polynomials is given by Eq. (4-13).

4-2 The algebraic expressions of Hermite polynomials $H_n(\rho)$ are given in Eq. (4-14) for n up to 5. Use the method of Box 4-1 to obtain similar expressions for $n = 6$ to 10.

4-3 Write a computer program to evaluate the numerical value of $H_n(\rho)$ for $n = 10$ to 15 and ρ in the interval $[0, 14]$ using the recursion relation of Eq. (4-15). For a given value of ρ, the starting point of the calculation is given by $H_0(\rho) = 1$ and $H_1(\rho) = 2\rho$. Compare the accuracies in the results with those calculated using the algebraic expressions obtained in Problem 4-2.

4-4 Show that the relation between spherical harmonics $Y_{\ell,m}(\theta,\phi)$ for $m = 0$ and Legendre polynomial $P_\ell(\cos\theta)$ given by Eq. (4-42) follows from the differential forms of $Y_{\ell,m}(\theta,\phi)$, given by Eq. (4-37), and $P_\ell(\cos\theta)$, given by

$$P_\ell(\eta) = \frac{1}{2^\ell \ell!}\frac{d^\ell}{d\eta^\ell}\left(\eta^2 - 1\right)^\ell$$

for $\eta \equiv \cos\theta$.

4-5 One way to generate spherical harmonics of all possible m values for a given ℓ is to start with $Y_{\ell,m}(\theta,\phi)$ for $m=\pm\ell$ and apply the angular momentum lowering and raising operator $\hat{\boldsymbol{L}}_\pm$ of Eq. (4-41). Derive the explicit algebraic forms of $Y_{\ell,m}(\theta,\phi)$ for $m=\pm\ell$ using Eq. (4-37) for $\ell=2$ and show that the results are consistent with those given in Eq. (4-38).

4-6 Verify that $Y_{2,2}(\theta,\phi)$ of Eq. (4-38) is the eigenfunction of $\hat{\boldsymbol{L}}^2$ and $\hat{\boldsymbol{L}}_z$ with the correct eigenvalues by calculating the results of the actions of these two operators on $Y_{2,2}(\theta,\phi)$.

4-7 The numerical values of the spherical Bessel function $j_\ell(\rho)$ for very high orders, such as $\ell=100$, are given in Abramowitz and Stegun[1] for $\rho=1$, 2, 5, 10, 50, and 100. Write a computer program to calculate these values and compare the results with the tabulated ones.

4-8 Evaluate the spherical Bessel function $j_5(\rho)$ for five different values of ρ in the interval $[0,10]$ using (a) the explicit algebraic form given in Table 4-3 and (b) by propagating from the values of $j_0(\rho)$ and $j_1(\rho)$ for these five arguments. Compare the accuracies of the two methods by checking the results against the tabulated values given in Abramowitz and Stegun.

4-9 Make a table of the values of $j_5(\rho)$ for different ρ in the interval $[0,10]$. Use the method of inverse interpolation in §3-6 to find all the zeros of the function in the interval. What is the main source of errors in the results?

4-10 Calculate the numerical values of $L_n(t)$ for $n=5$ and 6 for a number of t in the range $[0,5]$. Use the results for $n=6$ to obtain the value of dL_6/dt by numerical differentiation. Check the relation given by Eq. (4-83) for the numerical values obtained.

4-11 Evaluate the expectation value of ρ^2 for the three low-lying states of a hydrogen atom using the numerical values of the radial wave functions calculated with the expressions Eq. (4-93). Check the results against those calculated by evaluating the integrals analytically using the algebraic expressions for the wave functions.

4-12 Repeat Problem 4-11 for the isotropic harmonic oscillator using the radial wave functions given by Eq. (4-94).

4-13 Verify that the uncertainties in the values of error functions calculated using the approximation of Eq. (4-103) are $|\epsilon|\le 2.5\times10^{-5}$ by comparing the results with numerical integration using Eq. (4-100). Note that it is necessary to establish first that the error in numerical integration is less than 2.5×10^{-5}.

4-14 The integral representation of $\Gamma(\frac{1}{2})$ is

$$\Gamma\Big(\frac{1}{2}\Big)=\int_0^\infty \frac{e^{-t}}{\sqrt{t}}dt.$$

Choose a method to evaluate this integral numerically. Is this a good way to calculate the value of π?

4-15 Use the properties of $\Gamma(x)$ for half-integer arguments to show that

$$z!\left(z+\frac{1}{2}\right)! = \frac{\pi^{1/2}}{2^{2z+1}}(2z+1)!$$

Chapter 5

Matrices

With computers taking over many of the tedious tasks, matrix methods are becoming increasingly more popular in solving problems of interest. In this chapter, we shall be mainly concerned with some of the basic numerical calculations involving matrices. Later, we shall apply some of the techniques described here to solve other problems, such as differential equations.

5-1 System of linear equations

Consider a simple example of a direct-current circuit, such as that shown in Fig. 5-1. There are three resistors, labeled R_1, R_2 and R_3, and two sources of electromotive force, V_1 and V_2. One possible interest in the problem is to find the three currents, I_1, I_2, and I_3, flowing respectively through the three resistors, R_1, R_2, and R_3. By considering the voltages and currents in each one of the two circuit loops, we can set up the following three linear equations:

$$\begin{aligned} R_1 I_1 + R_2 I_2 \qquad\quad &= V_1 \\ - R_2 I_2 + R_3 I_3 &= V_2 \\ I_1 - \quad I_2 - \quad I_3 &= 0. \end{aligned} \tag{5-1}$$

There are several ways to solve this system of equations. The particular method we are interested in here is to make use of determinants.

A more compact notation is to write the system of linear equations given by Eq. (5-1) in terms of matrices,

$$\boldsymbol{R}\boldsymbol{I} = \boldsymbol{V}$$

where $\boldsymbol{R}$ is a (3×3) square matrix,

$$\boldsymbol{R} \equiv \begin{pmatrix} R_1 & R_2 & 0 \\ 0 & -R_2 & R_3 \\ 1 & -1 & -1 \end{pmatrix}.$$

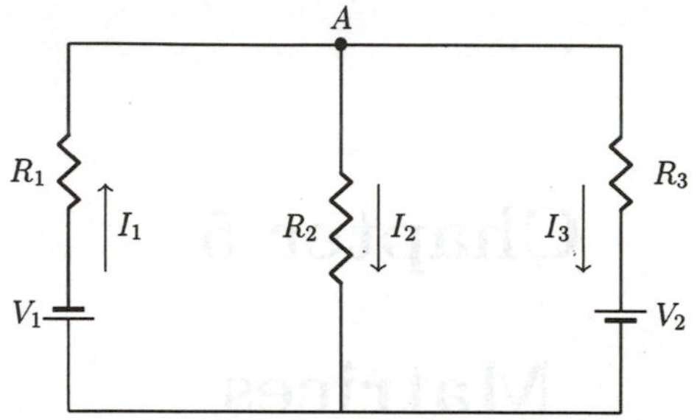

Figure 5-1: A direct-current circuit consisting of three resistors R_1, R_2, and R_3 and two sources of electromotive force V_1 and V_2.

The other two, $\boldsymbol{I}$ and $\boldsymbol{V}$, are column matrices with three elements each:

$$\boldsymbol{I} \equiv \begin{pmatrix} I_1 \\ I_2 \\ I_3 \end{pmatrix} \qquad \boldsymbol{V} \equiv \begin{pmatrix} V_1 \\ V_2 \\ 0 \end{pmatrix}.$$

In general, a system of n linear equations for n unknown quantities $x_1, x_2, \ldots, x_n$, may be written in the following form:

$$\sum_{j=1}^{n} a_{i,j}\, x_j = y_i \tag{5-2}$$

where $i = 1, 2, \ldots, n$. In matrix notation, the same relations are written in the compact form of

$$\boldsymbol{AX} = \boldsymbol{Y} \tag{5-3}$$

where the matrices $\boldsymbol{A}$, $\boldsymbol{X}$, and $\boldsymbol{Y}$ are given by

$$\boldsymbol{A} = \begin{pmatrix} a_{1,1} & a_{1,2} & \cdots & a_{1,n} \\ a_{2,1} & a_{2,2} & \cdots & a_{2,n} \\ \vdots & \vdots & \ddots & \vdots \\ a_{n,1} & a_{n,2} & \cdots & a_{n,n} \end{pmatrix} \qquad \boldsymbol{X} = \begin{pmatrix} x_1 \\ x_2 \\ \vdots \\ x_n \end{pmatrix} \qquad \boldsymbol{Y} = \begin{pmatrix} y_1 \\ y_2 \\ \vdots \\ y_n \end{pmatrix}.$$

Sometimes it is convenient to use a shorthand to indicate that the elements of matrix $\boldsymbol{A}$ are $a_{i,j}$ in the following way:

$$\boldsymbol{A} = \{a_{i,j}\}.$$

We shall do this where it is convenient.

When a matrix $\boldsymbol{A}$ is multiplied by a scalar quantity λ, it means that every element of the matrix is multiplied by λ. That is,

$$\lambda \boldsymbol{A} = \begin{pmatrix} \lambda a_{1,1} & \lambda a_{1,2} & \cdots & \lambda a_{1,n} \\ \lambda a_{2,1} & \lambda a_{2,2} & \cdots & \lambda a_{2,n} \\ \vdots & \vdots & \ddots & \vdots \\ \lambda a_{n,1} & \lambda a_{n,2} & \cdots & \lambda a_{n,n} \end{pmatrix}.$$

When a matrix $\boldsymbol{A}$ is multiplied by another matrix $\boldsymbol{B}$ to form matrix $\boldsymbol{C}$,

$$\boldsymbol{C} = \boldsymbol{A}\boldsymbol{B}$$

each element of $\boldsymbol{C}$ is given by a sum of the products of the elements of $\boldsymbol{A}$ and $\boldsymbol{B}$ in the following way:

$$c_{i,k} = \sum_j a_{i,j} b_{j,k}.$$

For the multiplication to have meaning, the number of columns of $\boldsymbol{A}$ must equal to the number of rows of $\boldsymbol{B}$. If the matrix $\boldsymbol{A}$ has n rows and m columns, and $\boldsymbol{B}$ has m rows and ℓ columns, the product matrix $\boldsymbol{C}$ has n rows and ℓ columns. Different types of matrices commonly encountered in problems are listed in Table 5-1.

Determinant In terms of determinants, the solution of a system of equations given by Eq. (5-3) may be written as

$$x_k = \frac{\det \begin{vmatrix} a_{1,1} & a_{1,2} & \cdots & a_{1,k-1} & y_1 & a_{1,k+1} & \cdots & a_{1,n} \\ a_{2,1} & a_{2,2} & \cdots & a_{2,k-1} & y_2 & a_{2,k+1} & \cdots & a_{2,n} \\ \vdots & \vdots & \ddots & \vdots & \vdots & \vdots & \ddots & \vdots \\ a_{n,1} & a_{n,2} & \cdots & a_{n,k-1} & y_n & a_{n,k+1} & \cdots & a_{n,n} \end{vmatrix}}{\det \begin{vmatrix} a_{1,1} & a_{1,2} & \cdots & a_{1,n} \\ a_{2,1} & a_{2,2} & \cdots & a_{2,n} \\ \vdots & \vdots & \ddots & \vdots \\ a_{n,1} & a_{n,2} & \cdots & a_{n,n} \end{vmatrix}}. \tag{5-4}$$

where the denominator is the determinant $\boldsymbol{A}$ itself and the numerator is that of $\boldsymbol{A}$ with the elements in column k replaced by those of $\boldsymbol{Y}$. Our problem is now reduced to one of evaluating determinants.

The value of the determinant of a square matrix $\boldsymbol{A}$ of order n, that is, one with both the number of columns and the number of rows equal to n, is given by

$$\det \boldsymbol{A} = \sum_{i,j,k,\ldots} \epsilon_{i,j,k,\ldots} a_{1,i} a_{2,j} a_{3,k} \cdots. \tag{5-5}$$

Here, $\epsilon_{i,j,k,\ldots}$ are the Levi-Civita symbols. It is equal to $+1$ for any even permutations of the n subscripts $\{1, 2, 3, \ldots, n\}$, -1 for odd permutations, and zero if any two

Table 5-1: Properties of various types of matrices.

Name	Symbol	Matrix element	Property
Null	$\mathbf{0}$	$c_{i,j} = 0$	$\mathbf{0A} = \mathbf{A0} = \mathbf{0}$
Unit	$\mathbf{1}$	$c_{i,j} = \delta_{i,j}$	$\mathbf{1A} = \mathbf{A1} = \mathbf{A}$
Diagonal	$\mathbf{D}$	$c_{i,j} = c_{i,i}\delta_{i,j}$	
Complex conjugate	$\mathbf{A}^*$	$c_{i,j} = a^*_{i,j}$	
Real		$a_{i,j} = a^*_{i,j}$	$\mathbf{A} = \mathbf{A}^*$
Pure imaginary		$a_{i,j} = -a^*_{i,j}$	$\mathbf{A} = -\mathbf{A}^*$
Inverse	$\mathbf{A}^{-1}$		$\mathbf{A}^{-1}\mathbf{A} = \mathbf{A}\mathbf{A}^{-1} = \mathbf{1}$
Transpose	$\widetilde{\mathbf{A}}$	$c_{i,j} = a_{j,i}$	
Symmetric		$a_{i,j} = a_{j,i}$	$\mathbf{A} = \widetilde{\mathbf{A}}$
Skew-symmetric		$a_{i,j} = -a_{j,i}$	$\mathbf{A} = -\widetilde{\mathbf{A}}$
Hermitian adjoint	$\mathbf{A}^\dagger$	$c_{i,j} = a^*_{j,i}$	
Hermitian		$a_{i,j} = a^*_{j,i}$	$\mathbf{A} = \mathbf{A}^\dagger$
Skew-Hermitian		$a_{i,j} = -a^*_{j,i}$	$\mathbf{A} = -\mathbf{A}^\dagger$
Unitary			$\mathbf{A}^{-1} = \mathbf{A}^\dagger$
Trace	$\mathrm{Tr}\mathbf{A}$	$\sum_i a_{i,i}$	
Addition	$\mathbf{A}+\mathbf{B}=\mathbf{C}$	$c_{i,j} = a_{i,j} + b_{i,j}$	
Multiplication	$\mathbf{C} = \lambda\mathbf{A}$	$c_{i,j} = \lambda a_{i,j}$	
	$\mathbf{C} = \mathbf{AB}$	$c_{i,j} = \sum_k a_{i,k}b_{k,j}$	

indices are the same. For example, for $n = 3$,

$$\begin{aligned}\det \mathbf{A} &= \det \begin{vmatrix} a_{1,1} & a_{1,2} & a_{1,3} \\ a_{2,1} & a_{2,2} & a_{2,3} \\ a_{3,1} & a_{3,2} & a_{3,3} \end{vmatrix} \\ &= a_{1,1}a_{2,2}a_{3,3} - a_{1,1}a_{2,3}a_{3,2} + a_{1,3}a_{2,1}a_{3,2} - a_{1,2}a_{2,1}a_{3,3} \\ &\quad + a_{1,2}a_{2,3}a_{3,1} - a_{1,3}a_{2,2}a_{3,1}. \end{aligned} \tag{5-6}$$

From this we see that if any two rows, or any two columns, are interchanged the value of the determinant changes sign. By the same token, if any two rows, or any two columns, are identical, the value of the determinant vanishes.

One method to evaluate a determinant is based on the Laplace expansion theorem, which states that the value of a determinant is given by the sum of the products of all the elements of an arbitrary row j (or a column) with their corresponding minors:

$$\det \mathbf{A} = \sum_i (-1)^{i+j} a_{i,j} M_{i,j}. \tag{5-7}$$

The minor $M_{i,j}$ is defined as a determinant obtained from $\boldsymbol{A}$ by removing row i and column j. In place of minors, one may use cofactors

$$F_{i,j} = (-1)^{i+j} M_{i,j}. \tag{5-8}$$

Our $n = 3$ example given in Eq. (5-6) may now be written in the form

$$\begin{aligned} \det \boldsymbol{A} &= a_{1,1}M_{1,1} - a_{2,1}M_{2,1} + a_{3,1}M_{3,1} \\ &= a_{1,1}F_{1,1} + a_{2,1}F_{2,1} + a_{3,1}F_{3,1} \end{aligned} \tag{5-9}$$

where

$$M_{1,1} = F_{1,1} = \begin{vmatrix} a_{2,2} & a_{2,3} \\ a_{3,2} & a_{3,3} \end{vmatrix} \qquad M_{2,1} = -F_{2,1} = \begin{vmatrix} a_{1,2} & a_{1,3} \\ a_{3,2} & a_{3,3} \end{vmatrix}$$

$$M_{3,1} = F_{3,1} = \begin{vmatrix} a_{1,2} & a_{1,3} \\ a_{2,2} & a_{2,3} \end{vmatrix}.$$

This method, however, becomes very cumbersome if n is large.

Gauss-Jordan elimination method As an algorithm for computer calculations, we shall take a somewhat different approach that depends on a special property of determinants, namely its value is unchanged if a column is replaced by a linear combination of itself and other columns. For example,

$$\det \begin{vmatrix} a_{1,1} & a_{1,2} & \cdots & a_{1,n} \\ a_{2,1} & a_{2,2} & \cdots & a_{2,n} \\ \vdots & \vdots & \ddots & \vdots \\ a_{n,1} & a_{n,2} & \cdots & a_{n,n} \end{vmatrix} = \det \begin{vmatrix} a_{1,1} + \lambda a_{1,2} & a_{1,2} & \cdots & a_{1,n} \\ a_{2,1} + \lambda a_{2,2} & a_{2,2} & \cdots & a_{2,n} \\ \vdots & \vdots & \ddots & \vdots \\ a_{n,1} + \lambda a_{n,2} & a_{n,2} & \cdots & a_{n,n} \end{vmatrix}. \tag{5-10}$$

This is also true for rows. We can make use of this property to transform a determinant such that all the off-diagonal elements vanish. For example, by subtracting $(a_{2,1}/a_{1,1})$ times each of the elements in the first row from the corresponding ones in the second row,

$$a'_{2,j} = a_{2,j} - a_{1,j} \times \frac{a_{2,1}}{a_{1,1}} \qquad \text{for} \qquad j = 1, 2, \ldots, n$$

the $(i,j) = (2,1)$ element of the resulting determinant vanishes. If this process is carried out for all the rows beyond the first one, the result is that the determinant is changed into a form with all the off-diagonal elements in the first column equal zero,

$$\det \begin{vmatrix} a_{1,1} & a_{1,2} & \cdots & a_{1,n} \\ a_{2,1} & a_{2,2} & \cdots & a_{2,n} \\ \vdots & \vdots & \ddots & \vdots \\ a_{n,1} & a_{n,2} & \cdots & a_{n,n} \end{vmatrix} = \det \begin{vmatrix} a_{1,1} & a_{1,2} & \cdots & a_{1,n} \\ 0 & a'_{2,2} & \cdots & a'_{2,n} \\ \vdots & \vdots & \ddots & \vdots \\ 0 & a'_{n,2} & \cdots & a'_{n,n} \end{vmatrix}. \tag{5-11}$$

Here

$$a'_{i,j} = a_{i,j} - a_{1,j} \times \frac{a_{i,1}}{a_{1,1}} \qquad \text{for} \quad j = 1, 2, \ldots, n$$
$$\text{and} \quad i = 2, 3, \ldots, n. \tag{5-12}$$

More generally, a transformation of the form

$$a'_{i,j} = a_{i,j} - a_{k,j} \times \frac{a_{i,k}}{a_{k,k}} \qquad \text{for} \quad j = k, k+1, \ldots, n$$
$$\text{and} \quad i = j+1, j+2, \ldots, n \tag{5-13}$$

reduces all the elements below the diagonal in column k to zero.

On carrying out this for $k = 1, 2, \ldots, (n-1)$, all the elements below the diagonal vanish. The result is an upper triangular matrix having the same value for the determinant as the original one:

$$\det \begin{vmatrix} a_{1,1} & a_{1,2} & \cdots & a_{1,n} \\ a_{2,1} & a_{2,2} & \cdots & a_{2,n} \\ \vdots & \vdots & \ddots & \vdots \\ a_{n,1} & a_{n,2} & \cdots & a_{n,n} \end{vmatrix} = \det \begin{vmatrix} a_{1,1} & a_{1,2} & a_{1,3} & \cdots & a_{1,n} \\ 0 & a'_{2,2} & a'_{2,3} & \cdots & a'_{2,n} \\ 0 & 0 & a''_{3,3} & \cdots & a''_{3,n} \\ \vdots & \vdots & \vdots & \ddots & \vdots \\ 0 & 0 & 0 & \cdots & a^{(n-1)}_{n,n} \end{vmatrix}.$$

As we shall see soon, the value of the determinant may be easily evaluated in this way and the method is known as the Gaussian elimination method.

In principle, we can carry out a back substitution to eliminate all the off-diagonal elements in the upper triangle as well. In fact, as we shall see soon, it is not necessary to perform the back substitution if we are only interested in the value of the determinant. However, for pedagogical reasons, we shall give a short discussion on this step. Let us start from the nth column and work backward one column at a time. In column n, we can use $a^{(n-1)}_{n,n}$ to reduce all $(n-1)$ off-diagonal elements to zero by subtracting suitable multiples of the last row from the other rows. Note that, since all the elements in the last row are zero except the diagonal one, the subtractions do not change the values of the elements in the earlier rows except those in column n. The only effect of this particular step in the back substitution is to reduce all the off-diagonal elements in column n to zero. Note also that the value of the diagonal element is not affected by the back substitution step.

We can now proceed to column $(n-1)$. Again, we have the situation that all the elements in row $(n-1)$ are zero except the diagonal one. As a result, we can subtract suitable multiples of this row from rows $(n-2)$, $(n-3)$, ..., and reduce all the remaining off-diagonal elements in column $(n-1)$ to zero. Again, no other elements, including the diagonal one in row $(n-1)$, are affected by this set of transformations.

Box 5-1 Function DETRM(A,N,NDMN)
Value of a determinant by Gauss-Jordan elimination

Argument list:
- A: Two-dimensional array for determinant $\boldsymbol{A}$.
- N: Order of the determinant.
- NDMN: Dimension of array A in the calling program.

1. Initialize the value of the determinant to be $\mathcal{D} = 1$.
2. For $i = 1$ to n, carry out the following steps:
   (a) Find $a_{j,k}$ with the largest absolute value for $j = i$ to n and $k = i$ to n.
      (i) If the value is zero, terminate the calculation and return zero for $\mathcal{D}$.
      (ii) If not, move this element to the diagonal position $a_{i,i}$:
      (iii) Reverse the sign of $\mathcal{D}$ for each interchange of rows or columns.
   (b) Multiply the diagonal matrix element to $\mathcal{D}$.
   (c) Eliminate the rest of the elements in the ith row using Eq. (5-13).
3. Return $\mathcal{D}$ as the value of the determinant.

We can continue this process until the entire determinant is diagonal,

$$\det\begin{vmatrix} a_{1,1} & a_{1,2} & \cdots & a_{1,n} \\ a_{2,1} & a_{2,2} & \cdots & a_{2,n} \\ \vdots & \vdots & \ddots & \vdots \\ a_{n,1} & a_{n,2} & \cdots & a_{n,n} \end{vmatrix} = \det\begin{vmatrix} a_{1,1} & 0 & 0 & \cdots & 0 \\ 0 & a'_{2,2} & 0 & \cdots & 0 \\ 0 & 0 & a''_{3,3} & \cdots & 0 \\ \vdots & \vdots & \vdots & \ddots & \vdots \\ 0 & 0 & 0 & \cdots & a^{(n-1)}_{n,n} \end{vmatrix}.$$

Since the only nonvanishing elements are along the diagonal, the value of such a determinant is simply the product of the diagonal ones,

$$\det \boldsymbol{A} = a_{1,1}a'_{2,2}a''_{3,3}\cdots a^{(n-1)}_{n,n}. \tag{5-14}$$

Furthermore, since the diagonal elements are not changed, the matrix elements that enter on the right side of Eq. (5-14) are exactly the same ones as those before the back substitution step. This means that we could have obtained the value of the determinant without actually carrying out the back substitution.

A slight refinement of the method is needed before using it as an algorithm for calculating the value of an arbitrary determinant. The basic operation in Gaussian elimination is to reduce the off-diagonal element at position (i,j) to zero by subtracting from it multiples of a diagonal element. Such a diagonal element is usually referred to as the *pivot*, for obvious reasons. This method fails if, for any reason, one of the pivots vanishes. By the same token, the numerical accuracy of the result will be poor if some of the pivots used are much smaller compared with the values of the off-diagonal elements involved in the same calculation. To prevent this from happening, it is advantageous to use as pivot the element with the largest possible absolute value. In practice, this principle is applied in the following way. Consider the situation that the determinant is transformed to the stage that all the off-diagonal matrix elements in the lower half-triangle up to row k are reduced to zero. A search is made among the $(n-k) \times (n-k)$ remaining elements to locate the one with the largest absolute value. If this element is not at the position of the next pivot, the diagonal element at position (k, k), we shall permute the rows and columns of the determinant so as to bring it to position (k, k). When the refinement of pivoting is included as a part of Gaussian elimination, the method is known as Gauss-Jordan elimination. An algorithm based on it is summarized in Box 5-1.

Solution of a system of linear equations by elimination Instead of evaluating determinants, it is also possible to solve a system of linear equations directly by Gaussian elimination. The fact that we can add one row of a determinant to a multiple of another one without changing the value is equivalent to the statement that we can add a multiple of one member of a system of linear equations to another without affecting the solution. In practice, this means that a set of linear equations can be solved by carrying such operations on both sides at the same time.

It is more instructive to illustrate the method by a simple example. Consider the following system of three linear equations:

$$\begin{aligned} x_1 + 2x_2 + 3x_3 &= 14 \\ 3x_1 + x_2 + 2x_3 &= 11 \\ 2x_1 + 3x_2 + x_3 &= 11. \end{aligned} \tag{5-15}$$

If we subtract three times the first equation from the second one and twice the first one from third one, we eliminate x_1 from the second and third member of the system:

$$\begin{aligned} x_1 + 2x_2 + 3x_3 &= 14 \\ -5x_2 - 7x_3 &= -31 \\ -x_2 - 5x_3 &= -17. \end{aligned} \tag{5-16}$$

We can eliminate x_2 from the third member of this new set by subtracting from it one-fifth times the second member,

$$\begin{aligned} x_1 + 2x_2 + 3x_3 &= 14 \\ -5x_2 - 7x_3 &= -31 \\ -\frac{18}{5}x_3 &= -\frac{54}{5}. \end{aligned} \tag{5-17}$$

In matrix notation, the left side is a product of an upper triangular matrix with a column matrix. From the third member of Eq. (5-17), we obtain the solution $x_3 = 3$. We can now carry out the equivalent of back substitutions to obtain the solutions for x_2 and x_3. On inserting the value of x_3 into the second member, we have the result $x_2 = 2$. By putting both values into the first one, we get $x_1 = 1$.

Box 5-2 Subroutine LNEQN(A,Y,N,DET,NDMN)
Solve a system of linear equation $\sum_{j=1}^{n} a_{j,i}x_j = y_i$.
Using Gauss-Jordan elimination with pivoting

Argument list:
- A: Two-dimensional array for determinant $\boldsymbol{A}$.
- Y: Array for column matrix $\boldsymbol{Y}$.
- N: Number of linear equations.
- DET: Value of the determinant on output.
- NDMN: Dimension of the arrays A and Y in the calling program.

Initialization:
- (a) Start with the value of the determinant DET=1.
- (b) Zero an auxiliary array to keep track of the rows transformed.

1. Carry out steps 2 to 7 for i equal to 1 to n.
2. Find the next pivot:
   (a) Locate the largest element in absolute value among rows not yet transformed.
   (b) If the value is zero, return DET=0 to signal a singular case.
3. Multiply DET by the value of the pivot found.
4. If necessary, move the pivoting element to the diagonal position:
   (a) Interchange rows of $\boldsymbol{A}$ and $\boldsymbol{Y}$ to put the element at the diagonal position.
   (b) Change the sign of DET.
   (c) Record the interchanges using the auxiliary array.
5. Divide the elements in row i of $\boldsymbol{A}$ and $\boldsymbol{Y}$ by the pivot.
6. Mark the row as transformed in the auxiliary array.
7. Apply the transformation of Eq. (5-13) to the rest of $\boldsymbol{A}$ and $\boldsymbol{Y}$.
8. Return DET as the value of the determinant and Y as the solution for $\{x_i\}$.

This simple example shows that it is possible to find the solution of a set of n linear equations, such as that represented by Eq. (5-3) for $n = 3$, by taking appropriate linear combinations of two members and eliminating one of the independent variables between them. This is essentially the Gauss-Jordan elimination method for solving a system of linear equations and is outlined in Box 5-2. Similar to what we did earlier in calculating the value of a determinant, it includes also pivoting to avoid having diagonal elements with very small absolute values.

To check our results, we can substitute the solution obtained for $x_1, x_2, \ldots, x_n$ into the original linear equations and see if the sum of products $\sum_j a_{i,j}x_j$ is equal to y_i within the accuracies required. To do this in a computer program, it is necessary to save the matrix elements of $\boldsymbol{A}$ in a separate array, as the one that goes into the calculation of the solution is usually destroyed in the process of Gauss-Jordan elimination. For the purpose of testing the computer program, it is possible to make use of a random number generator to produce the input matrix elements $a_{i,j}$ and the components of the vector $\boldsymbol{Y}$. This approach has the advantage that a large number of test cases can be conducted with relatively little effort.

5-2 Matrix inversion and LU-decomposition

Instead of using determinants, we can also solve a system of linear equations by matrix inversion. If $\boldsymbol{A}$ is a matrix, its inverse $\boldsymbol{A}^{-1}$ is defined by

$$\boldsymbol{A}^{-1}\boldsymbol{A} = \boldsymbol{A}\boldsymbol{A}^{-1} = \mathbf{1}. \tag{5-18}$$

Here $\mathbf{1}$ is the unit matrix. As given in Table 5-1, it is a matrix with all the diagonal elements equal to 1 and all the off-diagonal elements 0. In terms of matrix elements, we can express Eq. (5-18) in the following way. Let

$$\boldsymbol{B} = \boldsymbol{A}^{-1}.$$

If $a_{i,k}$ and $b_{k,j}$ are the matrix elements of, respectively, $\boldsymbol{A}$ and $\boldsymbol{B}$, Eq. (5-18) is the same as

$$\sum_k b_{i,k}a_{k,j} = \sum_k a_{i,k}b_{k,j} = \delta_{i,j}. \tag{5-19}$$

For simplicity, we shall consider here only square matrices with the number of rows equal to that of columns, although many of the considerations below apply also to rectangular matrices with the number of rows not equal to that of columns.

Using the inverse matrix, we can solve the system of linear equations in Eq. (5-3) by multiplying both sides with $\boldsymbol{A}^{-1}$:

$$\boldsymbol{A}^{-1}\boldsymbol{A}\boldsymbol{X} = \boldsymbol{A}^{-1}\boldsymbol{Y}. \tag{5-20}$$

From the definition of the inverse given by Eq. (5-18), we obtain $\boldsymbol{X}$ in terms of $\boldsymbol{Y}$:

$$\boldsymbol{X} = \boldsymbol{A}^{-1}\boldsymbol{Y}. \tag{5-21}$$

In terms of matrix elements, the same relation appears in the form

$$x_i = \sum_j b_{i,j}\, y_j$$

where $b_{i,j}$ are the elements of $\boldsymbol{A}^{-1}$. If $\boldsymbol{A}^{-1}$ is known, we have a solution of x_i in terms of the elements of the column matrix $\boldsymbol{Y}$.

In principle, we can find the inverse of a matrix by making use of the definition given by Eq. (5-19). For a given j, the second form of the definition may be regarded as a system of linear equations for the n unknown quantities $b_{1,j}, b_{2,j}, \ldots, b_{n,j}$:

$$\sum_k a_{i,k}b_{k,j} = \delta_{i,j}.$$

This is identical to Eq. (5-2) with $b_{k,j}$ taking on the role of the unknown x_i and $\delta_{i,j}$ occupying the position of the known input y_i. On solving this set of equations, we obtain the values of all n elements in column j of $\boldsymbol{A}^{-1}$. For the complete inverse matrix, we must carry out the operation for all n columns, with $j = 1, 2, \ldots, n$. As a result, we have to solve a total of n^2 equations to find the n^2 unknown elements $b_{i,j}$ in this approach.

We shall regard this merely as a statement that the problem has a solution. The actual calculations involved are much simpler. In practice, the solution requires only a single application of a slightly modified form of Gauss-Jordan elimination, as we shall see next.

Gauss-Jordan elimination In §5-1, we have used the method of Gauss-Jordan elimination to evaluate the determinant of an arbitrary matrix $\boldsymbol{A}$ by reducing it to a diagonal form. For our discussion here, the operations may be expressed symbolically in terms of an operator,

$$\hat{\mathcal{O}}\boldsymbol{A} = \det|\boldsymbol{A}|\,\mathbf{1} \tag{5-22}$$

where the operator $\hat{\mathcal{O}}$ represents all the steps required to reduce $\boldsymbol{A}$ to a unit matrix, multiplied by a constant given by the value of the determinant of $\boldsymbol{A}$. What happens if the same operations are carried out on a unit matrix? The answer is

$$\hat{\mathcal{O}}\mathbf{1} = \boldsymbol{A}^{-1}\det|\boldsymbol{A}|. \tag{5-23}$$

The proof may be constructed in the following way. Formally, we can write

$$\boldsymbol{A}^{-1} = \frac{\mathbf{1}}{\boldsymbol{A}}.$$

If, on the right side of the equation, we apply the same set of operations to both the matrices in the numerator and the denominator, the ratio is unchanged. In other words,

$$\boldsymbol{A}^{-1} = \frac{\hat{\mathcal{O}}\mathbf{1}}{\hat{\mathcal{O}}\boldsymbol{A}}.$$

This is essentially the same as Eq. (5-23), since the denominator is the product of the determinant of $\boldsymbol{A}$ and a unit matrix, as we saw earlier in Eq. (5-22).

From a practical point of view, we can interpret Eq. (5-23) in the following way. The inverse of a matrix $\boldsymbol{A}$ may be found by the same steps as used above to reduce it to a unit matrix. To store the steps, we can apply the operations to a unit matrix at the same time. Except for a normalization constant, given by the value of the determinant, $\boldsymbol{A}$ is transformed at the end into a unit matrix, while the unit matrix is transformed into the inverse of $\boldsymbol{A}$. This is essentially what we have done in the previous paragraph in proving Eq. (5-22).

An example using a three-dimensional matrix will serve to illustrate this point. At the start, we have the matrix $\boldsymbol{A}$ we wish to invert and a unit matrix $\boldsymbol{U}$,

$$\boldsymbol{A} = \begin{pmatrix} a_{1,1} & a_{1,2} & a_{1,3} \\ a_{2,1} & a_{2,2} & a_{2,3} \\ a_{3,1} & a_{3,2} & a_{3,3} \end{pmatrix} \qquad \boldsymbol{U} = \begin{pmatrix} 1 & 0 & 0 \\ 0 & 1 & 0 \\ 0 & 0 & 1 \end{pmatrix}. \tag{5-24}$$

Using steps similar to those carried out in Eq. (5-12), we can reduce the second and third elements of the first row of $\boldsymbol{A}$ to zero. Upon normalizing the diagonal element in the first row to unity, these two matrices are changed to

$$\boldsymbol{B} = \begin{pmatrix} 1 & 0 & 0 \\ \frac{a_{2,1}}{a_{1,1}} & a_{2,2} - a_{2,1}\frac{a_{1,2}}{a_{1,1}} & a_{2,3} - a_{2,1}\frac{a_{1,3}}{a_{1,1}} \\ \frac{a_{3,1}}{a_{1,1}} & a_{3,2} - a_{3,1}\frac{a_{1,2}}{a_{1,1}} & a_{3,3} - a_{3,1}\frac{a_{1,3}}{a_{1,1}} \end{pmatrix} \qquad \boldsymbol{V} = \begin{pmatrix} \frac{1}{a_{1,1}} & -\frac{a_{1,2}}{a_{1,1}} & -\frac{a_{1,3}}{a_{1,1}} \\ 0 & 1 & 0 \\ 0 & 0 & 1 \end{pmatrix}. \tag{5-25}$$

Next, we can reduce the first and third elements of the second row of $\boldsymbol{B}$ to zero. This gives us the results

$$\boldsymbol{C} = \begin{pmatrix} 1 & 0 & 0 \\ 0 & 1 & 0 \\ b_{3,1} - b_{3,2}\frac{b_{2,1}}{b_{2,2}} & \frac{b_{2,3}}{b_{2,2}} & b_{3,3} - b_{3,2}\frac{b_{2,3}}{b_{2,2}} \end{pmatrix}$$

$$\boldsymbol{W} = \begin{pmatrix} v_{1,1} - v_{1,2}\frac{b_{2,1}}{b_{2,2}} & \frac{v_{1,2}}{b_{2,2}} & v_{1,3} - v_{1,2}\frac{b_{2,3}}{b_{2,2}} \\ -\frac{v_{2,1}}{b_{2,2}} & \frac{1}{b_{2,2}} & -\frac{b_{2,3}}{b_{2,2}} \\ 0 & 0 & 1 \end{pmatrix}$$

where, to simplify the notation, we have used the matrix elements of $\boldsymbol{B}$ and $\boldsymbol{V}$ of Eq. (5-25) instead of those of the original matrix $\boldsymbol{A}$.

The final step in reducing the left side to a unit matrix requires the first and second elements in the last row of $\boldsymbol{C}$ to be transformed to zeros and the third element normalized to unity. In terms of the matrix elements of $\boldsymbol{C}$ and $\boldsymbol{W}$, we have the result

$$\boldsymbol{D} = \begin{pmatrix} 1 & 0 & 0 \\ 0 & 1 & 0 \\ 0 & 0 & 1 \end{pmatrix} \qquad \boldsymbol{X} = \begin{pmatrix} w_{1,1} - w_{1,3}\frac{c_{3,1}}{c_{3,3}} & w_{1,2} - w_{1,3}\frac{c_{3,2}}{c_{3,3}} & \frac{w_{1,3}}{c_{3,3}} \\ w_{2,1} - w_{2,3}\frac{c_{3,1}}{c_{3,3}} & w_{2,2} - w_{2,3}\frac{c_{3,2}}{c_{3,3}} & \frac{w_{2,3}}{c_{3,3}} \\ -\frac{c_{3,1}}{c_{3,3}} & -\frac{c_{3,2}}{c_{3,3}} & \frac{1}{c_{3,3}} \end{pmatrix}.$$

Since the original matrix is now transformed to $\boldsymbol{D}$, a unit matrix, the inverse of $\boldsymbol{A}$ is given by the matrix $\boldsymbol{X}$. In terms of the matrix elements of $\boldsymbol{A}$, the results may be expressed as

$$\boldsymbol{A}^{-1} = \frac{1}{\det|\boldsymbol{A}|}\begin{pmatrix} a_{2,2}a_{3,3} - a_{2,3}a_{3,2} & a_{1,3}a_{3,2} - a_{1,2}a_{3,3} & a_{1,2}a_{2,3} - a_{1,3}a_{2,2} \\ a_{2,3}a_{3,1} - a_{2,1}a_{3,3} & a_{1,1}a_{3,3} - a_{1,3}a_{3,1} & a_{1,3}a_{2,1} - a_{1,1}a_{2,3} \\ a_{2,1}a_{3,2} - a_{2,2}a_{3,1} & a_{1,2}a_{3,1} - a_{1,1}a_{3,2} & a_{1,1}a_{2,2} - a_{1,2}a_{2,1} \end{pmatrix}. \tag{5-26}$$

We can check that it is, indeed, the inverse matrix by multiplying it by $\boldsymbol{A}$.

Two refinements are required for any practical application of the method described above. The first is in reducing the storage required for the elements. The operations outlined above require two $(n \times n)$ arrays to store the matrices, for example, $\boldsymbol{A}$ and $\boldsymbol{U}$ of Eq. (5-24). However, in each step, the number of newly generated elements of $\boldsymbol{U}$, as well as their locations in the matrix, correspond exactly to the

Box 5-3 Subroutine MATIV(A,N,DET,NDMN)
Matrix inversion by Gauss-Jordan elimination

Argument list:

A: Input, matrix $\boldsymbol{A}$; stores the inverse of $\boldsymbol{A}$ on output.
N: Input, dimension of the matrix.
DET: Output, value of the determinant.
NDMN: Dimension of the array A in the calling program.

Initialization:

(a) Set up two arrays to record row and column numbers in the original matrix.
(b) Let the value of the determinant be DET = 1.

1. For $k = 1$ to n:
   (a) Find the pivot by locating the largest element remaining:
      (i) If the pivot is zero, the matrix is singular. Terminate the inversion.
      (ii) Otherwise:
         Store the positions of the row and column of the pivot.
         Multiply DET with the pivot.
         Put the pivot at diagonal position by column and row interchanges.
         Reverse the sign of DET for each interchange.
   (b) Replace the elements in column k by those of the inverse.
   (c) Transformation the rest of the matrix except row and column k.
2. Normalize the elements of the row k.
3. Restore the inverse matrix to the original order using the auxiliary arrays.
8. Return the value of DET and the inverse matrix.

number and locations of elements in $\boldsymbol{A}$ reduced to zero. Furthermore, there is really no need to store the unit matrix at the beginning and to have the unit matrix at the end. As a result, the two matrices at all the intermediate stages of the calculations can occupy the same memory locations in the computer. In other words, it is possible to make both matrices coexist in the same array and only a single one needs to be allocated. This complicates the programming somewhat, but is worth doing especially for large matrices.

The second refinement is to apply pivoting, as we did in the previous section. To incorporate this improvement, it is necessary to interchange two columns and two rows of the matrix whenever the element we wish to use as the pivot does not fall on the diagonal. Since interchanging all the elements in two rows or columns is a time-consuming process, methods have been designed to avoid the actual operations, as we did earlier in Box 5-2. However, we shall not carry it out here. We are already using the same two-dimensional array to store two different matrices at the same time — it would make the program even more difficult to understand if we introduce any further complications. For this reason, explicit row and column interchanges are applied in the algorithm of Box 5-3, if necessary, at every step of reduction. To restore the inverse matrix to the original order, it is necessary to carry out the same set of interchanges in reverse at the end.

The inverse matrix produced may be checked by multiplying it with the original matrix and observing if the conditions of Eq. (5-19) are satisfied. Again, to test the computer program, we can use a random number generator to produce all the elements of the test matrices. Since the original matrix $\boldsymbol{A}$ is replaced by its inverse at the end of the calculation, a copy of $\boldsymbol{A}$ must be saved for the purpose of checking.

Lower and upper triangular matrices We have seen in the previous section that, in evaluating a determinant by Gauss-Jordan elimination, the back substitution step is relative easy. The simplification comes from the fact that the matrix is already reduced to an upper triangular form, with zeros as the elements below the diagonal. The same is true if the matrix is reduced to a lower triangular form, with all the elements above the diagonal equal zero. We shall now try to take advantage of the properties of triangular matrices in solving a system of linear equations of the form

$$\boldsymbol{AX} = \boldsymbol{Y}$$

given earlier in Eq. (5-3). To achieve this, we shall write the matrix $\boldsymbol{A}$ as a product of a lower triangular matrix $\boldsymbol{L}$ and an upper triangular matrix $\boldsymbol{U}$,

$$\boldsymbol{A} = \boldsymbol{LU} \tag{5-27}$$

where

$$\boldsymbol{L} = \begin{pmatrix} \ell_{1,1} & 0 & 0 & \cdots & 0 \\ \ell_{2,1} & \ell_{2,2} & 0 & \cdots & 0 \\ \ell_{3,1} & \ell_{3,2} & \ell_{3,3} & \cdots & 0 \\ \vdots & \vdots & \vdots & \ddots & \vdots \\ \ell_{n,1} & \ell_{n,2} & \ell_{n,3} & \cdots & \ell_{n,n} \end{pmatrix} \qquad \boldsymbol{U} = \begin{pmatrix} u_{1,1} & u_{1,2} & u_{1,3} & \cdots & u_{1,n} \\ 0 & u_{2,2} & u_{2,3} & \cdots & u_{2,n} \\ 0 & 0 & u_{1,3} & \cdots & u_{3,n} \\ \vdots & \vdots & \vdots & \ddots & \vdots \\ 0 & 0 & 0 & \cdots & u_{n,n} \end{pmatrix}. \tag{5-28}$$

We shall see next why *LU-decomposition* can help us to solve a system of linear equations.

In terms of $\boldsymbol{L}$ and $\boldsymbol{U}$, the relation expressed by Eq. (5-3) may be written as

$$\boldsymbol{AX} = \boldsymbol{L}(\boldsymbol{UX}) = \boldsymbol{Y}. \tag{5-29}$$

The solution is obtained in two stages. First, we write $(\boldsymbol{UX})$ as a column matrix $\boldsymbol{Z}$ of dimension n and put Eq. (5-29) in the form

$$\boldsymbol{LZ} = \boldsymbol{Y}. \tag{5-30}$$

We can solve these n equations for the elements of $\boldsymbol{Z}$. Once this is done, we can move to the second stage and make use of $\boldsymbol{Z}$ as the right side of the system of linear equations:

$$\boldsymbol{UX} = \boldsymbol{Z}. \tag{5-31}$$

The solution gives us the elements of $\boldsymbol{X}$, our original aim of solving Eq. (5-3).

Let us defer till later the question of how to reduce $\boldsymbol{A}$ to a product of $\boldsymbol{L}$ and $\boldsymbol{U}$. Here, we shall consider first the problem of solving Eqs. (5-30) and (5-31). Since each of the two stages in the previous paragraph involves only a triangular matrix, the calculations are fairly simple. For the first stage, Eq. (5-30) may be written in terms of matrix elements in the form

$$\boldsymbol{L}\boldsymbol{Z} = \begin{pmatrix} \ell_{1,1} & 0 & 0 & \cdots & 0 \\ \ell_{2,1} & \ell_{2,2} & 0 & \cdots & 0 \\ \ell_{3,1} & \ell_{3,2} & \ell_{3,3} & \cdots & 0 \\ \vdots & \vdots & \vdots & \ddots & \vdots \\ \ell_{n,1} & \ell_{n,2} & \ell_{n,3} & \cdots & \ell_{n,n} \end{pmatrix} \begin{pmatrix} z_1 \\ z_2 \\ z_3 \\ \vdots \\ z_n \end{pmatrix} = \begin{pmatrix} y_1 \\ y_2 \\ y_3 \\ \vdots \\ y_n \end{pmatrix}. \tag{5-32}$$

In terms of y_i, we obtain from the first row of the matrices the result

$$z_1 = \frac{y_1}{\ell_{1,1}}.$$

Next we can substitute forward to find z_2 in terms of y_2 and z_1

$$z_2 = \frac{1}{\ell_{2,2}}(y_2 - \ell_{2,1}z_1).$$

By continuing the forward substitution to the kth row, we obtain the general result

$$z_k = \frac{1}{\ell_{k,k}}\Big(y_k - \sum_{j=1}^{k-1} \ell_{k,j}z_j\Big). \tag{5-33}$$

Once the values of all the elements of $\boldsymbol{Z}$ are found, they can be used in Eq. (5-31) to find the elements of $\boldsymbol{X}$ by back substitution. The steps are similar to the forward substitution above, and we shall leave it as an exercise to derive the general formula.

***LU*-decomposition** We shall now address the question of decomposing a matrix into the product of an upper and a lower triangular matrix so as to take advantage of the simplicities discussed above. The method is basically provided by Eq. (5-27). In terms of the matrix elements of $\boldsymbol{L}$, $\boldsymbol{U}$, and $\boldsymbol{A}$, we have the relation

$$\begin{pmatrix} \ell_{1,1} & 0 & 0 & \cdots & 0 \\ \ell_{2,1} & \ell_{2,2} & 0 & \cdots & 0 \\ \ell_{3,1} & \ell_{3,2} & \ell_{3,3} & \cdots & 0 \\ \vdots & \vdots & \vdots & \ddots & \vdots \\ \ell_{n,1} & \ell_{n,2} & \ell_{n,3} & \cdots & \ell_{n,n} \end{pmatrix} \begin{pmatrix} u_{1,1} & u_{1,2} & u_{1,3} & \cdots & u_{1,n} \\ 0 & u_{2,2} & u_{2,3} & \cdots & u_{2,n} \\ 0 & 0 & u_{3,3} & \cdots & u_{3,n} \\ \vdots & \vdots & \vdots & \ddots & \vdots \\ 0 & 0 & 0 & \cdots & u_{n,n} \end{pmatrix}$$

$$= \begin{pmatrix} a_{1,1} & a_{1,2} & a_{1,3} & \cdots & a_{1,n} \\ a_{2,1} & a_{2,2} & a_{2,3} & \cdots & a_{2,n} \\ a_{3,1} & a_{3,2} & a_{3,3} & \cdots & a_{3,n} \\ \vdots & \vdots & \vdots & \ddots & \vdots \\ a_{n,1} & a_{n,2} & a_{n,3} & \cdots & a_{n,n} \end{pmatrix}. \tag{5-34}$$

We may regard this matrix equation as a system of n^2 equations. However, the number of unknowns is (n^2+n), since $\boldsymbol{L}$ and $\boldsymbol{U}$ each has $n(n+1)/2$ nonvanishing elements. This means that we have the freedom to assign values to n matrix elements among those of $\boldsymbol{L}$ and $\boldsymbol{U}$.

We shall make use of the freedom to simplify the decomposition as much as possible. A common choice is to give the value 1 to all the diagonal elements of $\boldsymbol{L}$. That is, for $i = 1, 2, \ldots, n$, we have the assignment

$$\ell_{i,i} = 1. \tag{5-35}$$

Once this is done, Eq. (5-34) may be used to calculate all the other $n(n-1)/2$ elements $\ell_{i,j}$, for $j = 1$ to $(i-1)$ and $i = 1$ to n, as well as the $n(n+1)/2$ elements $u_{i,j}$, for $i = 1$ to n and $j = i$ to n. In principle, we can obtain them using one of the methods for a system of linear equations discussed earlier. However, such an approach negates the original purpose of using an LU-decomposition to simplify the solution for a system of linear equations. In fact, we shall soon see that, with the proper method, the LU-decomposition is much faster on a computer.

Because of the particular shapes of the matrices $\boldsymbol{L}$ and $\boldsymbol{U}$, solutions of $\ell_{i,j}$ and $u_{i,j}$ in terms of $a_{i,j}$ can be worked out in a simple way. The equality expressed by Eq. (5-34) may be stated in terms of individual matrix elements as

$$\sum_m \ell_{i,m} u_{m,j} = a_{i,j}. \tag{5-36}$$

For $i = j = 1$, all $\ell_{i,m} = 0$ except for $m = 1$. As a result, the sum has only one term,

$$\ell_{1,1} u_{1,1} = a_{1,1}.$$

Since $\ell_{1,1}$ is chosen to be 1 in Eq. (5-35), we obtain at once

$$u_{1,1} = a_{1,1}.$$

In fact, this relation applies to all $u_{i,j}$ for $i = 1$ and $j = 1, 2, \ldots, n$,

$$u_{1,j} = a_{1,j}. \tag{5-37}$$

This gives us all the elements in the first row of $\boldsymbol{U}$.

At this moment we shall only make use of the value of $u_{1,1}$. For the convenience of pivoting, to which we shall return later, the calculations will be carried out one complete column at a time for all the elements of both $\boldsymbol{U}$ and $\boldsymbol{L}$. Let us now consider the elements in the first column. There is only one nonvanishing element for $\boldsymbol{U}$, and it is already given by Eq. (5-37). For $j = 1$, Eq. (5-36) reduces to the form

$$\ell_{i,1} u_{1,1} = a_{i,1} \qquad \text{for} \qquad i = 2, 3, \ldots, n.$$

This gives us a solution for all the elements in the first column of $\boldsymbol{L}$

$$\ell_{i,1} = \frac{1}{u_{1,1}} a_{i,1} \qquad \text{for} \qquad i = 2, 3, \ldots, n. \tag{5-38}$$

Note that we have no further use for the elements $a_{j,1}$ for $j = 1$ to n in the first column of $\boldsymbol{A}$. As a result, we can use their locations in the computer memory to store the elements in the first columns of both $\boldsymbol{U}$ and $\boldsymbol{L}$. The only exception is the diagonal element of $\boldsymbol{L}$. Since it is chosen to be unity, there is no need to have its value retained in the computer. This leaves exactly n nonvanishing elements to be kept, one for $\boldsymbol{U}$ and $(n-1)$ for $\boldsymbol{L}$. As we shall see later, this is useful in minimizing the storage requirement to carry out LU-decomposition.

With all the elements in the first column known, we can proceed to solve for those in the second column. First, we shall apply Eq. (5-37) to calculate the elements in the rest of the first row of $\boldsymbol{U}$. Once this is done, we can apply Eq. (5-36) for $i = 2$ to produce the equation for all the elements in the second row of $\boldsymbol{U}$:

$$u_{2,j} = a_{2,j} - \ell_{2,1}u_{1,j} \qquad \text{for} \qquad j = 2, 3, \ldots, n. \tag{5-39}$$

Similarly, by applying Eq. (5-36) for $j = 2$, we obtain the elements of the second column of $\boldsymbol{L}$:

$$\ell_{i,2} = \frac{1}{u_{2,2}}(a_{i,2} - \ell_{i,1}u_{1,2}) \qquad \text{for} \qquad i = 3, 4, \ldots, n. \tag{5-40}$$

The values for all the elements in the second column of both $\boldsymbol{L}$ and $\boldsymbol{U}$ are now available.

Let us work out explicitly the equations for one more column before giving the general result. For the elements of $\boldsymbol{U}$ in the third row, we can again invoke Eq. (5-36) with $i = 3$. This gives us the result

$$u_{3,j} = a_{3,j} - (\ell_{3,1}u_{1,j} + \ell_{3,2}u_{2,j}) \tag{5-41}$$

for $j = 3, 4, \ldots, n$. For now, we need only to make use of $u_{3,3}$. The elements in the third column of $\boldsymbol{L}$ may be obtained from Eq. (5-36) with $j = 3$. That is,

$$\ell_{i,3} = \frac{1}{u_{3,3}}\{a_{i,3} - (\ell_{i,1}u_{1,3} + \ell_{i,2}u_{2,3})\}$$

for $i = 4, 5, \ldots, n$. Besides $a_{i,3}$, the input required for $\ell_{i,3}$ are $\ell_{i,1}$, already found from Eq. (5-38), $\ell_{i,2}$ from (5-40), $u_{1,3}$ from (5-37), and $u_{2,3}$ from (5-39).

It is now clear what the general result should look like. We shall maintain the procedure of completing the calculation for all the elements one column at a time for both $\boldsymbol{U}$ and $\boldsymbol{L}$, starting from the first column onward. By extending the arguments leading to Eq. (5-37) for $i = 1$, (5-39) for $i = 2$, and (5-41) for $i = 3$, we obtain from Eq. (5-36) the general result for the elements of $\boldsymbol{U}$ in column k:

$$u_{\alpha,k} = a_{\alpha,k} - \sum_{m=1}^{\alpha-1} \ell_{\alpha,m}u_{m,k} \tag{5-42}$$

for $\alpha = 1, 2, \ldots, k$. To conform with Eq. (5-37) for $k = 1$, the sum must vanish if the upper limit is less than 1.

To calculate $u_{i,k}$, we need as input the values of $\ell_{i,m}$ for $m < k$ (since $m < i$ and $i \leq k$) and $u_{m,k}$ for $m < i$, in addition to $a_{i,k}$. All the elements $\ell_{i,j}$ for $j < k$ are already available from the calculations of the previous rows. As for the necessary input from the elements of $\boldsymbol{U}$, we note that none is needed to obtain $u_{1,k}$. Once $u_{1,k}$ is available, we have enough information to calculate $u_{2,k}$. This, in turn, allows us to obtain $u_{3,k}$, and so on. In this way we always have all the necessary input if the calculations are carried out in the correct order, one complete column for both $\boldsymbol{U}$ and $\boldsymbol{L}$ before proceeding to the next one. Within each column, the calculation starts with first element of $\boldsymbol{U}$ and proceeds down the column one element at a time until we come to the end at the diagonal element.

Once all the elements of $\boldsymbol{U}$ in column k are known, we move to those of $\boldsymbol{L}$ in the same column. Again, by substituting α for i and k for j in Eq. (5-36), we obtain the equation for the elements of $\boldsymbol{L}$ in column k:

$$\ell_{\alpha,k} = \frac{1}{u_{k,k}}\Big(a_{\alpha,k} - \sum_{m=1}^{k-1} \ell_{\alpha,m}u_{m,k}\Big) \qquad \text{for} \quad \alpha = 1, 2, \ldots, (k-1). \tag{5-43}$$

As in Eq. (5-42), it is necessary that the sum on the right side vanish if the upper limit is less than 1 in order for the expression to agree with Eq. (5-38). Since $m < k$, all the elements of $\boldsymbol{L}$ required as input to the calculation are those in the earlier columns. Similarly, the required elements from $\boldsymbol{U}$ are in the present column and they have been calculated already, as we saw in the previous paragraph.

So far, we have assumed that it is possible to achieve *LU*-decomposition of a matrix $\boldsymbol{A}$ as long as it is not singular. In fact, such a condition may be relaxed. A proof of the existence theorem for *LU*-decomposition can be found on page 154 of Dahlquist and Björck.[15]

Pivoting It is important to note here that each step of the calculation requires the result of the previous step. For example, in Eq. (5-42), the element $u_{\alpha,k}$ is found from the values of $u_{m,k}$ for $m < \alpha$, as well as the elements of $\boldsymbol{L}$ in the earlier columns. Similarly, to calculate $\ell_{\alpha,k}$ in Eq. (5-43), we need the values of all the elements of $\boldsymbol{U}$ in the same column, as well as those of $\boldsymbol{L}$ in the previous columns. In calculations of this type, the stability of the solution can be a problem and the errors in each step become cumulative in such a way that hardly any significant figures are left at the end. Pivoting is one way to improve the accuracy and to stabilize the solution.

In applying the Gauss-Jordan elimination earlier to reduce a determinant, we have selected the element with the largest absolute value in the remaining part of the determinant as the pivot. This was an efficient approach since the pivot appears in the denominator as, for example, in Eq. (5-13). In *LU*-decomposition, we have an analogous situation. In Eq. (5-43), a diagonal element of $\boldsymbol{U}$ appears in the denominator. As a result, it may be tempting to choose the largest element of $\boldsymbol{U}$ in a column or row as the pivot. Before we decide to do this, two further considerations must be taken into account.

The first is that the value of a particular element of $\boldsymbol{U}$ can be increased or decreased by multiplying a constant factor to a whole row without changing the solution

Box 5-4 Subroutine LUDCP(A,N,DET,L_ROW,NDMN)
LU* decomposition of matrix *A

Argument list:

- A: $\boldsymbol{A}$ on input; on output, $\boldsymbol{U}$ in upper triangle and $\boldsymbol{L}$, lower triangle.
- N: Order of the matrix.
- DET: Value of the determinant.
- L_ROW: List of row order.
- NDMN: Dimension of the arrays.

Initialization:

(a) Set the value of the determinant to be DET $= 1$.
(b) Zero L_ROW as the original positions of rows.
(c) Store the element with the largest absolute value in each row as A_MAX.

1. For $k = 1$ to n, carry out the following for all n columns:
   (a) Calculate all the elements of $\boldsymbol{U}$ in column k except the diagonal one:
      (i) Obtain the value of $u_{i,k}$ using Eq. (5-42).
      (ii) Store the result in the location for $a_{i,k}$.
   (b) Calculate the elements of $\boldsymbol{L}$ in column k:
      (i) Calculate $S_{i,k} = \sum_{m=1}^{k-1} \ell_{i,m} u_{m,k}$ for $i = k$ to n.
      (ii) Store $a_{i,k} - S_{i,k}$ in the location for $a_{i,k}$.
      (iii) Use as pivot the largest $|S_{i,k}|$ weighted by the largest element of the row.
      (iv) If the pivot is zero, the matrix is singular. Return DET $= 0$.
      (v) Bring the pivot to the diagonal by interchanging rows.
      (vi) Change the sign of the determinant for each row interchange.
      (vii) Store the original row number of the pivot in L_ROW(k).
      (viii) Multiply the value of the pivot by the determinant.
      (ix) Divide the sums obtained in (a) by $u_{k,k}$ to get the elements of $\boldsymbol{L}$.
2. Return DET, $\boldsymbol{L}$ and $\boldsymbol{U}$ stored in A, and the order of the rows in L_ROW.

of the system of linear equations. As a result, the criterion for selecting a pivoting element depends not only on the magnitude of the element itself but also on its size relative to the other elements. One way to incorporate the relative size into the selection criterion is to choose the largest element in each row as the weighting factor. We shall see soon how to put this point into practice.

The second consideration concerns efficiency – to find the pivot without having to do a large amount of work. Since the calculation is sequential, we do not know the value of the next element of $\boldsymbol{U}$ without first completing all the calculations on both $\boldsymbol{L}$ and $\boldsymbol{U}$. Thus, the weighting factor for selecting a pivot changes as each element is processed. As a result, the optimum solution requires the pivot to be selected essentially from scratch for each element to be reduced. This is a rather time-consuming process and a compromise must be made.

A good way is to select the pivot only from the elements in the column. The choice of calculating by column rather than by row also has the advantage that the upper limit of the summation in Eq. (5-43) is the same for all the elements of $\boldsymbol{L}$ in the column. Furthermore, for column k, the quantity inside the parentheses on the right side of the equation is the same as that of $u_{k,k}$ if it is the diagonal element. As a result, if we limit our choice of the pivot for column k to be among the elements in the lower part of the column, $i \geq k$ where i is the row index, very little additional work is required. In other words, for column k, all the elements $u_{i,k}$ up to $i = (k-1)$ are calculated using Eq. (5-42). Next we calculate the values inside the parentheses of the right side of Eq. (5-43) for $i = k$ to n. Among the latter $(n-k+1)$ quantities, the one with the largest absolute value, weighted by the element with largest absolute value in row k of $\boldsymbol{A}$, is taken as the pivot for this column. This element is now put in the diagonal position of column k as the element $u_{k,k}$. To achieve this, it may be necessary to interchange all the elements of two rows. Note also that, since the search for the pivot takes place only among elements $u_{i,k}$ for $i > k$, row interchanges do not affect any elements with $i < k$. This is important in the way the elements of $\boldsymbol{U}$ and $\boldsymbol{L}$ are stored, as we shall see in the next paragraph. The only work remaining now is to use the pivot as the denominator for all the quantities inside the parentheses on the right side of Eq. (5-43) we calculated earlier and make the quotients as the elements of $\boldsymbol{L}$ for column k.

Storage considerations We shall now turn to the question of storage. As mentioned earlier, once all the elements of a particular column of $\boldsymbol{U}$ and $\boldsymbol{L}$ are calculated, the elements in the same column of $\boldsymbol{A}$ are no longer needed. Since the calculation is carried out in the order from left to right one column at a time, and for each column, from the top to the bottom one element at a time, the elements of $\boldsymbol{A}$ can be discarded one after another, as the reduction of $\boldsymbol{A}$ to $\boldsymbol{L}$ and $\boldsymbol{U}$ progresses. The array for $\boldsymbol{A}$ may therefore be used to store both $\boldsymbol{U}$ and $\boldsymbol{L}$. Since one is an upper triangular matrix and the other is a lower triangular matrix, the only conflict arises at the diagonal. This is resolved by not storing the diagonal elements of $\boldsymbol{L}$, which, as we have seen earlier, are taken to be unity. One efficient way to arrange the storage is to use the upper triangle of $\boldsymbol{A}$, the locations for $a_{i,j}$ for $j \geq i$, for the elements of $\boldsymbol{U}$ and the rest, the locations for $a_{i,j}$ for $j > i$, for the elements of $\boldsymbol{L}$. The only additional information required is the order of the rows of $\boldsymbol{L}$, as they may be modified because of the interchanges made for pivoting purposes. It is simplest to put this information in a separate one-dimensional array and to take care of the actual permutations at the stage of making use of the $\boldsymbol{L}$ and $\boldsymbol{U}$ matrices. The algorithm is given in Box 5-4.

Forward and backward substitution Once the matrices $\boldsymbol{L}$ and $\boldsymbol{U}$ are found, we need to apply them to solve the original system of linear equations we are interested in. As we have seen earlier, two separate steps are required. The first is the forward substitution of Eq. (5-30) to find the auxiliary vector $\boldsymbol{Z}$ from $\boldsymbol{L}$ and the input vector $\boldsymbol{Y}$. With the elements of $\boldsymbol{Z}$, we can use Eq. (5-31) to solve for $\boldsymbol{X}$ by back substitution. The algorithm is given in Box 5-5. It is basically a straightforward calculation except for the interchanges made on the rows of $\boldsymbol{L}$. This is taken care of by permuting the order of the elements of $\boldsymbol{Y}$ before applying Eq. (5-31).

Box 5-5 Subroutine FBSBS(A,L_ROW,Y,N,NDMN)
Forward and back substitution after *LU* decomposition

Argument list:

A: Upper triangle for $\boldsymbol{U}$, lower triangle (except diagonal) for $\boldsymbol{L}$.
L_ROW: Order of rows in $\boldsymbol{L}$ and $\boldsymbol{U}$.
Y: Input matrix $\boldsymbol{Y}$.
N: Number of linear equations.
NDMN: Dimension of the arrays.

1. Forward substitution for $\boldsymbol{LZ} = \boldsymbol{Y}$ of Eq. (5-30). For $i = 1$ to n:
   (a) Use L_ROW to interchange elements of $\boldsymbol{Y}$ to the permuted order.
   (b) Calculate the elements of $\boldsymbol{Z}$ using Eq. (5-33).
   (c) Store the value of z_i in the location for y_i.
2. Back substitution for $\boldsymbol{UX} = \boldsymbol{Z}$ of Eq. (5-25). For $i = n$ backward to 1:
   (a) For $i = n$, let $x_n = z_n$. For $i < n$, let $x_i = (z_i - \sum_{m=i+1}^{n} u_{i,m} z_m)/u_{i,i}$.
   (b) Divide the elements by the diagonal and store them in the locations for y_i.
3. Return the array $\boldsymbol{Y}$ as the solution $\boldsymbol{X}$.

The computer program for LU-decomposition may be checked against by multiplying $\boldsymbol{L}$ with $\boldsymbol{U}$ and observing if we can reconstruct the original matrix $\boldsymbol{A}$ to sufficient accuracy. Alternatively, we can combine it with the forward and back substitution step and check the resulting solution of the linear equation in the same way as we did in the previous section.

5-3 Matrix approach to the eigenvalue problem

In quantum mechanics, the starting point for many problems is often the time-independent Schrödinger equation,

$$\left(-\frac{\hbar^2}{2\mu}\nabla^2 + V\right)\psi_\alpha = E_\alpha \psi_\alpha \tag{5-44}$$

where the first term on the left side represents the kinetic energy of the system with μ as the reduced mass. The second term comes from the potential energy, with V describing the interaction among different components of the system. Both the eigenvalue E_α and the eigenfunction ψ_α are to be found by solving the equation.

There are basically three ways to approach the eigenvalue problem. The first was given earlier in Chapter 4. When the potential has a simple form, such as in the case of a harmonic oscillator, the solution may be expressed in terms of some known function, such as the Laguerre polynomials in §4-4. In practice, there are only a limited number of cases where such analytic solutions are possible. A second approach is to solve Eq. (5-44) as a differential equation using numerical methods, and we shall discuss this type of approach in Chapter 8. A third way is solve the

Schrödinger equation by matrix methods. This is our primary goal for the next few sections of this chapter.

Basis states The first step in taking a matrix approach to solve Eq. (5-44) is to find a complete set of states $\phi_1, \phi_2, \ldots, \phi_n$ such that any function of interest in the problem may be expressed as a linear combination of these basis states. In particular, an eigenvector of Eq. (5-44) may be written in the form

$$\psi_\alpha = \sum_{i=1}^{n} C_{\alpha,i}\,\phi_i \tag{5-45}$$

where the coefficients $C_{\alpha,i}$ express the eigenvector ψ_α in terms of ϕ_i. For mathematical convenience, we shall choose the basis states $\{\phi_i\}$ such that each one is normalized to unity and different states are orthogonal to each other. That is,

$$\int \phi_i^* \phi_j \, d\tau = \delta_{i,j} \tag{5-46}$$

where ϕ_i^* is the complex conjugate of ϕ_i. The integral is taken over all the independent variables and

$$\delta_{i,j} = \begin{cases} 1 & \text{for } i = j \\ 0 & \text{otherwise} \end{cases}$$

is the Kronecker delta. To accommodate variables that are discrete, such as angular momentum, we can use the bra ($\langle$) and ket ($\rangle$) notation of Dirac and express the integration as

$$\langle \phi_i | \phi_j \rangle = \delta_{i,j}.$$

In principle, the total number of basis states n can be infinite; in practice, if n is too large, the problem is not tractable numerically. In addition to restrictions imposed by the size of the available memory, numerical accuracy and computational time also constrain the size of matrices we can handle.

To obtain realistic solutions in a restricted space, it is useful to choose the basis based on reasonable physical grounds. For example, if we wish to use a slightly more realistic potential than the one-dimensional harmonic oscillator of Eq. (4-1), we can include a correction term that is proportional to the fourth power of the displacement. The potential takes on the form

$$V(x) = \frac{1}{2}\mu\omega^2 x^2 + \epsilon\hbar\omega\Big(\frac{\mu\omega}{\hbar}\Big)^2 x^4 \tag{5-47}$$

where the constant ϵ is usually a small, dimensionless quantity. For such an *anharmonic oscillator*, the eigenfunctions are no longer those of Eq. (4-9), namely

$$\phi_m(\rho) = \frac{1}{\sqrt{2^m m! \sqrt{\pi}}}\, e^{-\rho^2/2} H_m(\rho) \tag{5-48}$$

where $H_m(\rho)$ is the Hermite polynomial of degree m and $\rho = x\sqrt{\mu\omega/\hbar}$. This comes from the fact that

$$\int_{-\infty}^{+\infty} \phi_k(x)\, x^4 \phi_\ell(x)\, dx \neq 0 \qquad \text{for} \qquad |k-\ell| = 0, 2, 4, \ldots. \tag{5-49}$$

On the other hand, $\{\phi_i(x)\}$ remains to be a complete set of states in the one-dimensional space of interest to us. As a result, the eigenfunctions of an anharmonic oscillator may be expressed as linear combinations of $\phi_\alpha(x)$.

For small ϵ, each eigenfunction $\psi_\alpha(\rho)$ is likely to be dominated by a single component of ϕ_i. Let us label ψ_α in such a way that the main contribution to it comes from ϕ_α. In this way, the primary effect of the anharmonic terms ϵx^4 is to introduce small admixtures with $i = \alpha \pm 2$, as we can see from Eq. (5-49). Other components, with $i = \alpha \pm 4$, $\alpha \pm 6, \ldots$, are also possible but their contributions are likely to be even smaller. If our interest is in ψ_α, it may be possible to ignore these higher-order terms and approximate the solution using a basis consisting of three states, $\phi_\alpha(x)$, $\phi_{\alpha-2}(x)$, and $\phi_{\alpha+2}(x)$. The problem of the anharmonic oscillator is now reduced to a fairly simple one, consisting of finding a linear combination of these three basis states that satisfies the Schrödinger equation.

The example illustrates the common situation where a part of the Hamiltonian can be solved easily, as we have done for the harmonic oscillator part of the example above. If the remainder of the Hamiltonian is small, the complete problem can be approximated to high accuracies by diagonalizing the matrix in only a small part of the complete space.

The eigenvalue problem Once a basis is selected, the eigenvalue problem is reduced essentially to one of finding the expansion coefficients $C_{\alpha,i}$ of Eq. (5-45). This we can see by having both sides of Eq. (5-44) multiplied by ϕ_j^* and integrating over all the independent variables. In bra-ket notation , the result may be written in the form

$$\langle \phi_j | \hat{H} | \psi_\alpha \rangle = E_\alpha \langle \phi_j | \psi_\alpha \rangle \tag{5-50}$$

where $\hat{H}$ is the Hamiltonian operator,

$$\hat{H} = -\frac{\hbar^2}{2\mu}\nabla^2 + V.$$

Both the eigenvalue E_α and the corresponding eigenvector ψ_α are not known at this stage of calculation. Since E_α is a number (rather than an operator), it may be taken outside the integration, as we have done on the right side of Eq. (5-50).

To put Eq. (5-50) into a matrix equation, we can make use of Eq. (5-45) and replace ψ_α by a linear combination of basis states ϕ_i. On applying the orthogonality relation of Eq. (5-46), we obtain an algebraic equation for the expansion coefficients:

$$\sum_{i=1}^{n} H_{j,i} C_{\alpha,i} = E_\alpha C_{\alpha,j} \qquad \alpha = 1, 2, \ldots, n \tag{5-51}$$

where we have made use of the Kronecker delta to eliminate the summation on the right side and

$$H_{j,i} \equiv \langle \phi_j | \hat{H} | \phi_i \rangle.$$

In matrix notation, we have

$$\begin{pmatrix} H_{1,1} & H_{1,2} & \cdots & H_{1,n} \\ H_{2,1} & H_{2,2} & \cdots & H_{2,n} \\ \vdots & \vdots & \ddots & \vdots \\ H_{n,1} & H_{n,2} & \cdots & H_{n,n} \end{pmatrix} \begin{pmatrix} C_{\alpha,1} \\ C_{\alpha,2} \\ \vdots \\ C_{\alpha,n} \end{pmatrix} = E_\alpha \begin{pmatrix} C_{\alpha,1} \\ C_{\alpha,2} \\ \vdots \\ C_{\alpha,n} \end{pmatrix}. \tag{5-52}$$

This can be put into the form of a system of n linear equations for $C_{\alpha,i}$,

$$\begin{pmatrix} (H_{1,1} - E_\alpha) & H_{1,2} & \cdots & H_{1,n} \\ H_{2,1} & (H_{2,2} - E_\alpha) & \cdots & H_{2,n} \\ \vdots & \vdots & \ddots & \vdots \\ H_{n,1} & H_{n,2} & \cdots & (H_{n,n} - E_\alpha) \end{pmatrix} \begin{pmatrix} C_{\alpha,1} \\ C_{\alpha,2} \\ \vdots \\ C_{\alpha,n} \end{pmatrix} = 0. \tag{5-53}$$

Since the right side is zero, a solution exits only if

$$\det \begin{vmatrix} (H_{1,1} - E_\alpha) & H_{1,2} & \cdots & H_{1,n} \\ H_{2,1} & (H_{2,2} - E_\alpha) & \cdots & H_{2,n} \\ \vdots & \vdots & \ddots & \vdots \\ H_{n,1} & H_{n,2} & \cdots & (H_{n,n} - E_\alpha) \end{vmatrix} = 0.$$

This is known as the characteristic equation and the eigenvalues are the roots of this equation. Once the value of E_α is known, the expansion coefficients $C_{\alpha,1}$, $C_{\alpha,2}, \ldots$, $C_{\alpha,n}$ may be found, for example, by solving the system of linear equations Eq. (5-53).

Matrix diagonalization A different way of approaching the problem posed by Eq. (5-52) is to find a transformation matrix $\boldsymbol{U}$ such that the similarity transformation

$$\boldsymbol{U}^{-1}\boldsymbol{H}\boldsymbol{U} = \boldsymbol{E} \tag{5-54}$$

results in a diagonal matrix $\boldsymbol{E}$. The process of reducing a matrix $\boldsymbol{H}$ to a diagonal form is called *diagonalization*.

In this section, we shall be dealing only with real symmetric matrices. This stems from the fact that quantum mechanical operators associated with physical observables are Hermitian. As a result, their matrices are also Hermitian. Such matrices can always be put into a form that is real and symmetric. The Hamiltonian operator is a member of this class, having the property

$$\boldsymbol{H} = \boldsymbol{H}^\dagger \qquad \text{or} \qquad H_{i,j} = H^*_{j,i}.$$

It is relatively easy to solve problems involving real symmetric matrices. However, the power of matrix methods extends far beyond such simple cases, and we shall return later at the end of this chapter to examine briefly a few of the more advanced topics on matrices.

Another way of describing the eigenvalue problem of Eq. (5-44) is to think hypothetically first in terms of a set of basis states made of the eigenvectors $\{\psi_\alpha\}$. In this basis, the Hamiltonian matrix is diagonal, with all off-diagonal elements equal to zero and the diagonal elements consisting of the eigenvalues. This can be seen by left multiplying both sides of Eq. (5-44) by the complex conjugate of an arbitrary eigenvector ψ_β. On integrating over all the independent variables, the result may be expressed in Dirac notation as

$$\langle\psi_\beta|\hat{H}|\psi_\alpha\rangle = \langle\psi_\beta|E_\alpha|\psi_\alpha\rangle. \tag{5-55}$$

Again, we can take E_α outside the matrix element, as it is not an operator. Furthermore, if we assume that the eigenvectors are normalized to unity and orthogonal to each other, the right side is reduced to

$$\langle\psi_\beta|E_\alpha|\psi_\alpha\rangle = E_\alpha\langle\psi_\beta|\psi_\alpha\rangle = E_\alpha\delta_{\alpha,\beta}.$$

The left side of Eq. (5-55) cannot be simplified further here since $\hat{H}$ is an operator. The relation given by the equation now takes on the form

$$\langle\psi_\beta|\hat{H}|\psi_\alpha\rangle = E_\alpha\,\delta_{\alpha,\beta}. \tag{5-56}$$

That is, in a basis made of eigenvectors, the Hamiltonian matrix is diagonal with the eigenvalues as the diagonal matrix elements. Such a basis is known as the eigenvector representation.

Our goal in matrix diagonalization is to find a transformation that takes us from an arbitrary basis $\{\phi_i\}$ to $\{\psi_\alpha\}$, made of eigenvectors. In terms of matrices, this may be stated as

$$\boldsymbol{U}\boldsymbol{\phi} = \boldsymbol{\psi} \tag{5-57}$$

where $\boldsymbol{U}$ is the transformation matrix and

$$\boldsymbol{\phi} = \begin{pmatrix} \phi_1 \\ \phi_2 \\ \phi_3 \\ \vdots \\ \phi_n \end{pmatrix} \qquad\qquad \boldsymbol{\psi} = \begin{pmatrix} \psi_1 \\ \psi_2 \\ \psi_3 \\ \vdots \\ \psi_n \end{pmatrix}.$$

From Eq. (5-45), it is easy to see that the elements of the transformation matrix $\boldsymbol{U}$ are the expansion coefficients of ψ_α in terms of ϕ_i:

$$\boldsymbol{U} = \begin{pmatrix} C_{1,1} & C_{1,2} & C_{1,3} & \cdots & C_{1,n} \\ C_{2,1} & C_{2,2} & C_{2,3} & \cdots & C_{2,n} \\ \vdots & \vdots & \vdots & \ddots & \vdots \\ C_{n,1} & C_{n,2} & C_{n,3} & \cdots & C_{n,n} \end{pmatrix}.$$

In other words, we have the relation $U_{\alpha,\beta} = C_{\alpha,\beta}$. For this reason, we can also say that $\boldsymbol{U}$ is made of all the eigenvectors in the space. We shall defer to books on quantum mechanics to show that $\boldsymbol{U}$ is unitary

$$\boldsymbol{U}^{\dagger} = \boldsymbol{U}^{-1} \tag{5-58}$$

and Eq. (5-57) is equivalent to Eq. (5-54).

Two-dimensional rotations The most straightforward way to diagonalize a matrix is the Jacobi method. The basic operation in this approach consists of a series of rotations among two columns and two rows. Consider first a real symmetric (2×2) matrix of the form

$$\boldsymbol{A} = \begin{pmatrix} a_{1,1} & a_{1,2} \\ a_{2,1} & a_{2,2} \end{pmatrix}$$

where $a_{2,1} = a_{1,2}$. A two-dimensional similarity transformation may be used to reduce the two off-diagonal elements to zero.

Because of the unitary requirement of Eq. (5-58), there is only one parameter in the transformation. We shall take this to be θ, which may be interpreted as the angle of rotation in the two-dimensional space. The transformation matrix may be expressed in terms of trigonometry functions in the following way:

$$\boldsymbol{T} = \begin{pmatrix} \cos\theta & \sin\theta \\ -\sin\theta & \cos\theta \end{pmatrix} \equiv \left\{ \begin{matrix} c & s \\ -s & c \end{matrix} \right\} \tag{5-59}$$

where, to shorten the notation, we have used the abbreviations $c \equiv \cos\theta$ and $s \equiv \sin\theta$. Since $\boldsymbol{T}^{-1} = \boldsymbol{T}^{\dagger} = \widetilde{\boldsymbol{T}}$, a similarity transformation of $\boldsymbol{A}$ using $\boldsymbol{T}$ gives us the result

$$\begin{aligned} \boldsymbol{A}' &= \boldsymbol{T}^{-1}\boldsymbol{A}\boldsymbol{T} \\ &= \begin{pmatrix} c & -s \\ s & c \end{pmatrix} \begin{pmatrix} a_{1,1} & a_{1,2} \\ a_{2,1} & a_{2,2} \end{pmatrix} \begin{pmatrix} c & s \\ -s & c \end{pmatrix} \\ &= \begin{pmatrix} c^2 a_{1,1} + s^2 a_{2,2} - 2cs a_{1,2} & cs(a_{1,1} - a_{2,2}) + (c^2 - s^2) a_{1,2} \\ cs(a_{1,1} - a_{2,2}) + (c^2 - s^2) a_{1,2} & s^2 a_{1,1} + c^2 a_{2,2} + 2cs a_{1,2} \end{pmatrix}. \end{aligned} \tag{5-60}$$

Our goal here is to find a value of θ such that the two off-diagonal matrix elements of $\boldsymbol{A}'$ vanish. From Eq. (5-60), we see that such a requirement is equivalent to the condition

$$\frac{cs}{c^2 - s^2} = \frac{a_{1,2}}{a_{2,2} - a_{1,1}}. \tag{5-61}$$

In terms of the rotation angle θ, the same condition may be written as

$$\tan 2\theta = \frac{2\cos\theta\sin\theta}{\cos^2\theta - \sin^2\theta} = \frac{2a_{1,2}}{a_{2,2} - a_{1,1}}. \tag{5-62}$$

Although the above equations are expressed in terms of trigonometry functions, the actual relations between the various quantities are algebraic. In other words, the sine, cosine, and tangent functions do not need to appear explicitly in the numerical calculations. This is important from an efficiency point of view, as trigonometry functions are, in general, time consuming to calculate.

Jacobi method The same type of rotation used in Eq. (5-60) for a two-dimensional matrix may also be applied to larger ones. Consider a real symmetric matrix $\boldsymbol{A}$ of some finite dimension n. If we wish to put to zero a particular pair of off-diagonal matrix elements, for example $a_{k,\ell}$ and $a_{\ell,k}$, we can apply a rotation between columns and rows k and ℓ, using a transformation similar to that given by Eq. (5-59). The major difference is that, instead of (2×2), the transformation matrix $\boldsymbol{T}$ is $(n \times n)$, as given in Eq. (5-63) below.

To bring the complete matrix to a diagonal form, we need to make a series of such rotations, each of which annihilates a pair of off-diagonal matrix elements. Let us assume that after $(m-1)$ such rotations the original matrix is transformed into $\boldsymbol{A}(m-1)$ with matrix elements $\{a_{i,j}^{(m-1)}\}$. Since all the transformations are unitary, the matrix $\boldsymbol{A}(m-1)$ remains real and symmetric. Let us further assume that, in the next rotation, the pair of off-diagonal elements we wish to annihilate are $a_{k,\ell}^{(m-1)}$ and $a_{\ell,k}^{(m-1)}$ (with $\ell \neq k$). The transformation matrix to achieve this goal may be written in the form

$$\boldsymbol{T}(m) = \begin{pmatrix} 1 & 0 & 0 & \cdots & 0 & \cdots & 0 & \cdots & 0 \\ 0 & 1 & 0 & \cdots & 0 & \cdots & 0 & \cdots & 0 \\ 0 & 0 & 1 & \cdots & 0 & \cdots & 0 & \cdots & 0 \\ \vdots & \vdots & \vdots & \ddots & \vdots & \ddots & \vdots & \ddots & 0 \\ 0 & 0 & 0 & \cdots & \cos\theta^{(m)} & \cdots & \sin\theta^{(m)} & \cdots & 0 \\ \vdots & \vdots & \vdots & \ddots & \vdots & \ddots & \vdots & \ddots & 0 \\ 0 & 0 & 0 & \cdots & -\sin\theta^{(m)} & \cdots & \cos\theta^{(m)} & \cdots & 0 \\ \vdots & \vdots & \vdots & \ddots & \vdots & \ddots & \vdots & \ddots & 0 \\ 0 & 0 & 0 & \cdots & 0 & \cdots & 0 & \cdots & 1 \end{pmatrix}. \tag{5-63}$$

Similar to Eq. (5-62), if we choose the angle $\theta^{(m)}$ such that

$$\tan 2\theta^{(m)} = \frac{2a_{k,\ell}^{(m-1)}}{a_{\ell,\ell}^{(m-1)} - a_{k,k}^{(m-1)}} \tag{5-64}$$

elements of the matrix

$$\boldsymbol{A}(m) = \boldsymbol{T}^{-1}(m)\boldsymbol{A}(m-1)\boldsymbol{T}(m) \tag{5-65}$$

take on the values given in the next paragraph.

First, the elements outside columns and rows k and ℓ are unchanged by the transformation, meaning

$$a_{i,j}^{(m)} = a_{i,j}^{(m-1)} \qquad \text{for} \quad i \neq k \neq \ell \quad \text{and} \quad j \neq k \neq \ell. \tag{5-66}$$

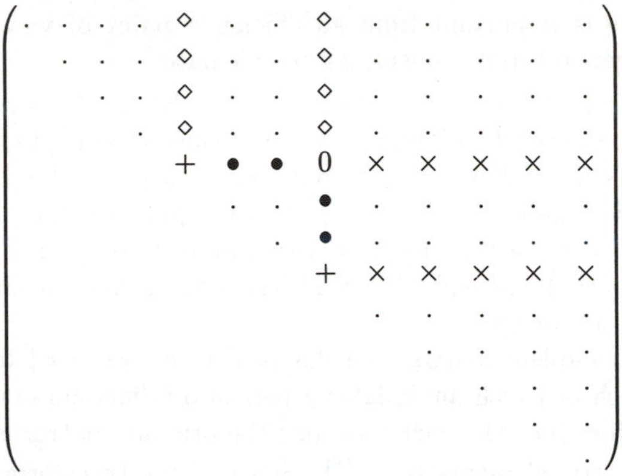

Figure 5-2: Three categories of elements in the upper triangle of a real symmetric matrix after a rotation between columns and rows k and ℓ in the Jacobi method. Elements outside k and ℓ $(\cdot)$ are unchanged [Eq. (5-63)]; diagonal elements for k and ℓ $(+)$ are transformed by Eq. (5-67); and elements in columns and rows k and ℓ $(\diamond, \bullet, \text{and } \times)$ are transformed according to Eq. (5-68).

Second, for the matrix elements in columns and rows k and ℓ, we have the analogous situation as Eq. (5-60):

$$\begin{aligned} a_{k,\ell}^{(m)} &= a_{\ell,k}^{(m)} = 0 \\ a_{k,k}^{(m)} &= c^2 a_{k,k}^{(m-1)} + s^2 a_{\ell,\ell}^{(m-1)} - 2cs a_{k,\ell}^{(m-1)} \\ a_{\ell,\ell}^{(m)} &= s^2 a_{k,k}^{(m-1)} + c^2 a_{\ell,\ell}^{(m-1)} + 2cs a_{k,\ell}^{(m-1)}. \end{aligned} \tag{5-67}$$

Here the symbols c and s are cosine and sine functions of the angle $\theta^{(m)}$,

$$c \equiv \cos\theta^{(m)} \qquad\qquad s \equiv \sin\theta^{(m)}.$$

Third, since the transformation affects all the elements in columns k and ℓ, as well as in rows k and ℓ, the elements in these columns and rows, other than those already given above, are transformed in the following way:

$$\begin{aligned} a_{i,k}^{(m)} &= a_{k,i}^{(m)} = c a_{i,k}^{(m-1)} - s a_{i,\ell}^{(m-1)} \\ a_{i,\ell}^{(m)} &= a_{\ell,i}^{(m)} = s a_{i,k}^{(m-1)} + c a_{i,\ell}^{(m-1)} \qquad \text{for} \qquad i \neq k \neq \ell. \end{aligned} \tag{5-68}$$

These three types of matrix elements may be distinguished by the aid of the diagram shown in Fig. 5-2.

Because of this third group of transformations, the Jacobi method must be an iterative one. This we can see from the fact that off-diagonal matrix elements that vanished in an earlier transformation may become nonzero by a subsequent transformation. The process can, however, be shown to be convergent, as successive transformations increase the absolute values of the diagonal elements and decrease those of the off-diagonal ones. We shall leave the proof of this point to texts on numerical analysis.

Eigenvectors Let us turn our attention to the transformation matrix that gives us the eigenvectors. Symbolically, we may represent a Jacobi diagonalization as the following similarity transformation

$$\boldsymbol{D} = \boldsymbol{V}^{-1}\boldsymbol{A}\boldsymbol{V} \tag{5-69}$$

where the transformation matrix $\boldsymbol{V}$ is the product of a series of rotations, each of which has the form given in Eq. (5-63). The total number of rotations required depends on the dimension of the matrix as well as the accuracy we wish to achieve. In principle, the transformation should start with the pair of off-diagonal elements having the largest magnitude and then proceed to the pair with the next largest value. In practice, this is seldom carried out, as the search itself is a time-consuming process. If the matrix dimension is not too large, it is fairly efficient just to carry out the reduction row by row (or column by column), starting from the first off-diagonal element in each row (or column). Since the matrix is symmetric, it is only necessary to examine either the upper or the lower triangle. When we encounter an element whose absolute value is larger than some small number, determined by the accuracy we wish to achieve in the calculation, a rotation is applied to reduce it (and the corresponding one in the other half of the matrix) to zero. After all the $n(n-1)/2$ off-diagonal elements in the matrix are examined in this way, we must go back to the beginning and repeat the process. The diagonalization is completed if all the off-diagonal elements can be taken as zero within the preset tolerance for error. In general, several iterations are necessary.

The actual calculations of the elements of $\boldsymbol{V}$ in Eq. (5-69) are very similar to that carried out in Eqs. (5-67) and (5-68). Let us write the form of $\boldsymbol{V}$ after applying m rotations to $\boldsymbol{A}$ as

$$\boldsymbol{A}(m) = \boldsymbol{V}^{-1}(m)\boldsymbol{A}\,\boldsymbol{V}(m) \tag{5-70}$$

where $\boldsymbol{A}$ is the original matrix we wish to diagonalize. Using Eq. (5-65), the same relation may also be expressed in the form

$$\boldsymbol{A}(m) = \boldsymbol{T}^{-1}(m)\big\{\boldsymbol{V}^{-1}(m-1)\boldsymbol{A}\boldsymbol{V}(m-1)\big\}\boldsymbol{T}(m)$$

where $\boldsymbol{T}(m)$ is a rotation given by Eq. (5-63). Comparing this with Eq. (5-70), we obtain the result

$$\boldsymbol{V}(m) = \boldsymbol{V}(m-1)\,\boldsymbol{T}(m). \tag{5-71}$$

In terms of matrix elements, this is equivalent to

$$\begin{aligned} v_{i,j}^{(m)} &= v_{i,j}^{(m-1)} \qquad\qquad \text{for} \quad j \neq k \neq \ell \\ v_{i,k}^{(m)} &= c v_{i,k}^{(m-1)} - s v_{i,\ell}^{(m-1)} \\ v_{i,\ell}^{(m)} &= s v_{i,k}^{(m-1)} + c v_{i,\ell}^{(m-1)} \end{aligned} \tag{5-72}$$

where $i = 1, 2, \ldots, n$. The starting point of the iterative relation given by Eq. (5-71) is the unit matrix

$$\boldsymbol{V}(0) \equiv \mathbf{1}$$

with elements $v_{i,j}^{(0)} = \delta_{i,j}$.

Improving efficiency For computational efficiency, it is advantageous to avoid calculating the trigonometry functions explicitly. We can achieve this by the following manipulation. What we need in Eqs. (5-67), (5-68), and (5-72) for a rotation are the values of $\cos\theta^{(m)}$ and $\sin\theta^{(m)}$ with the angle $\theta^{(m)}$ given by Eq. (5-64). To simplify the notation, we shall use R to represent the inverse of the ratio of matrix elements on the right side of Eq. (5-64). That is,

$$R \equiv \frac{a_{\ell,\ell}^{(m-1)} - a_{k,k}^{(m-1)}}{2a_{k,\ell}^{(m-1)}}.$$

Using the trigonometry relation

$$\tan 2\theta^{(m)} = \frac{2\tan\theta^{(m)}}{1 - \tan^2\theta^{(m)}}$$

we obtain an equation for $\tan\theta^{(m)}$ in terms of R,

$$\frac{2\tan\theta^{(m)}}{1 - \tan^2\theta^{(m)}} = \frac{1}{R}$$

or

$$\tan^2\theta^{(m)} + 2R\tan\theta^{(m)} - 1 = 0. \tag{5-73}$$

The two solutions for such a quadratic equation for $\tan\theta$ are

$$\tan\theta^{(m)} = -R \pm \sqrt{R^2 + 1}.$$

Alternatively, we can change Eq. (5-73) into a quadratic equation for $\cot\theta$ and obtain the solution

$$\cot\theta^{(m)} = R \pm \sqrt{R^2 + 1}$$

or

$$\tan\theta^{(m)} = \frac{1}{R \pm \sqrt{R^2 + 1}}.$$

Any one of these possibilities satisfies Eq. (5-64), and a choice among them can be made based on numerical accuracy considerations.

In general, the accuracy of such calculations is best if, in step m, we can manage to transform the off-diagonal matrix element $a_{k,\ell}^{(m-1)}$ to zero with only minimum changes in the other elements. To achieve this, we must take $\cos\theta^{(m)}$ to be as close to 1, and $\sin\theta^{(m)}$ as close to 0, as possible. The advantages of such choices are quite obvious if we examine Eqs. (5-67) and (5-68). In terms of the rotation angle, we want the absolute value of $\theta^{(m)}$ to be as close to 0 as possible. The optimum choice is then

$$\tan\theta^{(m)} = \frac{\mathcal{S}(R)}{|R| + \sqrt{R^2+1}} \tag{5-74}$$

where

$$\mathcal{S}(R) = \begin{cases} +1 & \text{if } R > 0 \\ -1 & \text{if } R < 0. \end{cases}$$

From the value of $\tan\theta^{(m)}$, we obtain those for $\cos\theta^{(m)}$ and $\sin\theta^{(m)}$ that actually enter into the calculations for the rotation as

$$\cos\theta^{(m)} = \frac{1}{\sqrt{1+\tan^2\theta^{(m)}}} \qquad\qquad \sin\theta^{(m)} = \cos\theta^{(m)}\tan\theta^{(m)}. \tag{5-75}$$

In this approach, two square roots are used to replace the slower calculations of an arc tangent, a sine, and a cosine.

To improve the numerical accuracy further, we shall reformulate slightly the three equations, i.e., (5-67), (5-68), and (5-72). Analogous to Eq. (5-61), we find that

$$\frac{a_{k,\ell}^{(m-1)}}{a_{\ell,\ell}^{(m-1)} - a_{k,k}^{(m-1)}} = \frac{cs}{c^2 - s^2}$$

where we have returned to the use of the abbreviations of $c = \cos\theta^{(m)}$, $s = \sin\theta^{(m)}$, and $t = \tan\theta^{(m)}$. When we substitute this relation into the second member of Eq. (5-67) and make use of the fact that $c^2 + s^2 = 1$, we obtain

$$\begin{aligned} a_{k,k}^{(m)} &= a_{k,k}^{(m-1)} + s^2\{a_{\ell,\ell}^{(m-1)} - a_{k,k}^{(m-1)}\} - 2cs a_{k,\ell}^{(m-1)} \\ &= a_{k,k}^{(m-1)} + \Big\{s^2\frac{c^2-s^2}{cs} - 2cs\Big\} a_{r,s}^{(m-1)} \\ &= a_{k,k}^{(m-1)} - t\, a_{k,\ell}^{(m-1)}. \end{aligned} \tag{5-76}$$

Similarly, the third member of Eq. (5-67) may be put into the form

$$a_{\ell,\ell}^{(m)} = a_{\ell,\ell}^{(m-1)} + t\, a_{k,\ell}^{(m-1)}. \tag{5-77}$$

The idea here is that the off-diagonal matrix elements are small in general, especially during the later iterations. Since we have already kept the absolute value of t as

Box 5-6 Subroutine JCBDG(A,V,E,N,NDMN,ACC)
Jacobi diagonalization of a real symmetric matrix

Argument list:

- A: Input matrix, upper triangle destroyed in the calculation.
- V: Array for output eigenvectors $\boldsymbol{V}$.
- E: Array for output eigenvalues $\boldsymbol{E}$.
- N: Number of rows and columns of the matrices.
- NDMN: Dimension of the arrays $\boldsymbol{A}$ and $\boldsymbol{V}$ in the calling program.
- ACC: Size of matrix element below which it is considered to be 0.

Initialization:

- (a) Set the maximum number of iterations allowed to 50.
- (b) Set $\boldsymbol{E}$ equal to the diagonal matrix elements of $\boldsymbol{A}$.
- (c) Set $\boldsymbol{V}$ equal to a unit matrix.
- (d) Zero the iteration counter I_t.

1. Increase I_t by 1 and set the requirement for further iteration to .FALSE..
2. If any $|a_{i,j}| >$ ACC in the upper triangle, then:
   - (a) Set the condition for needing further iteration to .TRUE..
   - (b) Calculate t, c, s, and τ in Eqs. (5-74), (5-75), and (5-78).
   - (c) Modify the diagonal matrix elements using Eqs. (5-76) and (5-77).
   - (d) Put the off-diagonal matrix element $a_{i,j}$ to zero.
   - (e) Change other off-diagonal elements in columns and rows i and j using Eqs. (5-78) and (5-79) by three loops to protect diagonal and elements below:
     - (i) Elements in columns i and j ($\diamond$ in Fig. 5-2).
     - (ii) Remaining off-diagonal elements in row i between columns i and j, and in column j between rows i and j ($\bullet$ in Fig. 5-2).
     - (iii) Remaining elements in rows i and j ($\times$ in Fig. 5-2).
   - (f) Update the matrix $\boldsymbol{V}$ using Eq. (5-80).
3. Check for the need of a further iteration:
   - (a) If not, return eigenvalues in $\boldsymbol{E}$ and eigenvectors in $\boldsymbol{V}$.
   - (b) If so, go back to step 2 up to the maximum number of iterations allowed.

small as possible, the above two equations represent small adjustments of the diagonal matrix elements.

The other elements modified by the same rotation are also in columns k and ℓ and in rows k and ℓ. It is useful to rearrange the terms in the first member of Eq. (5-68) into the following form:

$$\begin{aligned} a_{i,k}^{(m)} = a_{k,i}^{(m)} &= a_{i,k}^{(m-1)} - s a_{i,\ell}^{(m-1)} - (1-c) a_{i,k}^{(m-1)} \\ &= a_{i,k}^{(m-1)} - s\left\{ a_{i,\ell}^{(m-1)} + \frac{(1-c)}{s} a_{i,k}^{(m-1)} \right\} \\ &= a_{i,k}^{(m-1)} - s\{ a_{i,\ell}^{(m-1)} + \tau a_{i,k}^{(m-1)} \} \end{aligned} \tag{5-78}$$

where

$$\tau \equiv \frac{1-c}{s} = \frac{s}{1+c} = \tan\frac{1}{2}\theta^{(m)}.$$

Similarly, the second member of Eq. (5-68) may be put into the form

$$a_{i,\ell}^{(m)} = a_{\ell,i}^{(m)} = a_{i,\ell}^{(m-1)} + s\{a_{i,k}^{(m-1)} - \tau a_{i,\ell}^{(m-1)}\}. \tag{5-79}$$

Since τ is kept as small as possible, similar to s, the second terms of both Eqs. (5-78) and (5-79) are small in general. By the same token, the calculations required to obtain the matrix elements of $\boldsymbol{V}$ in the last two members of Eq. (5-72) may be put into similar forms:

$$\begin{aligned} v_{i,k}^{(m)} &= v_{i,k}^{(m-1)} - s\{v_{i,\ell}^{(m-1)} + \tau v_{i,k}^{(m-1)}\} \\ v_{i,\ell}^{(m)} &= v_{i,\ell}^{(m-1)} + s\{v_{i,k}^{(m-1)} - \tau v_{i,\ell}^{(m-1)}\}. \end{aligned} \tag{5-80}$$

These are the steps carried out in the algorithm for the Jacobi method of matrix diagonalization given in Box 5-6. Many other improvements are possible, but we shall not incorporate them here so as to keep the method relatively simple and still reasonably efficient.

To check the computer program, we can start by finding out whether the eigenvectors are normalized to unity and orthogonal to each other, namely satisfying the relation

$$\boldsymbol{V}^{-1}\boldsymbol{V} = \boldsymbol{V}^{\dagger}\boldsymbol{V} = \mathbf{1}.$$

In terms of matrix elements, this is the same as

$$\sum_k v_{k,i}\, v_{k,j} = \delta_{i,j}.$$

These results provide us with an idea of the accuracy achieved in the calculation. A more stringent test is to apply the transformations given by $\boldsymbol{V}$ to the original matrix $\boldsymbol{A}$ and see if Eq. (5-69) is satisfied within the numerical accuracy required. This we can do by multiplying the original matrix $\boldsymbol{A}$ on the left by $\boldsymbol{V}^{-1}$ and on the right by $\boldsymbol{V}$. The result should be a diagonal matrix with the values of the diagonal elements equal to the eigenvalues produced by the diagonalization process. Because of the unitary property, there is no need to invert $\boldsymbol{V}$ to obtain $\boldsymbol{V}^{-1}$. Using the fact that $\boldsymbol{V}^{-1} = \widetilde{\boldsymbol{V}}$, the relation in terms of matrix elements takes on the form

$$(\boldsymbol{V}^{-1}\boldsymbol{A}\boldsymbol{V})_{i,j} = \sum_{k=1}^{n}\sum_{\ell=1}^{n} v_{k,i}\, a_{k,\ell} v_{\ell,j}.$$

Note the order of the subscripts in the first v. They are the matrix elements of $\widetilde{\boldsymbol{V}}$ and are given by the relation $(\widetilde{\boldsymbol{V}})_{i,j} = v_{j,i}$.

5-4 Tridiagonalization method

The Jacobi method of matrix diagonalization suffers from the fact that the method is an iterative one. In general, the number of iterations required to reduce a symmetric matrix of dimension n to a diagonal one is proportional to n. Since in each iteration the number of off-diagonal elements to be annihilated is proportional to n^2, the method requires the order of n^3 operations, thus making it rather impractical for large matrices.

The method of Givens The main weakness in the Jacobi method lies in the fact that each two-dimensional rotation that annihilates one pair of off-diagonal matrix elements affects the values of others in the same column and row. As a result, matrix elements put to zero in one operation may become nonzero at a later stage when another pair is being reduced. In the method of Givens, each rotation remains to be between two columns and two rows. However, they are arranged in such a way that the elements already reduced to zero remain zero. The transformation matrix for each rotation retains the form of Eq. (5-62) for the Jacobi method. However, for later discussions, we shall now label each rotation matrix $\boldsymbol{T}$ by the indices of the two columns and two rows it affects:

$$\boldsymbol{T}(k,\ell) = \begin{array}{c} \\ \\ \\ \\ k \\ \\ \ell \\ \\ \\ \end{array} \overset{\qquad\qquad\qquad\quad k \qquad\quad \ell}{\begin{pmatrix} 1 & 0 & 0 & \cdots & 0 & \cdots & 0 & \cdots & 0 \\ 0 & 1 & 0 & \cdots & 0 & \cdots & 0 & \cdots & 0 \\ 0 & 0 & 1 & \cdots & 0 & \cdots & 0 & \cdots & 0 \\ \vdots & \vdots & \vdots & \ddots & \vdots & \ddots & \vdots & \ddots & 0 \\ 0 & 0 & 0 & \cdots & c & \cdots & s & \cdots & 0 \\ \vdots & \vdots & \vdots & \ddots & \vdots & \ddots & \vdots & \ddots & 0 \\ 0 & 0 & 0 & \cdots & -s & \cdots & c & \cdots & 0 \\ \vdots & \vdots & \vdots & \ddots & \vdots & \ddots & \vdots & \ddots & 0 \\ 0 & 0 & 0 & \cdots & 0 & \cdots & 0 & \cdots & 1 \end{pmatrix}}.$$

All the diagonal matrix elements of $\boldsymbol{T}(k,\ell)$ equal 1 except

$$t_{k,k} = t_{\ell,\ell} = c.$$

The off-diagonal matrix elements are given by the following relations:

$$t_{i,j} = \begin{cases} s & \text{for } i = k \text{ and } j = \ell \\ -s & \text{for } i = \ell \text{ and } j = k \\ 0 & \text{otherwise.} \end{cases}$$

The values of c and s are chosen such that the similarity transformation

$$\boldsymbol{A}' = \boldsymbol{T}^{-1}(k,\ell)\boldsymbol{A}\,\boldsymbol{T}(k,\ell) \tag{5-81}$$

reduces the off-diagonal element $a_{k-1,\ell}$ to zero (and through symmetry, $a_{\ell,k-1} = 0$) by taking a linear combination of $a_{k-1,\ell}$ with the element $a_{k-1,k}$. The major difference

$$
\boldsymbol{J} = \begin{pmatrix}
d_1 & f_2 & 0 & 0 & 0 & \cdots & 0 & 0 & 0 \\
f_2 & d_2 & f_3 & 0 & 0 & \cdots & 0 & 0 & 0 \\
0 & f_3 & d_3 & f_4 & 0 & \cdots & 0 & 0 & 0 \\
0 & 0 & f_4 & d_4 & f_5 & \cdots & 0 & 0 & 0 \\
\vdots & \vdots & \vdots & \vdots & \vdots & \ddots & \vdots & \vdots & \vdots \\
0 & 0 & 0 & 0 & 0 & \cdots & d_{n-2} & f_{n-1} & 0 \\
0 & 0 & 0 & 0 & 0 & \cdots & f_{n-1} & d_{n-1} & f_n \\
0 & 0 & 0 & 0 & 0 & \cdots & 0 & f_n & d_n
\end{pmatrix}
$$

Figure 5-3: A symmetric, tridiagonal matrix with all the elements zero except n diagonal matrix elements $d_1, d_2, \ldots, d_n$, and $(n-1)$ off-diagonal elements $f_2, f_3, \ldots, f_n$, just above and below the diagonal.

from the Jacobi method is that the reduction makes use of the superdiagonal element rather than the diagonal element. As a result, it is not possible to bring the matrix to a diagonal form. Instead, the end result is a tridiagonal matrix, having the form shown in Fig. 5-3. As we shall see in the next section, it is a relatively simple matter to diagonalize a matrix once it is brought to the tridiagonal form.

A more efficient method to bring a matrix into a tridiagonal form is that of Householder. Instead of operating on one element at a time, a transformation is carried out in such a way that all the elements in a row $a_{k,\ell}$, for $\ell = k+2, k+3, \ldots, n$, are reduced to zero. Although the method applies to any Hermitian matrices, we shall restrict our interest here to real symmetric matrices.

Inner product and outer product Before going into the details of Householder method, it is useful to differentiate between two types of matrix multiplications, inner and outer products. A column vector with n elements

$$
\boldsymbol{a} = \begin{pmatrix} a_1 \\ a_2 \\ a_3 \\ \vdots \\ a_n \end{pmatrix}
$$

may be thought as a matrix having only a single column. The transpose of $\boldsymbol{a}$ is a row vector, meaning a matrix consisting of a single row of n elements

$$
\widetilde{\boldsymbol{a}} = \begin{pmatrix} a_1 & a_2 & a_3 & \cdots & a_n \end{pmatrix}.
$$

There are two different kinds of matrix product we can form and they are best illustrated using a column matrix and a row matrix. Let $\boldsymbol{b}$ be a column vector also of dimension n. According to the rule of matrix multiplication, the following product is a scalar quantity,

$$\tilde{\boldsymbol{a}}\boldsymbol{b} = \begin{pmatrix} a_1 & a_2 & a_3 & \cdots & a_n \end{pmatrix} \begin{pmatrix} b_1 \\ b_2 \\ b_3 \\ \vdots \\ b_n \end{pmatrix} = \sum_{i=1}^{n} a_i b_i.$$

That is, it is invariant under a transformation of the basis. Such a product of two vectors is known as the scalar or *inner product.* The inner product of $\tilde{\boldsymbol{a}}$ with $\boldsymbol{a}$,

$$\|\boldsymbol{a}\|^2 = \tilde{\boldsymbol{a}}\boldsymbol{a} = \sum_{i=1}^{n} a_i^2 \tag{5-82}$$

gives the square of the norm of the vector $\boldsymbol{a}$.

We can form another kind of product between a column matrix and a row matrix. If $\boldsymbol{a}$ is a column matrix with m elements and $\boldsymbol{b}$ one with n elements, the product

$$\boldsymbol{a}\tilde{\boldsymbol{b}} = \begin{pmatrix} a_1 \\ a_2 \\ a_3 \\ \vdots \\ a_m \end{pmatrix} \begin{pmatrix} b_1 & b_2 & b_3 & \cdots & b_n \end{pmatrix} = \begin{pmatrix} a_1b_1 & a_1b_2 & a_1b_3 & \cdots & a_1b_n \\ a_2b_1 & a_2b_2 & a_2b_3 & \cdots & a_2b_n \\ a_3b_1 & a_3b_2 & a_3b_3 & \cdots & a_3b_n \\ \vdots & \vdots & \vdots & \ddots & \vdots \\ a_mb_1 & a_mb_2 & a_mb_3 & \cdots & a_mb_n \end{pmatrix}$$

is a rectangular matrix of m rows and n columns. To distinguish from inner products, this type of matrix product is known as the *outer product.*

The method of Householder Consider the following n-dimensional, square matrix

$$\boldsymbol{P} = \boldsymbol{1} - \beta \boldsymbol{u}\tilde{\boldsymbol{u}} \tag{5-83}$$

where $\boldsymbol{1}$ is a unit matrix and β is a constant to be determined later. The column vector $\boldsymbol{u}$ of dimension n is constructed out of another one, $\boldsymbol{a}$, in the following way,

$$\boldsymbol{u} = \boldsymbol{a} \pm \|\boldsymbol{a}\| \boldsymbol{\epsilon}_1 \tag{5-84}$$

where

$$\boldsymbol{\epsilon}_1 = \begin{pmatrix} 1 \\ 0 \\ 0 \\ \vdots \\ 0 \end{pmatrix}$$

is a column matrix with only the first element nonvanishing and equal to 1 [and the other $(n-1)$ elements equal to 0]. In other words,

$$\boldsymbol{u} = \begin{pmatrix} a_1 \pm \|\boldsymbol{a}\| \\ a_2 \\ a_3 \\ \vdots \\ a_n \end{pmatrix}.$$

By making a suitable choice of the value of β, it is possible to show that the column vector formed out of the product of $\boldsymbol{P}$ with $\boldsymbol{a}$ is a column vector with only the first element nonzero. This applies to both the plus and the minus signs in Eq. (5-84). As we shall see later, the choice of the sign can be used to our advantage in designing an algorithm to transform a matrix into tridiagonal form.

We can express $\tilde{\boldsymbol{u}}$, the transpose of $\boldsymbol{u}$, in terms of $\tilde{\boldsymbol{a}}$ and $\tilde{\boldsymbol{\epsilon}}_1$. This gives us

$$\boldsymbol{P}\boldsymbol{a} = \boldsymbol{a} - \beta\boldsymbol{u}(\tilde{\boldsymbol{a}} \pm \|\boldsymbol{a}\|\tilde{\boldsymbol{\epsilon}}_1)\boldsymbol{a} = \boldsymbol{a} - \beta\boldsymbol{u}(\|\boldsymbol{a}\|^2 \pm \|\boldsymbol{a}\|a_1). \tag{5-85}$$

Because of the definition of $\boldsymbol{u}$ given by Eq. (5-84), it is easy to see that the quantity inside the parentheses is half the square of the norm of $\boldsymbol{u}$:

$$\|\boldsymbol{u}\|^2 = (\tilde{\boldsymbol{a}} \pm \|\boldsymbol{a}\|\tilde{\boldsymbol{\epsilon}}_1)(\boldsymbol{a} \pm \|\boldsymbol{a}\|\boldsymbol{\epsilon}_1) = 2(\|\boldsymbol{a}\|^2 \pm \|\boldsymbol{a}\|a_1).$$

As a result, if we take

$$\beta = \frac{2}{\|\boldsymbol{u}\|^2} \tag{5-86}$$

we find that

$$\boldsymbol{P}\boldsymbol{a} = \mp\|\boldsymbol{a}\|\boldsymbol{\epsilon}_1. \tag{5-87}$$

Since all the elements of $\boldsymbol{\epsilon}_1$ are zero except the first one $(= 1)$, only the first element of the column vector $\boldsymbol{P}\boldsymbol{a}$ is different from zero. Similarly, the row vector

$$\tilde{\boldsymbol{a}}\tilde{\boldsymbol{P}} = \mp\|\boldsymbol{a}\|\tilde{\boldsymbol{\epsilon}}_1 \tag{5-88}$$

has only one nonvanishing element. This particular property of $\boldsymbol{P}$ may be used to transform all the elements of a row or a column of a matrix to a tridiagonal form.

Reduction of one row and one column to tridiagonal form The property of $\boldsymbol{P}$ given by Eqs. (5-87) and (5-88) allows us to perform a similarity transformation on an arbitrary real symmetric matrix and change one of its columns and rows into a tridiagonal form. To reduce the whole matrix $\boldsymbol{A}$ of dimension n to a tridiagonal form, it takes $(n-1)$ such transformations. To introduce the method, it is convenient to carry out the transformations starting with the first row and column. Similar to Eq. (5-83), the transformation matrix has the form

$$\boldsymbol{T}(1) = \mathbf{1} - \beta\boldsymbol{w}^{(1)}\tilde{\boldsymbol{w}}^{(1)} \tag{5-89}$$

where the column vector

$$\boldsymbol{w}^{(1)} = \begin{pmatrix} 0 \\ a_{2,1} \pm \alpha \\ a_{3,1} \\ a_{4,1} \\ \cdots \\ a_{n,1} \end{pmatrix} \tag{5-90}$$

is essentially made of the elements of the first column (or row) of $\boldsymbol{A}$. The only exceptions are that the first element is zero and that the second element contains an additional factor α given by

$$\alpha^2 = \sum_{i=2}^{n} a_{i,1}^2.$$

The structure of $\boldsymbol{w}^{(1)}$ is similar to that of $\boldsymbol{u}$ in Eq. (5-84), except that the vector is made of all the elements of the first column of the matrix $\boldsymbol{A}$ below the diagonal.

If we choose β in Eq. (5-89) in the same way as we did in Eq. (5-86) by taking

$$\beta = \frac{2}{\|\boldsymbol{w}^{(1)}\|^2} \tag{5-91}$$

the matrix $\boldsymbol{T}(1)$ is unitary:

$$\boldsymbol{T}^{-1}(1) = \widetilde{\boldsymbol{T}}(1) = \boldsymbol{T}(1).$$

This can be shown in the following way. First, we see that

$$\widetilde{\boldsymbol{T}}(1) = \widetilde{\mathbf{1}} - \beta\widetilde{\boldsymbol{w}^{(1)}\widetilde{\boldsymbol{w}}^{(1)}} = \mathbf{1} - \beta\widetilde{\widetilde{\boldsymbol{w}}}^{(1)}\widetilde{\boldsymbol{w}}^{(1)} = \mathbf{1} - \beta\boldsymbol{w}^{(1)}\widetilde{\boldsymbol{w}}^{(1)} = \boldsymbol{T}(1)$$

as $\widetilde{\widetilde{\boldsymbol{w}}}^{(1)}$, the transpose of $\widetilde{\boldsymbol{w}}^{(1)}$, is the same as $\boldsymbol{w}^{(1)}$. Note also that the product of $\widetilde{\boldsymbol{w}}^{(1)}$ with $\boldsymbol{w}^{(1)}$ is an outer product, as required by the definition of $\boldsymbol{T}(1)$ in Eq. (5-89). Second, we find that

$$\boldsymbol{T}^2(1) = \mathbf{1} - 2\beta\boldsymbol{w}^{(1)}\widetilde{\boldsymbol{w}}^{(1)} + \beta^2\boldsymbol{w}^{(1)}\widetilde{\boldsymbol{w}}^{(1)}\boldsymbol{w}^{(1)}\widetilde{\boldsymbol{w}}^{(1)}.$$

For the last term on the right, we can take first the product of the $\widetilde{\boldsymbol{w}}^{(1)}$ with the $\boldsymbol{w}^{(1)}$ on its right. This is an inner product, as $\boldsymbol{T}^2(1)$ is an inner product by definition. The result is a scalar quantity, the square of the norm of $\boldsymbol{w}^{(1)}$. Because of the choice of β made in Eq. (5-91), we have

$$\widetilde{\boldsymbol{w}}^{(1)}\boldsymbol{w}^{(1)} = \frac{2}{\beta}.$$

As a result, $\boldsymbol{T}^2(1) = \mathbf{1}$.. This means that we can write the similarity transformation of $\boldsymbol{A}$ using $\boldsymbol{T}(1)$ in the form $\boldsymbol{T}(1)\boldsymbol{A}\boldsymbol{T}(1)$.

If we think of $\boldsymbol{A}$ as a collection of n column vectors, we find that, by comparing with Eq. (5-87),

$$\begin{aligned}
\boldsymbol{A}' &= \boldsymbol{T}(1)\boldsymbol{A} \\
&= \begin{pmatrix} 1 & 0 & 0 & \cdots & 0 \\ 0 & 1-\beta w_2 w_2 & -\beta w_2 w_3 & \cdots & -\beta w_2 w_n \\ 0 & -\beta w_3 w_2 & 1-\beta w_3 w_3 & \cdots & -\beta w_3 w_n \\ 0 & \vdots & \vdots & \ddots & \vdots \\ 0 & -\beta w_n w_2 & -\beta w_n w_3 & \cdots & 1-\beta w_n w_n \end{pmatrix} \begin{pmatrix} a_{1,1} & a_{1,2} & a_{1,3} & \cdots & a_{1,n} \\ a_{2,1} & a_{2,2} & a_{2,3} & \cdots & a_{2,n} \\ a_{3,1} & a_{3,2} & a_{3,3} & \cdots & a_{3,n} \\ \vdots & \vdots & \vdots & \ddots & \vdots \\ a_{n,1} & a_{n,2} & a_{n,3} & \cdots & a_{n,n} \end{pmatrix} \\
&= \begin{pmatrix} a_{1,1} & a_{1,2} & a_{1,3} & \cdots & a_{1,n} \\ \mp\alpha & a'_{2,1} & a'_{2,3} & \cdots & a'_{2,n} \\ 0 & a'_{3,2} & a'_{3,3} & \cdots & a'_{3,n} \\ \vdots & \vdots & \vdots & \ddots & \vdots \\ 0 & a'_{n,2} & a'_{n,3} & \cdots & a'_{n,n} \end{pmatrix}
\end{aligned} \tag{5-92}$$

where w_k are the matrix elements of $\boldsymbol{w}^{(1)}$. We shall return later to the forms of the matrix elements $a'_{i,j}$ for both i and j greater than 1. For the moment, the important point is that the elements in the first row of the product matrix $\boldsymbol{A}'$ remain unchanged from those of $\boldsymbol{A}$, arising from the fact that the first element of $\boldsymbol{w}$ is zero. Note that the elements of the first column are now in a tridiagonal form.

By considering $\boldsymbol{A}'$ as a collection of n row vectors, we obtain, using Eq. (5-89),

$$\boldsymbol{A}'' = \boldsymbol{A}'\boldsymbol{T}(1) = \begin{pmatrix} a_{1,1} & \mp\alpha & 0 & \cdots & 0 \\ \mp\alpha & a''_{2,1} & a''_{2,3} & \cdots & a''_{2,n} \\ 0 & a''_{3,2} & a''_{3,3} & \cdots & a''_{3,n} \\ \vdots & \vdots & \vdots & \ddots & \vdots \\ 0 & a''_{n,2} & a''_{n,3} & \cdots & a''_{n,n} \end{pmatrix}. \tag{5-93}$$

Again, the first columns of $\boldsymbol{A}''$ and $\boldsymbol{A}'$ are the same because of the structure of $\boldsymbol{w}^{(1)}$.

The net result obtained from the previous two paragraphs is that the transformation $\boldsymbol{T}(1)\boldsymbol{A}\boldsymbol{T}(1)$ reduces $\boldsymbol{A}$ to that on the right side of Eq. (5-93), with the first column and row in the standard form of a tridiagonal matrix. We can now proceed to apply a second transformation. Instead of Eq. (5-90), we shall use

$$\boldsymbol{w}^{(2)} = \begin{pmatrix} 0 \\ 0 \\ a''_{3,2} \pm \alpha \\ a''_{4,2} \\ a''_{5,2} \\ \cdots \\ a''_{n,2} \end{pmatrix} \qquad \alpha = \sum_{i=3}^{n} (a''_{i,2})^2. \tag{5-94}$$

The new transformation matrix $\boldsymbol{T}(2)$, made of $\boldsymbol{w}^{(2)}$ in an analogous way as Eq. (5-89), reduces $\boldsymbol{A}''$ to a form with one more tridiagonal column and row.

We can generalize the procedure to one that transforms a particular row and column of an arbitrary real symmetric matrix $\boldsymbol{A}$ to tridiagonal form. If we apply such transformations $(m-1)$ times, starting from the first column and row as prescribed in Eqs. (5-92) and (5-93), the resulting matrix is a partially tridiagonal one,

$$\boldsymbol{A}(m-1) = \left(\begin{array}{cccccc|cccc} d_1 & f_2 & 0 & \cdots & 0 & 0 & 0 & 0 & \cdots & 0 \\ f_2 & d_2 & f_3 & \cdots & 0 & 0 & 0 & 0 & \cdots & 0 \\ 0 & f_3 & d_3 & \cdots & 0 & 0 & 0 & 0 & \cdots & 0 \\ \vdots & \vdots & \vdots & \ddots & \vdots & \vdots & \vdots & \vdots & \ddots & \vdots \\ 0 & 0 & 0 & \cdots & f_{m-1} & d_{m-1} & f_m & 0 & \cdots & 0 \\ \hline 0 & 0 & 0 & \cdots & 0 & f_m & a^{(m-1)}_{m,m} & a^{(m-1)}_{m,m+1} & \cdots & a^{(m-1)}_{m,n} \\ 0 & 0 & 0 & \cdots & 0 & 0 & a^{(m-1)}_{m+1,m} & a^{(m-1)}_{m+1,m+1} & \cdots & a^{(m-1)}_{m+1,n} \\ \vdots & \vdots & \vdots & \ddots & \vdots & \vdots & \vdots & \vdots & \ddots & \vdots \\ 0 & 0 & 0 & \cdots & 0 & 0 & a^{(m-1)}_{n,m} & a^{(m-1)}_{n,m+1} & \cdots & a^{(m-1)}_{n,n} \end{array}\right) \tag{5-95}$$

with $(m-1)$ columns and rows in tridiagonal form. To reduce this matrix to one with one more tridiagonal column and row, we follow Eq. (5-89) and use the transformation matrix

$$\boldsymbol{T}(m) = \mathbf{1} - \beta \boldsymbol{w}^{(m)} \widetilde{\boldsymbol{w}}^{(m)} \tag{5-96}$$

where the elements of the column vector $\boldsymbol{w}^{(m)}$ have the form

$$w_i^{(m)} = \begin{cases} 0 & \text{for } i \leq m \\ a_{m+1,m}^{(m-1)} \pm \alpha & \text{for } i = m+1 \\ a_{m,i}^{(m-1)} & \text{for } m+1 < i \leq n \end{cases} \tag{5-97}$$

with

$$\alpha = \sum_{j=m+1}^{n} \left\{ a_{m,j}^{(m-1)} \right\}^2.$$

Using Eqs. (5-95) and (5-96), it is easy to show that

$$\boldsymbol{T}(m)\boldsymbol{A}(m-1)\,\boldsymbol{T}(m) = \boldsymbol{A}(m) \tag{5-98}$$

where $\boldsymbol{A}(m)$ is a matrix with the first m columns and rows tridiagonal.

We can take Eq. (5-98) as a recursion relation to transform an arbitrary real symmetric matrix $\boldsymbol{A}$. The starting point of such a process is $\boldsymbol{A}(0) = \boldsymbol{A}$. The transformation matrix $\boldsymbol{T}(1)$ is given by Eq. (5-89), and it generates $\boldsymbol{A}(1)$ with the first column and row tridiagonal, as shown on the right side of Eq. (5-93). From $\boldsymbol{A}(1)$ we can produce $\boldsymbol{A}(2)$ using $\boldsymbol{T}(2)$ made of $\boldsymbol{w}^{(2)}$ of Eq. (5-94), and so on. After $(n-1)$ such similarity transformations, each having the form of Eq. (5-96), we arrive at a tridiagonal equivalent of the original $\boldsymbol{A}$.

It is useful to rewrite Eq. (5-98) in a form that is more convenient for practical calculations. Using Eq. (5-96), we have

$$\begin{aligned} \boldsymbol{A}(m) &= \left(\mathbf{1} - \beta \boldsymbol{w}^{(m)} \widetilde{\boldsymbol{w}}^{(m)}\right) \boldsymbol{A}(m-1) \left(\mathbf{1} - \beta \boldsymbol{w}^{(m)} \widetilde{\boldsymbol{w}}^{(m)}\right) \\ &= \boldsymbol{A}(m-1) - \beta \boldsymbol{A}(m-1) \boldsymbol{w}^{(m)} \widetilde{\boldsymbol{w}}^{(m)} - \beta \boldsymbol{w}^{(m)} \widetilde{\boldsymbol{w}}^{(m)} \boldsymbol{A}(m-1) \\ &\qquad + \beta^2 \boldsymbol{w}^{(m)} \widetilde{\boldsymbol{w}}^{(m)} \boldsymbol{A}(m-1) \boldsymbol{w}^{(m)} \widetilde{\boldsymbol{w}}^{(m)} \\ &= \boldsymbol{A}(m-1) - \boldsymbol{p} \widetilde{\boldsymbol{w}}^{(m)} - \boldsymbol{w}^{(m)} \widetilde{\boldsymbol{p}} + \beta \boldsymbol{w}^{(m)} \widetilde{\boldsymbol{p}} \boldsymbol{w}^{(m)} \widetilde{\boldsymbol{w}}^{(m)} \end{aligned} \tag{5-99}$$

where we have defined a column matrix

$$\boldsymbol{p} = \beta \boldsymbol{A}(m-1) \boldsymbol{w}^{(m)}. \tag{5-100}$$

Since $\boldsymbol{A}(m)$ is a symmetric matrix, we have

$$\widetilde{\boldsymbol{p}} = \beta \widetilde{\boldsymbol{w}}^{(m)} \widetilde{\boldsymbol{A}}(m-1) = \beta \widetilde{\boldsymbol{w}}^{(m)} \boldsymbol{A}(m-1).$$

Let us define another column matrix

$$\boldsymbol{q} = \boldsymbol{p} - \frac{1}{2}\beta\left(\widetilde{\boldsymbol{p}} \boldsymbol{w}^{(m)}\right) \boldsymbol{w}^{(m)} = \boldsymbol{p} - \frac{1}{2}\beta \boldsymbol{w}^{(m)} \left(\widetilde{\boldsymbol{p}} \boldsymbol{w}^{(m)}\right) \tag{5-101}$$

where $(\tilde{\boldsymbol{p}}\boldsymbol{w}^{(m)})$ is a scalar product. Using this, the final result of Eq. (5-99) may be written as

$$\begin{aligned}\boldsymbol{A}(m) &= \boldsymbol{A}(m-1) - \left\{\boldsymbol{p} - \frac{1}{2}\beta\boldsymbol{w}^{(m)}(\tilde{\boldsymbol{p}}\boldsymbol{w}^{(m)})\right\}\tilde{\boldsymbol{w}}^{(m)} - \boldsymbol{w}^{(m)}\left\{\tilde{\boldsymbol{p}} - \frac{1}{2}\beta(\tilde{\boldsymbol{p}}\boldsymbol{w}^{(m)})\tilde{\boldsymbol{w}}^{(m)}\right\} \\ &= \boldsymbol{A}(m-1) - \boldsymbol{q}\tilde{\boldsymbol{w}}^{(m)} - \boldsymbol{w}^{(m)}\tilde{\boldsymbol{q}}. \qquad (5\text{-}102)\end{aligned}$$

This is the form commonly employed in computer programs for Householder tridiagonalization .

It is also possible to start the tridiagonalization process from the last column and row, instead of the first column and row as we did above. In the place of Eq. (5-90), we can use a column matrix of the form

$$\boldsymbol{w}^{[n]} = \begin{pmatrix} a_{1,n} \\ a_{2,n} \\ \cdots \\ a_{n-2,n} \\ a_{n-1,n} \pm \alpha \\ 0 \end{pmatrix} \qquad \alpha = \sum_{i=1}^{n-1} a_{i,n}^2.$$

Let us name the transformation matrix constructed from $\boldsymbol{w}^{[n]}$ as $\boldsymbol{T}[n]$:

$$\boldsymbol{T}[n] = 1 - \beta\boldsymbol{w}^{[n]}\tilde{\boldsymbol{w}}^{[n]}.$$

It transforms the matrix $\boldsymbol{A}$ into one with column and row n tridiagonal,

$$\boldsymbol{A}[n] = \boldsymbol{T}[n]\boldsymbol{A}\boldsymbol{T}[n] = \begin{pmatrix} a_{1,1}^{[n]} & a_{2,1}^{[n]} & \cdots & a_{n-1,1}^{[n]} & 0 \\ a_{2,1}^{[n]} & a_{2,2}^{[n]} & \cdots & a_{n-1,2}^{[n]} & 0 \\ \vdots & \vdots & \ddots & \vdots & \vdots \\ a_{1,n-1}^{[n]} & a_{2,n-1}^{[n]} & \cdots & a_{n-1,n-1}^{[n]} & \mp\alpha \\ 0 & 0 & \cdots & \mp\alpha & a_{n,n}^{[n]} \end{pmatrix}.$$

The definition of β is the same as that given by Eq. (5-91). Next we apply a transformation using $\boldsymbol{T}[n-1]$, constructed in a similar way as $\boldsymbol{T}[n]$ with the two last elements in $\boldsymbol{w}^{[n-1]}$ equal to zero and using the elements of $\boldsymbol{A}[n]$ for the rest. The result is $\boldsymbol{A}[n-1]$ with columns and rows n and $(n-1)$ tridiagonal.

We can proceed in this way to reduce the matrix elements in column and row $(n-2)$, and so on, until the whole matrix is tridiagonal. The final result must be the same as that discussed earlier where the reduction is carried forward starting from column and row 1. The choice between these two approaches is usually based on the ground of numerical accuracy.

It is obvious from the discussions that, if there are large off-diagonal matrix elements in $\boldsymbol{A}$, it will be better if we try to reduce them to zeros as late as possible.

This stems from the fact that cancelations among large numbers have a tendency to reduce the numerical accuracy. Since every transformation modifies the values of all the elements in the part of the matrix that is not yet tridiagonal, the absolute magnitudes of the large elements will likely be reduced. As a result, the need for reducing large elements to zero may not materialize in practice. For this reason, it may be advantageous to interchange the columns and rows of $\boldsymbol{A}$ before the tridiagonalization step in such a way that the large off-diagonal matrix elements are, as far as possible, in the later columns and rows.

Most of the published matrix diagonalization programs involving tridiagonalization start the reduction from the last column and row. That is, if the dimension of the matrix is n, the first column and row to be reduced are the nth. In this case, the large elements should be put in the leftmost columns and uppermost rows, if possible. Alternatively, if the reduction starts from the first column and row, it is preferable to have the large elements, as far as possible, in the rightmost columns and bottommost rows. To distinguish between these two approaches, we have been using brackets to indicate the steps in transforming $\boldsymbol{A}$ to the tridiagonal form $\boldsymbol{J}$ starting from column and row n, and parentheses for those starting from column and row 1.

To improve the numerical accuracy further, it is useful to scale the matrix elements that enter into the calculation of $\boldsymbol{w}^{[m]}$. This comes from the observation that some of the off-diagonal matrix elements can be small, especially after the initial few steps. Since $\boldsymbol{w}^{[m]}$ depends only on the relative size of $a_{m,j}^{[m-1]}$ for $j = 1, 2, \ldots, (m-1)$, the transformation matrix $\boldsymbol{T}^{[m]}$ is not changed if we divide all $a_{m,j}^{[m-1]}$ by a scale factor s. One choice is to take s to be the sum over j of the absolute values of $a_{m,j}^{[m-1]}$. This is included as a part of the algorithm given in Box 5-7.

Transformation matrix So far we have concerned ourselves only with the process of changing the matrix $\boldsymbol{A}$ to a tridiagonal form. Our ultimate goal is to obtain the eigenvalues and the eigenvectors. To this end, it is also necessary to diagonalize the tridiagonal matrix. In the approach used here, the transformation matrix that brings the original matrix $\boldsymbol{A}$ to a diagonal one consists of two parts. The first part reduces $\boldsymbol{A}$ to a tridiagonal matrix $\boldsymbol{J}$ and the second part, to which we shall return in the next section, turns $\boldsymbol{J}$ into a diagonal matrix. Let us define a transformation matrix $\boldsymbol{R}$ that accomplishes the first part. That is,

$$\widetilde{\boldsymbol{R}}\boldsymbol{A}\boldsymbol{R} = \boldsymbol{J}. \tag{5-103}$$

From the discussions above, it is clear that, if we carry out the reduction backwards, starting from column and row n,

$$\boldsymbol{R} = \boldsymbol{T}[n]\boldsymbol{T}[n-1]\cdots\boldsymbol{T}[2]. \tag{5-104}$$

Alternatively, if we carry out the reduction starting from column and row 1, we have

$$\boldsymbol{R} = \boldsymbol{T}(1)\boldsymbol{T}(2)\cdots\boldsymbol{T}(n-1).$$

Box 5-7 Subroutine TRIDG(A,N,D,F,NDMN)
Householder tridiagonalization of a real symmetric matrix

Argument list:

A: Input real symmetric matrix $\boldsymbol{A}$. Output matrix $\boldsymbol{R}$ of Eq. (5-103).
N: Matrix dimension n.
D: Output diagonal elements of tridiagonal matrix $\boldsymbol{J}$.
F: Output superdiagonal matrix elements of $\boldsymbol{J}$.
NDMN: Dimension of the arrays in the calling program.

1. If $n = 1$, return $a_{1,1}$ as the eigenvalue and 1.0 as the eigenvector.
   If $n = 2$, use the Jacobi rotation of Eq. (5-60).
2. For $n > 2$, carry out the following steps for $m = n$ backwards to 2:
   (a) Find the scale factor $s = \sum_{j=1}^{m-1} |a_{m,j}|$.
   If $s = 0$, set $d_m = a_{m,m}$, $f_m = a_{m,m-1}$, $\boldsymbol{w}^{[m]} = 0$, and skip steps (b) to (h).
   (b) Scale the row of matrix elements and find the norm.
   (c) Calculate $\alpha^2 = \sum_{j=1}^{m-1} a_{m,j}^2$ and take the sign of α to be that of $-a_{m,m-1}$.
   (d) Construct $\boldsymbol{w}^{[m]}$ with $w_j = a_{m,j}$ for $j = 1$ to $(m-2)$,
   $w_{m-1} = a_{m,m-1} - \alpha$, $w_j = 0$ for $j = m$ to n, and $\|\boldsymbol{w}^{[m]}\|^2 = \alpha^2 - \alpha a_{m,m-1}$.
   (e) Store $\{w_i\}$ in array D.
   (f) Construct column matrix $\boldsymbol{p}$ of Eq. (5-100) and store the elements in array F.
   (g) Construct column matrix $\boldsymbol{q}$ of Eq. (5-101) and replace p_j by q_j in array F.
   (h) Reduce column m and row m of $\boldsymbol{A}$ to tridiagonal form using Eq. (5-102).
   (i) Store $a_{m,m}$ as d_m and $s * a_{m,m-1}$ as f_m.
3. Construct the transformation matrix $\boldsymbol{R}$ of Eq. (5-103) from $\boldsymbol{w}^{[m]}$:
   (a) Start with $m = 1$ and $r_{1,1} = 1$.
   (b) Obtain $\boldsymbol{R}[m]$ from $\boldsymbol{R}[m-1]$ and $\boldsymbol{w}^{[m]}$. Calculate only matrix elements $r_{i,j}$ for $i = 1$ to $(m-1)$ and $j = 1$ to $(m-1)$ using Eq. (5-106).
   (c) Let $r_{m,m} = 1$ and $r_{m,j} = r_{j,m} = 0$ for $j = 1$ to $(m-1)$.
   (d) Repeat steps (b) and (c) until $m = n$.
4. Return d_i in D, f_i in F, and $\boldsymbol{R}$ in A.

Note that there are only $(n-1)$ steps in either approach, as the matrix is tridiagonal by the end of step $(n-1)$. The matrix $\boldsymbol{R}$ is, however, not the eigenvector matrix, since we still have to diagonalize the tridiagonal matrix.

Storage considerations It is a relatively simple matter to obtain the transformation matrix $\boldsymbol{R}$ from the product of $\boldsymbol{T}[n], \boldsymbol{T}[n-1], \ldots, \boldsymbol{T}[2]$. However, to store $\boldsymbol{R}$, we need, in principle, a two-dimensional array of size n^2. The computer program will, then, require two such arrays, one to store the matrix $\boldsymbol{A}$ and the other for $\boldsymbol{R}$. As we shall see in the following discussion, it is possible to reduce the storage requirement to only one such array. The reduction by almost a factor of 2 is important for large n.

In the tridiagonalization process, the matrix $\boldsymbol{T}[k]$ is usually discarded at the end of step k. If we wish to construct the transformation matrix $\boldsymbol{R}$, the information on

$\boldsymbol{T}[k]$ must be stored before the next step in the reduction is carried out. At the same time, we also notice that, as each column and row of $\boldsymbol{A}$ are reduced to a tridiagonal form, only two nonzero matrix elements are left. The locations used to store the other off-diagonal matrix elements are no longer needed and may be used to keep information required to build the transformation matrix $\boldsymbol{R}$. Consider now step k of the reduction. The crucial information on $\boldsymbol{R}$ is contained in the column vector $\boldsymbol{w}^{[k]}$. The number of nonvanishing elements in $\boldsymbol{w}^{[k]}$ is $(k-1)$, exactly the same as the number of off-diagonal matrix elements annihilated by $\boldsymbol{T}[k]$. As a result, we have the correct amount of storage for $\boldsymbol{w}^{[k]}$ in the space "discarded" by the off-diagonal matrix elements of $\boldsymbol{A}[k-1]$. This is true in every step of the tridiagonalization process.

The space vacated by the discarded matrix elements of $\boldsymbol{A}$ is not suitable to store the transformation matrix $\boldsymbol{R}$ itself. Since the transformation of $\boldsymbol{A}$ into a tridiagonal matrix is carried out row by row (or column by column), there is not enough space to store $\boldsymbol{R}$ at the end of each step. This comes from the simple fact that $\boldsymbol{T}[i]$ is a square matrix of dimension n and, consequently, takes up more space than that vacated by the matrix elements of $\boldsymbol{A}$ in the transformation. However, all we need is $\boldsymbol{w}^{[k]}$, from which it is possible to reconstruct $\boldsymbol{R}$ at the end of the tridiagonalization step.

The reconstruction can also be carried out using the space originally occupied by $\boldsymbol{A}$. Let us use $\boldsymbol{R}[k]$ to represent the product of the first k of the $(n-1)$ number of $\boldsymbol{T}[i]$. In terms of $\boldsymbol{R}[k]$, Eq. (5-104) may be put into the form of a recursion relation

$$\boldsymbol{R}[k] = \boldsymbol{T}[k]\boldsymbol{R}[k-1]. \tag{5-105}$$

The starting point and the final result are, respectively,

$$\boldsymbol{R}[1] = \mathbf{1} \qquad\qquad \boldsymbol{R} = \boldsymbol{R}[n-1].$$

Since $\boldsymbol{T}[k]$ has the form

$$\begin{aligned}\boldsymbol{T}[k] &= \mathbf{1} - \beta \boldsymbol{w}^{[k]}\tilde{\boldsymbol{w}}^{[k]} \\ &= \begin{pmatrix} 1-\beta w_1^{[k]}w_1^{[k]} & -w_1^{[k]}w_2^{[k]} & -w_1^{[k]}w_3^{[k]} & \cdots & -w_1^{[k]}w_{k-1}^{[k]} & 0 & \cdots & 0 \\ -w_2^{[k]}w_1^{[k]} & 1-\beta w_2^{[k]}w_2^{[k]} & -w_2^{[k]}w_3^{[k]} & \cdots & -w_2^{[k]}w_{k-1}^{[k]} & 0 & \cdots & 0 \\ -w_3^{[k]}w_1^{[k]} & -w_3^{[k]}w_2^{[k]} & 1-\beta w_3^{[k]}w_3^{[k]} & \cdots & -w_3^{[k]}w_{k-1}^{[k]} & 0 & \cdots & 0 \\ \vdots & \vdots & \vdots & \ddots & \vdots & \vdots & \ddots & \vdots \\ -w_{k-1}^{[k]}w_1^{[k]} & -w_{k-1}^{[k]}w_2^{[k]} & -w_{k-1}^{[k]}w_3^{[k]} & \cdots & 1-\beta w_{k-1}^{[k]}w_{k-1}^{[k]} & 0 & \cdots & 0 \\ 0 & 0 & 0 & \cdots & 0 & 1 & \cdots & 0 \\ \vdots & \vdots & \vdots & \ddots & \vdots & \vdots & \ddots & \vdots \\ 0 & 0 & 0 & \cdots & 0 & 0 & \cdots & 1 \end{pmatrix}\end{aligned}$$

it is clear that the essential information of $\boldsymbol{R}[k]$, that is, matrix elements that are neither zero nor one, is confined to the upper-left square corner of a matrix with dimension $(k-1)$. This means we can build $\boldsymbol{R}$ in the same two-dimensional array used to store the $(n-1)$ column matrices $\boldsymbol{w}^{[k]}$. The useful information for $\boldsymbol{R}[k]$ may be stored without destroying the matrix elements $w^{[i]}$ for $i = k, k+1, \ldots, n$.

In terms of $\boldsymbol{T}[k]$ given by the analogous form to Eq. (5-96), we can write $\boldsymbol{R}[k]$ as

$$\boldsymbol{R}[k] = \left(\mathbf{1} - \beta \boldsymbol{w}^{[k]}\widetilde{\boldsymbol{w}}^{[k]}\right)\boldsymbol{R}[k-1] = \boldsymbol{R}[k-1] - \beta \boldsymbol{R}'$$

where

$$\boldsymbol{R}' = \boldsymbol{w}^{[k]}\widetilde{\boldsymbol{w}}^{[k]}\boldsymbol{R}[k-1].$$

In terms of matrix elements

$$r_{i,j}^{[k]} = \sum_{\ell=1}^{k-1}\left(\boldsymbol{w}^{[k]}\widetilde{\boldsymbol{w}}^{[k]}\right)_{i,\ell} r_{\ell,j}^{[k-1]} = w_i^{[k]}\sum_{\ell=1}^{k-1} w_\ell^{[k]} r_{\ell,j}^{[k-1]} \tag{5-106}$$

where $r_{i,j}^{[k]}$ are the matrix element of $\boldsymbol{R}[k]$ at position (i,j). The algorithm based on such an approach is outlined in Box 5-7, and the transformation matrix $\boldsymbol{R}$ produced is used in the next section to generate the eigenvectors of $\boldsymbol{A}$.

We can test a computer program for tridiagonalization by checking whether the transformation matrix $\boldsymbol{R}$ produced satisfies the relation

$$\widetilde{\boldsymbol{R}}\boldsymbol{A}\boldsymbol{R} = \boldsymbol{J}$$

given by Eq. (5-103). Here $\boldsymbol{J}$ is a tridiagonal, symmetric matrix with diagonal matrix elements $d_1, d_2, \ldots, d_n$ and superdiagonal elements $f_2, f_3, \cdots, f_n$.

5-5 Eigenvalues and eigenvectors of a tridiagonal matrix

The most straightforward method to find the eigenvalues of a real symmetric tridiagonal matrix $\boldsymbol{J}$ is to solve the characteristic equation

$$\boldsymbol{J} - \lambda \mathbf{1} = \mathbf{0} \tag{5-107}$$

where $\mathbf{0}$ is the null matrix having the same dimension as $\boldsymbol{J}$ but with all the elements zero. Since the matrix $\boldsymbol{J}$ has the simple form shown in Fig. 5-3, it is fairly easy to find the roots.

Sturm sequence of polynomials Consider first the case of dimension $n = 2$. The characteristic equation may be written as a second-order polynomial in λ,

$$p_2(\lambda) = \det\begin{vmatrix} d_1 - \lambda & f_2 \\ f_2 & d_2 - \lambda \end{vmatrix} = (d_1 - \lambda)(d_2 - \lambda) - f_2^2. \tag{5-108}$$

For $n = 3$, the characteristic polynomial for a tridiagonal matrix has the form

$$p_3(\lambda) = \det\begin{vmatrix} d_1 - \lambda & f_2 & 0 \\ f_2 & d_2 - \lambda & f_3 \\ 0 & f_3 & d_3 - \lambda \end{vmatrix} = (d_3 - \lambda)p_2(\lambda) - f_3^2 p_1(\lambda)$$

where we have defined

$$p_1(\lambda) = d_1 - \lambda \tag{5-109}$$

and the form of the second-order polynomial $p_2(\lambda)$ was given earlier in Eq. (5-108). In fact, if we also define a polynomial of order zero,

$$p_0 = 1$$

we can express $p_2(\lambda)$ of Eq. (5-108) as

$$p_2(\lambda) = (d_2 - \lambda)p_1(\lambda) - f_2^2 p_0. \tag{5-110}$$

By continuing with this line of argument, it is not difficult to see that, for tridiagonal matrices, the characteristic polynomials are given by the following recursion relation;

$$p_r(\lambda) = (d_r - \lambda)p_{r-1}(\lambda) - f_r^2 p_{r-2}(\lambda) \tag{5-111}$$

where $2 \leq r \leq n$. For a tridiagonal matrix of dimension n, the characteristic equation is then given by

$$p_n(\lambda) = (d_n - \lambda)p_{n-1}(\lambda) - f_n^2 p_{n-2}(\lambda) = 0 \tag{5-112}$$

a polynomial of order n in λ. A set of polynomials, $p_0(\lambda)$, $p_1(\lambda)$, $p_2(\lambda), \ldots, p_n(\lambda)$, satisfying the recursion relation Eq. (5-111) is said to form a Sturm sequence. There are many interesting properties of such a sequence of polynomials and we shall make use of them in designing an algorithm to find the eigenvalues of a tridiagonal matrix.

For all practical purposes, we need to consider only tridiagonal matrices without any of the superdiagonal (and subdiagonal) elements $f_2, f_3, \ldots, f_n$ equal to zero. This can be seen from the following argument. If any one of the matrix elements f_i is zero, the matrix can be split into two separate submatrices. For example, if $f_m = 0$, we have one matrix of dimension $(m-1)$, consisting of diagonal elements $d_1, d_2, \ldots, d_{m-1}$ and superdiagonal (and subdiagonal) elements $f_2, f_3, \ldots, f_{m-1}$, and another one of dimension $(n-m+1)$ consisting of diagonal elements $d_m, d_{m+1}, \ldots, d_n$ and superdiagonal (and subdiagonal) elements $f_{m+1}, f_{m+2}, \ldots, f_n$. Similarly, the basis states of the tridiagonal matrix also split into two groups, one consisting of states with indices $1, 2, \ldots, (m-1)$ and the other consisting of states $m, (m+1), \ldots, n$. Since there are no matrix elements joining these two blocks of states, transformations within one group do not affect the other. As a result, the eigenvalues of one submatrix are independent of those in the other. Conversely, the characteristic equation of $\boldsymbol{J}$ cannot be separated into two or more equations unless some of the superdiagonal (and subdiagonal) matrix elements vanish.

A simple example is the case of a tridiagonal matrix with dimension $n = 4$,

$$\boldsymbol{J} = \begin{pmatrix} d_1 & f_2 & 0 & 0 \\ f_2 & d_2 & f_3 & 0 \\ 0 & f_3 & d_3 & f_4 \\ 0 & 0 & f_4 & d_4 \end{pmatrix}.$$

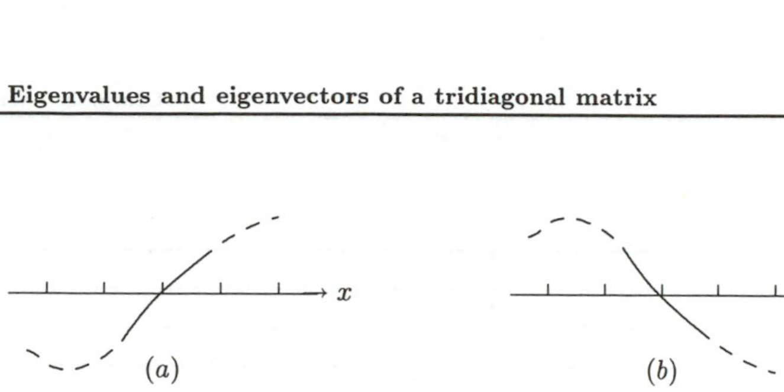

Figure 5-4: Behavior of a polynomial in an interval where there is only one root. The value must change sign, either from negative to positive as shown in case (a), or from positive to negative as shown in case (b).

If $f_3 = 0$, the matrix is equivalent to two $n = 2$ matrices. Transformations among columns (and rows) 1 and 2 do not affect the matrix elements d_3, d_4, and f_4.

Bisection method A simple way to find the roots of a polynomial is to use the method of bisection, as we did earlier in §3-6 for inverse interpolation. If, by some other means, we know that there is one but only one root of the polynomial $p_r(\lambda)$ in the interval $[v_\ell, v_h]$, the value of the polynomial at $\lambda = v_\ell$ must have a different sign from that at $\lambda = v_h$, as illustrated in Fig. 5-4. Without loss of generality, we can assume that $p_r(v_\ell) < 0$ and $p_r(v_h) > 0$, as illustrated in (a) of Fig. 5-4. We can now divide the interval into two equal parts and calculate the value of $p_r(\lambda)$ at the center of the interval. Let

$$v = \frac{1}{2}(v_\ell + v_h). \tag{5-113}$$

If $p_r(v) > 0$, the root is in the lower half of the interval. On the other hand, if $p_r(v) < 0$, the root lies in the upper half of the interval. In the former case, we can reduce the interval to half by replacing the upper limit v_h by v. Similarly, in the latter case, we can replace the lower limit v_ℓ by v. The process of bisection may be repeated as many times as we wish, until the difference between v_h and v_ℓ is less than the error in the root we can tolerate. The method is quite efficient, particularly if we want only a few of the roots of a polynomial of degree n. However, for this method to work, we must have a way to determine the interval within which each root lies.

Roots of polynomial For a polynomial of degree 1, there is only root,

$$\lambda^{(1)} = d_1$$

as can be seen from Eq. (5-109). The two roots of $p_2(\lambda)$ can be found by solving the quadratic equation given by $p_2(\lambda)$ of Eq. (5-110):

$$\lambda_i^{(2)} = \frac{1}{2}(d_1 + d_2) \mp \frac{1}{2}(d_1 - d_2)\sqrt{1 + \Big(\frac{2f_2}{d_1 - d_2}\Big)^2} \tag{5-114}$$

where the subscript $i = 1$ for the upper sign and 2 for the lower sign. Since f_2 is assumed to be nonzero, we have the situation that, for $d_1 > d_2$, and both $p_1(\lambda)$ and

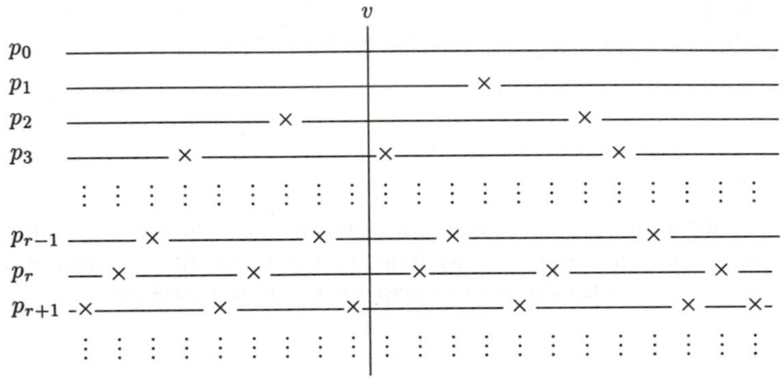

Figure 5-5: Relation between the roots of the polynomials in a Sturm sequence.

$p_2(\lambda)$ belong to the same Sturm sequence,

$$\lambda_1^{(2)} < \lambda^{(1)} < \lambda_2^{(2)} \tag{5-115}$$

as can be seen by inspection. For $d_2 > d_1$, the same inequality holds except that the directions are reversed.

The property expressed by Eq. (5-115) can be shown to be general for the polynomials in a Sturm sequence. That is, for $i = 1, 2, \ldots$, if $\lambda_i^{(r)}$ are the roots of $p_r(\lambda)$ and $\lambda_i^{(r+1)}$, those of $p_{r+1}(\lambda)$, the roots of the two polynomials satisfy the inequality

$$\lambda_1^{(r+1)} < \lambda_1^{(r)} < \lambda_2^{(r+1)} < \lambda_2^{(r)} < \lambda_3^{(r+1)} < \cdots < \lambda_m^{(r+1)} < \lambda_m^{(r)} < \lambda_{m+1}^{(r+1)} < \cdots < \lambda_r^{(r+1)} < \lambda_r^{(r)} < \lambda_{r+1}^{(r+1)}. \tag{5-116}$$

These relations imply also that two such polynomials, differing in degree by 1, and with $f_r \neq 0$, cannot have any common roots. We can see this from the recursion relation Eq. (5-111). Let us assume for the moment that λ_q is a root for both $p_{r+1}(\lambda)$ and $p_r(\lambda)$. For this to be true, it must also be the root of $p_{r-1}(\lambda)$ to satisfy Eq. (5-111). By the same token, if λ_q is the root of $p_r(\lambda)$ and $p_{r-1}(\lambda)$, it must also be a root of $p_{r-2}(\lambda)$. When we continue this line of reasoning down to $r = 2$, we reach an absurdity that λ_q is also the root of p_0. Since, by definition, p_0 is a constant and does not have a root, our original assumption must be wrong, and we must conclude that $p_{r+1}(\lambda)$ and $p_r(\lambda)$ cannot share a common root.

To prove the inequality relations of Eq. (5-116), we need also to show that there is only one root of $p_r(\lambda)$ between any two adjacent roots of $p_{r+1}(\lambda)$. We shall leave this part of the arguments to books on numerical analysis, such as Wilkinson.[56] Schematically, the relation given by Eq. (5-116) is shown in Fig. 5-5.

Except for the two end ones, Eq. (5-116) provides us with a way to find the lower and upper limits of all the roots. To complete the definition for all the ranges, we need further to show that all the eigenvalues E_i of a tridiagonal matrix $\boldsymbol{J}$ are bounded by the relation

$$|E_i| \leq \lambda_{\max} \tag{5-117}$$

where

$$\lambda_{\max} = \max\{|f_i| + |d_i| + |f_{i+1}|\} \qquad \text{for} \qquad i = 1, 2, \ldots, n.$$

That is, $\lambda_{\max}$ is given by the maximum value among sums of the absolute values of the three elements in a given row. Since the first row has only two nonvanishing elements, d_1 and f_2, the value of f_1 in the above expression is taken to be 0. Similarly, $f_{n+1} = 0$, as there are also only two nonvanishing elements in the last row. The quantity $\lambda_{\max}$ is called the maximum or infinite subordinate matrix norm of $\boldsymbol{J}$. A proof of this relation involves inequalities among matrix norms and can be found in Stoer and Bulirsch.[54]. Our interest here lies mainly in the possibility of making use of $\lambda_{\max}$ to provide us with the necessary range of values to start the bisection method for locating the roots of Eq. (5-112).

Since the largest possible eigenvalue is less than or equal to $\lambda_{\max}$, the roots of any of the polynomials in a Sturm sequence are less than $\lambda_{\max}$. As a result, for $v = \lambda_{\max} + \epsilon$, where ϵ is a small, positive quantity, we have the situation

$$p_r(v) \begin{cases} > 0 & \text{for } r = \text{even} \\ < 0 & \text{for } r = \text{odd}. \end{cases} \tag{5-118}$$

This relation implies that, for $v > \lambda^{(n)}_{\max}$, the sign changes between two polynomials differ by 1 in the degree. It is easy to convince ourselves that this is true by considering the limiting case of $v \to +\infty$, where every polynomial is dominated by the first term of Eq. (5-111) and

$$p_r(v) \xrightarrow[v \to \infty]{} (-v)^r. \tag{5-119}$$

Since there are no roots between $v = +\infty$ and $v = (\lambda_{\max} + \epsilon)$ for any of the polynomials, the signs of $p_r(v)$ cannot change and we obtain Eq. (5-118). As we decrease the value of v from the upper limit of $\lambda_{\max}$ until it is just less than the largest eigenvalue $\lambda^{(n)}_n$, the sign of $p_n(v)$ changes, since it passes through a root. However, the signs of the other polynomials are not changed, as the value of v is still larger than any of the roots of the other polynomials. As a result, the values of $p_n(\lambda)$ and $p_{n-1}(\lambda)$ have the same sign for λ in the vicinity of this particular value of v.

Upper and lower limits of eigenvalues Let us use s to record the number of agreements in sign at any given value of v between any two polynomials different in degree by 1. When $v = \lambda_{\max} + \epsilon$, no two adjacent polynomials have the same sign, as can be seen from Eq. (5-119). In this case, we have $s = 0$. This value of s is not changed until we come to $v = \lambda^{(n)}_n - \epsilon$. For small positive values of ϵ, we have $s = 1$ since $p_n(v)$ and $p_{n-1}(v)$ now have the same sign. The value of s remains 1 until v becomes less than $\lambda^{(n)}_{n-1}$, the second largest eigenvalue. This can be seen for the following reasons. As we decrease the value of v from $\lambda_n^{(n)}$, it will eventually become less than $\lambda_{n-1}^{(n-1)}$, the largest root of $p_{n-1}(\lambda)$. At this value, the sign of $p_{n-1}(v)$ changes but not that of $p_n(v)$ or any of the other polynomials, as $\lambda_{n-1}^{(n-1)}$ cannot be the root of any other polynomials. The sign of $p_{n-1}(v)$ now disagrees with that of $p_n(v)$ but is in agreement with that of $p_{n-2}(v)$. As a result, s remains 1. When $v = \lambda_{n-1}^{(n)} - \epsilon$, the sign of $p_n(v)$ changes and is now in agreement with that of $p_{n-1}(v)$. Since none of the other polynomials have a change of sign here, we have $s = 2$. If we continue this line of argument, we will come to the conclusion that the value of s is also equal to the number of eigenvalues greater than v. For example, when v is in the interval between $-\infty$ and $(\lambda_1^{(n)} - \epsilon)$, the smallest root of $p_n(\lambda)$ minus a small positive quantity, all the polynomials are positive and the number of sign agreement is $s = n$.

With both upper and lower limits of all the eigenvalues given by Eq. (5-117), and the connection between s and the number of eigenvalues above, we have a practical way to apply the method of bisection to find the eigenvalues of a tridiagonal matrix. For the convenience of discussion, we shall label all the eigenvalues in descending order according to their values. That is,

$$E_1 > E_2 > \cdots > E_n.$$

Furthermore, for the time being we shall ignore the possibility of degeneracies, that is, two or more eigenvalues equal in value. We shall return to this practical question later when we construct an algorithm to diagonalize a tridiagonal matrix.

Let us assume that we are interested in the ith largest eigenvalue E_i. In terms of the roots of the polynomial $p_n(v)$, it is equivalent to find the ith largest root $\lambda_i^{(n)}$. From Eq. (5-117), we see that the value must lie within the interval $[-\lambda_{\max}, +\lambda_{\max}]$. Using the bisection method, we have

$$v_\ell = -\lambda_{\max} \qquad\qquad v_h = \lambda_{\max}.$$

Next, we assign v as the average of v_ℓ and v_h, as we did earlier in Eq. (5-113). In addition to calculating the value of $p_n(v)$, we check also, for $r = 0$ to $(n-1)$, whether $p_{r+1}(v)$ agrees in sign with $p_r(v)$ and store the number of agreements as s. Very little additional work is involved in carrying out such a counting if we use the recursion relation Eq. (5-111) to evaluate $p_n(v)$. For $s \geq i$, we know that E_i is in the upper half of the interval, and for $s < i$, we find that E_i is in the lower half. The interval is now reduced to half, and the process is repeated until the interval is less than the tolerance of error. This is essentially the same procedure as that discussed earlier for the simple case of a single root in a given interval. The only exception is that the value of s is used here to determine which half of the interval to use for the next step.

A complication in the method comes when one of the polynomials becomes zero for some value of v. Let us assume that this takes place at $p_r(v)$. From Eq. (5-116), we know that both $p_{r-1}(v)$ and $p_{r+1}(v)$ cannot vanish. A moment of reflection with the help of Fig. 5-5 will convince us that $p_r(v) = 0$ should be considered as a disagreement in sign with $p_{r-1}(v)$ and an agreement in sign with $p_{r+1}(v)$. This completes the basic

Box 5-8 Subroutine BISCT(D,F,V,N,NDMN,ACC)
Eigenvalues of a tridiagonal matrix by bisection

Argument list:
- D: Input array for the diagonal matrix elements.
- F: Input array for the superdiagonal elements.
- V: Output array of eigenvalues.
- N: Matrix dimension.
- NDMN: Dimension of the arrays D, F, and V in the calling program.
- ACC: Accuracy required.

Initialization:
- (a) Set up arrays for f_i^2, and the two limits of each eigenvalue.
- (b) Store f_i^2, squares of the superdiagonal matrix elements.

1. Find the two limits of each eigenvalue:
   - (a) Calculate the infinite subordinate matrix norm $\lambda_{\max}$ of Eq. (5-117).
   - (b) Store $\lambda_{\max}$ as the upper and $-\lambda_{\max}$ as the lower limits of all eigenvalues.
2. Find the eigenvalues by bisection:
   - (a) Get the interval from the upper and lower limits stored.
   - (b) Calculate the value at the middle point.
   - (c) Count the number of sign agreements:
     - (i) Let p be the polynomial value for this order and p_{-1} the previous order.
     - (ii) If $p * p_{-1} > 0$, no sign change,
     - (iii) If $p = 0$, the sign is considered to be different, and
     - (iv) If $p_{-1} = 0$, the sign is considered to be in agreement.
   - (d) Bisect the interval.
   - (e) Update the limits for the other eigenvalues.
   - (f) Check if the interval is smaller than the error tolerance.
     - (i) If so, store the average of the two limits as the eigenvalue.
     - (ii) If not, go back to step (b) using the new interval.
3. Return the eigenvalues.

method for finding the eigenvalues of a tridiagonal matrix by bisection. Because of the possibility of searching for any one of the eigenvalues, the method is most useful if only a few are required.

A small improvement may be added to the method. For a given value of v, we know that there are s eigenvalues above v and $(n-s)$ eigenvalues below v. As a result, the value of v becomes the lower bound for all E_i with $i \leq s$. Similarly, v becomes the upper bound of the $(n-s)$ eigenvalues with subscript $i < s$. In this way, the value of v used in calculating the limits for any one of the eigenvalues can also serve either as the upper or lower limits of all the other unknown E_i. It is, therefore, possible to keep a list of the upper and lower limits of all the eigenvalues known so far in the calculation. Each time the value of s is calculated for a particular v, we can check the list and update it if v represents a better estimate of the limits for any of the unknown eigenvalues. This improves the efficiency of the method since the starting range of search for any eigenvalue decreases as more and more eigenvalues are found. The algorithm is summarized in Box 5-8.

It is relatively easy to test a computer program based on the bisection method to find the eigenvalues of a tridiagonal matrix. All we need to do is to substitute the eigenvalues back into the characteristic equation of Eq. (5-112) and see if they are, indeed, the roots of $p_n(v)$ within the specified numerical accuracy. That is, we can check how close the value $p_n(E_i)$ is to zero for all the eigenvalues found.

***QR*- and *QL*-transformations** The strength of the bisection method lies in its simplicity. However, as mentioned earlier, it is efficient only in cases where a few eigenvalues are needed. Furthermore, the method does not lend itself easily to calculating the corresponding eigenvectors. For these reasons, the QL- and QR-algorithms are preferred for diagonalizing tridiagonal matrices.

The principle of QL- and QR-methods applies to a wider class of matrices than the tridiagonal one of interest to us here. For this reason, we shall begin by considering symmetric matrices in general. It can be shown that any symmetric matrix $\boldsymbol{A}$ is reducible to a product of two matrices,

$$\boldsymbol{A} = \boldsymbol{Q}\boldsymbol{R} \tag{5-120}$$

where $\boldsymbol{Q}$ is a unitary matrix, with the property $\boldsymbol{Q}^{-1} = \boldsymbol{Q}^{\dagger}$, and $\boldsymbol{R}$ is an upper triangular matrix having nonvanishing matrix elements only in the upper triangle. This is similar in spirit to the LU-decomposition used in §5-2 to solve a system of linear equations.

Using the unitary property of $\boldsymbol{Q}$, we obtain from Eq. (5-120) the relation

$$\boldsymbol{Q}^{\dagger}\boldsymbol{A} = \boldsymbol{Q}^{-1}\boldsymbol{A} = \boldsymbol{R}.$$

Alternatively, we can define a new matrix $\boldsymbol{B}$ such that

$$\boldsymbol{B} \equiv \boldsymbol{R}\boldsymbol{Q} = \boldsymbol{Q}^{\dagger}\boldsymbol{A}\boldsymbol{Q}. \tag{5-121}$$

Since $\boldsymbol{B}$ is the unitary transformation of $\boldsymbol{A}$, their eigenvalues are identical to each other.

In general, $\boldsymbol{B}$ is not a triangular matrix, since multiplying the triangular matrix $\boldsymbol{R}$ by $\boldsymbol{Q}$ produces nonvanishing matrix elements in the other half of the triangle. However, if we define

$$\boldsymbol{A}(1) \equiv \boldsymbol{A}$$

and apply a series of unitary transformations given by the recursion relation

$$\boldsymbol{A}(k+1) = \boldsymbol{Q}^{\dagger}(k)\boldsymbol{A}(k)\boldsymbol{Q}(k) \tag{5-122}$$

for $k = 1, 2, \ldots$, using unitary matrices $\boldsymbol{Q}(1)$, $\boldsymbol{Q}(2), \ldots$ satisfying Eq. (5-120), it is possible to prove that

$$\boldsymbol{A}(k) \xrightarrow[k\to\infty]{} \boldsymbol{D}. \tag{5-123}$$

If $\boldsymbol{A}$ is a real symmetric matrix, $\boldsymbol{D}$ is a diagonal matrix and, as a result, the diagonal elements are the eigenvalues of $\boldsymbol{A}$. In the more general case of a nonsymmetric matrix, the diagonal matrix elements of $\boldsymbol{D}$ remain as the eigenvalues, but the matrix $\boldsymbol{D}$ is triangular instead, with vanishing elements only above the diagonal. This theorem is not obvious and readers interested in a proof should consult Wilkinson[56] or Stoer and Bulirsch.[54]

Similar to Eq. (5-120), we can also decompose $\boldsymbol{A}$ into a product of unitary matrix $\boldsymbol{Q}$ and a lower triangular matrix $\boldsymbol{L}$, with nonvanishing elements only in the lower half,

$$\boldsymbol{A} = \boldsymbol{QL}. \tag{5-124}$$

This is known as the QL-decomposition. Since there is nothing special about the upper half-triangle of a matrix, compared with the lower half, Eq. (5-123) applies here also. We have now a choice between QR- and QL-decomposition in diagonalizing a matrix. The selection may be made based on numerical accuracy considerations. As we shall see later, if the largest elements are in the lower-right corner of the matrix, the QL-decomposition is more likely to generate better results. On the other hand, if the largest elements are in the upper-left corner, the QR-method is preferred.

As a general method for matrix diagonalization, the QL- and QR-methods are not very useful because of the iterative nature of the procedure, as can be seen from Eq. (5-123). On the other hand, for tridiagonal matrices and band matrices, that is, matrices with only nonvanishing elements along and near the diagonal, these two methods are likely to be more efficient. As practical methods for diagonalizing real symmetric matrices, it is preferable to apply the QL- or QR-algorithm after the matrices are reduced to a tridiagonal form using, for example, the method of Householder given in the previous section. Furthermore, we shall choose the QL-approach here, following the common approach originated by Bowdler and coauthors.[6] For the remainder of this section, we shall therefore concentrate on the application of the QL-algorithm to tridiagonal matrices, in particular those obtained from real symmetric matrices.

Shift of the diagonal elements Before embarking on the steps to carry out a QL-decomposition, it is necessary to realize a special point concerning the question of convergence. In general, the rate at which an off-diagonal matrix element $a_{i,j}$ goes to zero in Eq. (5-123) is proportional to the ratio of the absolute values of the two eigenvalues λ_i and λ_j,

$$\text{rate}\{a_{i,j}^{(k)} \to 0\} \sim \left|\frac{\lambda_i}{\lambda_j}\right|^k. \tag{5-125}$$

As a result, the convergence of QL-algorithm can be slow if two eigenvalues happen to be very close to each other in their absolute values. To prevent this from happening, we can shift all the diagonal values by a constant η_k. That is, instead of $\boldsymbol{A}$, we work with the matrix

$$\boldsymbol{A}' = \boldsymbol{A} - \eta_k \boldsymbol{1} \tag{5-126}$$

where the value of the parameter η_k will be determined later.

The transformation of Eq. (5-122) is not affected by such a shift. We can see this in the following way. In step k, we have the relation

$$\boldsymbol{A}'(k) = \boldsymbol{Q}(k)\boldsymbol{L}(k)$$

instead of Eq. (5-124). Similar to Eq. (5-122), we have the recursion relation

$$\boldsymbol{A}'(k+1) = \boldsymbol{L}(k)\boldsymbol{Q}(k) = \boldsymbol{Q}^{\dagger}(k)\boldsymbol{A}'(k)\boldsymbol{Q}(k). \tag{5-127}$$

Since

$$\boldsymbol{Q}^{\dagger}(k)\mathbf{1}\boldsymbol{Q}(k) = \boldsymbol{Q}^{\dagger}(k)\boldsymbol{Q}(k) = \mathbf{1}$$

the result of Eq. (5-127) is equivalent to

$$\boldsymbol{A}(k+1) - \eta_k\mathbf{1} = \boldsymbol{L}(k)\boldsymbol{Q}(k)\boldsymbol{Q}^{\dagger}(k)\boldsymbol{A}'(k)\boldsymbol{Q}(k) = \boldsymbol{Q}^{\dagger}(k)\boldsymbol{A}(k)\boldsymbol{Q}(k) - \eta_k\mathbf{1}.$$

From this we see that the relation

$$\boldsymbol{A}(k+1) = \boldsymbol{Q}^{\dagger}(k)\boldsymbol{A}(k)\boldsymbol{Q}(k) \tag{5-128}$$

remains the same as Eq. (5-122), except that a QL-decomposition is used here.

Although the shift of Eq. (5-126) does not affect the transformation itself, the eigenvalues of $\boldsymbol{A}'$ are different and equal to $(\lambda_i - \eta_k)$, shifted in value by η_k from those of $\boldsymbol{A}$. As a result, the relation given by Eq. (5-125) becomes

$$\text{rate}\{a_{i,j}^{(k)} \to 0\} \sim \left|\frac{\lambda_i - \eta_k}{\lambda_j - \eta_k}\right|^k. \tag{5-129}$$

If $|\lambda_i|$ is not exactly equal to $|\lambda_j|$, we can choose the value of η_k to be as close to that of λ_i as possible and the ratio on the right side of Eq. (5-129) becomes small. The net result is that $a_{i,j}$ goes to zero in far fewer number of steps than without the shift.

What happens if we have degenerate eigenvalues? First, this cannot take place in a tridiagonal matrix unless one of the superdiagonal elements vanishes, as we saw earlier. As an illustration, let us consider again a two-dimensional case. The two eigenvalues of the matrix

$$\boldsymbol{C}_2 \equiv \begin{pmatrix} d_1 & f_2 \\ f_2 & d_2 \end{pmatrix}$$

are, as we saw in Eq. (5-114),

$$\lambda_i = \frac{1}{2}(d_1 + d_2) \pm \frac{1}{2}(d_1 - d_2)\sqrt{1 + \left(\frac{2f_2}{d_1 - d_2}\right)^2}$$

where the subscript i takes on values of either 1 or 2. Their difference is

$$\Delta \equiv |d_1 - d_2| = (d_1 - d_2)\sqrt{1 + \left(\frac{2f_2}{d_1 - d_2}\right)^2} = \sqrt{(d_1 - d_2)^2 + 4f_2^2}$$

and cannot be zero if $f_2 \neq 0$.

In practical terms, when the numerical accuracy of the computer cannot distinguish $(|d_1| + f_2)$ from $|d_1|$ or $(|d_2| + f_2)$ from $|d_2|$, the value of f_2 is immaterial. For the general case of an n-dimensional tridiagonal matrix, we can use the criterion

$$[d] + f_i = [d] \tag{5-130}$$

to determine whether f_i may be treated as zero. Here, $[d]$ is the larger of $|d_i|$ and $|d_{i+1}|$. When this condition is true, the tridiagonal matrix may be separated into two distinct blocks, the same as we saw earlier in relation with the bisection method.

In general, degenerate eigenvalues are not a problem in QL- and QR-methods. However, a practical algorithm for diagonalizing tridiagonal matrices must test for the possible existence of degeneracies and make allowance for them. We shall return to this point later when we come to the actual algorithm.

The unitary matrix $\boldsymbol{Q}$ We shall now turn our attention to the unitary matrix $\boldsymbol{Q}(k)$ in Eq. (5-124). For this purpose, it is more convenient to consider instead its Hermitian adjoint $\boldsymbol{Q}^\dagger(k)$ that changes $\boldsymbol{A}(k)$ into a lower triangular matrix $\boldsymbol{L}(k)$:

$$\boldsymbol{Q}^\dagger(k)\boldsymbol{A}(k) = \boldsymbol{L}(k).$$

If $\boldsymbol{A}(k)$ is a tridiagonal matrix of dimension n, this particular step may be achieved by $(n-1)$ rotations, each of which reduces one of the $(n-1)$ superdiagonal matrix elements of $\boldsymbol{A}(k)$ to zero. However, as we have seen earlier, several such passes may be necessary, as each rotation does not necessarily preserve the values of off-diagonal matrix elements annihilated earlier. Although the method applies to a wider class of matrices, we shall be mainly concerned with real symmetric tridiagonal ones.

Let us use $\boldsymbol{P}_i^\dagger(k)$ to denote one of these $(n-1)$ rotations, the one that transforms to zero the super-diagonal element $f_i^{(k)}$ at location $(i-1, i)$ of a tridiagonal matrix $\boldsymbol{J}$. The calculation involves taking a linear combination of elements in rows $(i-1)$ and i such that the element $(i-1, i)$ of $\boldsymbol{L}(k)$ vanishes. The matrix $\boldsymbol{P}_i^\dagger(k)$ may be expressed in terms of two parameters c_i and s_i in the following form:

$$\begin{array}{c} \\ \boldsymbol{P}_i^\dagger(k) = \end{array} \begin{array}{c} \qquad\qquad (i-1) \quad i \\ \begin{pmatrix} 1 & 0 & \cdots & 0 & 0 & 0 & \cdots & 0 \\ 0 & 1 & \cdots & 0 & 0 & 0 & \cdots & 0 \\ \vdots & \vdots & \ddots & \vdots & \vdots & \vdots & \ddots & \vdots \\ 0 & 0 & \cdots & c_i & -s_i & 0 & \cdots & 0 \\ 0 & 0 & \cdots & s_i & c_i & 0 & \cdots & 0 \\ 0 & 0 & \cdots & 0 & 0 & 1 & \cdots & 0 \\ \vdots & \vdots & \ddots & \vdots & \vdots & \vdots & \ddots & \vdots \\ 0 & 0 & \cdots & 0 & 0 & 0 & \cdots & 1 \end{pmatrix} \end{array} \begin{array}{c} \\ \\ \\ (i-1) \\ i \\ \\ \\ \\ \end{array}. \tag{5-131}$$

The values of c_i and s_i must satisfy the condition that $\boldsymbol{P}_i^\dagger(k)\boldsymbol{P}_i(k) = \boldsymbol{1}$. In terms of matrix elements, this is equivalent to the requirement that

$$c_i^2 + s_i^2 = 1. \tag{5-132}$$

A second condition for these two parameters is that the superdiagonal matrix element at location $(i-1, i)$ must vanish. That is,

$$c_i\,\bar{a}_{i,i+1} - s_i\,\bar{a}_{i+1,i+1} = 0 \tag{5-133}$$

where we have used bars on top of the matrix elements to remind us that they are not the same as $f_{i+1}^{(k)}$ and $d_{i+1}^{(k)}$ of the tridiagonal matrix $\boldsymbol{J}(k)$. As we shall see later, earlier rotations in the same step k may have modified their values. Our main concern at the moment is with the values of the two parameters c_i and s_i. The rotation induced by $\boldsymbol{P}_i^{\dagger}(k)$ is different from those used in the Jacobi or Givens method, since the off-diagonal matrix elements here are reduced to zero by the action of $\boldsymbol{P}_i(k)$ alone, rather than through a similarity transformation used in Eq. (5-60) or (5-81).

At the start of step k, we have a tridiagonal matrix $\boldsymbol{J}(k)$, having the form shown in Fig. 5-3. Being a real symmetric matrix, the subdiagonal elements of $\boldsymbol{J}(k)$ are equal to the corresponding superdiagonal ones. The matrix is completely specified by n diagonal elements and $(n-1)$ superdiagonal elements. For $\boldsymbol{J}(k)$, these may be taken as $d_1^{(k)}, d_2^{(k)}, \ldots, d_n^{(k)}$ and $f_2^{(k)}, f_3^{(k)}, \ldots, f_n^{(k)}$. To make use of a shift in the eigenvalues to improve the rate of convergence, we shall apply a rotation to the $\boldsymbol{J}(k)$ with the diagonal matrix elements shifted in value by a constant η_k:

$$\boldsymbol{J}'(k) = \begin{pmatrix} d_1^{(k)}-\eta_k & f_2^{(k)} & 0 & \cdots & 0 & 0 & 0 \\ f_2^{(k)} & d_2^{(k)}-\eta_k & f_3^{(k)} & \cdots & 0 & 0 & 0 \\ 0 & f_3^{(k)} & d_3^{(k)}-\eta_k & \cdots & 0 & 0 & 0 \\ \vdots & \vdots & \vdots & \ddots & \vdots & \vdots & \vdots \\ 0 & 0 & 0 & \cdots & d_{n-2}^{(k)}-\eta_k & f_{n-1}^{(k)} & 0 \\ 0 & 0 & 0 & \cdots & f_{n-1}^{(k)} & d_{n-1}^{(k)}-\eta_k & f_n^{(k)} \\ 0 & 0 & 0 & \cdots & 0 & f_n^{(k)} & d_n^{(k)}-\eta_k \end{pmatrix}.$$

The choice of η_k is not easy, as we have no knowledge of any of the eigenvalues at this point. The procedure usually recommended is the following. If we wish to reduce the superdiagonal element $f_i^{(k)}$ to zero, we can make an approximation and treat the submatrix formed by $d_{i-1}^{(k)}$, $d_i^{(k)}$, and $f_i^{(k)}$ as a (2×2) matrix. From Eqs. (5-76) and (5-77), the two eigenvalues of this matrix are

$$\lambda_1 = d_{i-1}^{(k)} - t f_i^{(k)} \qquad\qquad \lambda_2 = d_i^{(k)} + t f_i^{(k)} \tag{5-134}$$

where $t \equiv \tan\theta^{(i)}$ is given by Eq. (5-74). Alternatively, it can be expressed as

$$t = \frac{1}{|\xi| \pm \sqrt{1+\xi^2}} \qquad\qquad \xi = \frac{d_i^{(k)} - d_{i-1}^{(k)}}{2 f_i^{(k)}}. \tag{5-135}$$

The choice of the sign in the denominator on the right side depends on that of ξ. If ξ is positive, we take the plus sign and, if ξ is negative, we take the negative sign. By approximating η_k with the value of λ_1 calculated in this way, reasonable convergence rates are expected.

We shall start the $(n-1)$ rotations by trying first to zero the superdiagonal element $f_n^{(k)}$. For this purpose, the rotation matrix $P(k)$ takes on the form

$$\boldsymbol{P}_n^{\dagger}(k) = \begin{pmatrix} 1 & 0 & \cdots & 0 & 0 \\ 0 & 1 & \cdots & 0 & 0 \\ \vdots & \vdots & \ddots & \vdots & \vdots \\ 0 & 0 & \cdots & c_n & -s_n \\ 0 & 0 & \cdots & s_n & c_n \end{pmatrix}$$

with

$$c_n = \frac{d_n^{(k)} - \eta_k}{r_n} \qquad s_n = \frac{f_n^{(k)}}{r_n} \qquad r_n = \sqrt{(d_n^{(k)} - \eta_k)^2 + (f_n^{(k)})^2}.$$

Let us denote the matrix after this rotation as $\boldsymbol{L}_n(k)$, having the form

$$\boldsymbol{L}_n(k) = \begin{pmatrix} d_1^{(k)} - \eta_k & f_2^{(k)} & 0 & \cdots & 0 & 0 & 0 \\ f_2^{(k)} & d_2^{(k)} - \eta_k & f_3^{(k)} & \cdots & 0 & 0 & 0 \\ 0 & f_3^{(k)} & d_3^{(k)} - \eta_k & \cdots & 0 & 0 & 0 \\ \vdots & \vdots & \vdots & \ddots & \vdots & \vdots & \vdots \\ 0 & 0 & 0 & \cdots & d_{n-2}^{(k)} - \eta_k & f_{n-1}^{(k)} & 0 \\ 0 & 0 & 0 & \cdots & c_n f_{n-1}^{(k)} & p_{n-1} & 0 \\ 0 & 0 & 0 & \cdots & 0 & g_n & p_n \end{pmatrix}$$

where

$$\begin{aligned} p_{n-1} &= c_n(d_{n-1}^{(k)} - \eta_k) - s_n f_n^{(k)} \qquad & p_n &= s_n f_n^{(k)} + c_n(d_n^{(k)} - \eta_k) \\ g_n &= s_n(d_{n-1}^{(k)} - \eta_k) + c_n f_n^{(k)}. \end{aligned}$$

The matrix $\boldsymbol{L}_n(k)$ is not yet in a proper lower triangular form, as we have only performed one of the $(n-1)$ steps in multiplying $\boldsymbol{Q}^{\dagger}(k)$ with $\boldsymbol{J}(k)$. For the same reason, it is not yet possible to equate $\boldsymbol{L}_n(k)$ with $\boldsymbol{J}(k+1)$, since we have yet to multiply $\boldsymbol{L}(k)$ by $\boldsymbol{Q}(k)$ from the right.

Before we go on any further with the calculation to transform $\boldsymbol{J}(k)$ into $\boldsymbol{L}(k)$, it is important to recall that our ultimate goal in step k is not $\boldsymbol{L}(k)$ but $\boldsymbol{J}(k+1)$, given by the relation

$$\boldsymbol{J}(k+1) = \boldsymbol{Q}^{\dagger}(k)\boldsymbol{J}(k)\boldsymbol{Q}(k) = \boldsymbol{L}(k)\boldsymbol{Q}(k).$$

For this purpose, we must multiply $\boldsymbol{L}(k)$ obtained from Eq. (5-124) by $\boldsymbol{Q}(k)$. Since $\boldsymbol{Q}(k)$ is also made of a series of $(n-1)$ transformations, it is convenient to carry out each rotation in $\boldsymbol{Q}^{\dagger}(k)\boldsymbol{J}(k)$ in conjunction with the corresponding calculations required for $\boldsymbol{L}(k)\boldsymbol{Q}(k)$. In this way, it is possible to obtain $\boldsymbol{J}(k+1)$ without having to store the $(n-1)$ rotations that make up $\boldsymbol{Q}(k)$. This may be achieved in the following way. First, we note that the matrix element $d_n^{(k+1)}$ is not affected by any of the later rotations $\boldsymbol{P}_{n-1}^{\dagger}(k)$, $\boldsymbol{P}_{n-2}^{\dagger}(k)$, ... As a result, we can complete the calculation of $d_n^{(k+1)}$ by multiplying the last row of $\boldsymbol{L}_n(k)$ with the last column of $\boldsymbol{P}_n(k)$. This gives us

$$\begin{aligned} d_n^{(k+1)} - \eta_k &= g_n s_n + p_n c_n \\ &= s_n^2(d_{n-1}^{(k)} - \eta_k) + c_n^2(d_n^{(k)} - \eta_k) + 2c_n s_n f_n^{(k)} \\ &= (d_n^{(k)} - \eta_k) + s_n^2(d_n^{(k)} - \eta_k) + s_n^2(d_{n-1}^{(k)} - \eta_k). \end{aligned}$$

However, we cannot work out the value for $f_n^{(k+1)}$ yet, as the matrix elements in column $(n-1)$ will be modified by a later rotation in the $(n-2, n-1)$ plane. This is not a basic problem, as a procedure can be designed such that some quantities are kept from one rotation to the next so as to carry out the transformation for $f_n^{(k+1)}$ later. In this way, a fairly efficient scheme can be constructed to carry out all the rotations in a step.

We can now derive a set of general formulas to carry out the $(n-1)$ rotations for step k. After rotation $(j+1)$, the partial lower triangular matrix $\boldsymbol{L}_{j+1}(k)$, resulting from applying the rotations $\boldsymbol{P}_n^{\dagger}(k)$, $\boldsymbol{P}_{n-1}^{\dagger}(k), \ldots, \boldsymbol{P}_{j+1}^{\dagger}(k)$, takes on the form

$$\boldsymbol{L}_{j+1}(k) = \left(\begin{array}{cccccc|cccc} d_1^{(k)} - \eta_k & f_2^{(k)} & 0 & \cdots & 0 & 0 & 0 & 0 & 0 & \cdots \\ f_2^{(k)} & d_2^{(k)} - \eta_k & f_3^{(k)} & \cdots & 0 & 0 & 0 & 0 & 0 & \cdots \\ \vdots & \vdots & \vdots & \ddots & \vdots & \vdots & \vdots & \vdots & \vdots & \ddots \\ 0 & 0 & 0 & \cdots & f_{j-1}^{(k)} & d_{j-1}^{(k)} - \eta_k & f_j^{(k)} & 0 & 0 & \cdots \\ \hline 0 & 0 & 0 & \cdots & 0 & g_j & p_j & 0 & 0 & \cdots \\ 0 & 0 & 0 & \cdots & 0 & z_{j+1} & g_{j+1} & p_{j+1} & 0 & \cdots \\ 0 & 0 & 0 & \cdots & 0 & 0 & z_{j+2} & g_{j+2} & p_{j+2} & \cdots \\ \vdots & \vdots & \vdots & \ddots & \vdots & \vdots & \vdots & \vdots & \vdots & \ddots \end{array}\right) \begin{array}{c} \\ \\ \\ (j-1) \\ j \\ \\ \\ \\ \end{array}. \tag{5-136}$$

(columns marked: $(j-1)$, j)

Note that only the superdiagonal elements of $\boldsymbol{L}_{j+1}(k)$ beyond column j are zero, since we have yet to apply the operations $P_1(k)P_2(k)\cdots P_j(k)$ on the right to complete the similarity transformation on $\boldsymbol{J}$. For the same reason, the diagonal elements of $\boldsymbol{L}_{j+1}(k)$ beyond column j are not yet the eigenvalues of the matrix $\boldsymbol{A}$, and they are denoted as p_j, $p_{j+1}, \ldots, p_n$. At the same time, there are nonvanishing off-diagonal matrix elements z_{j+1}, $z_{j+2}, \ldots$ just below the subdiagonal elements. As we shall see below, these elements arise because of transformation $P_n^{\dagger}(k)$, $P_{n-1}^{\dagger}(k), \ldots, P_{j+1}^{\dagger}(k)$.

The aim of the next rotation is to annihilate the element $f_j^{(k)}$. For this purpose, the rotation matrix $\boldsymbol{P}_j^{\dagger}(k)$ takes on the form of Eq. (5-131). From the conditions given by Eqs. (5-132) and (5-133), we have the results

$$c_j = \frac{p_j}{r_j} \qquad s_j = \frac{f_j^{(k)}}{r_j} \qquad r_j = \sqrt{p_j^2 + (f_j^{(k)})^2}.$$

On applying $\boldsymbol{P}_j^\dagger(k)$ to $\boldsymbol{L}_{j-1}(k)$, we obtain the matrix

$$
\boldsymbol{L}_j(k) \;=\; \begin{array}{c}
\qquad\qquad\qquad\qquad\qquad (j-1)\;\; j \\
\left(\begin{array}{cccccc|cccc}
d_1^{(k)}-\eta_k & f_2^{(k)} & 0 & \cdots & 0 & 0 & 0 & 0 & 0 & \cdots \\
f_2^{(k)} & d_2^{(k)}-\eta_k & f_3^{(k)} & \cdots & 0 & 0 & 0 & 0 & 0 & \cdots \\
\vdots & \vdots & \vdots & \ddots & \vdots & \vdots & \vdots & \vdots & \vdots & \ddots \\
0 & 0 & 0 & \cdots & g_{j-1} & p_{j-1} & 0 & 0 & 0 & \cdots \\
\hline
0 & 0 & 0 & \cdots & z_j & g'_j & p'_j & 0 & 0 & \cdots \\
0 & 0 & 0 & \cdots & 0 & z_{j+1} & g_{j+1} & p_{j+1} & 0 & \cdots \\
0 & 0 & 0 & \cdots & 0 & 0 & z_{j+2} & g_{j+2} & p_{j+2} & \cdots \\
\vdots & \vdots & \vdots & \ddots & \vdots & \vdots & \vdots & \vdots & \vdots & \ddots
\end{array}\right)
\end{array}
\begin{array}{l}
\\ \\ \\ \\ \\ (j-1) \\ j \\ \\ \\ \\
\end{array} .
$$

A new nonzero matrix element z_j in row j and column $(j-2)$ is produced from the product of s_j in row j and column $(j-1)$ of $P_j^\dagger(k)$ and $f_{j-1}^{(k)}$ in row $(j-1)$ and column $(j-2)$ of $\boldsymbol{L}_{j+1}(k)$. We shall not be concerned with it here, as the final form of $\boldsymbol{J}(k+1)$ is tridiagonal, and any elements outside the diagonal and subdiagonal of $\boldsymbol{L}(k)$ must vanish when it is multiplied by $\boldsymbol{Q}(k)$.

There are four other changes in the matrix elements between $\boldsymbol{L}_{j+1}(k)$ and $\boldsymbol{L}_j(k)$ (in addition to $z_j = s_j f_{k-1}^{(k)}$, which we do not need to be concerned with):

$$
\begin{aligned}
p_{j-1} &= c_j(d_{j-1}^{(k)} - \eta_k) - s_j g_j & \qquad\qquad g_{j-1} &= c_j f_{j-1}^{(k)} \\
p'_j &= s_j f_j^{(k)} + c_j p_j = r_j & \qquad\qquad g'_j &= s_j(d_{j-1}^{(k)} - \eta_k) + c_j g_j .
\end{aligned}
\tag{5-137}
$$

The values of p_{j-1} and g_{j-1} are needed in the next rotation and must be stored temporarily. The values of g'_j and p'_j may be put to use right away to produce the values of $d_j^{(k+1)}$ and $f_j^{(k+1)}$ for the matrix $\boldsymbol{J}(k+1)$. For this purpose, we need to multiply with the matrix for the previous rotation $\boldsymbol{P}_{j+1}(k)$, involving matrix elements c_{j+1} and s_{j+1}, as well as $\boldsymbol{P}_j(k)$ of this rotation. From the matrix element at location $(j, j+1)$ in the product $\boldsymbol{L}_j(k)\boldsymbol{P}_{j+1}(k)$, we obtain

$$
f_{j+1}^{(k+1)} = s_{j+1} p'_j = s_{j+1} r_j .
$$

Similarly, the matrix element at location (j, j) now has the value $c_{j+1} r_j$. To calculate the matrix element $d_j^{(k+1)}$, we need to multiply from the right by $\boldsymbol{P}_j(k)$. This gives

$$
\begin{aligned}
d_j^{(k+1)} &= s_j g'_j + c_j c_{j+1} r_j \\
&= s_j\{s_j(d_{j-1}^{(k)} - \eta_k) + c_j g_j\} + c_{j+1}\, c_j\, r_j \\
&= s_j^2(d_{j-1}^{(k)} - \eta_k) + s_j\, c_j\, g_j + \zeta
\end{aligned}
\tag{5-138}
$$

where

$$
\zeta = c_{j+1}\, c_j\, r_j = c_{j+1}\, p_j .
$$

Box 5-9 Subroutine TRIQL(D,F,V,N,NDMN)
Eigenvalues and eigenvectors of a real symmetric matrix – *QL* algorithm

Argument list:

D: Input diagonal matrix elements. Eigenvalues on output.
F: Input superdiagonal matrix elements.
V: Input transformation matrix $\boldsymbol{R}$. Output eigenvector matrix.
N: Matrix dimension.
NDMN: Dimension of arrays in the calling programs.

Initialization:

Shift the location of f_i by one, (i.e., $f_{i-1} = f_i$). Let $f_n = 0$.

For $i = 1$ to n:

(a) Initialize the iteration counter for each f_i.
(b) Set up the criterion to test f_i according to Eq. (5-130).
(c) Break the matrix into two if any superdiagonal element is small.
(d) Shift the diagonal matrix elements:
  (i) Diagonalize the (2×2) matrix formed by d_i, d_{i+1}, and f_i.
  (ii) Use λ_1 of Eq. (5-134) as the shift parameter η_k.
  (iii) Store the sum of all the shifts so far.
  (iv) Shift all the diagonal elements not yet in diagonal form.
(e) QL-transformation:
  (i) Initialize $p = d_m$, $c = 1$, and $s = 0$.
  (ii) For $j = (m-1)$ to i in steps of -1: Calculate r_{j+1}, g_{j+1}, h_{j+1}, c_j, s_j, p_j, $f_{j+1}^{(k+1)}$, and $d_{j+1}^{(k+1)}$ of Eq. (5-139).
  (iii) Update the transformation matrix according to Eq. (5-142).
  (iv) Calculate the values of $d_i^{(k+1)}$ and $f_i^{(k+1)}$.
(f) Repeat step (e) if $f_i^{(k+1)}$ is not essentially zero.

Return the eigenvalues in D and the eigenvectors in A.

The final forms of both Eqs. (5-137) and (5-138) do not involve c_{j+1} and s_{j+1}. However, we must store the value of ζ before destroying c_{j+1}. Furthermore, it is also possible to arrange the calculations in such a way that the values of $f_{j+1}^{(k+1)}$ and $d_j^{(k+1)}$ are stored in the same locations as those for $f_{j+1}^{(k)}$ and $d_j^{(k)}$, saving the need of a separate array for $\boldsymbol{J}(k+1)$. The algorithm, modeled after that described in an article by Wilkinson and Reinsch,[57] and used also in lapack,[2] is given in Box 5-9.

The actual calculations required for step $(k+1)$ in a QL-transformation are summarized by Bowdler and coauthors[6] in the following way:

$$p_m = d_m^{(k)} - \eta_k \qquad c_{m+1} = 1 \qquad s_{m+1} = 0$$

$$
\text{for } \substack{j=m-1,\\ m-2,\ldots,\\ i+1,i} \begin{cases} r_{j+1} = \{p_{j+1}^2 + (f_{j+1}^{(k)})^2\}^{1/2} \\ g_{j+1} = c_{j+2} f_{j+1}^{(k)} \\ h_{j+1} = c_{j+2} p_{j+1} \\ f_{j+2}^{(k+1)} = s_{j+2} r_{j+1} \\ c_{j+1} = p_{j+1}/r_{j+1} \\ s_{j+1} = f_{j+1}^{(k)}/r_{j+1} \\ p_j = c_{j+1}(d_j^{(k)} - \eta_k) - s_{j+1} g_{j+1} \\ d_{j+1}^{(k+1)} = h_{j+1} + s_{j+1}\{c_{j+1} g_{j+1} + s_{j+1}(d_j^{(k)} - \eta_k)\} \end{cases}
$$

$$
f_{i+1}^{(k+1)} = s_{i+1} p_i \qquad d_i^{(k+1)} = c_{i+1} p_i. \tag{5-139}
$$

Usually, more than one iteration is required to bring a superdiagonal matrix elements f_i to sizes sufficiently small compared with $|d_{i-1}|$ and $|d_i|$ to be considered as zero. However, the number of iterations required is seldom large, four to five for small values of i if the matrix dimension is $n = 30$ or so, and much less later (larger values of i) when the matrix has undergone several QL-transformations and is partially diagonal already.

Eigenvectors To obtain the eigenvectors, we must start with the transformation matrix $\boldsymbol{R}$ of the previous section that reduces the original real symmetric matrix $\boldsymbol{A}$ to a tridiagonal form $\boldsymbol{J}$:

$$
\widetilde{\boldsymbol{R}}\boldsymbol{A}\boldsymbol{R} = \boldsymbol{J}.
$$

The matrix $\boldsymbol{V}$ that transforms $\boldsymbol{A}$ to a diagonal matrix $\boldsymbol{D}$ may be defined as

$$
\widetilde{\boldsymbol{V}}\boldsymbol{A}\boldsymbol{V} = \boldsymbol{D}. \tag{5-140}
$$

If we use $\boldsymbol{Q}$ to represent the transformation that reduces a tridiagonal matrix $\boldsymbol{J}$ to a diagonal matrix $\boldsymbol{D}$, we have the relation

$$
\boldsymbol{V} = \boldsymbol{R}\boldsymbol{Q}.
$$

Since $\boldsymbol{Q}$ is the product of many rotations, each having the form $\boldsymbol{P}(k)$ given by Eq. (5-131), $\boldsymbol{V}$ is obtained from a recursion relation in the form

$$
\boldsymbol{V}(k) = \boldsymbol{V}(k-1)\boldsymbol{P}(k). \tag{5-141}
$$

The starting point is $\boldsymbol{V}(1) = \boldsymbol{R}$. If the transformation coefficients are c_j and s_j in $\boldsymbol{P}(k)$ in step k, the matrix elements of $\boldsymbol{V}(k)$ are obtained from those of $\boldsymbol{V}(k-1)$ in the following way:

$$
v_{\ell,j+1}^{(k+1)} = c_j v_{\ell,j+1}^{(k)} + s_j v_{\ell,j}^{(k)} \qquad v_{\ell,j}^{(k+1)} = c_j v_{\ell,j}^{(k)} - s_j v_{\ell,j+1}^{(k)} \tag{5-142}
$$

where j and $(j+1)$ are the two rows where the rotation takes place. Both ℓ and j take on values $1, 2, \ldots, n$.

5-6 Lanczos method of constructing matrices

We have seen in the previous two sections that an efficient way to diagonalize a real symmetric matrix is to bring it first to a tridiagonal form. Once this is done, it is a relatively simple matter to obtain the eigenvalues and eigenvectors. We have not been concerned with the basis states in which the matrix is constructed. This comes from the principle that the final result should be independent of the basis. The usual practice is to take a set that is convenient for carrying out calculations of operators we are interested in.

For some problems it is possible to construct a basis, such that the matrix is tridiagonal to start with. Usually, it takes some extra work to find such a set of states. However, the effort may well be worthwhile, as there are two major advantages in using a tridiagonal basis. The first is the obvious ease in diagonalizing such a matrix. The second is in the amount of computer memory required to store the matrix. As we saw earlier, an n-dimensional tridiagonal matrix requires only n locations to store the diagonal elements $d_1, d_2, \ldots, d_n$, and another $(n-1)$ locations to store the off-diagonal elements $f_2, f_3, \ldots, f_n$. This is to be contrasted with the amount $n(n+1)/2$ required to store a real symmetric matrix in general. As a result, the tridiagonal basis offers the possibility of handling cases with much larger dimensions.

For eigenvalue problems it is possible to use the Lanczos method to construct a tridiagonal basis for the Hamiltonian operator. An extensive description of the various aspects of this method can be found in Cullum and Willoughby.[14] We will briefly introduce the method.

Construction of a tridiagonal basis Let us again start with the Schrödinger equation. Consider an arbitrary normalized wave function $|\Phi_1\rangle$. We shall assume that it is not an eigenvector of the Hamiltonian. As a result, the action of the Hamiltonian operator $\hat{H}$ on $|\Phi_1\rangle$ produces a function $|\boldsymbol{U}_1\rangle$ that is different from $|\Phi_1\rangle$,

$$\hat{H}|\Phi_1\rangle = |\boldsymbol{U}_1\rangle. \tag{5-143}$$

In general, $|\boldsymbol{U}_1\rangle$ is not normalized. Let us use N_1 to represent the normalization constant. That is,

$$\langle \boldsymbol{U}_1|\boldsymbol{U}_1\rangle = N_1^2. \tag{5-144}$$

A normalized vector $|\Psi_1\rangle$ may be defined by the following relation:

$$|\Psi_1\rangle = N_1^{-1}|\boldsymbol{U}_1\rangle.$$

Since $|\Phi_1\rangle$ is not an eigenvector of the Hamiltonian, $|\Psi_1\rangle$ and $|\Phi_1\rangle$ are not the same.

The difference between them may be expressed by saying that $|\Psi_1\rangle$ is a linear combination of $|\Phi_1\rangle$ and another function $|\Phi_2\rangle$. It is convenient to define $|\Phi_2\rangle$ as a normalized function orthogonal to $|\Phi_1\rangle$. That is,

$$\langle \Phi_2|\Phi_2\rangle = 1 \qquad\qquad \langle \Phi_2|\Phi_1\rangle = 0.$$

In terms of $|\Phi_1\rangle$ and $|\Phi_2\rangle$, we can write $|\Psi_1\rangle$ as

$$|\Psi_1\rangle = \alpha_1|\Phi_1\rangle + \beta_1|\Phi_2\rangle \tag{5-145}$$

with

$$\alpha_1^2 + \beta_1^2 = 1. \tag{5-146}$$

On multiplying both sides of Eq. (5-143) by $\langle\Phi_1|$ and integrating over all the independent variables, we obtain

$$\langle\Phi_1|\hat{H}|\Phi_1\rangle = N_1\langle\Phi_1|\Psi_1\rangle = \alpha_1 N_1. \tag{5-147}$$

Since the value of N_1 is known from Eq. (5-144), we obtain α_1 by calculating the diagonal matrix element of $\hat{H}$ for $|\Phi_1\rangle$. From this and Eq. (5-145), we obtain

$$|\Phi_2\rangle = \frac{1}{\beta_1}\Big(|\Psi_1\rangle - \alpha_1|\Phi_1\rangle\Big).$$

The value of β_1 may be found from Eq. (5-146) or from the normalization requirement of $|\Phi_2\rangle$. This is basically the same as the Gram-Schmidt orthogonalization method of constructing a set of orthonormal vectors from an arbitrary set. Before we proceed further, it is convenient to define two other quantities:

$$\begin{aligned} d_1 &\equiv \langle\Phi_1|\hat{H}|\Phi_1\rangle = \alpha_1 N_1 \\ f_2 &\equiv \langle\Phi_2|\hat{H}|\Phi_1\rangle = N_1\langle\Phi_2|\Psi_1\rangle = \beta_1 N_1. \end{aligned} \tag{5-148}$$

We shall see later that these are the first diagonal and superdiagonal elements of the tridiagonal matrix we are after.

What happens if we apply the Hamiltonian operator on $|\Phi_2\rangle$? Again, in general, $|\Phi_2\rangle$ cannot be an eigenvector of $\hat{H}$. As a result,

$$\hat{H}|\Phi_2\rangle = N_2|\Psi_2\rangle \tag{5-149}$$

where $|\Psi_2\rangle$ is a normalized state, similar to $|\Psi_1\rangle$, and N_2 is a constant. In general, $|\Psi_2\rangle$ cannot be a linear combination of $|\Phi_1\rangle$ and $|\Phi_2\rangle$ alone. We shall come back to the reasons for this later. For now, we shall take this premise for granted and, similar to Eq. (5-145), we shall express $|\Psi_2\rangle$ as a linear combination of $|\Phi_1\rangle$, $|\Phi_2\rangle$, and another function, which we shall label as $|\Phi_3\rangle$. That is,

$$|\Psi_2\rangle = \alpha_2|\Phi_1\rangle + \beta_2|\Phi_2\rangle + \gamma_2|\Phi_3\rangle. \tag{5-150}$$

In other words, any part of $|\Psi_2\rangle$ that cannot come from a linear combination of $|\Phi_1\rangle$ and $|\Phi_2\rangle$ is represented by the new state $|\Phi_3\rangle$. We shall choose $|\Phi_3\rangle$ to be a normalized function and orthogonal to both $|\Phi_1\rangle$ and $|\Phi_2\rangle$. This gives us the condition

$$\alpha_2^2 + \beta_2^2 + \gamma_2^2 = 1$$

obtained by squaring both sides of Eq. (5-150) and integrating over all the independent variables.

From the fact that the Hamiltonian is real and symmetric, we find that

$$\langle\Phi_1|\hat{H}|\Phi_2\rangle = \langle\Phi_2|\hat{H}|\Phi_1\rangle = f_2.$$

As we have already done in Eq. (5-148), this matrix element is labeled as f_2 because it is the first off-diagonal matrix element of $\hat{H}$. The convention we follow here is the same as that adopted in the previous sections in representing the superdiagonal elements of a tridiagonal matrix. On the other hand, from Eqs. (5-149) and (5-150), we obtain

$$\langle\Phi_1|\hat{H}|\Phi_2\rangle = N_2\langle\Phi_1|\Psi_2\rangle = N_2\alpha_2.$$

This gives us the value of α_2,

$$\alpha_2 = f_2/N_2.$$

With $|\Phi_2\rangle$, we obtain

$$d_2 \equiv \langle\Phi_2|\hat{H}|\Phi_2\rangle.$$

The value of β_2 is then given by

$$\langle\Phi_2|\hat{H}|\Phi_2\rangle = N_2\langle\Phi_2|\Psi_2\rangle = N_2\beta_2$$

or

$$\beta_2 = d_2/N_2.$$

Similar to Eq. (5-147), we have

$$|\Phi_3\rangle = \frac{1}{\gamma_2}\Big(|\Psi_2\rangle - \alpha_2|\Phi_1\rangle - \beta_2|\Phi_2\rangle\Big).$$

It is useful to define here

$$f_3 \equiv \langle\Phi_3|\hat{H}|\Phi_2\rangle$$

in the same spirit as we did in Eq. (5-148). Note also that the matrix element

$$\langle\Phi_3|\hat{H}|\Phi_1\rangle = 0. \tag{5-151}$$

This comes from the fact that, by the definition given in Eq. (5-145), the state produced by the action of $\hat{H}$ on $|\Phi_1\rangle$ is only a linear combination of $|\Phi_1\rangle$ and $|\Phi_2\rangle$, and both of them are orthogonal to $|\Phi_3\rangle$.

We shall continue this line of argument for one more step before giving the general result. When the Hamiltonian operator is applied to $|\Phi_3\rangle$, we obtain the result

$$\hat{H}|\Phi_3\rangle = N_3|\Psi_3\rangle$$

in analogy with Eq. (5-149). Similarly, we shall assume that $|\Psi_3\rangle$ is normalized to unity. Because of Eq. (5-151) and the fact that the Hamiltonian is Hermitian, the matrix is symmetric. As a result, we have

$$\langle\Phi_1|\hat{H}|\Phi_3\rangle = 0.$$

This comes from the fact that, by construction, the only nonzero off-diagonal matrix element of $\hat{H}$ acting on $|\Phi_1\rangle$ is $\langle\Phi_2|\hat{H}|\Phi_1\rangle$. It also implies that $|\Psi_3\rangle$ is orthogonal to $|\Phi_1\rangle$. In other words, $|\Psi_3\rangle$ can only be a linear combination of $|\Phi_2\rangle$, $|\Phi_3\rangle$, and a new basis state $|\Phi_4\rangle$:

$$|\Psi_3\rangle = \alpha_3|\Phi_2\rangle + \beta_3|\Phi_3\rangle + \gamma_3|\Phi_4\rangle. \tag{5-152}$$

Similar to what we have done earlier, $|\Phi_4\rangle$ is normalized and orthogonal to $|\Phi_2\rangle$ and $|\Phi_3\rangle$. Furthermore, since $|\Psi_3\rangle$ is orthogonal to $|\Phi_1\rangle$, the new state $|\Phi_4\rangle$ must also be orthogonal to $|\Phi_1\rangle$.

The relation represented by Eq. (5-152) is general. This can be seen in the following way: If we continue the process of finding new basis vector $|\Phi_1\rangle$, $|\Phi_2\rangle$, ..., until we come to $|\Phi_k\rangle$, we have the result

$$\hat{H}|\Phi_k\rangle = N_k|\Psi_k\rangle \tag{5-153}$$

in analogy to Eq. (5-149). It is always possible to express the new state $|\Psi_k\rangle$ produced as a linear combination of three components,

$$|\Psi_k\rangle = \alpha_k|\Phi_{k-1}\rangle + \beta_k|\Phi_k\rangle + \gamma_k|\Phi_{k+1}\rangle \tag{5-154}$$

as we have demonstrated for $|\Psi_3\rangle$ in Eq. (5-152). The new vector $|\Phi_{k+1}\rangle$ can be made to be orthogonal to all the known basis vectors, $|\Phi_1\rangle$, $|\Phi_2\rangle$, ..., $|\Phi_k\rangle$. Furthermore, we see that the Hamiltonian matrix is tridiagonal in the basis formed by $|\Phi_1\rangle$, $|\Phi_2\rangle$, ...

An exception to Eq. (5-154) happens when the dimension of the Hilbert space n is finite. In this case, the total number of linearly independent basis states we can construct is limited to n. When we reach $k = n$ in our basis state construction, all the required states have been found and no new ones can be generated. As a result,

$$|\Phi_{n+1}\rangle = 0.$$

This means that the process of finding more basis states, as expected, cannot be continued beyond $k = n$.

What happens if $\gamma_k = 0$ at some stage for $k < n$ of our construction for the basis states? Obviously, we cannot continue the procedure if this takes place, as it implies that the action of $\hat{H}$ on $|\Psi_k\rangle$ does not contain any new component that is not already in the basis states already found. Except for reasons of poor numerical accuracy, this can occur only if the Hilbert space for the problem consists of two or more independent subspaces. Earlier in §5-5 we encountered the inverse problem. There, we saw that a tridiagonal matrix can be separated into two or more independent submatrices if one or more of the superdiagonal elements vanish. It is obvious that the two topics, $\gamma_k = 0$ and $f_k = 0$, are closely related to each other. We shall delay any further discussion on this point until we come to the question of actually constructing a tridiagonal basis.

Because of truncation errors in the numerical calculations, it may be difficult in practice to judge if γ_k is essentially zero for large n. However, as we shall see next, the strength of the Lanczos method lies in that, for eigenvectors of interest, it may be quite adequate to generate only a small fraction of the total number of the tridiagonal basis states in the space, long before any problem of numerical accuracy can dominate the calculations.

Let us label the elements of a tridiagonal matrix in the same way as in the previous sections. That is,

$$d_k = \langle\Phi_k|\hat{H}|\Phi_k\rangle \qquad\qquad f_k = \langle\Phi_{k-1}|\hat{H}|\Phi_k\rangle. \tag{5-155}$$

It is easy to see that, in general,

$$\hat{H}|\Phi_k\rangle = f_k|\Phi_{k-1}\rangle + d_k|\Phi_k\rangle + f_{k+1}|\Phi_{k+1}\rangle. \tag{5-156}$$

Since it is relatively a simple matter to diagonalize such a matrix, we can afford to carry out the operation as we construct the basis states. This is a meaningful thing to do since, in most eigenvalue problems, we are interested only in a very small fraction of the total number of eigenstates in the complete Hilbert space. In fact, the usual situation is that we are only concerned with a few of the low-lying ones near the ground state. In such cases, if the starting state is well chosen, all the basis states that contribute significantly to the low-lying states emerge at relatively early stages in generating the tridiagonal basis. Once we have the main components, any additional states cannot affect very much the eigenvectors of interest. As a result, there cannot be much loss in the accuracy of the final results if we do not include any of the basis states that are yet to be found. In other words, good approximations of the results of interest to us may be obtained without having to complete the construction of all the possible tridiagonal basis states in the Hilbert space. This is one of the strengths of the Lanczos approach. What we need now is a practical way to implement the scheme by having a method that recognizes when a sufficient number of basis states is reached.

Ground state energy illustration As an example, let us consider the problem of finding the ground state energy of a system. Using intuition, we can usually pick a starting state $|\Phi_1\rangle$ for our tridiagonal basis that is close to the ground state wave function. Let the value of the diagonal matrix element calculated with this state be $d_1 = \langle\Phi_1|\hat{H}|\Phi_1\rangle$. In general, d_1 cannot be the ground state energy, as $|\Phi_1\rangle$ is unlikely to be the eigenfunction. In fact, from a variational principle point of view, we know that the expectation value of the Hamiltonian in any state in the same Hilbert space must be higher than that in the ground state. If we represent the true ground state energy as $\mathcal{E}_1$, we expect that $d_1 \geq \mathcal{E}_1$.

We can construct a second basis state $|\Phi_2\rangle$ using Eqs. (5-143) and (5-145). With $|\Phi_1\rangle$ and $|\Phi_2\rangle$, we can find d_2 and f_2. In this enlarged "active" space of two tridiagonal basis states, we expect to produce an eigenvector that is a better approximation of the ground state. Let λ_1 be the lower one of the two eigenvalues of this (2×2) matrix, formed of d_1, d_2, and f_2. Since it represents a better approximation to the ground state energy, we expect it to be lower in value than d_1 but most likely still higher than $\mathcal{E}_1$.

We can add more tridiagonal basis states to the calculation and enlarge the active space. Each time we add one more state, we expect the lowest eigenvalue, obtained from diagonalizing the matrix constructed so far, to be lowered. Eventually, the value will converge to $\mathcal{E}_1$, similar to a variational calculation. If the convergence is fast, and this seems to be true in many physical problems, the ground state energy may be obtained in an active space that is only a small fraction of the complete Hilbert space for the problem.

There are two ways to recognize when convergence is achieved. The first is to make use of the lowest eigenvalue obtained in the calculation. If the value is not changed in any significant manner when more tridiagonal basis states are added to the space, it is likely that the calculation has converged. The second is to make use of the eigenvector. When we get to the stage that additional basis states in the active space do not make any essential contributions to the wave function of the lowest state, it may be safe to assume that any further additions will not change the result in any significant way either and the calculation has converged. In either case, one must make sure that the result is a genuine indication of convergence rather than a fictitious one caused by truncation errors. One way to get some insurance against possible erroneous conclusions caused by poor numerical accuracy is to take the calculation some distance beyond convergence. Alternatively, we can try to start with a different $|\Phi_1\rangle$ and see if the calculation converges to the same result.

The most obvious exception to the practical method described above in recognizing convergence is when the Hilbert space consists of two or more parts that have only very small off-diagonal matrix elements connecting the states in the different parts. Under such circumstances, it may happen that the starting state chosen is in the wrong part. In this case, the calculation may converge only to the lowest state in that part of the space, but not necessarily to the ground state of the system as a whole. Such pathological cases can usually be detected on physical grounds, and a different starting state can be used.

Realization of tridiagonal basis states So far, we have discussed the tridiagonal basis states $|\Phi_i\rangle$ in abstract. In actual calculations, it is necessary for us to express $|\Phi_i\rangle$ in terms of some known functions. For this purpose, it is often more convenient to have a complete set of orthonormal functions $|\phi_j\rangle$, for $j = 1, 2, \ldots, n$, and expressed each tridiagonal state in terms of a linear combination of $|\phi_j\rangle$:

$$|\Phi_i\rangle = \sum_{j=1}^{n} c_{i,j}|\phi_j\rangle. \tag{5-157}$$

For a given set of $|\phi_j\rangle$, the function $|\Phi_i\rangle$ is completely specified by coefficients $c_{i,j}$.

The choice for $|\phi_j\rangle$ should be made both on physical grounds and for mathematical convenience. For example, the Hamiltonian may be separated into two parts:

$$\hat{H} = \mathcal{H}_0 + \mathcal{H}'.$$

If the eigenfunctions of $\mathcal{H}_0$ are known, such as those for a particle in a harmonic oscillator well of §4-1, we may use this set of functions as our $\{|\phi_i\rangle\}$.

Once a complete set of $|\phi_j\rangle$ is selected, we can proceed to find the starting tridiagonal basis state $|\Phi_1\rangle$. As mentioned earlier, this should be based on some physical notion we have of the system. Let

$$|\Phi_1\rangle = \sum_{j=1}^{n} c_{1,j}|\phi_j\rangle.$$

Once the values of $c_{1,j}$ for $j = 1, 2, \ldots, n$ are chosen, the rest of the calculations are quite mechanical and can be easily put on a computer. For example, we can express $|\boldsymbol{U}_1\rangle$ of Eq. (5-143) in terms of $|\phi_j\rangle$ by invoking the closure property for a complete set of states:

$$|\boldsymbol{U}_1\rangle = \hat{H}|\Phi_1\rangle = \sum_{j=1}^{n} c_{1,j}\hat{H}|\phi_j\rangle = \sum_{j=1}^{n} c_{1,j}\sum_{k=1}^{n}|\phi_k\rangle\langle\phi_k|\hat{H}|\phi_j\rangle = \sum_{k=1}^{n} g_{1,k}|\phi_k\rangle \tag{5-158}$$

where

$$g_{1,k} = \sum_{j=1}^{n} c_{1,j}\langle\phi_k|\hat{H}|\phi_j\rangle = \sum_{j=1}^{n} c_{1,j}H_{k,j}.$$

To simplify the notation, we have adopted the shorthand

$$H_{i,j} \equiv \langle\phi_i|\hat{H}|\phi_j\rangle \tag{5-159}$$

and we shall continue with this practice for the rest of this section.

In terms of $H_{i,j}$, the value of the first diagonal matrix element may be written as

$$d_1 = \langle\Phi_1|\hat{H}|\Phi_1\rangle = \sum_{i,j} c_{1,i}c_{1,j}H_{i,j}.$$

Since all the coefficients $c_{1,i}$ are available and the matrix elements $H_{i,j}$ can be calculated, there is no difficulty in obtaining the value of d_1. Once this is done, we shall proceed to find

$$|\Phi_2\rangle = \sum_{j=1}^{n} c_{2,j}|\phi_j\rangle$$

by calculating the coefficients $c_{2,j}$ for $j = 1, 2, \ldots, n$. Using the relations given by Eqs. (5-156) and (5-158), we have

$$\hat{H}|\Phi_1\rangle = d_1|\Phi_1\rangle + f_2|\Phi_2\rangle = \sum_{j=1}^{n} g_{1,j}|\phi_j\rangle.$$

Since the basis states $|\phi_j\rangle$ are independent of each other, we obtain a relation between the coefficients in the form

$$f_2\, c_{2,j} = g_{1,j} - d_1c_{1,j}.$$

Using the fact that $|\Phi_2\rangle$ is normalized to unity, $\sum_j c_{2,j}^2 = 1$, we obtain the result,

$$f_2^2 = \sum_{j=1}^{n}(g_{1,j} - d_1c_{1,j})^2.$$

With f_2, we can calculate the values of the coefficients

$$c_{2,j} = \frac{1}{|f_2|}(g_{1,j} - d_1 c_{1,j}) \tag{5-160}$$

for $j = 1, 2, \ldots, n$. There is an overall ambiguity in sign for all the coefficients $c_{2,1}$, $c_{2,2}, \ldots, c_{2,n}$ that cannot be determined. This affects the sign of f_2, as we can see from the relation

$$f_2 = \langle \Phi_1 | \hat{H} | \Phi_2 \rangle = \sum_{i,j} c_{1,i}\, c_{2,j} H_{i,j}.$$

However, it does not have any effect on the diagonal matrix element

$$d_2 = \langle \Phi_2 | \hat{H} | \Phi_2 \rangle = \sum_{i,j} c_{2,i}\, c_{2,j} H_{i,j}.$$

We shall see later that such an overall sign is not important physically.

To derive the equations for the general case, let us assume that we have already found k tridiagonal basis states $|\Phi_1\rangle$, $|\Phi_2\rangle, \ldots, |\Phi_k\rangle$. This means that all the coefficients $c_{i,j}$ for $i = 1, 2, \ldots, k$ and $j = 1, 2, \ldots, n$ are known. The input quantities required to find the next state,

$$|\Phi_{k+1}\rangle = \sum_{j=1}^{n} c_{k+1,j} |\phi_j\rangle$$

are the kth diagonal and superdiagonal elements, d_k and f_k, and the coefficients $c_{k-1,j}$ and $c_{k,j}$ for $j = 1, 2, \ldots, n$ related to the tridiagonal basis states $|\Phi_{k-1}\rangle$ and $|\Phi_k\rangle$. Let

$$|\boldsymbol{U}_k\rangle = \hat{H}|\Phi_k\rangle = \sum_{j=1}^{n} g_{k,j} |\phi_j\rangle \tag{5-161}$$

where, similar to Eq. (5-158), we have

$$g_{k,j} = \sum_{\ell=1}^{n} c_{k,\ell} H_{j,\ell}.$$

On the other hand, using Eq. (5-156), we have

$$\hat{H}|\Phi_k\rangle = f_k|\Phi_{k-1}\rangle + d_k|\Phi_k\rangle + f_{k+1}|\Phi_{k+1}\rangle.$$

At this stage, f_{k+1} and the expansion coefficient $c_{k+1,j}$ are still unknown to us.

From the fact that Eq. (5-156) holds for an arbitrary set of basis states, we obtain the following relation using Eq. (5-161):

$$f_{k+1} c_{k+1,j} = g_{k,j} - f_k c_{k-1,j} - d_k c_{k,j}. \tag{5-162}$$

Since $|\Phi_{k+1}\rangle$ is normalized to unity, $\sum_j c_{k+1,j}^2 = 1$, we have

$$f_{k+1}^2 = \sum_{j=1}^{n} (g_{k,j} - f_k c_{k-1,j} - d_k c_{k,j})^2.$$

This gives us the value of f_{k+1} up to a sign. Using this result, we can calculate the values of the coefficients that express $|\Phi_{k+1}\rangle$ in terms of $|\phi_i\rangle$,

$$c_{k+1,j} = \frac{1}{|f_{k+1}|}(g_{k,j} - f_k c_{k-1,j} - d_k c_{k,j}) \tag{5-163}$$

similar to what we have done in Eq. (5-160). From this, we can calculate d_{k+1} using the relation

$$d_{k+1} = \sum_{i,j} c_{k+1,i}\, c_{k+1,j} H_{i,j}. \tag{5-164}$$

This completes all the calculations associated with the new tridiagonal basis state $|\Phi_{k+1}\rangle$. We can now proceed to find $|\Phi_{k+2}\rangle$, if we wish, by substituting $(k+1)$ for k in Eqs. (5-161) to (5-164). In principle, we can repeat the process as many times as we wish until the total number of basis states in the Hilbert space is exhausted. However, as mentioned earlier, this is not the usual aim of the Lanczos method.

As soon as a reasonable number of diagonal and superdiagonal elements of the tridiagonal matrix is available, we can use either bisection or QL-method given in the previous section to diagonalize the matrix obtained so far. By comparing the results of diagonalizing matrices of successively larger dimensions, it is possible to reach a conclusion on whether we have enough tridiagonal basis states in the active space.

Note that in Eq. (5-164), if the overall sign of $c_{k+1,j}$ is reversed, the sign of f_{k+1} is changed but not that of d_{k+1}, as noted earlier. As a result, such a sign change does not affect the results of the next tridiagonal basis vector. This can be seen by looking at Eq. (5-163). If we change the overall sign of all the coefficients $c_{k-1,j}$, for $j = 1$ to n, of the basis vector $|\Phi_{k-1}\rangle$, the sign of f_k is changed, but the sign of the product $f_k\, c_{k-1,j}$ remains the same. Similarly, if we change the overall sign of $c_{k,j}$ for all possible values of j, both f_k and $g_{k,j}$ change sign, but not d_k. The net result is a sign change for all the coefficients $c_{k+1,j}$, for $j = 1$ to n, corresponding to an overall sign change in the tridiagonal basis state. Since such a sign difference cannot be observed in any measurements, it has no physical consequence.

Anharmonic oscillator example In §4-1, we have seen that analytical solutions are available for a quantum mechanical particle inside a one-dimensional, harmonic oscillator. The Hamiltonian operator has the form

$$\mathcal{H}_0 = -\frac{\hbar^2}{2\mu}\frac{d^2}{dx^2} + \frac{1}{2}\mu\omega^2 x^2.$$

In more realistic situations, we may also include an *anharmonic* term in the potential:

$$V_{\rm anh}(x) = \epsilon\hbar\omega\left(\frac{\mu\omega}{\hbar}\right)^2 x^4.$$

Here, ϵ is a dimensionless parameter describing the strength of the correction to the simple harmonic potential, as we saw earlier in Eq. (5-47) of §5-3.

Again, we shall write the variable x in terms of the dimensionless quantity

$$\rho = \sqrt{\frac{\mu\omega}{\hbar}}x.$$

The complete Hamiltonian takes on the form

$$\hat{H} = \mathcal{H}_0 + \epsilon\rho^4. \tag{5-165}$$

Similar to Eq. (4-8), the Schrödinger equation is reduced to the differential equation:

$$\frac{d^2\psi}{d\rho^2} + (\lambda - \rho^2 - \epsilon\rho^4)\psi(\rho) = 0. \tag{5-166}$$

Because of the additional term in the potential, the equation can no longer be solved by the method outlined in §4-1. If the anharmonic term is small, the solution is often found using perturbation techniques, such as the one used as illustrative example in §5-3. Here we shall use the Lanczos method to obtain the ground state. The advantage over the perturbative approach is that the parameter ϵ is not limited to small values; however, the convergence rate is faster for smaller ϵ.

Let us assume that the potential is still dominated by the harmonic term. As a result, it is convenient to use the harmonic oscillator wave functions

$$\psi_m(\rho) = \frac{1}{\sqrt{2^m m! \sqrt{\pi}}} e^{-\rho^2/2} H_m(\rho)$$

given in Eq. (4-9) as our set of functions $|\phi_i\rangle$. Here $H_m(\rho)$ is the Hermite polynomials of degree m defined in §4-1. Note that $\psi_m(\rho)$ is the eigenfunction of our $\mathcal{H}_0$ here, satisfying the relation

$$H_0\psi_m = \left(m + \frac{1}{2}\right)\psi_m \tag{5-167}$$

with the energy measured in units of $\hbar\omega$.

To match the notations used in this section with those adopted for §4-1, we shall define

$$\phi_i \equiv \psi_{i-1}(\rho)$$

so that ϕ_1 is the same as the ground state harmonic oscillator wave function $\psi_0(\rho)$, ϕ_2 that with principal quantum number 1, and so on. For our calculations here, we do not need explicitly the wave functions. Only the matrix elements

$$H_{i,j} \equiv \langle\phi_i|\hat{H}|\phi_j\rangle = \langle\psi_{i-1}|\mathcal{H}_0 + \epsilon\rho^4|\psi_{j-1}\rangle \tag{5-168}$$

of Eq. (5-159) enter into the various terms. Since the functions ϕ_i are the eigenfunctions of $\mathcal{H}_0$, we have the relation

$$\langle\phi_i|\mathcal{H}_0|\phi_j\rangle = \langle\psi_{i-1}|\mathcal{H}_0|\psi_{j-1}\rangle = \left(i - \frac{1}{2}\right)\delta_{i,j}$$

in units of $\hbar\omega$. This is essentially the same result as Eq. (5-167) except that it is written in the notation of this section. In terms of the matrix elements of $\hat{H}$, we have

$$\langle\psi_{i-1}|\mathcal{H}_0 + \epsilon\rho^4|\psi_{j-1}\rangle = \left(i - \frac{1}{2}\right)\delta_{i,j} + \epsilon\langle\psi_{i-1}|\rho^4|\psi_{j-1}\rangle.$$

For $m = \min(i, j) - 1$, the matrix elements of ρ^4 between harmonic oscillator wave functions may be expressed as

$$\langle\psi_{i-1}|\rho^4|\psi_{j-1}\rangle = \begin{cases} \dfrac{3}{2}(m^2 + m + \frac{1}{2}) & \text{for } i = j \\ (m + \dfrac{3}{2})\sqrt{(m+1)(m+2)} & \text{for } i = j \pm 2 \\ \dfrac{1}{4}\sqrt{(m+1)(m+2)(m+3)(m+4)} & \text{for } i = j \pm 4 \\ 0 & \text{otherwise.} \end{cases} \tag{5-169}$$

The derivation of these results is left as an exercise (see Problem 5-6).

A reasonable starting state to construct the tridiagonal basis is the ground state of the simple harmonic oscillator itself,

$$|\Phi_1\rangle = |\phi_1\rangle \equiv \psi_0(\rho).$$

In terms of the coefficients of Eq. (5-157), the same condition may be stated as

$$c_{1,j} = \begin{cases} 1 & \text{for } j = 1 \\ 0 & \text{otherwise.} \end{cases}$$

The first diagonal matrix element in our tridiagonal basis is then

$$d_1 = \langle\Phi_1|\hat{H}|\Phi_1\rangle = \langle\psi_0|\mathcal{H}_0 + \epsilon\rho^4|\psi_0\rangle = \frac{1}{2} + \frac{3}{4}\epsilon$$

in units of $\hbar\omega$. To obtain $|\Phi_2\rangle$, we use Eq. (5-158) to get

$$g_{1,k} = \sum_{j=1}^{n} c_{1,j}H_{k,j} = H_{k,1} = \begin{cases} d_1 & \text{for } k = 1 \\ \dfrac{3}{\sqrt{2}}\epsilon & \text{for } k = 3 \\ \sqrt{\dfrac{3}{2}}\epsilon & \text{for } k = 5 \\ 0 & \text{otherwise.} \end{cases}$$

From these, we obtain the values

$$f_2^2 = \epsilon^2\left(\frac{9}{2} + \frac{3}{2}\right) = 6\epsilon^2.$$

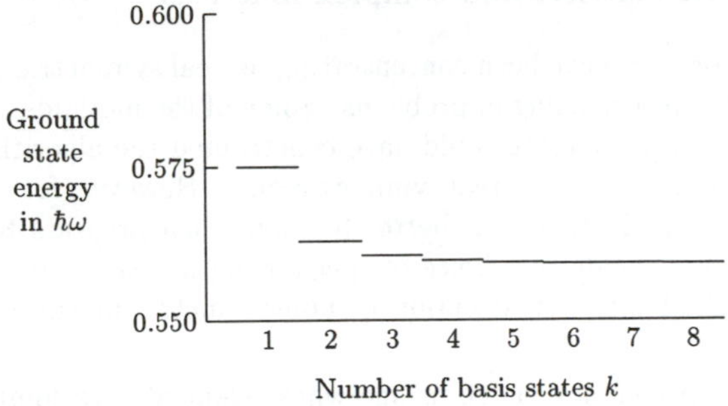

Figure 5-6: Ground state energy obtained with the Lanczos method for the anharmonic oscillator of (5-130) as a function of the number of tridiagonal basis states. For $\epsilon = 0.1$, convergence at $\sim 10^{-3}\hbar\omega$ is achieved with eight basis states.

This, in turn, gives us the value of $c_{2,j}$ using Eq. (5-160):

$$c_{2,j} = \left\{ 0,\ 0,\ \sqrt{\frac{3}{4}},\ 0,\ \sqrt{\frac{1}{4}},\ 0,\ 0, \cdots,\ 0 \right\}$$

for $j = 1, 2, \cdots, n$. With coefficients $c_{2,j}$, we can calculate d_2 according to Eq. (5-164), and so on.

The rest of the calculations may be carried out iteratively. For $k \geq 2$, we can calculate d_k by the relation

$$d_k = \sum_{i,j} c_{k,i} c_{k,j} H_{i,j}.$$

Similarly, f_k may be obtained using Eq. (5-162). This completes the calculations for all the elements in our tridiagonal basis with k states in the active space. We can now diagonalize the matrix obtained so far. Since we are interested only in the eigenvalues, the simpler bisection method may be used. Let us label the lowest eigenvalue as E_k, and this is to be compared with E_g, the lowest eigenvalue of a previous iteration. If the difference is larger than the accuracy required, we replace the value of E_g with that of E_k and proceed to obtain $|\Phi_{k+1}\rangle$. This is achieved by calculating $g_{k+1,j}$ using Eq. (5-161), f_{k+1} using (5-162), and then $c_{k+1,j}$ using (5-163). The step is completed by finding d_{k+1} using Eq. (5-164). For our illustrative example, we shall diagonalize the triangular matrix after each new basis state is added to the space. If we consider a change in the ground state energy of less than 10^{-3} in units of $\hbar\omega$ as the criterion, the calculation converges at $k = 4$ for a small anharmonic term of $\epsilon = 0.01$ and at $k = 8$ for a larger anharmonic term of $\epsilon = 0.1$. The rate of convergence for the latter case is shown in Fig. 5-6.

5-7 Nonsymmetric matrices and complex matrices

In the previous sections, we have been concentrating on real symmetric matrices, as it is the more common type occurring in problems. Some of the methods apply also to a more general class. In principle, we could have constructed the algorithms to handle a wider variety of matrices, not just real symmetric ones. However, for computational efficiency and ease of application, it is better to tailor each program to a particular type, and we have so far made the choice of specializing in real symmetric matrices. In this section, we shall turn our attention to nonsymmetric matrices and complex matrices.

Complex matrices A complex matrix is one whose elements are complex numbers. For example, the matrix element (k, ℓ) of a complex matrix $\boldsymbol{C}$ has the form

$$c_{k,\ell} = a_{k,\ell} + ib_{k,\ell}$$

where both $a_{k,\ell}$ and $b_{k,\ell}$ are real numbers, and $i^2 = -1$. Using the definition of matrix addition given in Table 5-1, we can write $\boldsymbol{C}$ as the sum of two real matrices $\boldsymbol{A}$ and $\boldsymbol{B}$ in an analogous way,

$$\boldsymbol{C} = \boldsymbol{A} + i\boldsymbol{B}. \tag{5-170}$$

Let $\boldsymbol{v}$ be an eigenvector of $\boldsymbol{C}$ with eigenvalue λ, i.e.,

$$\boldsymbol{C}\boldsymbol{v} = \lambda \boldsymbol{v}. \tag{5-171}$$

In general, the elements of $\boldsymbol{v}$ are also complex numbers and we can express $\boldsymbol{v}$ itself as a sum of two real vectors with complex coefficients:

$$\boldsymbol{v} = \boldsymbol{x} + i\boldsymbol{y} \tag{5-172}$$

similar to what we have done in Eq. (5-170).

In term of the real matrices $\boldsymbol{A}$, $\boldsymbol{B}$, $\boldsymbol{x}$, and $\boldsymbol{y}$, the complex eigenvalue problem of Eq. (5-171) may be written in the following way:

$$(\boldsymbol{A}\boldsymbol{x} - \boldsymbol{B}\boldsymbol{y}) + i(\boldsymbol{A}\boldsymbol{y} + \boldsymbol{B}\boldsymbol{x}) = \lambda(\boldsymbol{x} + i\boldsymbol{y}).$$

The equality must hold for the real and imaginary parts separately. If $\boldsymbol{C}$ is Hermitian, the eigenvalue λ is real. In this case, the relation may be expressed in terms of the following two for real matrices:

$$\begin{aligned} \boldsymbol{A}\boldsymbol{x} - \boldsymbol{B}\boldsymbol{y} &= \lambda \boldsymbol{x} \\ \boldsymbol{A}\boldsymbol{y} + \boldsymbol{B}\boldsymbol{x} &= \lambda \boldsymbol{y} \end{aligned}$$

or

$$\begin{pmatrix} \boldsymbol{A} & -\boldsymbol{B} \\ \boldsymbol{B} & \boldsymbol{A} \end{pmatrix} \begin{pmatrix} \boldsymbol{x} \\ \boldsymbol{y} \end{pmatrix} = \lambda \begin{pmatrix} \boldsymbol{x} \\ \boldsymbol{y} \end{pmatrix}. \tag{5-173}$$

In other words, the complex eigenvalue problem of Eq. (5-171) is equivalent to the following problem for real matrices:

$$\boldsymbol{C}'\boldsymbol{v}' = \lambda \boldsymbol{v}'$$

with

$$\boldsymbol{C}' = \begin{pmatrix} \boldsymbol{A} & -\boldsymbol{B} \\ \boldsymbol{B} & \boldsymbol{A} \end{pmatrix} \qquad \boldsymbol{v}' = \begin{pmatrix} \boldsymbol{x} \\ \boldsymbol{y} \end{pmatrix}. \tag{5-174}$$

If $\boldsymbol{C}$ is a complex matrix of dimension n, the dimension of the real matrix $\boldsymbol{C}'$ is $2n$. Similarly, $\boldsymbol{v}'$ is a column matrix with twice the number of elements as $\boldsymbol{v}$.

Since $\boldsymbol{C}$ is Hermitian, its matrix elements have the property

$$c^*_{k,\ell} = c_{\ell,k}. \tag{5-175}$$

In terms of real numbers, this is equivalent to

$$a_{k,\ell} - ib_{k,\ell} = a_{\ell,k} + ib_{\ell,k}.$$

From this, we conclude that submatrix $\boldsymbol{A}$ of Eq. (5-174) is a real symmetric matrix. Similarly, the submatrix $\boldsymbol{B}$ is a skew-symmetric matrix, with elements

$$b_{k,\ell} = -b_{\ell,k}.$$

As a result, $\boldsymbol{C}'$ is a real symmetric matrix. In this way, the eigenvalue problem for a complex, Hermitian matrix is reduced to one for a real symmetric matrix of twice the dimension.

If $\boldsymbol{v}$ of Eq. (5-172) is an eigenvector of the complex Hermitian matrix $\boldsymbol{C}$ with (real) eigenvalue λ, then $\boldsymbol{v}'$ of Eq. (5-174) is an eigenvector of the real matrix $\boldsymbol{C}'$ with the same eigenvalue. In fact, the eigenvalue equation for λ is also satisfied by

$$\boldsymbol{u}' = \begin{pmatrix} -\boldsymbol{y} \\ \boldsymbol{x} \end{pmatrix}.$$

We can see this by writing down the same equation as Eq. (5-173) for $\boldsymbol{u}'$

$$\begin{pmatrix} \boldsymbol{A} & -\boldsymbol{B} \\ \boldsymbol{B} & \boldsymbol{A} \end{pmatrix} \begin{pmatrix} -\boldsymbol{y} \\ \boldsymbol{x} \end{pmatrix} = \lambda \begin{pmatrix} -\boldsymbol{y} \\ \boldsymbol{x} \end{pmatrix}.$$

Since this is identical to Eq. (5-173), we find that $\boldsymbol{v}'$ and $\boldsymbol{u}'$ are linearly independent of each other as far as $\boldsymbol{C}'$ is concerned (but not $\boldsymbol{C}$). As a result, there are two degenerate eigenvectors, $\boldsymbol{v}'$ and $\boldsymbol{u}'$, for each eigenvalue λ. In other words, the number of unique eigenvalues for $\boldsymbol{C}'$ remains the same as that for $\boldsymbol{C}$. If we rewrite the eigenvectors in terms of complex numbers as

$$\boldsymbol{v} = \boldsymbol{x} + i\boldsymbol{y} \qquad \boldsymbol{u} = -\boldsymbol{y} + i\boldsymbol{x}$$

we find that

$$\boldsymbol{u} = i\boldsymbol{v}.$$

As a result, $\boldsymbol{u}$ is not independent of $\boldsymbol{v}$, and we find that the number of independent eigenvectors obtained from diagonalizing a complex matrix $\boldsymbol{C}$ of dimension n is exactly the same as diagonalizing an equivalent real matrix $\boldsymbol{C}'$ of dimension $2n$.

In terms of the efficiency in carrying out the calculations on a computer, it takes twice the number of locations to store $\boldsymbol{C}'$ than $\boldsymbol{C}$. This conclusion is arrived at in the following way. Ignoring for the moment any symmetry properties of the matrices, there are, in general, n^2 complex matrix elements in $\boldsymbol{C}$ and it takes $2n^2$ locations to store them (n^2 real and n^2 imaginary). In contrast, there are $(2n)^2$ real matrix elements in $\boldsymbol{C}'$. In principle, the factor of 2 increase in the number of matrix elements makes it inefficient to work with $\boldsymbol{C}'$ instead of $\boldsymbol{C}$. This conclusion can also be reached simply by counting the number of off-diagonal matrix elements that must be reduced to zero. However, this is not the entire story. Since it is usually more time consuming to carry out complex arithmetic on a computer, diagonalization routines written in terms of complex operations are often slower in practice. In fact, efficient diagonalization packages for complex matrices are often written in terms of real number operations. For this reason, it is preferable in many cases to use real symmetric matrix routines to handle complex Hermitian matrices.

If matrix $\boldsymbol{C}$ is not Hermitian, the eigenvalues are complex in general. Furthermore, the symmetry relations between matrix elements given by Eq. (5-175) are no longer true. As a result, the matrix $\boldsymbol{C}'$ is, in general, a nonsymmetric one, and methods to "diagonalize" nonsymmetric matrices must be used. However, all the other considerations discussed above for Hermitian matrices apply here.

Left and right eigenvectors Before we start on a discussion of the method to diagonalize a real nonsymmetric matrix, we shall examine first some of the differences from a real symmetric matrix. For a Hermitian matrix, both real and complex, the eigenvalues are real. On the other hand, for a real nonsymmetric matrix, some of the eigenvalues can be complex. Complex eigenvalues, if present, appear always in pairs for such matrices, with each member of a pair the conjugate to the other.

So far, we have always written the eigenvalue problem in the form

$$\boldsymbol{H}\boldsymbol{v}_R = \lambda \boldsymbol{v}_R \tag{5-176}$$

where λ is the eigenvalue and $\boldsymbol{v}_R$ the corresponding eigenvector. We shall call eigenvectors, defined in the way above for $\boldsymbol{v}_R$, right eigenvectors. Alternatively, the eigenvalue equation can also be written in the form with the eigenvector appearing on the left of the matrix $\boldsymbol{H}$,

$$\boldsymbol{v}_L \boldsymbol{H} = \lambda' \boldsymbol{v}_L. \tag{5-177}$$

The eigenvectors $\boldsymbol{v}_L$ are known as left eigenvectors. By transposing both sides of the equation, we have

$$\widetilde{\boldsymbol{v}_L \boldsymbol{H}} = \widetilde{\boldsymbol{H}}\widetilde{\boldsymbol{v}}_L = \lambda' \widetilde{\boldsymbol{v}}_L.$$

Since the value of the determinant of a matrix is equal to that of its transpose,

$$\det|\boldsymbol{H}| = \det|\widetilde{\boldsymbol{H}}|$$

we find that

$$\det|\boldsymbol{H}-\lambda\mathbf{1}| = \det|\widetilde{\boldsymbol{H}}-\lambda'\mathbf{1}|.$$

From this, we see that λ and λ' are the roots of two equivalent characteristic equations. As a result,

$$\lambda = \lambda'.$$

That is, the eigenvalues are the same for both left and right eigenvectors.

For a real symmetric matrix $\boldsymbol{H} = \widetilde{\boldsymbol{H}}$, we have the result

$$\boldsymbol{v}_L = \boldsymbol{v}_R$$

obtained by comparing Eq. (5-176) with (5-177). For this reason, we did not have to make any distinctions between left and right eigenvectors up to now. For a complex Hermitian matrix, $\boldsymbol{H} = \boldsymbol{H}^\dagger$, we have, instead, the relation

$$\boldsymbol{v}_L = \boldsymbol{v}_R^\dagger.$$

For a real nonsymmetric matrix, the situation is slightly more complicated, as we shall see below.

Consider the transformation matrix $\boldsymbol{V}_R$ made of all the right eigenvectors. By definition,

$$\boldsymbol{V}_R^{-1}\boldsymbol{H}\boldsymbol{V}_R = \boldsymbol{\lambda} \tag{5-178}$$

where $\boldsymbol{\lambda}$ is a diagonal matrix consisting of all the eigenvalues

$$\boldsymbol{\lambda} = \begin{pmatrix} \lambda_1 & 0 & 0 & \cdots & 0 \\ 0 & \lambda_2 & 0 & \cdots & 0 \\ \vdots & \vdots & \vdots & \ddots & \vdots \\ 0 & 0 & 0 & \cdots & \lambda_n \end{pmatrix}.$$

Since $\boldsymbol{\lambda}$ is diagonal, we may also write Eq. (5-178) in the form

$$\boldsymbol{H}\boldsymbol{V}_R = \boldsymbol{V}_R\boldsymbol{\lambda}. \tag{5-179}$$

Similarly, for the transformation matrix consisting of all the left eigenvectors, we have

$$\boldsymbol{V}_L\boldsymbol{H} = \boldsymbol{\lambda}\boldsymbol{V}_L. \tag{5-180}$$

Multiplying $\boldsymbol{V}_L$ from the left to both sides of Eq. (5-179), we obtain the result

$$\boldsymbol{V}_L\boldsymbol{H}\boldsymbol{V}_R = \boldsymbol{V}_L\boldsymbol{V}_R\boldsymbol{\lambda}.$$

Similarly, we can multiply $\boldsymbol{V}_R$ from the right to both sides of Eq. (5-180) and obtain the result

$$\boldsymbol{V}_L\boldsymbol{H}\boldsymbol{V}_R = \boldsymbol{\lambda}\boldsymbol{V}_L\boldsymbol{V}_R. \tag{5-181}$$

Comparing the last two equations, we find that

$$\boldsymbol{V}_L\boldsymbol{V}_R\boldsymbol{\lambda} = \boldsymbol{\lambda}\boldsymbol{V}_L\boldsymbol{V}_R.$$

For this to be true, it is necessary that the product $\boldsymbol{V}_L\boldsymbol{V}_R$ be a diagonal matrix. We can choose the normalization for each of the eigenvectors $\boldsymbol{v}_L$ and $\boldsymbol{v}_R$ in such a way that

$$\boldsymbol{V}_L\boldsymbol{V}_R = \boldsymbol{1}. \tag{5-182}$$

That is, the orthogonality relation among the eigenvectors of a real nonsymmetric matrix is between left and right eigenvectors. In fact, there is no corresponding relation among the left eigenvectors themselves or among the right eigenvectors. Furthermore, if there are degenerate eigenvalues, the left and right eigenvectors may not be orthogonal to each other to start with. In this case, it is necessary to take linear combinations of the degenerate eigenvectors to form the left and right eigenvectors that are orthogonal to each other.

Hessenberg form and $\boldsymbol{QL}$-transformation If our interest is in the eigenvalues alone, it is not necessary to diagonalize a nonsymmetric matrix completely — it is adequate to bring it to an upper or lower triangular form, a matrix with either all the elements below the diagonal or above the diagonal vanish. The diagonal elements are now the eigenvalues, as can be seen from the roots of the characteristic equation $\det|\boldsymbol{H} - \lambda\boldsymbol{1}| = 0$ for such a matrix.

Similarly, a Householder transformation of the type of Eq. (5-96) cannot reduce a nonsymmetric matrix $\boldsymbol{H}$ to a tridiagonal form. This is different from a symmetric matrix discussed in §5-4. In general, the most we can hope to achieve here is to reduce to zero all the elements in the lower (or upper) triangle, with the exception of the subdiagonal (or superdiagonal) ones. The transformed matrix takes on the form

$$\boldsymbol{H}' = \begin{pmatrix} h_{1,1} & h_{1,2} & h_{1,3} & \cdots & h_{1,n-1} & h_{1,n} \\ h_{2,1} & h_{2,2} & h_{2,3} & \cdots & h_{2,n-1} & h_{2,n} \\ 0 & h_{3,2} & h_{3,3} & \cdots & h_{3,n-1} & h_{3,n} \\ 0 & 0 & h_{4,3} & \cdots & h_{4,n-1} & h_{4,n} \\ \vdots & \vdots & \vdots & \ddots & \vdots & \vdots \\ 0 & 0 & 0 & \cdots & h_{n,n-1} & h_{n,n} \end{pmatrix}.$$

This is known as the upper Hessenberg form. From this, a QL-algorithm may be used to reduce all the subdiagonal elements to zero and transform the matrix into an upper triangular form. Alternatively, we can use a combination of Householder transformation and the QR-algorithm to achieve the same goal and reduce the matrix to a lower Hessenberg form. The details of the theory are given in Stoer and Bulirsch[54] and Wilkinson and Reinsch.[57] Computer programs to diagonalize real nonsymmetric matrices using such an approach can be found in the lapack[2] and in Press and coauthors.[44]

Jacobi method of diagonalizing a real nonsymmetric matrix To illustrate the differences between diagonalizing a real nonsymmetric matrix from that of a real symmetric one, it is instructive for us to adopt the simpler, but less efficient, Jacobi method. Our discussion is based on the paper by Eberlein and Boothroyd[20] that is also reprinted in Wilkinson and Reinsch.[57]

Our aim is to find a matrix $\boldsymbol{T}$ such that a similarity transformation of the form

$$\boldsymbol{T}^{-1}\boldsymbol{H}\boldsymbol{T} = \boldsymbol{\lambda}$$

reduces the real nonsymmetric $\boldsymbol{H}$ to a matrix $\boldsymbol{\lambda}$ containing the eigenvalues. Similar to the Jacobi method for real symmetric matrices, the transformation matrix $\boldsymbol{T}$ consists of the product of a series of transformations in the form

$$\boldsymbol{T} = \boldsymbol{T}_1\boldsymbol{T}_2\cdots\boldsymbol{T}_i\cdots$$

each of which is a rotation between two columns and two rows. Since the matrix is no longer symmetric, a rotation $\boldsymbol{T}_i$ in such a subspace, in general, can transform only one of the two off-diagonal elements to zero. As a result, it is necessary to add a shear, or complex rotation, in the same two-dimensional subspace to ensure convergence.

Each step of the transformation, represented symbolically by $\boldsymbol{T}_i$, is now a product of two operations, a rotation $\boldsymbol{R}_i$ and a shear $\boldsymbol{S}_i$. If the rotation is taking place in the subspace consisting of columns and rows k and ℓ with $k < \ell$, the elements of these two transformation matrices are given by

$$\begin{aligned} r_{k,k} = \quad r_{\ell,\ell} = \quad \cos\theta \qquad & s_{k,k} = s_{\ell,\ell} = \cosh p \\ r_{k,\ell} = -r_{\ell,k} = -\sin\theta \qquad & s_{k,\ell} = s_{\ell,k} = -\sinh p. \end{aligned} \tag{5-183}$$

For $i \neq k \neq \ell$ and $j \neq k \neq \ell$, the corresponding matrix elements are

$$r_{i,j} = \delta_{i,j} \qquad\qquad s_{i,j} = \delta_{i,j}. \tag{5-184}$$

Analogous to Eq. (5-64), the rotation angle here is selected to be

$$\tan 2\theta = \frac{h_{k,\ell} + h_{\ell,k}}{h_{k,k} - h_{\ell,\ell}}$$

where $h_{i,j}$ are the elements of the real nonsymmetric matrix whose eigenvalues and eigenvectors we wish to find. The shear parameter p in Eq. (5-183) has the value

$$\tanh p = \frac{ed - h}{g + 2(e^2 + d^2)}$$

where

$$\begin{aligned} e &= h_{k,\ell} - h_{\ell,k} \\ d &= (h_{k,k} - h_{\ell,\ell})\cos 2\theta + (h_{k,\ell} + h_{\ell,k})\sin 2\theta \\ g &= \sum_{i\neq k,\ell} (h_{k,i}^2 + h_{i,k}^2 + h_{\ell,i}^2 + h_{i,\ell}^2) \\ h &= \cos 2\theta \sum_{i\neq k,\ell} (h_{k,i}h_{\ell,i} - h_{i,k}h_{i,\ell}) - \frac{1}{2}\sin 2\theta \sum_{i\neq k,\ell} (h_{k,i}^2 + h_{i,k}^2 - h_{\ell,i}^2 - h_{i,\ell}^2). \end{aligned} \tag{5-185}$$

Box 5-10 Subroutine JNSYM(A,T,N,NDMN,L_R)
Eigenvalues and eigenvectors of real nonsymmetric matrices using a Jacobi-like algorithm

Argument list:

A:	Array for the input matrix.
T:	Array for the output transformation matrix.
N:	Matrix dimension.
NDMN:	Dimension of the arrays in the calling program.
L_R:	Input, +1 right eigenvector, −1 left eigenvector, 0 no eigenvector. Returns the number of iterations taken.

Initialization:
(a) Set up two criteria for small numbers, EPS $= 10^{-8}$ and EP $= \sqrt{\text{EPS}}$.
(b) Let the maximum number of iterations allowed be 50.
(c) Zero the iteration counter.
(d) Use logical variable MARK to indicate if further iterations needed.

1. Let T equal to the identity matrix.
2. Carry out a Jacobi-like transformation to $\boldsymbol{H}$:
   (a) Apply the transformation $\boldsymbol{R}_i$ and $\boldsymbol{S}_i$ according to Eq. (5-183):
      (i) Calculate the parameters e, d, g, and h of Eq. (5-185).
      (ii) Calculate rotation matrix elements $r_{i,j}$.
      (iii) Calculate the shear matrix elements $s_{i,j}$.
      (iv) Form the transformation matrix $\boldsymbol{T}_i = \boldsymbol{R}_i \times \boldsymbol{S}_i$.
   (b) Skip the transformation if $\boldsymbol{T}_i$ is an identity matrix. Otherwise
      (i) Set MARK to .FALSE..
      (ii) Apply the transformation to $\boldsymbol{H}$.
      (iii) Update $\boldsymbol{T}$ depending on left, right, or no eigenvectors are required.
3. Go back to step 2 unless one of the following is true:
   (a) If the maximum of iterations is exceeded, or
   (b) If MARK = .TRUE. (no transformation in the previous iteration), or
   (c) For all $i \neq j$, if $|h_{i,j} + h_{j,i}|$, $|h_{i,j} - h_{j,i}|$, and $|h_{i,i} - h_{j,j}| \leq$EPS.
4. Return eigenvalues in A, eigenvectors in T, and number of iterations in L_R.

The values are chosen to minimize the (Euclidean) norm of the matrix

$$\boldsymbol{H}(i+1) = (\boldsymbol{T}_1\boldsymbol{T}_2\cdots\boldsymbol{T}_i)^{-1}\boldsymbol{H}(\boldsymbol{T}_1\boldsymbol{T}_2\cdots\boldsymbol{T}_i).$$

Similar to Jacobi diagonalization of a real symmetric matrix, the calculation is iterative. Several passes, each going through all the nonvanishing off-diagonal matrix elements $h_{k,\ell}$ and $h_{\ell,k}$ for $k = 1, 2, \ldots, n$ and $\ell = k+1, k+2, \ldots, n$, are necessary before the matrix is reduced to a diagonal form.

If all the eigenvalues are real the transformed matrix $\boldsymbol{\lambda}$ is diagonal

$$\lambda_{k,\ell} = 0 \qquad \text{for } k \neq \ell$$

and the diagonal matrix elements are the eigenvalues. In general, some of the eigenvalues may be complex and, for these components of the matrix $\boldsymbol{\lambda}$, the elements have the form

$$\lambda_{k,\ell} = -\lambda_{\ell,k} \qquad \text{and} \qquad \lambda_{k,k} = \lambda_{\ell,\ell} \qquad \text{for } k \neq \ell.$$

Each complex eigenvalue has the form $\lambda_{k,k} \pm i\lambda_{k,\ell}$. For convenience, we can arrange each pair of such columns (and their corresponding rows), k and ℓ, to be adjacent to each other so that the complex eigenvalues appear as $\lambda_{k,k} \pm i\lambda_{k,k+1}$.

For real eigenvalues, the columns of the transformation matrix $\boldsymbol{T}$ are the right eigenvectors and the rows of $\boldsymbol{T}^{-1}$ are the left eigenvectors. For a complex eigenvalue $\lambda_{k,k} \pm i\lambda_{k,k+1}$, the corresponding eigenvector is also complex. The right eigenvector is $t_{j,k} \pm it_{j,k+1}$ for $j = 1, 2, \ldots, n$, where $t_{j,k}$ and $t_{j,k+1}$ are the matrix elements in columns k and $(k+1)$ of the jth row of $\boldsymbol{T}$. The left eigenvector is a similar linear combination of the matrix elements in rows k and $(k+1)$ of $\boldsymbol{T}^{-1}$.

The relation given by Eq. (5-181) may also be used to check the computer program for diagonalization. Because of Eq. (5-182), we have

$$\boldsymbol{V}_L \boldsymbol{H} \boldsymbol{V}_R = \boldsymbol{\lambda}. \tag{5-186}$$

In other words, we can use a pair of left and right vectors to transform $\boldsymbol{H}$. If the two vectors do not correspond to the same eigenvalue, the result should be zero. On the other hand, for both left and right vectors belonging to the same eigenvalue, a value λ is expected. To carry out this check, a copy of the original matrix must be saved. At the same time, both the left eigenvector matrix $\boldsymbol{V}_L$ and the right eigenvector matrix $\boldsymbol{V}_R$ must be calculated.

For matrices with relatively small dimensions, algorithms based on the Jacobi method are quite efficient. A simple way to implement it is given in Box 5-10. However, similar to the case of real symmetric matrices, the amount of computation increases as the matrix dimension cubed. As a result, methods based on Householder and QL-transformations are preferred for matrices with large dimensions.

Problems

5-1 Find the values of the three currents I_1, I_2, and I_3 for the circuit of Fig. 5-1 using $R_1 = 5\ \Omega$, $R_2 = 10\ \Omega$, $R_3 = 15\ \Omega$, and $V_1 = V_2 = 1.5$ V.

5-2 For a system of three linear equations with $\boldsymbol{A}$ given by Eq. (5-24), verify that the solution obtained using the inverse of $\boldsymbol{A}$, given by Eq. (5-20), is identical to that of the Gauss-Jordan elimination method of §5-1.

5-3 Construct an algorithm to obtain the value of a determinant of arbitrary order n by the method of decomposition into minors. Check the method by writing a computer program based on it. Test the program with the help of a random number generator.

5-4 If all the elements of the two matrices $\boldsymbol{U}$ and $\boldsymbol{Z}$ in Eq. (5-31) are known, solve for $\boldsymbol{X}$ by expressing each of the matrix elements, $x_1, x_2, \ldots, x_n$, in terms of $\ell_{i,j}$ and z_k. Compare the relations with those given by Eq. (5-33).

5-5 Obtain analytically the eigenvalues and eigenvectors of an anharmonic oscillator in the basis space consisting of three states, ϕ_0, ϕ_2, and ϕ_4, defined by Eq. (5-48). Use the potential V given by Eq. (5-47) and assume that ϵ is small. Express the results as a power series in ϵ.

5-6 Evaluate the matrix element $\langle\phi_m|\rho^4|\phi_{m'}\rangle$ using the harmonic oscillator wave functions given by Eq. (5-48). Check that the results are consistent with those given by Eq. (5-169).

5-7 One way to obtain an approximate solution of the ground state wave function for the anharmonic oscillator of Eq. (5-47) is to solve it as an eigenvalue problem in the Hilbert space of three harmonic oscillator states $\phi_0(\rho)$, $\phi_2(\rho)$, and $\phi_4(\rho)$. Construct the Hamiltonian matrix for $\epsilon = 0.1$ and compare the lowest eigenvalue obtained with that shown in Fig. 5-6, obtained by the Lanczos method.

5-8 Show that the trace of a matrix $\boldsymbol{M}$ is unchanged by a unitary transformation, that is, a similarity transformation $\boldsymbol{U}^{-1}\boldsymbol{M}\boldsymbol{U}$ with the transformation matrix $\boldsymbol{U}$ satisfying Eq. (5-58).

5-9 Construct all the polynomials in the Sturm sequence for the characteristic equation of the following tridiagonal matrix:

$$\boldsymbol{A} = \begin{pmatrix} 10 & 5 & 0 \\ 5 & -1 & -1 \\ 0 & -1 & 10 \end{pmatrix}.$$

Find the eigenvalues using the bisection method.

Chapter 6

Methods of Least Squares

A large part of the activity in science and engineering centers around taking measurements and comparing the results with theoretical expectations. This is a relatively easy task if the data are precise and the theories are well understood. However, such ideal situations occur only rarely. For most of the time, we are more likely to be dealing with cases where physical phenomena are not clear and measurements are imperfect. Under such circumstances, it may be necessary to make conjectures or models of the situation. To compensate for our incomplete knowledge, the models may contain a number of parameters that must be determined empirically. The uncertainties in the data, caused either by the limitations of the instruments used or by the nature of the physical quantities themselves, may contrive to make it difficult to determine these parameters from the measured values. If the number of independent pieces of data is larger than that of free parameters, it is possible to make progress with the help of statistical analyses. Among those commonly used, the method of least squares is by far the most important.

6-1 Statistical description of data

If we take two separate measurements of a physical object, there is a high probability that the results will be different from each other. As an example, let us think in terms of determining the distance between the curb on the north side of a city block and the one on the south side. There are many possible reasons for the discrepancy between the two results. For example, the measurements may have been carried out by counting the number of paces it takes to walk from one end of the block to the other and, as a result, it may not be possible to obtain the distance better than roughly to one-tenth of a pace, on the order of centimeters.

Other possible sources of discrepancies may also be present and they may have nothing to do with the accuracy of the "apparatus" used for the task. Our city block is not a sharply defined object. As a result, even if we replace our pace with a high-precision measuring device, there will still be differences, perhaps on the order of millimeters, between two measurements. The source of uncertainty in this case is related to the nature of the object we wish to measure. In physics, there are also quantities that cannot be measured precisely, such as those governed by the Heisenberg uncertainty principle.

We can also imagine another possible source of discrepancy between different measurements of the same quantity. If two persons with different strides are used to measure the length of a city block, the results in terms of the number of paces will be different. This is an obvious case of systematic error, resulting from a certain bias in the instruments used. If the sources of systematic error are known, it may be possible to make corrections for them. However, there are occasions where this is impossible and the uncertainties associated with systematic errors become an integral part of the data.

Distribution in the measured results Given the uncertainties associated with any measurements, how can we find the "true" value of a physical quantity? Let us begin the discussion by assuming that there is such a thing as the true value. Even where such an assumption is correct, there can still be differences in the results of any two measurements. Which one of the two results corresponds to the true value? The most probable situation is neither. Furthermore, it is not possible to argue that the average of the two measurements is the true value either. We can arrive at this conclusion in the following way. If we make a third measurement, it is unlikely that the result is exactly equal to the average value of the first two. As a result, the average of all three measurements is different from that of the first two. Since the average value changes, it is improbable that the average of any two measurements corresponds to the true value. On the other hand, if we make a large number of measurements, the average will approach some definite value, and additional measurements will not change it in any significant way. As a result, there is a high probability that the average value in this case is the same as the true value.

The notion that the true value of a quantity is given by the average of a large number of measurements is based on statistics. If the uncertainties associated with the measuring process are random, the values obtained will most likely be scattered around the true value with some definite distribution. In general, it may not be possible to deduce the distribution by theoretical considerations. On the other hand, if enough measurements are carried out under the same conditions, it is possible to map out this distribution to the desired precision. In the absence of any bias in the measurements, such as those caused by systematic errors, the probability of getting a particular measured value is purely statistical. In this way, the distribution reflects the stochastic nature of the uncertainties associated with the measurements and the nature of the quantity itself. Furthermore, if all the measured values are scattered randomly around the true value, the average of the measurements must approach the true value when the number of measurements is large.

It is also possible that in many circumstances there may not be a definite value associated with a particular quantity. For example, a diatomic molecule may be thought of as two spheres connected together by a spring. At finite temperatures, the two atoms vibrate around the center of mass with some finite amplitude. For simplicity, we shall consider the vibration to be along the line joining the two atoms. As a result, the distance between the centers of the two atoms is a function of time. In this case, there is no single value for the distance and we must make some reasonable definition for the "true" value of such a quantity. A good choice in this case is to use the average measured on many similar molecules, as this is the quantity that best characterizes the distance between the two atoms at the particular temperature. The distribution of the measured values around the average is therefore a part of the quantity itself and cannot be reduced by any improvements in the apparatus used.

Another common example of such quantities is the energy of an excited state in an atom or a nucleus. Because of the finite lifetime Δt associated with the state, the energy has a "natural" width ΔE that is given by the Heisenberg uncertainty principle $\Delta E \Delta t \approx \hbar$. In this case, there is an uncertainty ΔE associated with the excitation energy even when we use apparatus of infinite precision to carry out the measurements.

Probability distribution The distribution $p(x)$ of a quantity x may be characterized by a number of parameters, such as where it is located, how broad it is, and what shape it has. If the value obtained in measurement i is x_i, the *mean*, or *centroid*, μ of the distribution is given by the average value in the limit where the number of measurements is infinite:

$$\mu = \lim_{N\to\infty} \frac{1}{N} \sum_{i=1}^{N} x_i. \tag{6-1}$$

If the distribution is a continuous one and the probability of finding a value x in the interval dx is $p(x)\,dx$, the mean is given, instead, by the integral

$$\mu = \int_a^b x\,p(x)\,dx.$$

Here, we follow the generally accepted practice to normalize $p(x)$ to unity:

$$\int_a^b p(x)\,dx = 1 \tag{6-2}$$

with a and b as, respectively, the upper and lower limits of the possible values of x.

Two other quantities are also often used to indicate the location of a distribution. The *median*, usually denoted as $\mu_{1/2}$, is defined as the value where the probabilities of finding x below and above it are the same. That is

$$p(x < \mu_{1/2}) = p(x > \mu_{1/2}). \tag{6-3}$$

The *most probable value*, which we shall write as $\mu_{\max}$, is given by the value of x where the probability distribution attains the maximum value:

$$p(x = \mu_{\max}) \;\geq\; p(x \neq \mu_{\max}). \tag{6-4}$$

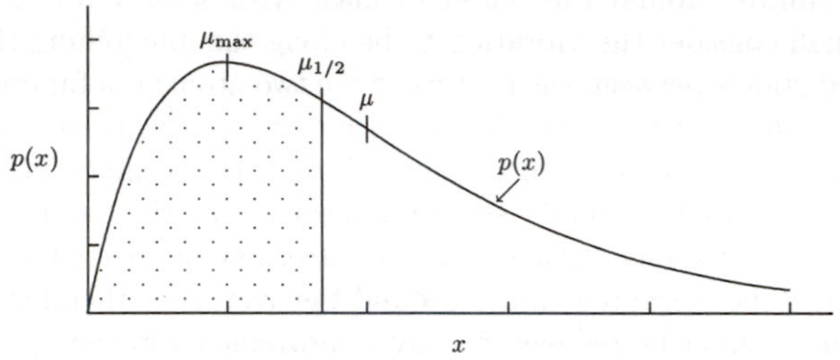

Figure 6-1: Most probable value $\mu_{\max}$, median $\mu_{1/2}$, and mean μ of a distribution.

In general, these three quantities, μ, $\mu_{1/2}$, and $\mu_{\max}$, are not equal to each other, as illustrated in Fig. 6-1 by the distribution

$$p(x)dx = xe^{-x}dx \qquad \text{for} \qquad x = [0, \infty). \tag{6-5}$$

Mathematically, the mean is the most convenient quantity in many applications, and we shall use it to characterize the location of a distribution for the most part.

A measure of the possible deviation of a member from the mean is given by the *variance* σ^2. For a discrete distribution, it is defined as

$$\sigma^2 = \lim_{N\to\infty} \frac{1}{N} \sum_{i=1}^{N} (x_i - \mu)^2. \tag{6-6}$$

For a continuous distribution in the interval $x = [a, b]$,

$$\sigma^2 = \int_a^b (x - \mu)^2 p(x)\, dx.$$

The square root of the variance, σ, is known as the standard deviation. In physics, it is often referred to as the *half-width* of the distribution. Strictly speaking, σ is equal to half of Γ, the full width of the distribution at half maximum, only in the case of a Lorentzian distribution:

$$p(x)dx = \frac{1}{\pi} \frac{\Gamma/2}{(x - \mu)^2 + (\Gamma/2)^2}\, dx. \tag{6-7}$$

Another quantity related to σ is the *probable error* s, defined by

$$\int_{\mu-s}^{\mu+s} p(x)\, dx = \frac{1}{2}. \tag{6-8}$$

It provides the range of the values in which we can find x with a probability of half. For a normal or Gaussian distribution,

$$p(x)dx = \frac{1}{\sigma\sqrt{2\pi}} e^{-(x-\mu)^2/2\sigma^2}\, dx \tag{6-9}$$

we have $s = 0.6745\sigma$ and the full width at half maximum is $\Gamma = 2 \times 1.177\sigma$.

To characterize the shape of a distribution in more detail, we can make use of the central moments:

$$\mu_r = \int_a^b (x - \mu)^r p(x)\, dx. \tag{6-10}$$

The $r = 0$ moment is unity from the normalization condition Eq. (6-2), and the first moment, $r = 1$, is zero since we are taking moments about the mean. The second moment is the same as the variance σ^2 and, as mentioned earlier, it provides an idea of how broad the distribution is. An alternate form of the third moment μ_3 is the *skewness*:

$$\gamma_3 = \mu_3/\sigma^3.$$

It gives a measure of the asymmetry of the distribution about the mean. Similarly, the fourth moment μ_4 is related to the *excess* or *kurtosis*:

$$\gamma_4 = \mu_4/\sigma^4 - 3.$$

For a distribution that is close to normal, a positive γ_4 means it is sharper in the central region than that given by Eq. (6-9) and a negative γ_4 means it is flatter. Usually, it is not very meaningful to compare the higher-order moments of two distributions unless their lower-order ones are roughly the same.

The definitions given by Eqs. (6-1) and (6-2) imply that an infinite number of measurements is required to determine the distribution. Since this is impossible, the distribution must be postulated in some way. Following the convention used in statistics, we shall call such a hypothetical distribution $p(x)$ the *parent distribution*. The aim of a statistical analysis is to arrive at the parent distribution from a finite sample of the values and to have an estimate of the likely errors, or uncertainties, incurred as a result of our incomplete knowledge of the system. Each measurement we make constitutes one of the infinite number required to define the distribution. After a number of samples are taken, we can, for instance, find the mean of the results obtained so far. The value may or may not correspond to the mean of the parent distribution, but it provides an estimate for it. As the number of samples grows, we can be confident that the sample mean is getting closer and closer to the mean of the parent distribution. At the same time, the distribution of our sample results becomes a better and better representation of the parent distribution. Before discussing some of the ways to estimate the parent distribution, we shall see first a few of the commonly used probability distributions.

Binomial distribution Consider a sample of N radioactive nuclei. If, in a given time interval dt, the probability for any one nucleus to decay to the ground state is $p\,dt$, what is the probability $P_B(N, n, p)\,dt$ of observing n decays among the N nuclei in the same time interval? For the purpose of illustration, we shall assume that the errors associated with our timer and counter for measuring the decays are sufficiently small that we can ignore them. The only source of uncertainty then arises from the probabilistic nature of radioactive decay. Furthermore, the decay of any one nucleus is independent of the others. As a result, there is no way to predict which of the N nuclei in the sample will decay in the time interval. All we know is that the probability for any one to decay is p per unit time. The value of p is a property that characterizes the nucleus and is related to its half-life or decay constant.

For our derivation of $P_B(N, n, p)$, it is convenient to think first in terms of a situation in which each of the N nuclei is labeled by a number, 1, 2, ..., N. For nucleus i, the probability to decay in the given time interval is $p\,dt$. Consequently, the probability for two particular nuclei, i and j, to decay in the same time interval is p^2dt. By the same token, the probability of n particular nuclei to decay in the same time interval is $p^n dt$. However, this is not of interest to us — what we want to know is the probability of any n, and only n, of the N nuclei to decay. To arrive at this quantity, several corrections must be made on the result of $p^n dt$ obtained above. The first comes from the requirement that none of the remaining $(N - n)$ nuclei have decayed in the mean time. For any one nucleus, the probability not to decay in the time interval is, by definition, $(1 - p)\,dt$. The probability of finding nucleus i decayed and another nucleus k not decayed in the time interval is therefore the product $p(1 - p)\,dt$.

Our thought experiment on radioactive decay can proceed further along the following line. At the beginning of a measurement, we have a sample of N radioactive nuclei in a box. After the given time interval has elapsed, we take out each of them and see whether it has decayed to the ground state or remained in the excited state. The first one we take out of the box can be any of the N nuclei in the box and, since it is immaterial to us which of the N nuclei it is, there are N possible ways to pick the first one. The probability of finding the first nucleus to have decayed is therefore Np. The second one can be any of the remaining $(N - 1)$ nuclei, and the probability for this nucleus to be in the ground state is $(N - 1)p$. Combining the two arguments, we find that the probability for the first two nuclei to be in the ground state is

$$P(N, 2, p)\,dt = \frac{N(N-1)}{2}p^2 dt.$$

The factor of 2 in the denominator comes from the fact that we do not wish to distinguish between the two possibilities, one with nucleus i as the first one and j as the second one, and the other with nucleus j as the first one and i as the second one. All we need to know is that both the first one and the second one are in the ground state.

The natural extension of this line of argument to the probability for starting with a sample of N excited nuclei and finding the first n nuclei in the ground state at the end of a time interval dt is then

$$P(N, n, p)\,dt = \frac{N(N-1)\cdots(N-n+1)}{n!}\,p^n dt. \tag{6-11}$$

Here the factor $n!$ in the denominator on the right side comes from the fact that all we care about is that there are n nuclei in the ground state — the $n \times (n-1) \times \cdots \times 2 \times 1$ different possible orders for the n nuclei to come out of the box are immaterial.

The result given by Eq. (6-11) is still not the probability $P_B(N,n,p)\,dt$ we are looking for, as we have not yet imposed the condition that the other $(N-n)$ nuclei remain in the excited state. Since the probability for $(N-n)$ nuclei not decayed is $(1-p)^{N-n}dt$, we obtain the final result that

$$\begin{aligned} P_B(N,n,p)\,dt &= \frac{N(N-1)\cdots(N-n+1)}{n!}p^n(1-p)^{N-n}dt \\ &= \frac{N!}{n!(N-n)!}p^n(1-p)^{N-n}dt = \binom{N}{n} p^n(1-p)^{N-n}dt \end{aligned} \tag{6-12}$$

where the binomial coefficient $\binom{N}{n}$ is defined in Eq. (A-1). This is known as the binomial distribution. The name comes from the binomial theorem

$$(p+q)^n = \sum_{r=0}^{n} \binom{n}{r} p^r q^{n-r}.$$

In Problem 6-2, we shall see that the mean and variance of a binomial distribution are, respectively, Np and $Np(1-p)$.

Poisson distribution For most radioactive transitions, the probability p for a single nucleus to decay per unit time is small. On the other hand, the sample is usually large, with N far in excess of 10^6. As a result, the product Np remains a finite number. Let us define

$$\mu \equiv Np.$$

In the limit that $p \to 0$, $N \to \infty$, and μ remains constant, the binomial distribution approaches a Poisson distribution:

$$P_P(n,\mu) = \frac{\mu^n}{n!}\,e^{-n}. \tag{6-13}$$

In Problem 6-2, we see that both the mean and the variance of this distribution are μ. The distribution is therefore completely specified by a single parameter, commonly taken to be μ.

The relation between Poisson and binomial distributions may be seen from the following considerations. Let us start from the second form of $P_B(N,n,p)$ given in Eq. (6-12) and rewrite it as

$$P_B(N,n,p) = \frac{N!}{n!(N-n)!}\,p^n(1-p)^N(1-p)^{-n}.$$

For $p \to 0$, we have $n \ll N$. The first two factors on the right side may be approximated as

$$\frac{N!}{n!(N-n)!}p^n \xrightarrow[p\to 0]{} \frac{1}{n!}N^n p^n = \frac{1}{n!}\mu^n \tag{6-14}$$

where we have made use of the fact $\mu = Np$ for $P_B(N,n,p)$. The third factor may be rewritten in terms of the mean

$$(1-p)^N = \left\{(1-p)^{1/p}\right\}^{\mu} \xrightarrow[p\to 0]{} e^{-\mu}. \tag{6-15}$$

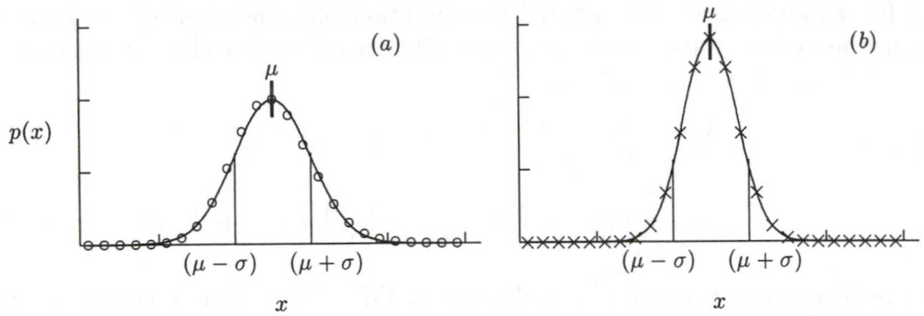

Figure 6-2: Comparison of Poisson (∘) and binomial distributions (×) with normal distributions (solid curve) of the same mean when there are large numbers of degrees of freedom.

This comes from the observation that, if we write $m = -p^{-1}$, we have

$$\begin{aligned}\left(1+\frac{1}{m}\right)^m &= 1 + m\frac{1}{m} + \frac{m(m-1)}{2!}\left(\frac{1}{m}\right)^2 + \cdots \\ &\quad + \frac{m(m-1)\cdots(m-r+1)}{r!}\left(\frac{1}{m}\right)^r + \cdots \\ &\xrightarrow[m\gg 1]{} 1 + \frac{1}{1} + \frac{1}{2!} + \cdots + \frac{1}{r!} + \cdots = e.\end{aligned}$$

When we put together the results of Eqs. (6-14) and (6-15) with the approximation $(1-p)^{-n} \to 1$, the Poisson distribution of Eq. (6-13) emerges.

Normal distribution The normal or Gaussian distribution of Eq. (6-9)

$$p(x)dx = \frac{1}{\sigma\sqrt{2\pi}} e^{-(x-\mu)^2/2\sigma^2} dx$$

is specified by two parameters, the mean μ and the variance σ^2. It occurs in a large variety of situations where many independent degrees of freedom are available to the system. The reason for its popularity comes from the central limit theorem in statistics. It states that, if a variable x has a large number of independent degrees of freedom, the distribution of x goes asymptotically to a normal distribution. The theorem is true without much regard to the distribution for each of the degrees of freedom. As an example, consider a number x that is made of the sum of n random numbers. If n is large, the distribution of x has the form of Eq. (6-9), essentially independent of the distribution of the n random numbers it is made of. Problem 6-3 provides an illustration of this point.

Because of the central limit theorem, the distribution of many quantities approaches that of a normal one when the number of degrees is large. Consider the case

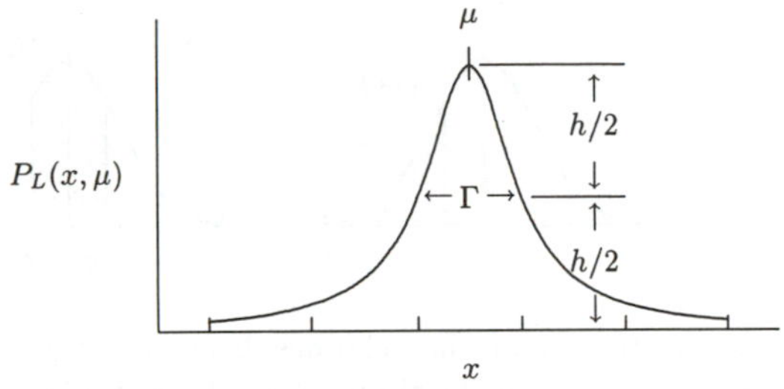

Figure 6-3: Lorentzian distribution given by Eq. (6-7). For this distribution, the full width at half-maximum is equal to twice the standard deviation.

of tossing coins. If the coins are unbiased, the probability of any one landing head up is $\frac{1}{2}$. If we flip two such coins, the probabilities of having two heads, one head, and no head are given by the binomial distribution with $N = 2$ and $p = (1 - p) = 0.5$. The results are, respectively, $P_B(2, 2, 0.5) = 0.25$, $P_B(2, 1, 0.5) = 0.5$, and $P_B(2, 0, 0.5) = 0.25$. Such a distribution is symmetric around the mean of one head, but the shape does not resemble that for a normal distribution. However, if instead of tossing two coins, we deal with a much larger number, such as $N = 50$, the resulting shape will be very close to that of a normal distribution. In the place of coins, we can also illustrate the same point using a large number of independent quantities, each of which has a Poisson distribution. The result is a normal distribution as shown in Fig. 6-2.

Lorentzian distribution The Lorentzian distribution of Eq. (6-7),

$$p(x)dx = \frac{1}{\pi}\frac{\Gamma/2}{(x-\mu)^2 + (\Gamma/2)^2}dx$$

is found in many natural phenomena, such as the power dissipated as a function of frequency across a resistor in an alternating-current circuit consisting of a capacitor, an inductor, and a resistor. The mechanical analog of the same type of situation is the power supplied by a source as a function of frequency to drive a harmonic oscillator.

The Lorentzian distribution gives also the *natural line shape* of the energy emitted when a particle decays from an excited state to a lower one. Because of the Heisenberg uncertainty principle, the energy of an unstable state cannot be infinitely sharp. Instead, it takes on the form shown in Fig. 6-3, with mean equal to μ and full width at half-maximum characterized by Γ. In terms of the lifetime of the state be τ, the time interval at the end of which only $1/e$ of the original number of excited nuclei remains,

$$\Gamma = \hbar/\tau$$

where $\hbar$ is the Planck's constant divided by 2π. As mentioned earlier, the standard deviation for a Lorentzian shape distribution is $\frac{1}{2}\Gamma$.

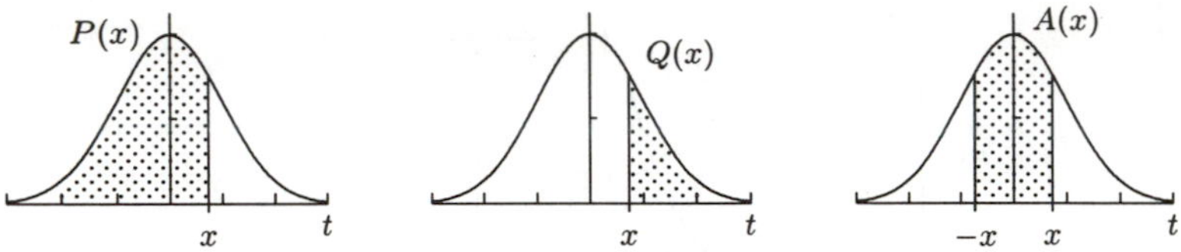

Figure 6-4: Schematic illustration of the relations between integrals $P(x)$ of Eq. (6-16) and integrals $Q(x)$ and $A(x)$ of Eq. (6-17).

Probability functions If the probability of finding a quantity with value t in an interval dt is given by $p(t)\,dt$, the total probability for finding the quantity with a value less or equal to x is given by the integral

$$P(x) = \int_{-\infty}^{x} p(t)\,dt.$$

For the normal distribution, the integral

$$P(x) = \frac{1}{\sqrt{2\pi}} \int_{-\infty}^{x} e^{-t^2/2} dt \tag{6-16}$$

is known sometimes as the *normal probability function.* It is related to the error function of Eq. (4-100) by the relation

$$\text{erf}(x) = 2P(\sqrt{2}\,x) - 1$$

as shown earlier in Eq. (4-102), and to the incomplete gamma function

$$\gamma\Big(\frac{1}{2}, x\Big) = \sqrt{\pi}\{2P(\sqrt{2x}) - 1\}$$

defined in Eq. (4-115).

Two other functions are also useful in probability studies,

$$Q(x) = \frac{1}{\sqrt{2\pi}} \int_{x}^{\infty} e^{-t^2/2} dt \qquad A(x) = \frac{1}{\sqrt{2\pi}} \int_{-x}^{x} e^{-t^2/2} dt. \tag{6-17}$$

Both may be written in terms of $P(x)$ of Eq. (6-16)

$$P(x) + Q(x) = 1 \qquad P(-x) = Q(x) \qquad A(x) = 2P(x) - 1. \tag{6-18}$$

The relations between $P(x)$, $Q(x)$, and $A(x)$ are illustrated in Fig. 6-4.

Earlier in §2-5, we used a method of evaluating $A(x)$ by Monte Carlo integration. An efficient rational polynomial approximation is given in Abramowitz and Stegun[1]

$$P(x) = 1 - \frac{1}{2}\frac{1}{(1 + d_1x + d_2x^2 + d_3x^3 + d_4x^4 + d_5x^5 + d_6x^6)^{16}} + \epsilon(x). \tag{6-19}$$

The six coefficients have the values

$$d_1 = 0.0498673470 \qquad d_2 = 0.0211410061 \qquad d_3 = 0.0032776263$$
$$d_4 = 0.0000380036 \qquad d_5 = 0.0000488906 \qquad d_6 = 0.0000053830.$$

The error quoted is $|\epsilon(x)| < 1.5 \times 10^{-7}$. Using the relations given in Eq. (6-18), the values of $Q(x)$ and $A(x)$ for a given x may be obtained from that of $P(x)$.

6-2 Uncertainties and their propagation

Since uncertainties are unavoidable in measurements, it is essential that we have some idea of the influence they have on any conclusion we wish to draw from data. Let us begin by considering the case of equating the true value of an object by the average obtained from a finite number of measurements.

Uncertainty associated with finite sample size Consider again the hypothetical experiment to determine the length ℓ of a city block. Let the values for $N = 25$ measurements be those listed in Table 6-1 in units of meters. The average of the 25 results is $\mu = 49.9$ m, and the root mean square deviation from the average is

$$\sigma = \left\{\frac{1}{N}\sum_{i=1}^{N}(\ell_i - \mu)^2\right\}^{1/2} = 1.0 \text{ m}.$$

The question we wish to ask ourselves is the following: How likely is the average value of 49.9 m to be the true length of the city block? Unless we are provided with some additional information, we have no reason to reject the sample average of $\mu = 49.9$ m as the mean of the parent distribution, the mean if the number of measurements N is infinity instead of 25. On the other hand, we have no reason to believe that the sample average is equal to the mean of the parent distribution either.

There can be a number of reasons why there are differences among the measured results. If the sources of error are numerous and independent of each other, we can invoke the central limit theorem and claim that the parent distribution of the measured values is a normal one. If the mean and variance of each source of error are known, it is possible to deduce the corresponding quantities for the parent distribution. However, this is impractical for most cases and we must find a way to make estimates of the two parameters required to specify the normal distribution for the parent. One way is to take the following approach. If the sample of 25 measurements represents a random selection among the members in the parent distribution, the measured values must also distribute normally with the mean and variance, μ and σ, that are good estimates of the corresponding values in the parent distribution.

Since we are assuming that the measured values follow a normal distribution, the probability for the mean of the parent distribution to be in the range $[\mu - \sigma, \mu + \sigma]$ is given by the integral $A(x)$ of Eq. (6-17) for $x = \sigma$:

$$A(\sigma) = \int_{\mu-\sigma}^{\mu+\sigma} P_G(\mu, \sigma, t)\, dt = 0.68. \tag{6-20}$$

Table 6-1: 25 measurements of the length of a city block in meters.

Exp. no.	Length	Exp. no.	Length	Exp. no.	Length
1	50.6	2	50.3	3	50.6
4	47.7	5	50.7	6	49.4
7	50.6	8	50.7	9	48.7
10	50.7	11	50.5	12	49.3
13	50.5	14	48.9	15	50.2
16	49.2	17	51.0	18	48.3
19	50.7	20	49.9	21	51.2
22	48.7	23	48.6	24	49.1
25	51.6				
				$\mu = 49.9$	$\sigma = 1.0$

In other words, if we perform the experiment of measuring the length of the same city block 100 times, each consisting of 25 measurements, we expect that the average value is between 48.9 m and 50.9 m for 68 of the 100 times. For the remaining 32 times, the average values are expected to be either below 48.9 m or above 50.9 m.

Before accepting this result, let us check whether it is correct for us to assume that the distribution for the measured values is a normal one. For this purpose we shall plot the results in the form of a histogram. Since we have only a sample size of $N = 25$, the bin size is chosen to be 1 m. A convenient way to make the plot is to include in the first bin all the measured results with $47 \leq \ell < 48$ m, the second bin, those with $48 \leq \ell < 49$ m, and so on. This is shown in Fig. 6-5. There is only one measured result in the range $47 \leq \ell < 48$ m and, as a result, the height of the first bin is 1. There are five results in the range $48 \leq \ell < 49$ m and the height of the second bin is 5, and so on. In the figure, we have also superimposed on the histogram a normal distribution of the same mean and variance. For ease of comparison, the normalization of the distribution is 25 instead of 1, as done in Eq. (6-9). We notice that in two of the bins, the one for $49 \leq \ell < 50$ m and the one for $50 \leq \ell < 51$ m, the deviations from the values expected of a normal distribution curve are quite noticeable.

Let us examine whether these two deviations are significant. The value in a given bin for a normal distribution with $N = 25$ and $(\mu, \sigma) = (49.9, 1.0)$ may be found from the integral

$$E(a,b) = \frac{N}{\sigma\sqrt{2\pi}} \int_a^b \exp\Big\{-\frac{1}{2}\Big(\frac{x-c}{\sigma}\Big)^2\Big\} dx. \tag{6-21}$$

For bin $49 \leq \ell < 50$ m, $a = 49$ and $b = 50$, the result is 8.9. This is to be compared with the value 5 for the height of the bin. For bin $50 \leq \ell < 51$ m, $a = 50$ and $b = 51$, the result is 8.1 and the height of the bin is 11. The differences are approximately 3 in both cases. Are these differences significant enough that we should question the

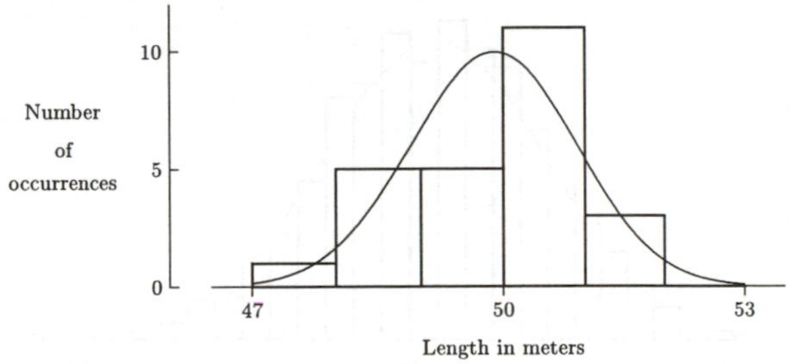

Figure 6-5: Histogram showing the distribution of the values obtained in the 25 measurements of a hypothetical city block listed in Table 6-1. A normal distribution of the same mean and variance (solid curve) is superimposed for comparison.

assumption of a normal distribution for our measured results?

To answer this question, we need to recognize that the occupancy in each bin must also follow some definite distribution. For our example, we have only one set of "measured" values, obtained by counting the number of occurrences of the value of ℓ in a given range in Table 6-1. Again, we must assume that this particular set of measured values is only one possible set among a large number of similar ones in the parent distribution. In other words, if we carry out another set of 25 measurements and plot the results as a histogram in the same way as we have done for our original one, the resulting plot is likely to be different. Now, if we repeat the process of taking 25 measurements many times and plot the histogram each time, we can find out the distribution of the number of occurrences for each one of the bins. In this way, we are able, at least in principle, to map out the distribution for the number of occurrences in each histogram bin among the measurements.

In reality, however, we have only one set of measurements and we wish to find out the significance of the set of values we have in hand. With only one value for each bin, the best we can do is to take it as the mean value for the distribution of the bin. Now we have exhausted the only piece of information in our possession. To make progress, we must do something else. A common approach is to assume the Poisson distribution of Eq. (6-13) as the parent distribution; that is, the distribution of the height of a particular histogram bin is Poisson in the limit of an infinite number of samples of 25 measurements of our city block. Since the variance of a Poisson distribution is equal to the mean, our distribution is completely specified by the one value we have available to us. Furthermore, a Poisson distribution approaches that of a normal distribution when the number of degrees of freedom is large. As a result, we can say that, if the occupancy is n in a particular bin, the standard deviation for the bin is $\sqrt{n}$. This means that there is a probability of 68%, as we can see from Eq. (6-20), that the true value for this bin is in the range $[n-\sqrt{n}, n+\sqrt{n}]$. In the

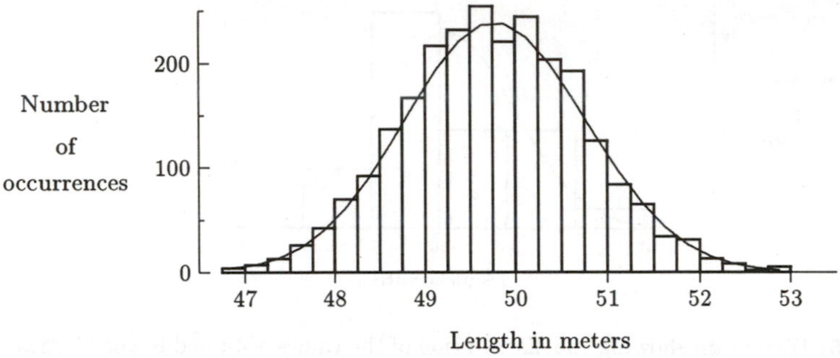

Figure 6-6: Histogram showing the distribution of the values of 2,500 measurements of the type given in Table 6-1. Compared with Fig. 6-5, we have here a much better agreement with a normal distribution (solid curve).

next section, we shall see that it is also possible to arrive at the same conclusion without having to make the assumption of a Poisson distribution.

Returning to histogram bin $50 \leq \ell < 51$ m, we have $n = 11$ and $\sqrt{11} = 3.3$. If the parent distribution for the height of this bin is Poisson, the uncertainty associated with the height is estimated to be $\sigma = 3.3$. The difference of the observed value from that expected of Eq. (6-21) is 2.9. Since this is within one standard deviation of the mean, it is probable (> 68%) that the sample result belongs to the same parent distribution we have assumed. Similarly for bin $49 \leq \ell < 50$ m, the measured value of n is 5 and the expected value from Eq. (6-21) for this bin is 8.8. The difference is therefore outside one standard deviation of $\sqrt{5} = 2.3$. For a normal distribution there is a probability of 32% for this to take place. Among a total of 6 bins in the histogram, we should therefore allow, on the average, roughly two bins to have differences larger than one standard deviation from the mean. As a result, we cannot treat the deviation from the expected value for this bin as significant. The net conclusion is that, although our histogram does not resemble a bell-shaped distribution, we cannot find any significant departure from one either.

In fact, the results in Table 6-1 are obtained from a computer simulation, produced using a random number generator with a normal distribution to simulate the uncertainties in the measured values. If we enlarge our sample size from 25 to a much larger number, the resulting histogram resembles more closely a normal distribution, as shown in Fig. 6-6. In the next section we shall develop more quantitative methods to test whether two distributions are similar to each other.

Propagation of error Consider a quantity y that is a function of independent variable x:

$$y = f(x).$$

If we vary x by Δx, the value of y changes by Δy. The relation between Δx and Δy may be found from a Taylor series expansion of $f(x \pm \Delta x)$ around $f(x)$:

$$y \pm \Delta y = f(x \pm \Delta x) = f(x) + \frac{df}{dx}(\pm \Delta x) + \cdots.$$

On truncating the series after the second term, we obtain the result

$$\Delta y = \frac{df}{dx}\Delta x. \tag{6-22}$$

If y is a function of several independent variables $u, v, w, \ldots,$

$$y = f(u, v, w, \ldots)$$

the relation analogous to Eq. (6-22) becomes

$$\Delta y = \frac{\partial f}{\partial u}\Delta u + \frac{\partial f}{\partial v}\Delta v + \frac{\partial f}{\partial w}\Delta w + \cdots. \tag{6-23}$$

We shall make use of this result to find the relation between uncertainties in the independent variables and that of a quantity derived from them.

The common practice is to associate the standard deviation σ_x in the distribution of a variable x as the uncertainty. That is,

$$\Delta x = \sigma_x.$$

This definition is particularly useful in dealing with experimental data. Since there are many independent sources of random errors associated with a measurement, the values obtained are likely to follow a normal distribution. The assignment of the standard deviation as the uncertainty has a statistical significance and can provide us with an estimate of the likelihood of obtaining values different from the mean, as we have seen earlier.

Using Eq. (6-23), we can relate the uncertainties in the independent variables u, v, $w, \ldots$, to that of a quantity calculated from them. If there are N measurements of u, v, $w, \ldots$, and the values obtained for them are u_i, v_i, $w_i, \ldots$, in measurement i, the variance in the distribution of y is given by

$$\begin{aligned}
\sigma_y^2 &= \lim_{N\to\infty}\frac{1}{N}\sum_{i=1}^{N}\left\{\frac{\partial f}{\partial u}(u_i-\mu_u)+\frac{\partial f}{\partial v}(v_i-\mu_v)+\frac{\partial f}{\partial w}(w_i-\mu_w)+\cdots\right\}^2 \\
&= \lim_{N\to\infty}\frac{1}{N}\sum_{i=1}^{N}\left\{\left(\frac{\partial f}{\partial u}\right)^2(u_i-\mu_u)^2+\left(\frac{\partial f}{\partial v}\right)^2(v_i-\mu_v)^2+\left(\frac{\partial f}{\partial w}\right)^2(w_i-\mu_w)^2+\cdots\right. \\
&\qquad \left. +2\left(\frac{\partial f}{\partial u}\right)\left(\frac{\partial f}{\partial v}\right)(u_i-\mu_u)(v_i-\mu_v)+2\left(\frac{\partial f}{\partial v}\right)\left(\frac{\partial f}{\partial w}\right)(v_i-\mu_v)(w_i-\mu_w)+\cdots\right\} \\
&= \left(\frac{\partial f}{\partial u}\right)^2\sigma_u^2+\left(\frac{\partial f}{\partial v}\right)^2\sigma_v^2+\left(\frac{\partial f}{\partial w}\right)^2\sigma_w^2+\cdots \\
&\qquad +2\left(\frac{\partial f}{\partial u}\right)\left(\frac{\partial f}{\partial v}\right)\sigma_{uv}^2+2\left(\frac{\partial f}{\partial v}\right)\left(\frac{\partial f}{\partial w}\right)\sigma_{vw}^2+\cdots
\end{aligned} \tag{6-24}$$

where μ_u, μ_v, μ_w, ..., are the mean values and σ_u^2, σ_v^2, σ_w^2, ..., the variances of the distributions of, respectively, u, v, w, ... Since there are several variables, there may be a correlation between any two of them. This is measured by the covariance

$$\sigma_{uv}^2 = \lim_{N\to\infty} \frac{1}{N} \sum_{i=1}^{N} \{(u_i - \mu_u)(v_i - \mu_v)\}. \tag{6-25}$$

If u and v are uncorrelated, the probability of obtaining a value u_i for u is completely independent of that for getting a value v_i for v and we have $\sigma_{uv}^2 = 0$.

It is useful to see some explicit examples of Eq. (6-24) involving simple functions. For instance, if y is a linear function of two independent variables u and v, such as

$$y = au \pm bv$$

where a and b are constants, the variance of y is given by

$$\sigma_y^2 = a^2\sigma_u^2 + b^2\sigma_v^2 \pm 2ab\,\sigma_{uv}^2. \tag{6-26}$$

This is also an example showing the importance of the covariance σ_{uv}^2. It is not difficult to think of a strongly correlated case where the error in measuring u is always compensated by that in measuring v. In this case, the value of σ_{uv}^2 must be such that σ_y^2 vanishes.

If y is the product of u and v, such as the function

$$y = auv$$

the partial derivatives of y are $\partial y/\partial u = av$ and $\partial y/\partial v = au$. The variance of y is then

$$\sigma_y^2 = (av)^2\sigma_u^2 + (av)^2\sigma_v^2 + 2a^2uv\,\sigma_{uv}^2.$$

This is more commonly written in the form

$$\frac{\sigma_y^2}{y^2} = \frac{\sigma_u^2}{u^2} + \frac{\sigma_v^2}{v^2} + 2\frac{\sigma_{uv}^2}{uv}. \tag{6-27}$$

In this case, any correlation between u and v always increases the uncertainty in y.

For the function

$$y = a\frac{u}{v}$$

the corresponding result is

$$\frac{\sigma_y^2}{y^2} = \frac{\sigma_u^2}{u^2} + \frac{\sigma_v^2}{v^2} - 2\frac{\sigma_{uv}^2}{uv}.$$

The difference in the sign of the covariance term from that of Eq. (6-27) comes from the fact that $\partial y/\partial v = -au/v^2$.. Derivations of similar relations for other standard functions are given as exercises in Problem 6-5.

6-3 The method of maximum likelihood

Let us, again, use the decay of an unstable particle as an example to introduce the method of maximum likelihood. Since the decay is purely statistical in nature, we can only predict the probability for the process to take place in a given time interval, but not the time when a particular particle actually undergoes the transformation. From this we obtain that the number of decays per unit time dN/dt at time t is proportional to $N(t)$, the number of radioactive nuclei present. That is,

$$\frac{dN}{dt} = -\frac{1}{\tau}N(t)$$

where we have written the proportional constant as $1/\tau$ for reasons that will become clear soon. The negative sign reflects the fact that each decay decreases the number of radioactive nuclei present.

The solution of this differential equation is the familiar exponential decay law

$$N(t) = N_0 e^{-t/\tau} \tag{6-28}$$

where N_0 is the number of radioactive nuclei at $t = 0$. When $t = \tau$, the number of radioactive nuclei is reduced to $1/e$ of the amount at $t = 0$. For this reason, τ is known as the lifetime or mean life for the decay. The value of τ is a property of the particle and equals the half-life $\tau_{1/2}$ divided by $\ln 2$ (= 0.693). From Eq. (6-28), we obtain the probability $p(t)$ of observing a particle to decay at time t

$$p(t) = \frac{1}{\tau}e^{-t/\tau} \tag{6-29}$$

where the factor $1/\tau$ on the right side is necessary to normalize $p(t)$ such that the integrated probability is unity in the time interval $[0, \infty)$. Let us see how we can find the value of τ by measuring the radioactivity as a function of time.

Because of the probabilistic nature of the process, it is necessary to observe a large number of decays. For simplicity, we shall assume for now that instrumental errors associated with the measurements are negligible; the only uncertainty in the results comes from the statistical nature of the process. The data then consist of N values, $t_1, t_2, \ldots, t_N$, each giving the time when one of the N particles is observed to decay.

If we take a different set of measurements of a similar collection of the same type of nucleus, the N values of time obtained will most likely be different from those in the set above. However, by definition, the value of τ underlying these N values must be the same in both sets, as we are, after all, measuring the same type of transition. The method we use to deduce the value of τ from the data must, therefore, be such that the result is, as far as possible, independent of the fluctuations from one set of data to another.

Likelihood function The probability $\mathcal{L}(\tau)$ for the set of measured results to come out the way we have recorded above is given by the product of the probabilities of Eq. (6-29) for the N particles to decay at times $t_1, t_2, \ldots, t_N$:

$$\mathcal{L}(\tau) = \prod_{i=1}^{N} p(t_i) = \prod_{i=1}^{N} \exp\Big\{-\frac{t_i}{\tau} - \ln\tau\Big\} = \exp\Big\{-\frac{1}{\tau}\sum_{i=1}^{N} t_i - N\ln\tau\Big\}. \tag{6-30}$$

This expression, however, cannot be used to evaluate $\mathcal{L}(\tau)$, as τ is not yet known. Some estimate must be made before we can proceed with the calculation. An optimum choice is one that maximizes $\mathcal{L}(\tau)$. As we shall see later, if $\mathcal{L}(\tau)$ has a sharply peaked distribution, the possibilities for any alternate choices are small. In fact, in the limit that $\mathcal{L}(\tau)$ is a delta function, the value of τ is uniquely determined. However, the limiting case is of no interest to us here, since it implies a completely deterministic situation and, as a result, there is really no need to construct the function $\mathcal{L}(\tau)$.

It is easy to see that the peak of the distribution $\mathcal{L}(\tau)$ occurs at the location that satisfies the condition

$$\frac{d\mathcal{L}(\tau)}{d\tau} = \Big\{\frac{1}{\tau^2}\sum_{i=1}^{N} t_i - \frac{N}{\tau}\Big\}\mathcal{L}(\tau) = 0.$$

From this, we obtain

$$\overline{\tau} = \frac{1}{N}\sum_{i=1}^{N} t_i \tag{6-31}$$

where we have used a bar over τ to distinguish it from the true value of the lifetime τ. The meaning of Eq. (6-31) is that the average of t_i is the value of τ that produces the maximum probability for the function $\mathcal{L}(\tau)$. This way of estimating the value of an unknown quantity is called the method of *maximum likelihood.* For this reason, $\mathcal{L}(\tau)$ is called the *likelihood function.* A more formal statement of the method can be found in standard references on probability and mathematical physics, such as Mathews and Walker.[38] It is important to realize here that this is not the only choice. In fact, several other criteria are also used in practice. Except for cases where the number of independent pieces of data is small or the uncertainty associated with each piece is large, the results obtained using different methods are essentially identical to each other. For our purpose, we shall concentrate on the method of maximum likelihood.

It is useful to have an estimate of the uncertainty associated with the value of $\overline{\tau}$ determined. If N is large, the shape of $\mathcal{L}(\tau)$ approaches that of a normal distribution,

$$\mathcal{L}(\tau) \xrightarrow[N\to\infty]{} \frac{1}{\sigma_\tau\sqrt{2\pi}} e^{-(\tau-\overline{\tau})^2/2\sigma_\tau^2}. \tag{6-32}$$

In this limit, the uncertainty of τ is given by the standard deviation σ_τ of this distribution. Using Eq. (6-32), the derivative of the logarithm of $\mathcal{L}(\tau)$ is

$$\frac{d}{d\tau}\ln\mathcal{L}(\tau) = -\frac{\tau-\overline{\tau}}{\sigma_\tau^2}.$$

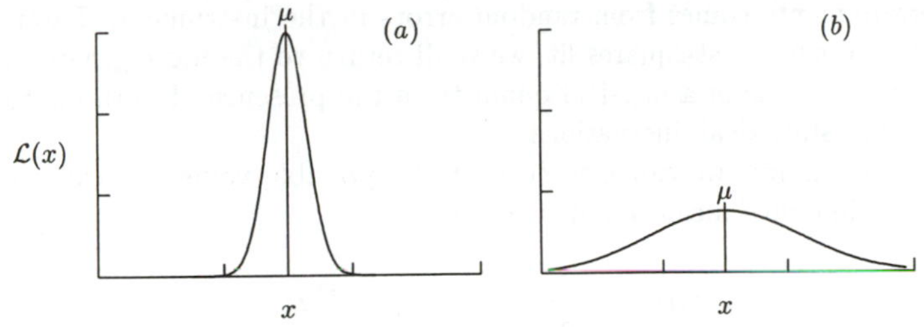

Figure 6-7: Mean value and uncertainty of likelihood function. If it is sharply peaked with a small uncertainty, as in (a), the mean μ is more effective in characterizing the quantity than that of a flat distribution shown in (b).

This gives us the relation

$$\sigma_\tau = \left\{ \frac{1}{\overline{\tau} - \tau} \frac{d}{d\tau} \ln \mathcal{L}(\tau) \right\}^{-1/2}. \tag{6-33}$$

To obtain an idea of the actual value of σ_τ for a given function, we can put $\mathcal{L}(\tau)$ of Eq. (6-30) into (6-33)

$$\sigma_\tau = \left\{ \frac{1}{\overline{\tau} - \tau} \left(\frac{1}{\tau^2} \sum_{i=1}^{N} t_i - \frac{N}{\tau} \right) \right\}^{-1/2} = \frac{\tau}{\sqrt{N}} \longrightarrow \frac{\overline{\tau}}{\sqrt{N}}. \tag{6-34}$$

This is the same result as the previous section in estimating the uncertainty of the height of a histogram, except there we had to invoke a Poisson distribution.

If N is not large enough for the likelihood function to be approximated by a normal distribution, the final result given by Eq. (6-34) may not be valid. In this case, one should make a plot of the likelihood function and see if $\mathcal{L}(\tau)$ has a narrow distribution. Unless $\mathcal{L}(\tau)$ is sharply peaked, the value $\overline{\tau}$ determined may not be very meaningful. This is illustrated schematically in Fig. 6-7.

Data points with unequal uncertainties In the previous example, we assumed that the source of uncertainty in a measurement comes purely from statistical fluctuations. In reality, it is inevitable that we must deal with instrumental uncertainties as well. Because of conditions beyond our control, the accuracy that can be achieved in one measurement may be quite different from that of another. For example, in dealing with a radioactive sample, there are many more decays in the beginning of the experiment and, as a result, we may not be able to measure the time as well as we can toward the end of the experiment when the activity is lower.

Earlier, we have seen in Eq. (6-24) how uncertainties σ_i in the measured quantity x_i are propagated to y that is a known function of x_i. Here we shall extend the notion to include relations between x_i and y determined through the maximum likelihood method. For simplicity, we shall consider here a system where the only difference between measurements comes from random errors in the instrument. Later, in the discussion of nonlinear least-squares fit, we shall return to the more general problem of determining the value of a physical quantity in the presence of both instrumental uncertainties and statistical fluctuations.

For each data point, we can assume that the possible values x_i are given by a normal distribution with mean μ and variance σ_i^2:

$$p(x_i)\,dx_i = \frac{1}{\sigma_i\sqrt{2\pi}} e^{-(x_i-\mu)^2/2\sigma_i^2}\,dx_i. \tag{6-35}$$

Since the only difference between the various data points here is due to instrumental errors, the mean μ is the same for all the data points $x_1, x_2, \ldots, x_N$. However, the variance for one point may be different from another. In general, it may not be easy to determine σ_i for each data point, but we shall not be concerned with this question here. For our purpose, we shall take the attitude that all the values of σ_i are given to us. Furthermore, we shall assume that there are many possible sources of instrumental error and, as a result, central limit theorem applies.

We can now construct the likelihood function for the mean μ of the parent distribution under the assumption that each data point follows the distribution given by Eq. (6-35). Analogous to Eq. (6-30), we have here a product of normal distribution functions:

$$\mathcal{L}(\mu) = \prod_{i=1}^{N}\Big\{\frac{1}{\sigma_i\sqrt{2\pi}} e^{-(x_i-\mu)^2/2\sigma_i^2}\Big\} = \Big\{\prod_{i=1}^{N}\frac{1}{\sigma_i\sqrt{2\pi}}\Big\}\exp\Big\{-\frac{1}{2}\sum_{i=1}^{N}\Big(\frac{x_i-\mu}{\sigma_i}\Big)^2\Big\}. \tag{6-36}$$

As we have seen earlier, the maximum of this function is given by the condition $d\mathcal{L}(\mu)/d\mu = 0$. This is equivalent to the requirement that

$$\frac{d}{d\mu}\Big(\frac{x_i-\mu}{\sigma_i}\Big)^2 = 0$$

or

$$\overline{\mu} = \Big(\sum_i \frac{1}{\sigma_i^2}\Big)^{-1}\sum_{i=1}^{N}\frac{1}{\sigma_i^2}\,x_i. \tag{6-37}$$

In other words, $\overline{\mu}$, the maximum likelihood value of the mean μ, is given by the average of the data points x_i, weighted by the inverse of the square of the uncertainty of each point. In the limit that the values of σ_i for different points are equal to each other, we obtain the same expression as the average value in the absence of any uncertainties for each data point.

Since $\overline{\mu}$ is a function of x_i, the uncertainty in μ is given by Eq. (6-24),

$$\sigma_\mu^2 = \sum_{i=1}^{N}\Big\{\sigma_i^2\Big(\frac{\partial\mu}{\partial x_i}\Big)^2\Big\}$$

where we have ignored any covariance between two data points, as we are assuming that each measurement is independent of the other. From Eq. (6-37), we obtain the value of the partial derivative as

$$\frac{\partial}{\partial x_i}\left\{\left(\sum_j \frac{1}{\sigma_j^2}\right)^{-1} \sum_{k=1}^{N} \frac{1}{\sigma_k^2} x_k\right\} = \left(\sum_j \frac{1}{\sigma_j^2}\right)^{-1} \frac{1}{\sigma_i^2}.$$

This gives us the result

$$\sigma_\mu^2 = \sum_{i=1}^{N}\left\{\sigma_i^2\left(\left[\sum_j \frac{1}{\sigma_j^2}\right]^{-1} \frac{1}{\sigma_i^2}\right)^2\right\} = \left(\sum_i \frac{1}{\sigma_i^2}\right)^{-1}. \tag{6-38}$$

If all the uncertainties are the same and equal to a constant σ, the expression reduces to $\sigma_\mu^2 = \sigma^2/N$. Thus the uncertainty due to instrument decreases if more independent measurements are taken in an experiment.

6-4 The method of least squares

We shall now apply the method of maximum likelihood to a more general situation. If, based on some theoretical grounds, we know that a quantity y is a function of x with m parameters a_1, a_2, ..., a_m,

$$y = f(a_1, a_2, \ldots, a_m; x) \tag{6-39}$$

what are the values of the parameters that give the best description to a set of measured values of y? This is the familiar problem of fitting a function to a set of data points. If $f(a_1, a_2, \ldots, a_m; x)$ is a linear function in the parameters, the method is known as a linear least-squares fit. In this section, we shall give an introduction to the method by considering the simple case of a straight-line relation between y and x. Later, in §6-6, we shall consider the general case of more than two parameters. The case of fitting with a nonlinear function is treated in §6-7.

Let y_i be the measured value of y at $x = x_i$. Because of uncertainties associated with the process, the value of y_i is likely to be different if we repeat the measurement. Based on the central limit theorem, the results are expected to follow a normal distribution if the measurements for y_i are carried out many times,

$$p(y_i)\,dy_i = \frac{1}{\sigma_i\sqrt{2\pi}}\, e^{-(y_i - f_i)^2/2\sigma_i^2}\, dy_i.$$

The mean of the distribution $p(y_i)\,dy_i$ is given by

$$f_i = f(a_1, a_2, \ldots, a_m; x_i).$$

The variance σ_i^2 of the distribution comes from a variety of sources, both statistical fluctuations and instrumental uncertainties. However, $p(y_i)$ may not be available to us, as it is usually impossible to repeat the measurement of y_i (for a given x_i) a large number of times. On the other hand, if the distribution is a normal one, all we need, in addition to the mean value f_i, is the variance to determine $p(y_i)$. For our interest here, we shall assume that σ_i^2 is provided to us somehow, as there is no way we can deduce its value without being given more information.

The likelihood function for the parameters $a_1, a_2, \ldots, a_m$ is given by the product of $p(y_1), p(y_2), \ldots, p(y_N)$. That is,

$$\begin{aligned}\mathcal{L}(a_1, a_2, \ldots, a_m) &= \prod_{i=1}^{N}\Big\{\frac{1}{\sigma_i\sqrt{2\pi}}e^{-(y_i-f_i)^2/2\sigma_i^2}\Big\} \\ &= \Big\{\prod_{i=1}^{N}\frac{1}{\sigma_i\sqrt{2\pi}}\Big\}\exp\Big\{-\frac{1}{2}\sum_{i=1}^{N}\Big(\frac{y_i-f_i}{\sigma_i}\Big)^2\Big\}.\end{aligned} \tag{6-40}$$

As we shall see soon, the sum in the argument of the exponential function

$$\chi^2 \equiv \sum_{i=1}^{N}\Big(\frac{y_i-f_i}{\sigma_i}\Big)^2 \tag{6-41}$$

is fundamental to the method of least squares and is given the name *chi-square*.

Similar to what we have done earlier, the maximum of the likelihood distribution $\mathcal{L}(a_1, a_2, \ldots, a_m)$ is obtained by requiring that its partial derivative with respect to each of the parameters vanish. This is equivalent to

$$\frac{\partial}{\partial a_k}\chi^2 = -2\sum_{i=1}^{N}\Big(\frac{y_i-f_i}{\sigma_i}\Big)\frac{\partial f_i}{\partial a_k} = 0 \tag{6-42}$$

for $k = 1, 2, \ldots, m$.

Least-squares fit to a straight line To make progress, we need the explicit functional dependence of f on the parameters $\{a_k\}$. For the simple case of a linear function involving two parameters, with $a_1 = a$ and $a_2 = b$, we have

$$f(a, b; x) = a + bx. \tag{6-43}$$

The partial derivatives of $f(a, b; x)$ with respect to parameters a and b are then

$$\frac{\partial f}{\partial a} = 1 \qquad\qquad \frac{\partial f}{\partial b} = x.$$

The maximum likelihood condition of Eq. (6-42) in this case becomes

$$\sum_{i=1}^{N}\Big(\frac{y_i-f_i}{\sigma_i^2}\Big) = 0 \qquad\qquad \sum_{i=1}^{N}\Big(\frac{y_i-f_i}{\sigma_i^2}\Big)x_i = 0$$

or

$$a\sum_{i=1}^{N}\frac{1}{\sigma_i^2} + b\sum_{i=1}^{N}\frac{x_i}{\sigma_i^2} = \sum_{i=1}^{N}\frac{y_i}{\sigma_i^2} \qquad\qquad a\sum_{i=1}^{N}\frac{x_i}{\sigma_i^2} + b\sum_{i=1}^{N}\frac{x_i^2}{\sigma_i^2} = \sum_{i=1}^{N}\frac{x_iy_i}{\sigma_i^2} \tag{6-44}$$

Table 6-2: A sample of measured values of y for different values of x.

i	x_i	y_i	σ_i	i	x_i	y_i	σ_i
1	0.25	0.86	0.27	6	3.64	8.84	0.66
2	1.05	2.18	1.16	7	3.92	8.71	0.98
3	2.25	4.84	1.14	8	4.94	11.98	0.93
4	2.88	5.80	0.93	9	5.92	12.40	0.60
5	2.97	6.99	0.31				

where N is the total number of pieces of data points. The values of a and b are obtained by solving this system of two linear equations. In statistics, fitting a sample of N data points to the relation given by Eq. (6-43) is known as *linear regression* analysis. We shall, however, use the more conventional name of *linear least-squares* fit to a straight line.

In matrix form, Eq. (6-44) may be written as

$$\begin{pmatrix} \alpha & \beta \\ \gamma & \delta \end{pmatrix} \begin{pmatrix} a \\ b \end{pmatrix} = \begin{pmatrix} \theta \\ \phi \end{pmatrix}.$$

The six elements in the equation are given by

$$\alpha = \sum_{i=1}^{N} \frac{1}{\sigma_i^2} \qquad \beta = \sum_{i=1}^{N} \frac{x_i}{\sigma_i^2} \qquad \gamma = \beta$$

$$\delta = \sum_{i=1}^{N} \frac{x_i^2}{\sigma_i^2} \qquad \theta = \sum_{i=1}^{N} \frac{y_i}{\sigma_i^2} \qquad \phi = \sum_{i=1}^{N} \frac{x_i y_i}{\sigma_i^2}. \tag{6-45}$$

The values of a and b are obtained by solving Eq. (6-44), a simple application of the method given by Eq. (5-4),

$$\begin{aligned} a &= \frac{1}{D} \det \begin{vmatrix} \theta & \beta \\ \phi & \delta \end{vmatrix} = \frac{1}{D} \{\theta\delta - \beta\phi\} \\ b &= \frac{1}{D} \det \begin{vmatrix} \alpha & \theta \\ \gamma & \phi \end{vmatrix} = \frac{1}{D} \{\alpha\phi - \theta\gamma\}. \end{aligned} \tag{6-46}$$

The value of the determinant D is

$$D = \det \begin{vmatrix} \alpha & \beta \\ \gamma & \delta \end{vmatrix} = \alpha\delta - \beta\gamma = \alpha\delta - \beta^2. \tag{6-47}$$

The calculations involved here are quite simple once the six quantities α, β, γ, δ, θ, and ϕ of Eq. (6-45) are obtained from the input values $\{x_i, y_i\}$ and the uncertainties $\{\sigma_i^2\}$ associated with $\{y_i\}$. As an example, a collection of nine pieces of hypothetical

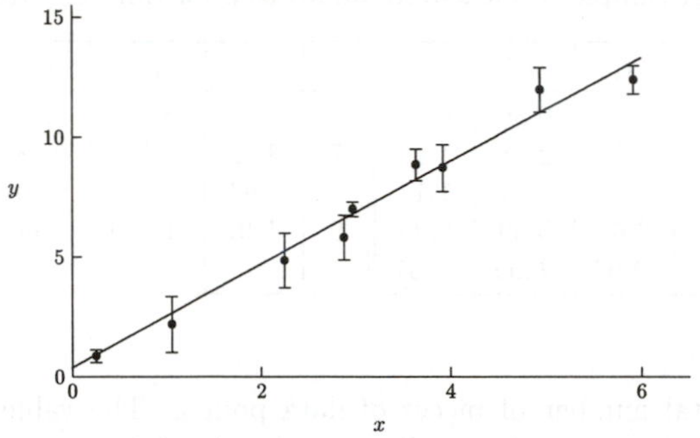

Figure 6-8: Linear least-squares fit to the nine data points in Table 6-2. The straight line $y = a + bx$ is plotted using the best-fit values of $a = 0.380$ and $b = 2.157$.

data, together with their uncertainties, is given in Table 6-2. The result of a straight-line fit is shown in Fig. 6-8.

Uncertainties in the parameters In addition to the maximum likelihood values of a and b, we also need to have estimates of their uncertainties arising from those in y_i. We can accomplish this task with Eq. (6-24) by treating both a and b as functions of $y_1, y_2, \ldots, y_N$. For simplicity, we shall assume that the uncertainties in different data points are uncorrelated and the covariance σ_{ij}^2 between any two of them vanishes. As a result, we have

$$\sigma_a^2 = \sum_{i=1}^{N}\left\{\left(\frac{\partial a}{\partial y_i}\right)^2 \sigma_i^2\right\} \qquad \sigma_b^2 = \sum_{i=1}^{N}\left\{\left(\frac{\partial b}{\partial y_i}\right)^2 \sigma_i^2\right\}. \tag{6-48}$$

Using Eq. (6-46), the partial derivatives involved are

$$\frac{\partial a}{\partial y_i} = \frac{1}{D}\left(\frac{1}{\sigma_i^2}\delta - \frac{x_i}{\sigma_i^2}\beta\right) \qquad \frac{\partial b}{\partial y_i} = \frac{1}{D}\left(\frac{x_i}{\sigma_i^2}\alpha - \frac{1}{\sigma_i^2}\gamma\right). \tag{6-49}$$

Inserting these results into Eq. (6-48), we obtain

$$\begin{aligned}
\sigma_a^2 &= \frac{1}{D^2}\sum_{i=1}^{N}\left\{\sigma_i^2\left(\frac{1}{\sigma_i^2}\delta - \frac{x_i}{\sigma_i^2}\beta\right)^2\right\} = \frac{1}{D^2}\sum_{i=1}^{N}\left\{\delta^2\frac{1}{\sigma_i^2} + \beta^2\frac{x_i^2}{\sigma_i^2} - 2\delta\beta\frac{x_i}{\sigma_i^2}\right\} \\
&= \frac{1}{D^2}\left\{\delta^2\alpha + \delta\beta^2 - 2\delta\beta^2\right\} = \frac{\delta}{D} \\
\sigma_b^2 &= \frac{1}{D^2}\sum_{i=1}^{N}\left\{\sigma_i^2\left(\frac{x_i}{\sigma_i^2}\alpha - \frac{1}{\sigma_i^2}\gamma\right)^2\right\} = \frac{1}{D^2}\sum_{i=1}^{N}\left\{\alpha^2\frac{x_i^2}{\sigma_i^2} + \gamma^2\frac{1}{\sigma_i^2} - 2\alpha\gamma\frac{x_i}{\sigma_i^2}\right\} \\
&= \frac{1}{D^2}\left\{\alpha^2\delta + \alpha\gamma^2 - 2\alpha\gamma\beta\right\} = \frac{\alpha}{D}
\end{aligned} \tag{6-50}$$

Box 6-1 Subroutine LLSQ(NPT,X,Y,SY,A,SA,B,SB)
Linear least-squares fit to a straight line $y = a + b * x$

Argument list:
- NPT: Number of data points.
- X: Array for the values of the independent variable x_i.
- Y: Array for the values of the dependent variable y_i.
- SY: Array for σ_i, the uncertainty of y_i.
- A: Output for coefficient a.
- SA: Standard deviation σ_a for a.
- B: Output for coefficient b.
- SB: Standard deviation σ_b for b.

1. Weight each point:
   (a) If $\sigma_i = 0$, $\omega_i = 1$.
   (b) If $\sigma_i \neq 0$, $\omega_i = 1/\sigma_i^2$.
2. Calculate α, β, δ, θ, and ϕ of Eq. (6-45) from x_i, y_i and ω_i.
3. Compute the value of the determinant D using Eq. (6-47).
4. Calculate the values of a and b using Eq. (6-46).
5. Calculate σ_a and σ_b using Eq. (6-48).
6. Return a, σ_a, b, σ_b, and $\sigma_{a,b}^2$.

where we have made use of the fact that $\beta = \gamma$.

One other quantity that can also provide some idea of the uncertainties in the values of a and b is the covariance between them. This is defined as

$$\sigma_{a,b}^2 = \sum_{i=1}^{N} \left\{ \frac{\partial a}{\partial y_i} \frac{\partial b}{\partial y_i} \sigma_i^2 \right\}.$$

Using the forms of $\partial a/\partial y_i$ and $\partial b/\partial y_i$ from Eq. (6-49), we find that

$$\sigma_{a,b}^2 = -\frac{\beta}{D}. \tag{6-51}$$

We shall make use of this quantity later. The calculations for σ_a and σ_b are incorporated as part of the algorithm for linear least-squares fit to data outlined in Box 6-1.

Correlation between y_i and x_i For our discussion of the method of least-squares fit to a straight-line relation, we have started with the premise that y is given as a linear function of x in the form of Eq. (6-43). As a part of the test of the significance of the fit, we shall also ask the question whether our data can actually be described by such a model. As usual in statistical analyses, we cannot hope to obtain a definite yes or no answer to this question. To get an idea of whether the fit is meaningful, we can calculate the correlation coefficient $r_{x,y}$ between $\{y_i\}$ and $\{x_i\}$:

$$r_{x,y} \equiv \frac{\sigma_{x,y}^2}{\sigma_x \sigma_y} = \frac{\sum_{i=1}^{N} \frac{1}{\sigma_i^2}(x_i - \mu_x)(y_i - \mu_y)}{\left\{\sum_{i=1}^{N} \frac{1}{\sigma_i^2}(x_i - \mu_x)^2\right\}^{1/2} \left\{\sum_{i=1}^{N} \frac{1}{\sigma_i^2}(y_i - \mu_y)^2\right\}^{1/2}} \tag{6-52}$$

where the mean values of y_i and x_i are given by

$$\begin{aligned} \mu_x &= \Big(\sum_{i=1}^{N} \frac{1}{\sigma_i^2}\Big)^{-1} \sum_{i=1}^{N} \frac{1}{\sigma_i^2} x_i = \frac{\beta}{\alpha} \\ \mu_y &= \Big(\sum_{i=1}^{N} \frac{1}{\sigma_i^2}\Big)^{-1} \sum_{i=1}^{N} \frac{1}{\sigma_i^2} y_i = \frac{\theta}{\alpha}. \end{aligned} \tag{6-53}$$

The denominator on the right side of Eq. (6-52) is proportional to the square root of the product of the variances for the distributions of $\{x_i\}$ and $\{y_i\}$:

$$\begin{aligned} \sigma_x^2 &= \Big(\sum_{i=1}^{N} \frac{1}{\sigma_i^2}\Big)^{-1} \sum_{i=1}^{N} \frac{1}{\sigma_i^2}(x_i - \mu_x)^2 \\ &= \frac{1}{\alpha}\Big\{\sum_{i=1}^{N} \frac{1}{\sigma_i^2} x_i^2 - 2\mu_x \sum_{i=1}^{N} \frac{1}{\sigma_i^2} x_i + \mu_x^2 \sum_{i=1}^{N} \frac{1}{\sigma_i^2}\Big\} = \frac{\delta}{\alpha} - \frac{\beta^2}{\alpha^2} = \frac{D}{\alpha^2} \\ \sigma_y^2 &= \Big(\sum_{i=1}^{N} \frac{1}{\sigma_i^2}\Big)^{-1} \sum_{i=1}^{N} \frac{1}{\sigma_i^2}(y_i - \mu_y)^2. \end{aligned} \tag{6-54}$$

The numerator is related to the covariance between x and y

$$\begin{aligned} \sigma_{x,y}^2 &= \Big(\sum_{i=1}^{N} \frac{1}{\sigma_i^2}\Big)^{-1} \sum_{i=1}^{N} \frac{1}{\sigma_i^2}(x_i - \mu_x)(y_i - \mu_y) \\ &= \frac{1}{\alpha}\Big\{\sum_{i=1}^{N} \frac{1}{\sigma_i^2} x_i y_i - \mu_x \sum_{i=1}^{N} \frac{1}{\sigma_i^2} y_i - \mu_y \sum_{i=1}^{N} \frac{1}{\sigma_i^2} x_i + \mu_x \mu_y \sum_{i=1}^{N} \frac{1}{\sigma_i^2}\Big\} \\ &= \frac{1}{\alpha}\Big\{\sum_{i=1}^{N} \frac{1}{\sigma_i^2} x_i y_i - \mu_x \mu_y \sum_{i=1}^{N} \frac{1}{\sigma_i^2}\Big\} \\ &= \frac{1}{\alpha}\phi - \frac{\beta\theta}{\alpha^2} = \frac{bD}{\alpha^2}. \end{aligned} \tag{6-55}$$

We can now express the square of the correlation coefficient $r_{x,y}$ of Eq. (6-52) in terms of σ_y^2 of Eq. (6-54) and that of $\sigma_{x,y}^2$ of Eq. (6-55):

$$r_{x,y}^2 = \frac{\sigma_{x,y}^4}{\sigma_x^2 \sigma_y^2} = \frac{\frac{bD}{\alpha^2}\sigma_{x,y}^2}{\frac{D}{\alpha^2}\sigma_y^2} = b\frac{\sigma_{x,y}^2}{\sigma_y^2}. \tag{6-56}$$

That is, $r^2_{x,y}$ is related to the slope of y_i as a function of x_i We shall make use of this result later in §6-6 to construct a correlation coefficient for the more general case.

If $\{x_i\}$ and $\{y_i\}$ are independent of each other, the numerator of the right side of Eq. (6-52) vanishes and the correlation coefficient $r_{x,y} = 0$. On the other hand, if y increases linearly as x, we have the result $r_{x,y} = 1$ regardless of the slope (given by b) of a plot of y versus x. Similarly, if y decreases linearly as x, we have the result $r_{x,y} = -1$. For this reason, the absolute value of the correlation coefficient may be used as one of the indicators of whether it is meaningful to carry out a linear least-squares fit for y_i. More sensitive tests can be constructed from the measures discussed in the next section.

6-5 Statistical tests of the results

In carrying out a least-squares fit, we usually start with a reasonable model on how the independent and dependent variables are related to each other. One of the purposes of the calculation is to find out whether our model is supported by the data. Because of the statistical nature of the least-squares analysis and the presence of uncertainties in the data themselves, it is usually impossible for us to reach a definitive answer to the question. Instead, we must rely on tests, or *statistics*, to arrive at an estimate of the probabilities for our model to be correct. A large amount of work in the field of statistics has gone into the design of these tests. We shall describe here three popular ones: Pearson's χ^2-test, Student's t-test, and Fisher's F-test.

χ^2-test In the previous section, we have seen that the method of maximum likelihood produces the most likely values for the parameters a and b in our linear model by minimizing the value of χ^2 defined in Eq. (6-41):

$$\chi^2 = \sum_{i=1}^{N}\left(\frac{y_i - f_i}{\sigma_i}\right)^2.$$

If the model used is an exact description of the data, we expect f_i, the value expected from our model, is the same as the measured value y_i for all the data points. In this case, the value of χ^2 vanishes. In practical situations, this is unlikely, either because of the deficiencies in the model or because of the uncertainties in the measurements, or both. However, if the fit is a good one, we expect the χ^2 to be small. Let us extend this idea into a quantitative test of the quality of the fit.

Since χ^2 is a sum, its value increases with the total number of data points N. To serve as an indicator for the quality of a fit, such a dependence is inconvenient. Our first reaction is to adopt, instead, an average value over N. This is also incorrect. We can see this using the linear model of the previous section as an example. Since there are only two parameters, a and b, an exact fit is obtained if there are only two data points ($N = 2$). Although the value of χ^2 is zero in this case, it is clear that the "fit" is not a meaningful one. This leads us to the concept of the number of *degrees of freedom*, defined as the number of independent measurements, N in our case, minus the number of constraints, m. That is, the number of degrees of freedom is defined as

$$\nu = N - m. \tag{6-57}$$

For our purpose, m is given by the number of free parameters in the model we adopt to fit the data. In the linear case of the previous section, $m = 2$.

A useful measure of how well a model fits the data is given by the value of the χ^2 per degree of freedom, also known as the *reduced chi-square*:

$$\chi_\nu^2 \equiv \frac{\chi^2}{\nu} = \frac{1}{N-n}\sum_{i=1}^{N}\left\{\frac{1}{\sigma_i^2}(y_i - f_i)^2\right\}. \tag{6-58}$$

We shall see later that the value of χ^2 follows a distribution with the mean equal to ν. As a result, a value of χ_ν^2 of around unity is considered to be a good fit.

To construct a more quantitative criterion, we can compare our model prediction with the result of using a set of N random numbers instead of $\{y_i\}$. In other words, if we substitute the measured results with random numbers, what is the probability $P(\chi^2|\nu)$ of obtaining a value that is equal to or larger than χ^2 if the number of degrees of freedom ν remains the same? Clearly, if our fit to the set of data points is not better than that to a set of random numbers, it is not meaningful.

In general, it is more convenient to ask the opposite question: what is the probability

$$Q(\chi^2|\nu) = 1 - P(\chi^2|\nu) \tag{6-59}$$

that our sample of data $\{y_i\}$ is given by the set of $\{f_i\}$ calculated using the parameters obtained from the fitting procedure. For this purpose, let us define a new variable

$$\xi_i = \frac{y_i - f_i}{\sigma_i}.$$

The value of χ^2 is then the sum of ξ_i^2. If each y_i is a random number with a normal distribution centered around f_i and having a variance σ_i^2, then ξ_i is a random number with a normal distribution centered around 0 and having a variance of unity. That is, the distribution of ξ_i is given by

$$p(\xi_i)d\xi_i = \frac{1}{\sqrt{2\pi}}e^{-\xi_i^2/2}d\xi_i.$$

As we have seen earlier, this probability is normalized to unity. From these two conditions, we find that the distribution of the variable $\zeta = \xi_i^2$ is given by

$$p(\zeta)\,d\zeta = \frac{1}{\sqrt{2\pi\zeta}}e^{-\zeta/2}d\zeta \tag{6-60}$$

with the normalization

$$\int_0^\infty p(\zeta)\,d\zeta = 1.$$

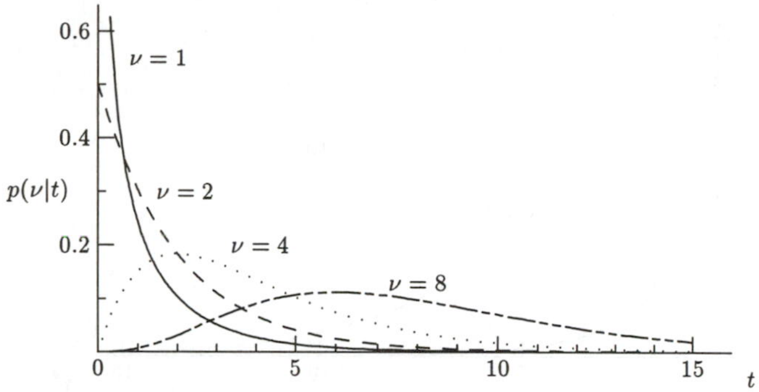

Figure 6-9: χ^2-distribution $p(\nu|t)$ for $\nu = 1$, 2, 4, and 8. The shape of $p(\nu|t)$ approaches that of a normal distribution for large values of ν.

This is known as the χ^2-distribution for a single degree of freedom.

For the purpose of testing our hypothesis, we are more interested in the distribution of the variable

$$t = \sum_{i=1}^{\nu} \xi_i^2 = \sum_{i=1}^{\nu} \left(\frac{y_i - f_i}{\sigma_i} \right)^2 \tag{6-61}$$

where each quantity ξ_i is an independent random variable with a normal distribution of zero mean and unit variance. The generalization of Eq. (6-60) is the χ^2-distribution for ν degrees of freedom,

$$p(t|\nu)\,dt = \frac{1}{2^{\nu/2}\Gamma(\nu/2)} t^{\frac{\nu}{2}-1} e^{-t/2} dt \tag{6-62}$$

where $\Gamma(x)$ is the gamma function of Eq. (4-104). The mean of $p(t|\nu)$ is ν and the variance is 2ν, as can be seen by explicit calculations. The shapes of $p(t|\nu)$ for a few low values of ν are shown in Fig. 6-9 for illustration.

The probability of having t less than or equal to some value χ^2 for a system of ν degrees of freedom is given by the probability integral of χ^2-distribution:

$$P(\chi^2|\nu) = \int_0^{\chi^2} \frac{1}{2^{\nu/2}\Gamma(\nu/2)} t^{\frac{\nu}{2}-1} e^{-t/2} dt. \tag{6-63}$$

We can identify this quantity as the probability we were looking for earlier to obtain a fit up to some χ^2-value in a least-squares calculation using a function with ν degrees of freedom. Comparing the right side of Eq. (6-63) with the incomplete gamma function $\gamma(a, x)$ of Eq. (4-115), we find that

$$P(\chi^2|\nu) = \frac{1}{\Gamma(\nu/2)} \gamma(a, x) \tag{6-64}$$

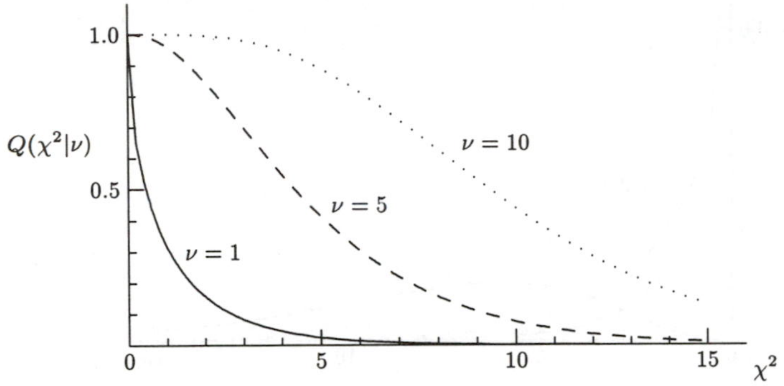

Figure 6-10: Value of $Q(\chi^2|\nu)$ for $\nu = 1$, 5, and 10 for $\chi^2 = [0, 15]$.

with

$$a = \frac{1}{2}\nu \qquad\qquad x = \frac{1}{2}\chi^2.$$

From Eq. (4-115), we have $\Gamma(a, x) = \Gamma(a) - \gamma(a, x)$. Using this relation, we obtain

$$Q(\chi^2|\nu) = 1 - \frac{1}{\Gamma(\nu/2)}\gamma(a, x) = \frac{1}{\Gamma(\nu/2)}\Gamma(a, x). \tag{6-65}$$

Values of $P(\chi^2|\nu)$ and $Q(\chi^2|\nu)$ are given in standard statistics tables. Alternatively, we can obtain $Q(\chi^2|\nu)$ from $\Gamma(x)$ calculated using the algorithm given in Box 4-7 and $\gamma(a, x)$, using the algorithm of Box 4-8. An idea of the variation of $Q(\chi^2|\nu)$ as a function of ν and χ^2 is given in Fig. 6-10.

As an example, we shall apply the χ^2-test to the histogram of Fig. 6-5 and see whether the distribution is indeed a normal one. The number of occurrences in each bin h_i is listed in the third column of Table 6-3 and the value n_i expected of a normal distribution is given in the fourth column. Again, we shall assume a Poisson distribution for the possible values in each histogram bin and take the uncertainty associated with h_i as given by the square root of the number in the bin. The contribution of each bin to the χ^2 is then

$$\chi_i^2 = (h_i - n_i)^2/n_i.$$

A small problem occurs in bin 6, where $n_i = 0$. To avoid an unphysical result of infinity as the contribution to the χ^2, we approximate the denominator by unity. The sum of χ_i^2 calculated in this way is 4.4.

Since the normal distribution assumed in the discussion above is adjusted to have the same mean and variance as the data, we have, in effect, imposed the equivalence

Table 6-3: χ^2-test for the histogram in Fig. 6-5 to be a normal distribution.

Bin	Range	Histogram	Normal distribution	χ_i^2
1	47–48	1	0.7	0.1
2	48–49	5	3.9	0.3
3	49–50	5	8.9	3.0
4	50–51	11	8.1	0.8
5	51–52	3	3.0	0.0
6	52–53	0	0.4	0.2
				$\sum \chi_i^2 = 4.4$

of two constraints on the system. With 6 bins, the number of degrees of freedom is, therefore, $\nu = (6 - 2) = 4$. The reduced χ^2 is then $4.4/4 = 1.1$. The probability of obtaining a χ^2 larger or equal to 4.4 is $Q(4.4|4) = 0.35$. The result, therefore, supports our assumption of a normal distribution. However, there is also a good chance that the assumption is incorrect. This is, essentially, the same conclusion we reached earlier in §5-2 — the difference is that we have a more quantitative conclusion here.

The χ^2-statistic is widely used to find the probability for a set of data to be in agreement with a model of the expected distribution. However, it is only one of the possible tests and, in many cases, cannot provide a clear answer without other corroborative evidence. Part of the difficulty in using a χ^2-test stems from the fact that two different sources contribute to its value at the same time. For a given set of $\{y_i, x_i\}$, there exists a functional relation between them, referred to as the *parent function* in the language of statistics. However, we have no knowledge of this function. In a least-squares calculation, we construct a model by making a guess of this function based on the best information available to us. Any differences between our model and the parent function increase the value of the χ^2. This is the first source contributing to the value of χ^2.

A second source comes from discrepancies between the data and the parent function. For example, some errors may have been introduced into the data, perhaps because of the limited precision we can achieve in the measurement. As a result, even if we make the correct guess for the parent function, we cannot have a perfect fit to the data points and the value of the χ^2 does not vanish. In a χ^2-test, we have no way to distinguish these two sources. For this reason, other statistical tests are needed. This is particularly important when the result from a χ^2-test is ambiguous and when the conclusion depends heavily on the outcome of the statistical analyses.

Student's *t*-test Consider again a sample of N data points, $y_1, y_2, \ldots, y_N$. The mean μ_y and variance σ^2 of the distribution of $\{y_i\}$ are given by

$$\mu_y = \frac{1}{N}\sum_{i=1}^{N} y_i \qquad\qquad \sigma^2 = \frac{1}{N-1}\sum_{i=1}^{N}(y_i - \mu_y)^2.$$

The Student's t-statistic[1] is a test based on the ratio between μ_y and σ. The reason that the denominator in the definition of σ^2 is $(N-1)$ is because that we are subtracting μ_y from each data point. Since μ_y is obtained from the data themselves, this is equivalent to imposing a constraint on the set of N variables.

If y_i is a random variable with a normal distribution of (unknown) mean μ and the same variance σ^2, then the ratio

$$t = \frac{\mu_y - \mu}{\sigma/\sqrt{N}} \tag{6-66}$$

is also a random variable distributed according to

$$p(t,\nu)\,dt = \frac{dt}{\nu^{1/2}B(\frac{1}{2},\frac{1}{2}\nu)(1+\frac{t^2}{\nu})^{\frac{1}{2}(\nu+1)}}$$

where $\nu = (N-1)$ is the number of degrees of freedom and $B(\frac{1}{2},\frac{1}{2}\nu)$ is the beta function

$$B(a,b) = \int_0^1 t^{a-1}(1-t)^{b-1}dt = \frac{\Gamma(b)\Gamma(a)}{\Gamma(a+b)}. \tag{6-67}$$

For our interest here, $a = \frac{1}{2}$ and $b = \frac{1}{2}\nu$. More detailed discussion of the Student's t-test and other statistics can be found in Cramer[13] and in Kendall and Stuart.[30] There are several ways to make use of the statistics. For example, we can construct a test based on the definition given by Eq. (6-66) to see if the sample mean μ_y obtained from the data is in agreement with the (unknown) parent mean μ.

Again, we are more interested in the integrated probability of t to be less than or equal to some value ζ. In terms of the incomplete beta function,

$$I_\zeta(a,b) = \frac{1}{B(a,b)}\int_0^\zeta t^{a-1}(1-t)^{b-1}dt \equiv \frac{B_\zeta(a,b)}{B(a,b)}. \tag{6-68}$$

The derivation of this result can be found in standard statistics textbooks. The values of $I_\zeta(a,b)$ or, alternatively, the values of

$$A(t|\nu) = 1 - I_{\frac{\nu}{\nu+t^2}}\left(\frac{\nu}{2},\frac{1}{2}\right) \tag{6-69}$$

are given in statistics tables. We shall first see how to evaluate this function before applying it to examples in least-squares fits to data in the next section.

[1] "Student" was the pen name of W.S. Gosset who published the statistic in 1908.

Incomplete beta function The most convenient way to calculate $I_\zeta(a,b)$ on a computer is to use the continued fraction form given on page 944 of Abramowitz and Stegun:[1]

$$I_x(a,b) = \frac{x^a(1-x)^b}{aB(a,b)}\left\{\frac{1}{1+}\frac{d_1}{1+}\frac{d_2}{1+}\cdots\right\} \tag{6-70}$$

where the coefficients d_i are given by

$$\begin{aligned} d_{2m+1} &= -\frac{(a+m)(a+b+m)}{(a+2m)(a+2m+1)}x \\ d_{2m} &= \frac{m(b-m)}{(a+2m-1)(a+2m)}x. \end{aligned} \tag{6-71}$$

The expression works best for

$$x < \frac{a-1}{a+b-2}. \tag{6-72}$$

For the other values of x, we can substitute $\xi = (1-x)$ in the integral of Eq. (6-68) and obtain the relation

$$\begin{aligned} B_x(a,b) &= \int_0^\zeta t^{a-1}(1-t)^{b-1}dt \\ &= -\int_1^{1-x}(1-\xi)^{a-1}\xi^{b-1}d\xi \\ &= \int_0^1 (1-\xi)^{a-1}\xi^{b-1}d\xi - \int_0^{1-x}(1-\xi)^{a-1}\xi^{b-1}d\xi \\ &= B(b,a) - B_{1-x}(b,a). \end{aligned}$$

In terms of the incomplete beta function, this is equivalent to the relation

$$I_x(a,b) = 1 - I_{1-x}(b,a). \tag{6-73}$$

If we apply the continued fraction of Eq. (6-70) to $I_{1-x}(b,a)$, the condition Eq. (6-72) becomes

$$(1-x) < \frac{b-1}{a+b-2}$$

or

$$x > \frac{a-1}{a+b-2}.$$

In other words, if x satisfies Eq. (6-72), we can use Eq. (6-71) to calculate $I_x(a,b)$. Otherwise, we can make use of Eq. (6-73) and calculate $I_{1-x}(b,a)$ with Eq. (6-71) instead. In this way, the continued fraction of Eq. (6-70) may be employed to calculate all the values of x in the range $[0,1]$.

To evaluate $I_x(a,b)$ using Eq. (6-70), we can put the relation into the standard form of Eq. (3-24) with partial numerators and denominators:

$$a_i = \begin{cases} 1 & \text{for } i=1 \\ d_{i-1} & \text{for } i>1 \end{cases} \qquad\qquad b_i = \begin{cases} 0 & \text{for } i=0 \\ 1 & \text{for } i>0. \end{cases}$$

Box 6-2 Function BETAI(x,a,b)
Incomplete beta function $I_x(a,b)$
Continued fraction approximation for $x < (a-1)/(a+b-2)$

Argument list:
- x: Upper limit of the integral in Eq. (6-68).
- a: The factor a in the integrand.
- b: The factor b in the integrand.

Subprogram used:
- GAMMA_LX: logarithm form of gamma function (cf. Box 4-7).

Initialization:
- (a) Set the maximum number of terms to be 100, and
- (b) Accuracy to be 5×10^{-7}.

1. Check for special values:
   - (a) $I_x(a,b) = 0$ for $x = 0$.
   - (b) $I_x(a,b) = 1$ for $x = 1$.
2. Calculate the starting values using Eq. (6-75) and set the scale factor $f = 1$.
3. For $m = 1$ to a maximum of 100 terms, carry out the following steps:
   - (a) Calculate d_{2m} using Eq. (6-71).
   - (b) Calculate A_{2m} and B_{2m} using Eq. (6-74).
   - (c) Divide A_{2m} and B_{2m} by f to prevent overflow and underflow.
   - (d) Calculate d_{2m+1} using Eq. (6-71) and divide the result by f.
   - (e) Calculate A_{2m+1} and B_{2m+1} using Eq. (6-74).
   - (f) If B_{2m+1} does not vanish, let $f = B_{2m+1}$. Otherwise, $f = 1$.
   - (g) Calculate $f_{2m+1} = A_{2m+1}/B_{2m+1}$.
     - (i) Stop the continued fraction if calculation converges.
     - (ii) Otherwise, add more terms by going to the next m value.
4. Evaluate $B(a,b)$ using Eq. (6-67):
   - (a) Calculate the gamma functions using the logarithmic form of Box 4-7.
   - (b) Obtain $B(a,b)$ from $\exp\{\ln(\Gamma(a)) + \ln(\Gamma(b)) - \ln(\Gamma(a+b))\}$.
5. Return $\frac{x^a(1-x)^b}{aB(a,b)} f_{2m+1}$.

In terms of d_{2m} and d_{2m+1} given in Eq. (6-71), the recursion relation of Eq. (3-24) may be written as

$$A_{2m} = A_{2m-1} + d_{2m}A_{2m-2} \qquad A_{2m+1} = A_{2m} + d_{2m+1}A_{2m-1}$$
$$B_{2m} = B_{2m-1} + d_{2m}B_{2m-2} \qquad B_{2m+1} = B_{2m} + d_{2m+1}B_{2m-1} \tag{6-74}$$

for $m = 1, 2, \ldots$. The starting point is

$$A_0 = B_0 = A_1 = 1 \qquad B_1 = 1 - \frac{a+b}{a+1}x. \tag{6-75}$$

Note that the indices for A_i and B_i are shifted by 1 from those used in Eq. (3-24).

This is done so that we can conveniently make use of the relation for d_i given by Eq. (6-71). The algorithm to evaluate $I_x(a, b)$ is given in Box 6-2.

Fisher's *F*-test The F-statistic is designed to test the probability of two least-squares fits with different χ^2-values being equivalent to each other. For example, in the previous section, we made a two-parameter fit to nine data points. To give a better fit, we may wish to added a third parameter associated with a quadratic term and use the method described in the next section to carry out the calculations. Since we have increased the number of parameters, the value of χ^2 obtained is likely to be smaller. Does this result imply that the parent function for the data contains a quadratic term? That is, we wish to find out whether the decrease in the value of χ^2 is significant enough for us to change our model by increasing the number of parameters.

The same kind of situation also occurs, for instance, if we carry out two separate sets of measurements for an experiment. If we apply the same least-squares fit to the two sets of results, there is a high probability that the two χ^2-values will be somewhat different from each other. If the difference is significant, it may imply that some underlying factors in the experiment have changed between the two measurements. The F-test provides us with a measure of the probability that the two sets are referring to different conditions.

In general, if we have two χ^2-values, χ_1^2 and χ_2^2, corresponding respectively to ν_1 and ν_2 degrees of freedom, the ratio

$$t \equiv \frac{\chi_1^2/\nu_1}{\chi_2^2/\nu_2} \tag{6-76}$$

is distributed according to

$$p(t; \nu_1, \nu_2)\, dt = \frac{\nu_1^{\nu_1/2} \nu_2^{\nu_2/2} t^{(\nu_1/2-1)}}{B(\frac{1}{2}\nu_1, \frac{1}{2}\nu_2)(\nu_1 t + \nu_2)^{(\nu_1+\nu_2)/2}}\, dt.$$

The range of possible values for t is from 0 to ∞. From the definition of t given by Eq. (6-76), it is obvious that t and $1/t$ have the same distribution except that the roles of ν_1 and ν_2 are interchanged.

The cumulative probability of t up to some value F is given by the integral

$$P(F|\nu_1, \nu_2) = \frac{\nu_1^{\nu_1/2} \nu_2^{\nu_2/2}}{B(\frac{1}{2}\nu_1, \frac{1}{2}\nu_2)} \int_0^F t^{(\nu_1/2-1)} (\nu_1 t + \nu_2)^{-(\nu_1+\nu_2)/2} dt.$$

More commonly, this is put in terms of the incomplete beta function $I_\varsigma(a, b)$,

$$Q(F|\nu_1, \nu_2) = 1 - P(F|\nu_1, \nu_2) = I_{\frac{\nu_2}{\nu_1 t + \nu_2}}(\frac{1}{2}\nu_2, \frac{1}{2}\nu_1). \tag{6-77}$$

When $F = 0$, the subscript in the incomplete beta function becomes 1 and

$$Q(0|\nu_1, \nu_2) = I_1(\frac{1}{2}\nu_2, \frac{1}{2}\nu_1) = 1.$$

Similarly, when $F = \infty$, the subscript becomes 0 and

$$Q(\infty|\nu_1, \nu_2) = I_0(\frac{1}{2}\nu_2, \frac{1}{2}\nu_1) = 0.$$

Thus, a value of $Q(F|\nu_1, \nu_2)$ near 0 or 1 implies that the two χ^2-values are quite different from each other. We shall see applications of the various tests in the examples of the next section.

6-6 Linear least-squares fit

In this section, we shall apply the method of least squares of §6-4 to functions with an arbitrary number of parameters, $a_1, a_2, \ldots, a_m$, subject only to the condition that the parameters appear linearly in the fitting function. That is, we wish to model the dependence of y on x by an expression of the form

$$y(x) = \sum_{k=1}^{m} a_k f_k(x). \tag{6-78}$$

At this moment, the functions $f_k(x)$ are arbitrary, except that they cannot involve any of the m parameters $a_1, a_2, \ldots, a_m$. In statistics, this way of analyzing data is known as *multiple regression.* The more conventional name is linear least-squares fit and we shall adopt it for the most part. Cases that fail to satisfy Eq. (6-78) fall into the subject of nonlinear least-squares methods and we shall return in the next section.

Similar to what we have done earlier, we shall denote as y_i the value of y measured at $x = x_i$. For each y_i, there is an uncertainty σ_i. If the total number of data points is N, the value of χ^2 defined in Eq. (6-41) is now

$$\chi^2 = \sum_{i=1}^{N} \frac{1}{\sigma_i^2}\Big\{y_i - \sum_{k=1}^{m} a_k f_k(x_i)\Big\}^2. \tag{6-79}$$

As we saw earlier in Eq. (6-42), the maximum of likelihood function $\mathcal{L}(a_1, a_2, \ldots, a_m)$ of Eq. (6-40) is given by

$$\frac{\partial\chi^2}{\partial a_\ell} = -2\sum_{i=1}^{N} \frac{1}{\sigma_i^2} f_\ell(x_i)\Big\{y_i - \sum_{k=1}^{m} a_k f_k(x_i)\Big\} = 0 \tag{6-80}$$

where $\ell = 1, 2, \ldots, m$, and we have made use of the fact $\partial f_k(x)/\partial a_\ell = 0$ for all k and ℓ.

Curvature matrix and covariance matrix The conditions given by Eq. (6-80) may be expressed as a set of m linear equations in the form

$$\sum_{k=1}^{m} a_k \sum_{i=1}^{N} \frac{1}{\sigma_i^2} f_k(x_i) f_\ell(x_i) = \sum_{i=1}^{N} \frac{1}{\sigma_i^2} f_\ell(x_i)\, y_i. \tag{6-81}$$

By defining a square matrix $\boldsymbol{F}$ and a column matrix $\boldsymbol{H}$, with elements

$$F_{\ell,k} = \sum_{i=1}^{N} \frac{1}{\sigma_i^2} f_k(x_i) f_\ell(x_i) = \frac{1}{2}\frac{\partial \chi^2}{\partial a_\ell \partial a_k} \tag{6-82}$$

$$H_\ell = \sum_{i=1}^{N} \frac{1}{\sigma_i^2} f_\ell(x_i) y_i \tag{6-83}$$

the m linear equations of Eq. (6-81) may be written as

$$\sum_{k=1}^{m} F_{\ell,k}\, a_k = H_\ell$$

where $\ell = 1, 2, \ldots, m$. In matrix notation, this set of equations may be written as

$$\boldsymbol{F}\boldsymbol{A} = \boldsymbol{H} \tag{6-84}$$

where $\boldsymbol{F}$ is known as the curvature matrix. The parameters $a_1, a_2, \ldots, a_m$ are the elements of the column matrix $\boldsymbol{A}$. Our aim here is to find their values by solving the system of linear equations.

As in Eq. (5-21), the solution may be found from the inverse of $\boldsymbol{F}$. Let

$$\boldsymbol{G} = \boldsymbol{F}^{-1}.$$

This is known as the covariance matrix. Using $\boldsymbol{G}$, the solution for $\boldsymbol{A}$ is given by

$$\boldsymbol{A} = \boldsymbol{G}\boldsymbol{H}. \tag{6-85}$$

In terms of matrix elements, the same result may be written as

$$\begin{aligned} a_k &= \sum_{\ell=1}^{m} G_{k,\ell} H_\ell \\ &= \sum_{\ell=1}^{m} G_{k,\ell} \sum_{i=1}^{N} \frac{1}{\sigma_i^2} f_\ell(x_i) y_i. \end{aligned} \tag{6-86}$$

In this approach, the main calculation involved is to obtain the covariance matrix $\boldsymbol{G}$ by inverting the curvature matrix $\boldsymbol{F}$. The algorithm is outlined in Box 6-3.

If, for some reason, a wrong choice is made on the functional form of $y(x)$, two or more of the parameters may become simply related to each other in essence. For example, one is a multiple of another. In this case, the curvature matrix becomes singular, the determinant of $\boldsymbol{F}$ vanishes, and the matrix cannot be inverted. There are two ways to correct this problem. The most obvious one is to modify our model so that the singular condition does not arise. This may not always be easy to do. The alternative is to use the method of singular value decomposition to solve the matrix inversion problem. This method is discussed in Stoer and Bulirsch[54] and in Press

Box 6-3 Subroutine MRGS(NPT,N_PARM,X,Y,SY,A,SA)

Multiple regression analysis for $y = \sum_{k=1}^{m} a_k f_k(x)$

Argument list:

NPT:	Number of data points.
N_PARM:	Number of parameters.
X:	Input array for independent variable $\{x_i\}$.
Y:	Input array for dependent variable $\{y_i\}$.
SY:	Input array for uncertainties $\{\sigma_i\}$.
A:	Output array for coefficients $\{a_j\}$.
SA:	Output array for the uncertainties in $\{a_j\}$.

Subprograms required:

FK:	Returns the value of $f_k(x)$ for a given set of $\{a_i\}$.
MATIV:	Matrix inversion of Box 5-3.

1. Zero the arrays for f_i and $h_{i,j}$.
2. From the input values of x_i, y_i, and σ_i, calculate:
   (a) If $\sigma_i = 0$, let the weighting factor $w_i = 1$, otherwise $w_i = 1/\sigma_i^2$.
   (b) Store the values of $f_k(x_i)$ in an auxiliary array.
   (c) Compute H_ℓ using Eq. (6-83).
   (d) Compute the lower half-triangle of $F_{\ell,k}$ using Eq. (6-82).
   (e) Complete the upper triangle of $F_{k,\ell}$ by symmetry.
3. Obtain the covariance matrix by inverting $\boldsymbol{F}$ using MATIV.
   Check if the determinant $\boldsymbol{F}$ is singular.
4. Compute a_k from the inverse of $\boldsymbol{F}$ using Eq. (6-86).
5. Calculate $\{\sigma_{a_i}\}$, the uncertainty of a_i, using Eq. (6-87).
6. Return the values of $\{a_i\}$ and $\{\sigma_{a_i}\}$.

and others.[44] Here we shall stay with the simpler approach and assume that we can take care of any singular situation that arises by changing the functional form of $y(x)$.

Example of fitting with Legendre polynomial As an illustration, we shall examine the angular correlation of two γ-rays emitted in the decay of an excited nucleus, one in the transition from an excited state with spin J_0 to an intermediate state with spin J_i, and the other from the intermediate state to the ground state with spin J_f. The angular correlation function $W(\theta)$, which expresses the probability of the second γ-ray to be emitted at angle θ with respect to the first, is given by

$$W(\theta) = 1 + \sum_{k=1}^{k_x} A_{2k} P_{2k}(\cos\theta)$$

where $P_\ell(\cos\theta)$ are the Legendre polynomials of §4-2. The upper limit of summation, k_x, is determined by the spins of the states involved. Because of the symmetry in the problem, only even-order Legendre polynomials enter the expression. The shape of the angular distribution is characterized by coefficients A_{2k}, and their values depend

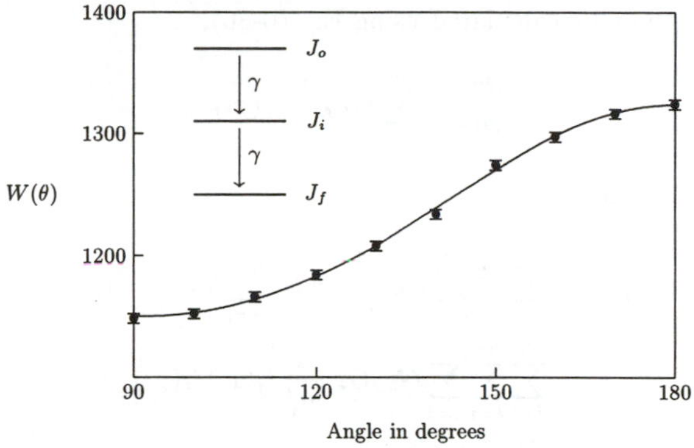

Figure 6-11: Angular correlation of γ-rays emitted in the decay of ^{60}Ni fitted to $C(\theta) = a_0 + a_2P_2(\cos\theta) + a_4P_4(\cos\theta)$ with $a_0 = 1200.3 \pm 1.6$, $a_2 = 110.5 \pm 3.3$, and $a_4 = 12.8 \pm 3.5$ (smooth curve). The data, with uncertainties shown in the form of error bars, are taken from Steffen.[53]

on the nature of the two γ-rays emitted and the spins of the three states involved. As a result, angular correlation studies constitute an important tool in determining the spin of a state. For our purpose here, we shall not be concerned with the connection between coefficients A_{2k} and the type of γ-rays emitted. Our interest is to find the values of A_{2k} that give the best fit to the measured correlation function $W(\theta)$.

To compare with the measured quantities, we shall rewrite the angular correlation function in terms of the number of γ-rays observed at angle θ with respect to another one emitted essentially at the same time,

$$C(\theta) = \sum_{k=0}^{m} a_{2k}P_{2k}(\cos\theta).$$

The experimental data are taken from the decay of ^{60}Ni measured by Steffen[53] and the fitted results are shown in Fig. 6-11. Note that, because of the symmetry of even-order Legendre polynomials, the angular distribution of $W(\theta)$ is symmetric around 90°. For this reason, only the backward angles are measured in many experiments of this type.

Uncertainties in the parameters The uncertainties in the parameters a_k determined in a linear least-squares calculation are given by the diagonal matrix elements of the covariance matrix $\boldsymbol{G}$. This can be seen from the following argument. From Eq. (6-24), we find that the uncertainty in the value of a_k is given by

$$\sigma_{a_k}^2 = \sum_{i=1}^{N} \sigma_i^2 \left(\frac{\partial a_k}{\partial y_i}\right)^2$$

if we ignore the covariance between any two data points. The partial derivative of a_k with respect to y_i may be calculated using Eq. (6-86),

$$\frac{\partial a_k}{\partial y_i} = \sum_{\ell=1}^{m} G_{k,\ell} \frac{1}{\sigma_i^2} f_\ell(x_i).$$

This gives us the result

$$\begin{aligned}
\sigma_{a_k}^2 &= \sum_{i=1}^{N} \sigma_i^2 \sum_{\ell=1}^{m} G_{k,\ell} \frac{1}{\sigma_i^2} f_\ell(x_i) \sum_{j=1}^{m} G_{k,j} \frac{1}{\sigma_i^2} f_j(x_i) \\
&= \sum_{i=1}^{N} \sum_{\ell=1}^{m} \sum_{j=1}^{m} G_{k,\ell} G_{k,j} \frac{1}{\sigma_i^2} f_j(x_i) f_\ell(x_i) \\
&= \sum_{\ell=1}^{m} \sum_{j=1}^{m} G_{k,\ell} G_{k,j} F_{j,\ell} = \sum_{\ell=1}^{m} G_{k,\ell} \delta_{k,\ell} = G_{k,k} \qquad \text{(6-87)}
\end{aligned}$$

where we have made use of Eq. (6-82) to sum over index i to obtain $F_{j,\ell}$ and the fact that $\boldsymbol{G}$ is the inverse of $\boldsymbol{F}$.

Chi-square test of the fit The minimum χ^2 is given by

$$\chi_{\min}^2 = \sum_{i=1}^{N} \frac{1}{\sigma_i^2} \Big\{ y_i - \sum_{k=1}^{m} a_k f_k(x_i) \Big\}^2$$

using the values of $\{a_k\}$ that minimize the χ^2. To have a feeling of the quality of the fit, we can find the probability to obtain a χ^2-value greater or equal to $\chi_{\min}^2$. This is given by the probability integral for χ^2-distribution $Q(\chi_{\min}^2|\nu)$ of Eq. (6-65). Here $\nu = (N - m)$ is the number of degrees of freedom, with N as the number of independent data points and m, the number of free parameters.

As an example, consider the 20 data points in Table 6-4. Let us begin by trying to fit them to a straight line as we did earlier in §6-4

$$y(a_1, a_2; x) = a_1 + a_2 x.$$

Since there are two parameters, the number of degrees of freedom is $\nu = 18$. The value of χ^2 at the best-fit values of $a_1 = 0.115 \pm 0.020$ and $a_2 = 2.282 \pm 0.011$ is $\chi_{\min}^2 = 168$, giving a reduced χ^2-value of $\chi_\nu^2 = 9.3$. The probability of obtaining a χ^2-value of 168 for 18 degrees of freedom is $Q(168|18) = 0.0$. Thus, we conclude that the fit is not a good one.

To find a better functional form, we include a quadratic term in $y(x)$. The fitting function now becomes

$$y(a_1, a_2; x) = a_1 + a_2 x + a_3 x^2.$$

With the addition of a parameter, the number of degrees of freedom is reduced by one to 17. The best-fit values of the parameters are now $a_1 = 0.0455 \pm 0.0206$,

Table 6-4: Twenty hypothetical measured values of y as a function of x.

i	x_i	y_i	$y_{\rm fit}$	σ_i	i	x_i	y_i	$y_{\rm fit}$	σ_i
1	2.48	6.04	6.259	0.900	11	4.09	9.29	9.072	1.030
2	2.92	7.57	7.123	0.340	12	2.31	5.89	5.906	0.050
3	1.41	4.03	3.858	0.690	13	2.00	5.47	5.235	0.580
4	0.04	0.16	0.164	0.020	14	2.38	5.94	6.053	0.640
5	0.30	1.61	0.918	0.730	15	3.88	8.76	8.760	0.040
6	0.04	0.49	0.164	0.320	16	0.67	1.74	1.949	0.650
7	4.07	9.14	9.043	0.910	17	3.03	6.80	7.328	0.770
8	1.90	4.60	5.011	0.600	18	0.82	2.18	2.353	0.190
9	0.64	2.51	1.868	0.200	19	3.21	7.94	7.654	1.260
10	1.49	3.79	4.052	1.540	20	0.39	0.96	1.174	0.310

$a_2 = 2.9656 \pm 0.0566$, and $a_3 = -0.1855 \pm 0.0151$. The value of χ^2 is reduced to $\chi^2_{\rm min} = 17$. Since $Q(17|17) = 0.45$, we conclude that the new fit is reasonable. The effect of the quadratic term is to put a curvature into the relation between x and y. From Fig. 6-12, we see that, even though the curvature is quite small, the refinement in the fit due to a small x^2-dependent term is quite significant.

The improvement produced by the quadratic term may prompt us to try to include even higher-order terms into the fitting function $y(x)$ to see if the value of the χ^2 can be further reduced. The addition of a cubic term lowers the χ^2-value only slightly to $\chi^2_{\rm min} = 16$. However, the smaller value does not necessarily imply a better fit. This can be seen, for example, from the fact that $Q(16|16) = 0.45$, unchanged from that of the quadratic fit. We shall see next that there are also other tests designed to check directly whether an additional parameter is significant.

Multiple correlation coefficient Another test of the quality of a fit is based on the correlation coefficient of Eq. (6-52). To make connection with the simple case of a straight-line fit, we shall assume that the first term in the sum on the right side of Eq. (6-78) is a constant term, independent of x. This is the equivalent of taking

$$f_1(x) = 1.$$

This approach has also the advantage that, when we limit ourselves to two parameters, the form matches that for a straight line used earlier in Eq. (6-43). Similar to Eq. (6-55), we can calculate the covariance between y and $f_k(x)$, the kth function in the sum on the right side of Eq. (6-78), in the following way:

$$\sigma^2_{k,y} \equiv \left(\sum_{i=1}^{N} \frac{1}{\sigma_i^2}\right)^{-1} \sum_{i=1}^{N} \frac{1}{\sigma_i^2}\{f_k(x_i) - \mu_k\}\{y_i - \mu_y\}$$

where μ_k is the mean of $f_k(x)$ and μ_y, that of y. These may be obtained in similar

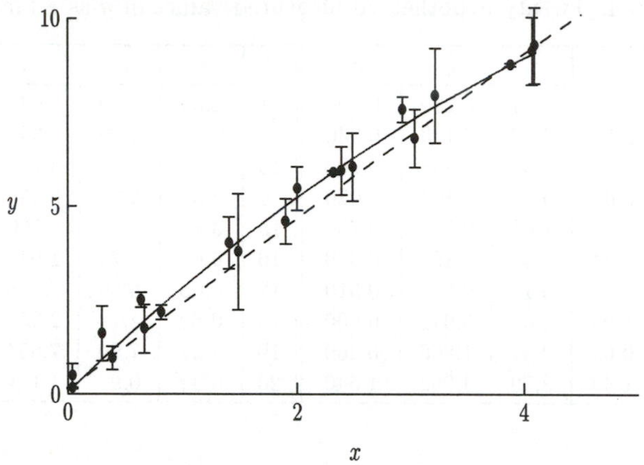

Figure 6-12: Linear least-squares fit to the 20 data points in Table 6-4 using a two-parameter form $y = a + bx$ with $a = 0.115 \pm 0.020$ and $b = 2.282 \pm 0.011$ (dashed line) and a three-parameter form $y = a_1 + a_2 x + a_3 x^2$ with $a_1 = 0.046 \pm 0.021$, $a_2 = 2.966 \pm 0.057$, and $a_3 = -0.186 \pm 0.015$ (solid curve).

ways as Eq. (6-53):

$$\mu_k = \left(\sum_{i=1}^{N} \frac{1}{\sigma_i^2}\right)^{-1} \sum_{i=1}^{N} \frac{1}{\sigma_i^2} f_k(x_i) \qquad\qquad \mu_y = \left(\sum_{i=1}^{N} \frac{1}{\sigma_i^2}\right)^{-1} \sum_{i=1}^{N} \frac{1}{\sigma_i^2} y_i.$$

Analogous to Eq. (6-56), we can define the square of a *multiple correlation coefficient* as

$$R^2 = \frac{1}{\sigma_y^2} \sum_{k=2}^{m} a_k \sigma_{k,y}^2 \tag{6-88}$$

where the variance of y is given by

$$\sigma_y^2 = \frac{1}{N-1} \sum_{i=1}^{N} (y_i - \mu_y)^2. \tag{6-89}$$

The summation on the right side of Eq. (6-88) starts with $k = 2$, as the $k = 1$ term is assumed to be a constant and therefore excluded from the summation.

We use $(N-1)$ rather than N in the denominator on the right side of Eq. (6-89) to define σ_y^2 for the following reason. If y_i is a random variable with a normal distribution, then σ_y^2 is a random variable with a χ^2-distribution, as can be seen by comparing Eq. (6-89) with (6-61). Since the mean μ_y is obtained by averaging over $\{y_i\}$, there are only $(N-1)$ degrees of freedom left among the N values of $(y_i - \mu_y)$.

The product $R^2\sigma_y^2$ is also a random variable with a χ^2-distribution. However, the number of degrees of freedom is only $(m-1)$. This is evident from the right side of Eq. (6-88). On multiplying both sides by σ_y^2, we obtain

$$R^2\sigma_y^2 = \sum_{k=2}^{m} a_k\sigma_{k,y}^2.$$

The right side is now simply a sum over $(m-1)$ squares of random variables. In terms of $R^2\sigma_y^2$, we can decompose σ_y^2 in the following way:

$$\sigma_y^2 = R^2\sigma_y^2 + (1-R^2)\sigma_y^2.$$

The left side of the equation consists of a random variable with a χ^2-distribution of $(N-1)$ degrees of freedom. Since the first term on the right side is a random variable with a χ^2-distribution of $(m-1)$ degrees of freedom, the second term must be a random variable with $(N-1)-(m-1)=(N-m)$ degrees of freedom. Physically, the first term is a measure of the spread in the dependent and independent variables in the data and the second term provides a feeling of the spread between the fit and the data. Since both quantities follow χ^2-distributions, we can use the F-test of §6-5 to supply a measure of the significance of the fit.

Using the definition of Eq. (6-76), we can construct an F-test for the multiple correlation coefficient R in terms of

$$F_R = \frac{R^2\sigma_y^2/(m-1)}{(1-R^2)\sigma_y^2/(N-m)} = \frac{R^2(N-m)}{(1-R^2)(m-1)}. \tag{6-90}$$

Similar to Eq. (6-77), the probability distribution of having a value of F_R larger or equal to some value F is given by the integral $Q(F|\nu_1,\nu_2)$, with $\nu_1=(m-1)$ and $\nu_2=(N-m)$. A large value of F_R here implies that the spread between the fit and the data is much smaller than the spread between the dependent and independent variables. This, in turn, is an indication that the fit is a good one.

Test of the need for additional parameters Earlier, we saw an example of putting additional terms into our fitting function. In general, we can expect a decrease in χ^2 as the number of parameters is increased. On the other hand, each additional term also decreases the number of degrees of freedom. As a result, the probability $Q(\chi^2|\nu)$ for obtaining the χ^2-value may not increase, indicating that the new term is not improving the fit. As a more quantitative measure, we can use F-test to tell us whether the additional term is significant or not.

For the convenience of discussion, we shall treat the additional term as the last one in our function $y(x)$, and there are m terms in total when the new one is included. Let the value of χ^2 obtained with m parameters be $\chi^2(m)$. If we carry out the least-squares calculation without the last term, we obtain, in general, a different value, which we shall label as $\chi^2(m-1)$. As we have seen earlier, $\chi^2(m)$ for N data points is a quantity following a χ^2-distribution of $(N - m)$ degrees of freedom, and $\chi^2(m-1)$ is one having $(N - m + 1)$ degrees of freedom. The difference between them,

$$\Delta\chi^2 = \chi^2(m-1) - \chi^2(m)$$

is therefore a quantity following a χ^2-distribution of $(N - m + 1) - (N - m) = 1$ degree of freedom. The ratio

$$F_\chi = \frac{\Delta\chi^2}{\frac{\chi^2(m)}{N-m}} = \frac{\chi^2(m-1) - \chi^2(m)}{\frac{\chi^2(m)}{N-m}} \tag{6-91}$$

is then a quantity following an F-distribution, as can be seen by comparing Eq. (6-91) with (6-76). If the value of F_χ is small, we conclude that the last term in our fitting function is not significant in improving the fit. More quantitatively, the probability for having a particular value of F in this case is given by the value of $Q(F|1, N-m)$ defined in Eq. (6-77).

Checking if two samples have the same distribution In an experiment, it may happen, for example, that the measurements are taken over long periods of time. How can we be certain that nothing in the setup has changed during the time interval? This is the same problem as the more general one of having several different sets of data and we wish to know whether they have the same distribution. Consider two such samples with N_1 and N_2 members in each. If both samples are analyzed using the same fitting function $y(x)$ of m parameters, we obtain two sets of values for the parameters. In general, they will not be identical to each other. On the other hand, because of the uncertainties associated with the values determined, it may not be easy to tell whether they are different from each other in a significant way. It is, therefore, useful to devise a statistical test to give us an estimate of the probability of two sets of parameter values describing the same distribution.

Let us represent the value of the kth parameter obtained from the first sample of N_1 data points as $a_k(N_1)$ and the corresponding quantity for the second set with N_2 data points as $a_k(N_2)$. The uncertainties associated with these two values are, respectively, $\sigma_k(N_1)$ and $\sigma_k(N_2)$. In each case, the uncertainty is to be interpreted as the standard deviation of the distribution in the value of the parameter if an infinite number of samples of the N_1, or N_2, data points were available and analyzed in the same way. Similarly, we shall assume that the mean value of its distribution is given by the value of the parameter itself. The question whether there are any significant differences between $a_k(N_1) \pm \sigma_k(N_1)$ and $a_k(N_2) \pm \sigma_k(N_2)$ is then equivalent to whether the distributions defined by these two sets of numbers are different from each other. To this end, we can construct a variable

$$t = \frac{a_k(N_1) - a_k(N_2)}{\frac{\sigma_k^2(N_1)}{N_1} + \frac{\sigma_k^2(N_2)}{N_2}}. \tag{6-92}$$

Comparing with the quantity defined by Eq. (6-66), we see that t here follows approximately a Student's t-distribution. The number of degrees of freedom is given

Table 6-5: Fitting the data in Table 6-4 as two groups of 10 points each.

Statistic	Set N_1	Set N_2	t	ν	$A(t\|\nu)$
N	10	10			
a_1	0.018 ± 0.024	-0.248 ± 0.274	2.56	6.09	0.96
a_2	3.650 ± 0.306	3.150 ± 0.206	3.59	10.5	1.00
a_3	-0.374 ± 0.104	-0.214 ± 0.036	3.85	7.42	0.99
ν	7	7			
χ^2	8.37	0.83			
$Q(\chi^2\|\nu)$	0.30	1.00			

by

$$\nu = \frac{\left(\frac{\sigma_k^2(N_1)}{N_1} + \frac{\sigma_k^2(N_2)}{N_2}\right)^2}{\frac{1}{N_1-1}\left(\frac{\sigma_k^2(N_1)}{N_1}\right)^2 + \frac{1}{N_2-1}\left(\frac{\sigma_k^2(N_2)}{N_2}\right)^2}. \tag{6-93}$$

In general, the value of ν here is not an integer. However, this does not cause any difficulty in evaluating the integral $A(t|\nu)$ of Eq. (6-69) that provides us with an estimate of the probability.

The possible range of values for t is $[0, \infty)$. Approaching the upper limit of $t = \infty$, it is obvious that $a_k(N_1)$ and $a_k(N_2)$ are quite different from each other. From Eq. (6-69), it is easy to see that as $t \to \infty$ we have $A(t|\nu) = 1$, as the value of the incomplete beta function vanishes in this limit. On the other hand, for $t = 0$, the value of the incomplete beta function becomes 1 and $A(t|\nu) = 0$. As a result, a value of $A(t|\nu)$ near 0 for t and ν defined, respectively, by Eqs. (6-92) and (6-93), implies that $a_k(N_1)$ and $a_k(N_2)$ are essentially the same.

As an example, we shall split the 20 data points given in Table 6-4 into two groups, with the first 10 data points in one group and the rest in the other. Again, let us use a three-parameter power series

$$y = a_1 + a_2 x + a_3 x^2$$

to fit each of the two groups of ten data points. The results are listed in Table 6-5. We can see from the values of $A(t|\nu)$ obtained that it is unlikely for the two sets of parameters to be describing the same distribution. This is not unexpected, as the differences between $a_k(N_1)$ and $a_k(N_2)$, for $k = 1$, 2 and 3, are in general larger than the uncertainties associated with each. The suggestion that the two groups are different from each other may also be seen from the values of the χ^2 obtained in the two cases. Such differences are not evident by examining directly the values of the data points in each group.

6-7 Nonlinear least-squares fit to data

All the methods of least squares we have discussed so far are restricted to fitting functions that are linear in the parameters a_1, a_2, ..., a_m. That is, the relation between the measured values y_i and the independent variable x_i is given by a function

$$y(x) = f(a_1, a_2, \ldots, a_m; x)$$

satisfying the condition

$$\frac{\partial^2 y}{\partial a_j \partial a_k} = 0 \tag{6-94}$$

for all j and $k = 1$, 2,..., m. (The same is also true for the higher-order partial derivatives, but we shall not be making any explicit use of the condition.) The usefulness of such a requirement can be seen from Eqs. (6-80) and (6-81). Because of Eq. (6-94), solutions can be obtained from a set of linear equations.

For fitting functions that are more complicated in form, nonlinear least-squares methods are needed. For example, if we wish to fit a set of data points y_i to a normal distribution with the normalization, mean, and standard deviation as adjustable parameters, we need to use a function of the form

$$f(a_1, a_2, a_3; x) = a_1 \frac{1}{a_3\sqrt{2\pi}} \exp\Big\{-\frac{1}{2}\Big(\frac{x - a_2}{a_3}\Big)^2\Big\}. \tag{6-95}$$

In this case, $y(x)$ is linear only in a_1. Both a_2 and a_3 do not satisfy the condition of Eq. (6-94). The method of maximum likelihood can still be applied and the value of χ^2 minimized by varying the parameters. The result is similar to Eq. (6-80) except that the equations are no longer linear in the unknown parameters.

An alternative is to go back to the primitive approach to locating the minimum of χ^2 by making a survey of the parameter space. Consider first the general case of m parameters. To simplify the discussion, let us divide the possible range of values for each parameter a_k into n equal parts. The complete m-dimensional space is then separated into n^m cells and we can calculate the value of χ^2 at the center of each cell. By comparing the values of χ^2 obtained, we can locate the minimum. The accuracy is then limited by the number of subdivisions we can afford to take for each parameter. Furthermore, the amount of computation can be large even for modest values of n and m. To improve the efficiency, we can replace the fixed grid with an iterative search. This is the basic principle behind most of the methods of nonlinear least squares, and we shall examine a few of the more commonly used ones.

General considerations of the parameter space For our discussion here, it is convenient to regard χ^2 as a function of m variables a_1, a_2,..., a_m, in the form

$$\chi^2(a_1, a_2, \ldots, a_m) = \sum_{i=1}^{N} \frac{1}{\sigma_i^2}\{y_i - f(a_1, a_2, \ldots, a_m; x_i)\}^2. \tag{6-96}$$

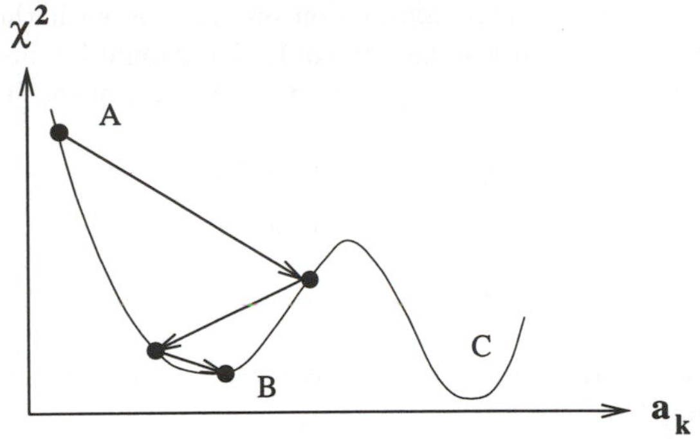

Figure 6-13: Schematic diagram showing the variation of χ^2 as a function of one of the parameters a_k in a multiparameter, nonlinear least-squares fit. A poor choice of the starting point A, for example, can lead to the search settling in a local minimum near B, rather than the absolute minimum at C.

Our interest is to find the location of its minimum. For $m = 2$, we can visualize χ^2 as a function of a_1 and a_2 by making an analogy to the view of the landscape observed from the top of a mountain. For example, we may identify a_1 with the north-south direction and a_2 with the east-west direction. Depending on the behavior of the function $\chi(a_1, a_2)$, the landscape may be quite flat or it may be full of hills and valleys. To make the discussion more quantitative, we shall define a coordinate system with a fixed origin. Any point in the two-dimensional space can then be specified by a vector $\boldsymbol{a}$ from the origin to the point. For $m = 2$, the vector is given by the values of a_1 and a_2. More generally,

$$\boldsymbol{a} \equiv \{a_1, a_2, \ldots, a_m\}$$

if there are m parameters.

The basic idea in most nonlinear least-squares methods is a simple one and relies on the local value of the slope of the χ^2-function. If the slope is pointing downhill, we know we must move forward in order to reach a lower value. Conversely, if the slope is pointing uphill, we must turn around and move in the opposite direction. A (local) minimum is indicated by the fact that the slope is positive in every direction.

In principle, we can start from any point in the parameter space and locate the minimum by following the slope from one point to another, much as water on land finds its way to the ocean. As a practical procedure, however, we must start from some initial estimates for all m parameters. Let us call this set $\boldsymbol{a}(0)$. Variations of one or more of the m values can be made by adding small increments $\delta\boldsymbol{a}(0)$ to $\boldsymbol{a}(0)$. This generates a new set of values

$$\boldsymbol{a}(1) = \boldsymbol{a}(0) + \delta\boldsymbol{a}(0).$$

By comparing the values of $\chi^2(\boldsymbol{a}(0))$ and $\chi^2(\boldsymbol{a}(1))$, we obtain the slope in the region in between $\boldsymbol{a}(0)$ and $\boldsymbol{a}(1)$. This information provides us with the direction and an estimate of the magnitude for the changes $\delta\boldsymbol{a}(1)$ that should be made to $\boldsymbol{a}(1)$ to lower $\chi^2(\boldsymbol{a})$ further. This process is repeated and the general approach for step ℓ may be stated as

$$\boldsymbol{a}(\ell+1) = \boldsymbol{a}(\ell) + \delta\boldsymbol{a}(\ell). \tag{6-97}$$

The end of the search is given by the condition

$$\Delta\chi^2 \equiv \chi^2(\boldsymbol{a}(\ell)+\delta\boldsymbol{a}) - \chi^2(\boldsymbol{a}(\ell)) \geq 0 \qquad \text{for} \quad \delta\boldsymbol{a} \to 0.$$

The major differences among the various nonlinear least-squares methods lie in their ways of estimating $\delta\boldsymbol{a}(\ell)$.

The main problem with the *local* approach outlined above is that there may be more than one minimum in the space. In the case of a linear model, $\chi^2(\boldsymbol{a})$ is a quadratic function of the parameters $\boldsymbol{a}$ and there is only a single minimum in the m-dimensional space. For a nonlinear model, this is no longer true and there can be a number of local minima in the landscape. In a least-squares fit, we are interested in the absolute minimum of the $\chi^2(\boldsymbol{a})$ function, as this is the point of maximum likelihood. A local search scheme, however, has no way to identify what type of minimum one is in. This is illustrated schematically in Fig. 6-13.

Parabolic approximation For small variations of the parameters, the value of $\chi^2(\boldsymbol{a}(\ell)+\delta\boldsymbol{a})$ may be expanded in terms of a Taylor series around $\chi^2(\boldsymbol{a}(\ell))$:

$$\begin{aligned}\chi^2(\boldsymbol{a}(\ell)+\delta\boldsymbol{a}) &= \chi^2(\boldsymbol{a}(\ell)) + \sum_{i=1}^{m} \delta a_i \frac{\partial\chi^2}{\partial a_i}\bigg|_{\boldsymbol{a}=\boldsymbol{a}(\ell)} \\ &\quad + \frac{1}{2!}\sum_{j=1}^{m}\sum_{i=1}^{m} \delta a_j \delta a_i \frac{\partial^2\chi^2}{\partial a_j \partial a_i}\bigg|_{\boldsymbol{a}=\boldsymbol{a}(\ell)} + \cdots.\end{aligned}$$

The change in $\chi^2(\boldsymbol{a})$ due to a small variation Δa_k in parameter a_k is given by

$$\chi^2(\boldsymbol{a}(\ell)+\Delta a_k) - \chi^2(\boldsymbol{a}(\ell)) = \Delta a_k \left\{ \frac{\partial\chi^2}{\partial a_k}\bigg|_{\boldsymbol{a}=\boldsymbol{a}(\ell)} + \sum_{j=1}^{m} \delta a_j \frac{\partial^2\chi^2}{\partial a_j \partial a_k}\bigg|_{\boldsymbol{a}=\boldsymbol{a}(\ell)} + \cdots \right\}.$$

The factor of $\frac{1}{2}$ in the second-order derivative term of the previous equation cancels with a factor of 2 coming from the fact that there are two summations involved and both δa_i and δa_j can be Δa_k. We can write the variations in the values of χ^2 due to small changes in a_k in the form of a derivative:

$$\begin{aligned}\frac{\Delta\chi^2}{\Delta a_k}\bigg|_{\boldsymbol{a}=\boldsymbol{a}(\ell)} &\equiv \frac{\chi^2(\boldsymbol{a}(\ell)+\Delta a_k) - \chi^2(\boldsymbol{a}(\ell))}{\Delta a_k} \\ &= \frac{\partial\chi^2}{\partial a_k}\bigg|_{\boldsymbol{a}=\boldsymbol{a}(\ell)} + \sum_{j=1}^{m} \delta a_j \frac{\partial^2\chi^2}{\partial a_j \partial a_k}\bigg|_{\boldsymbol{a}=\boldsymbol{a}(\ell)} + \cdots.\end{aligned} \tag{6-98}$$

The minimum of $\chi^2(\boldsymbol{a})$ is given by the condition

$$\frac{\Delta\chi^2}{\Delta a_k} = 0 \qquad \text{for} \qquad k = 1, 2, \ldots, m.$$

If the function $\chi^2(\boldsymbol{a})$ is approximated by a parabolic form and the series on the right side of Eq. (6-98) is truncated after the second term, we obtain a result that is similar to Eq. (6-80) for the linear case:

$$\left.\frac{\partial\chi^2}{\partial a_k}\right|_{\boldsymbol{a}=\boldsymbol{a}(\ell)} \approx -\sum_{j=1}^{m} \delta a_j \left.\frac{\partial^2\chi^2}{\partial a_j \partial a_k}\right|_{\boldsymbol{a}=\boldsymbol{a}(\ell)}. \qquad (6\text{-}99)$$

In terms of matrices, we can write these m equations as

$$\boldsymbol{\beta} = \boldsymbol{\alpha}\,\boldsymbol{\delta a}$$

where δa_j are the matrix elements of $\delta\boldsymbol{a}$. The elements of the square matrix $\boldsymbol{\alpha}$ and column matrix $\boldsymbol{\beta}$ are given, respectively, by

$$\alpha_{k,j} = \frac{1}{2}\frac{\partial^2\chi^2}{\partial a_j \partial a_k}\bigg|_{\boldsymbol{a}=\boldsymbol{a}(\ell)} \qquad \beta_k = -\frac{1}{2}\frac{\partial\chi^2}{\partial a_k}\bigg|_{\boldsymbol{a}=\boldsymbol{a}(\ell)}. \qquad (6\text{-}100)$$

For later convenience, a factor of $\frac{1}{2}$ is included in the definitions of both $\boldsymbol{\beta}$ and $\boldsymbol{\alpha}$. We see here also the reason for calling $\boldsymbol{\alpha}$ the *curvature matrix*, as it expresses the curvature of $\chi^2(\boldsymbol{a})$ in the parameter space.

On inverting the curvature matrix $\boldsymbol{\alpha}$, we obtain the solution for δa_k

$$\boldsymbol{\delta a} = \boldsymbol{\alpha}^{-1}\boldsymbol{\beta} \qquad (6\text{-}101)$$

where $\boldsymbol{\alpha}^{-1}$ is the covariance matrix. This is similar to Eq. (6-85) for the linear case where the solution gives directly the location of the minimum. Here, Eq. (6-99) is only an approximation. Furthermore, both $\boldsymbol{\alpha}$ and $\boldsymbol{\beta}$ are obtained using approximate values of $\boldsymbol{a}(\ell)$. As a result, the solution for $\delta\boldsymbol{a}$ does not locate the minimum of χ^2 for us, only the direction and an indication of the step size to look for it. As we shall soon see, the choice of step size is an important consideration for fast convergence.

A simplification can be introduced in the calculation of the curvature matrix. By taking the second-order partial derivative of $\chi^2(\boldsymbol{a})$ with respect to any two parameters, a_j and a_k, using the form given by Eq. (6-96), we obtain

$$\frac{\partial^2\chi^2}{\partial a_j \partial a_k} = 2\sum_{i=1}^{N}\frac{1}{\sigma_i^2}\left\{\frac{\partial f(\boldsymbol{a};x_i)}{\partial a_j}\frac{\partial f(\boldsymbol{a};x_i)}{\partial a_k} - [y_i - f(\boldsymbol{a};x_i)]\frac{\partial^2 f(\boldsymbol{a};x_i)}{\partial a_j \partial a_k}\right\}. \qquad (6\text{-}102)$$

If we ignore the second term on the right side, we are essentially approximating the function $f(\boldsymbol{a};x)$ as linear in the parameters $\boldsymbol{a}$, satisfying the condition given by Eq. (6-94). Elements of the curvature matrix reduce to the simple form

$$\boldsymbol{\alpha}_{k,j} = \frac{1}{2}\frac{\partial^2\chi^2}{\partial a_j \partial a_k} \approx \sum_{i=1}^{N}\frac{1}{\sigma_i^2}\left\{\frac{\partial f(\boldsymbol{a};x_i)}{\partial a_j}\frac{\partial f(\boldsymbol{a};x_i)}{\partial a_k}\right\} \qquad (6\text{-}103)$$

with the factor of $\frac{1}{2}$ in the definition given by Eq. (6-100) canceling the factor of 2 on the right side of Eq. (6-102). In practice, the second-order partial derivatives $\partial^2 f(\boldsymbol{a}; x_i)/\partial a_j \partial a_k$ in the second term of Eq. (6-102) have a tendency to make the numerical calculation unstable. For this reason also it is helpful to ignore it.

The parabolic approximation is equivalent to the assumption that the behavior of $\chi^2(\boldsymbol{a})$ can be described by a quadratic function in $\boldsymbol{a}$:

$$\chi^2(\boldsymbol{a}) \approx c_0 + \sum_{j,k} c_{j,k} a_j a_k$$

where the coefficients c_0 and $c_{j,k}$ are constants independent of $\boldsymbol{a}$. This is a familiar way to approximate a function near its minimum, as we have seen, for example, in Fig. 4-1. For the same reason, the approximation is a good one when the search is near a minimum in the parameter space of $\chi^2(\boldsymbol{a})$.

Gradient Method On the other hand, the parabolic approximation is slow in regions far away from a minimum, as it tends to search in small steps. As an alternative, the gradient method is often used. In this approach, the step size $\delta\boldsymbol{a}(\ell)$ of Eq. (6-97) is taken to be proportional to the gradient $\nabla\chi^2(\boldsymbol{a})$, the slope of the $\chi^2(\boldsymbol{a})$ surface at the point $\boldsymbol{a} = \boldsymbol{a}(\ell)$,

$$\delta\boldsymbol{a} \propto -\nabla\chi^2(\boldsymbol{a}(\ell)). \tag{6-104}$$

From Eq. (6-100), we find that

$$-\nabla\chi^2(\boldsymbol{a}(\ell)) = 2\boldsymbol{\beta}. \tag{6-105}$$

The gradient method is efficient, since the search follows the path of steepest descent of χ^2 as a function of the parameters. This is especially useful in regions far away from a minimum where the function usually varies smoothly. To make use of Eq. (6-104) to find $\delta\boldsymbol{a}$, we need to assign a value to the constant of proportionality. Unfortunately, there is no easy way to construct one. In general, the method tends to overshoot near a minimum and, as a result, the rate of convergence slows down as we approach one.

The method of Marquardt The optimum method for a nonlinear least-squares fit is to use a combination of gradient and parabolic approximations. This idea is attributed to Marquardt. Before implementing the method, we need a practical way of determining the step size in the gradient method. A hint can be obtained from a dimensional analysis of the problem. Since $\chi^2(\boldsymbol{a})$ is a dimensionless quantity, we see from Eq. (6-100) that the dimension of β_k, the kth element of $\boldsymbol{\beta}$, must be the inverse of that of a_k. (Note that different parameters have, in general, different dimensions, as each of them has a different physical meaning.) As a result, we find from Eq. (6-101) that the dimension of the matrix elements of $\boldsymbol{\alpha}$ must be given by

$$[\alpha_{j,k}^{-1}] = [a_j][a_k]$$

where the symbol $[q]$ represents the dimension of the quantity q.

We have now a way to implement the gradient method. When Eq. (6-105) is substituted into Eq. (6-104), we obtain

$$\delta a_k = \eta_k \beta_k$$

where the constant of proportionality η_k must have the dimension $[a_k]^2$. The dimensional analysis above tells us that the only quantity with the correct dimension for η_k is the inverse of $\alpha_{k,k}$, the kth diagonal element of the curvature matrix. This gives us the result

$$\delta a_k = \frac{1}{\lambda \alpha_{k,k}} \beta_k \tag{6-106}$$

where λ is a dimensionless constant of proportionality that must be determined in some other way.

It is possible to find alternate forms for Eq. (6-106). The particular choice made above has the advantage that we can incorporate the gradient approximation of Eq. (6-106) into a general approach of the parabolic method given by Eq. (6-101). For this purpose, we shall define a modified curvature matrix $\boldsymbol{\alpha}'$ whose elements are given by

$$\alpha'_{j,k} = \begin{cases} (1+\lambda)\alpha_{j,j} & \text{for } k = j \\ \alpha_{j,k} & \text{otherwise.} \end{cases} \tag{6-107}$$

In other words, the diagonal elements of the curvature matrix are increased by a factor $(1+\lambda)$ from the value given by Eq. (6-100). When λ is small, $\boldsymbol{\alpha}' = \boldsymbol{\alpha}$ and we have the parabolic approximation together with all its advantage near the minimum. When λ is large, the modified curvature matrix is dominated by the diagonal matrix elements and the value of the off-diagonal elements may be treated essentially as zero. In this limit, the elements of the covariance matrix may be written as

$$(\boldsymbol{\alpha}'^{-1})_{j,k} \approx \frac{1}{(1+\lambda)\alpha_{j,j}} \delta_{j,k}$$

where $\delta_{j,k}$ is the Kronecker delta and Eq. (6-101) becomes

$$\delta a_k \approx \frac{1}{(1+\lambda)\alpha_{k,k}} \beta_k.$$

This is essentially the same result as Eq. (6-106) for $\lambda \gg 1$.

The algorithm for carrying out a nonlinear least-squares calculation using the method of Marquardt is outlined in Box 6-4. It starts with an arbitrary set of values $\boldsymbol{a}(\ell)$, with $\ell = 0$, for the m parameters and an initial value of λ chosen to be 0.001. The calculations consist of the following steps:

(1) Calculate the value of $\chi^2(\boldsymbol{a})$ with this $\boldsymbol{a}(\ell)$.

(2) Construct $\boldsymbol{\alpha}'$ from $\boldsymbol{\alpha}$ of Eq. (6-100) using the value of λ given.

Box 6-4 Subroutine NLNFT(NPT,X,Y,SY,MP,A,SA,AMBDA,CHI2,COV,MD)
Nonlinear least-squares fit with Marquardt method

Argument list:

- NPT: Number of input data points.
- X: Input array of the values of $\{x_i\}$.
- Y: Input array of the values of $\{y_i\}$.
- SY: Input array of the uncertainty in $\{y_i\}$.
- MP: Number of parameters.
- A: Array for parameters $\{a_i\}$. Input estimates, output final values.
- SA: Array for the uncertainties in a_i.
- AMBDA: Value of λ of Eq. (6-106).
- CHI2: Output value of χ^2 for the parameter set.
- COV: Output array for the covariance matrix.
- MD: Dimension of the arrays in the calling program.

Subprograms required:

- FUNC: Returns the value of $f(\boldsymbol{a};x)$.
- DERIV: Returns dy/da_i for $i = 1$ to m.
- MATIV: Matrix inversion of Box 5-3.

1. Check the dimension of the auxiliary arrays and zero χ^2, $\boldsymbol{\beta}$, and covariance.
2. Carry out the following calculations using the input values $\{x_i, y_i, \sigma_i\}$:
   (a) If $\sigma_i = 0$, let the weighting factor $w_i = 1$; otherwise $w_i = 1/\sigma_i^2$.
   (b) Use FUNC and DERIV to calculate $f(\boldsymbol{a}, x_i)$ and $\partial f/\partial a_k$.
   (c) Compute the matrix elements of $\boldsymbol{\beta}$ according to Eq. (6-100).
   (d) Compute the matrix elements of $\boldsymbol{\alpha}$ according to Eq. (6-103).
3. Calculate the value of χ^2 using Eq. (6-96).
4. Construct the normalized, modified curvature matrix $\mathcal{A}$ using Eq. (6-108).
5. Invert the curvature matrix $\mathcal{A}$.
6. Calculate the next set of the values of $\boldsymbol{a}$:
   (a) Obtain $\boldsymbol{\alpha}'^{-1}$ using Eq. (6-109).
   (b) Use Eq. (6-101) to get $\delta\boldsymbol{a}$.
   (c) Find a new set of the parameters using Eq. (6-97).
7. Evaluate the new $\chi^2(\boldsymbol{a}')$.
8. Compare the new $\chi^2(\boldsymbol{a}')$ with the old $\chi^2(\boldsymbol{a})$:
   (a) If $\chi^2(\boldsymbol{a}') \geq \chi^2(\boldsymbol{a})$, increase λ by a factor of 10 and go back to step 5.
   (b) If $\chi^2(\boldsymbol{a}') < \chi^2(\boldsymbol{a})$, decrease the value of λ by a factor of 10 and return:
      (i) The value of $\chi^2(\boldsymbol{a}')$.
      (ii) $\boldsymbol{a}'$ as the new values of the parameters.
      (iii) $\sqrt{\alpha'_{k,k}}$ as the uncertainty of a_k.
      (iv) $\mathcal{A}^{-1}$ and diagonal elements of $\boldsymbol{\alpha}$ to construct the covariance matrix.
9. Call the subroutine again if the new χ^2 value differs significantly from the old.

(3) Invert $\boldsymbol{\alpha}'$ and calculate $\boldsymbol{\delta a}(\ell)$ with $\boldsymbol{\alpha}'^{-1}$ in the place of $\boldsymbol{\alpha}^{-1}$ in Eq. (6-101).

(5) Calculate $\boldsymbol{a}(\ell+1)$ using Eq. (6-97) and then the value of $\chi^2(\boldsymbol{a}(\ell+1))$.

(6) Compare $\chi^2(\boldsymbol{a}(\ell+1))$ with $\chi^2(\boldsymbol{a}(\ell))$:

(i) If $\chi^2(\boldsymbol{a}(\ell+1)) \geq \chi^2(\boldsymbol{a}(\ell))$, we are moving away from a minimum. Increase the value of λ by a large factor, such as 10, and go back to step (2).

(ii) If $\chi^2(\boldsymbol{a}(\ell+1)) < \chi^2(\boldsymbol{a}(\ell))$, we are approaching a minimum. Decrease λ by a factor of 10 and go back to step (2) with this new value of λ and increase ℓ by 1.

Convergence is reached when the difference between $\chi^2(\boldsymbol{a}(\ell+1))$ and $\chi^2(\boldsymbol{a}(\ell))$ is smaller than some predetermined value.

Numerical stability considerations In general, different parameters in a nonlinear function $f(\boldsymbol{a};x)$ play different roles and, as a result, each may have quite different behavior. This is demonstrated, for example, by the three parameters a_1, a_2, and a_3 in Eq. (6-95). For this reason, we expect that the values of the elements of the curvature matrix can be quite different from each other. This is undesirable from the point of view of numerical accuracy. To improve the situation, we can scale the elements of the modified curvature matrix and define a new matrix $\mathcal{A}$ in the following way:

$$\mathcal{A}_{j,k} = \frac{\alpha'_{j,k}}{\sqrt{\alpha_{j,j}\alpha_{k,k}}}. \tag{6-108}$$

All the diagonal matrix elements now have the value

$$\mathcal{A}_{j,j} = 1 + \lambda$$

as can be seen by substituting the values of $\alpha'_{j,j}$ from Eq. (6-107) into (6-108). Instead of $\boldsymbol{\alpha}'$, the matrix $\mathcal{A}$ is inverted. The relation between the inverse of $\mathcal{A}$ and the covariance matrix is given by

$$\alpha'^{-1}_{j,k} = \frac{\mathcal{A}^{-1}_{j,k}}{\sqrt{\alpha_{j,j}\alpha_{k,k}}}. \tag{6-109}$$

In this approach, the diagonal elements of $\boldsymbol{\alpha}$ must be saved before carrying out the transformation of Eq. (6-108). They may be regarded as the normalization factors that improve the numerical accuracy in the calculation. No additional storage space is required here, since the matrix α is saved anyway to allow variations on the value of λ as outlined earlier.

Radioactivity example As an example, let us consider a hypothetical experiment measuring the radioactivity of an unknown sample as a function of the energy of the radiation emitted. The 50 data points are plotted in Fig. 6-14 as dots, together with an error bar for each point to indicate the uncertainty in the measured number of counts (vertical axis) in each energy interval (horizontal axis). In this experiment, we expect two components in the radiation detected: an exponential background, coming from a variety of sources of no direct interest to the decay, and the decay of a particular state with a Lorentzian distribution. The counting rate $y(x)$ as a function of energy x is, therefore, conjectured to take on the form

$$y(x) = A_1 e^{-\gamma x} + \frac{A_2}{\pi}\frac{\frac{1}{2}\Gamma}{(x-\mu)^2 + (\frac{1}{2}\Gamma)^2} \tag{6-110}$$

where A_1 is the strength of the background and γ its decay constant. The Lorentzian distribution is centered at μ with a width Γ and strength A_2. All five quantities, A_1, A_2, γ, Γ, and μ, are treated as unknown parameters.

We shall now carry out a least-squares fit to find the values of these five quantities from the 50 measurements. For the fitting function, we shall adopt the form

$$f(a_1, a_2, a_3, a_4, a_5; x) = a_1 e^{-a_2 x} + \frac{a_3}{(x-a_4)^2 + a_5}$$

where we have, for mathematical convenience, defined

$$a_3 = A_2 \frac{\Gamma}{2\pi} \qquad\qquad a_5 = \left(\frac{1}{2}\Gamma\right)^2.$$

The other three parameters, $a_1 = A_1$, $a_2 = \gamma$, and $a_4 = \mu$, have direct relations with their counterparts in Eq. (6-110). The five partial derivatives of $f(\boldsymbol{a}; x)$ with respect to the parameters $a_1, a_2, \ldots, a_5$ are

$$\frac{\partial f}{\partial a_1} = e^{-a_2 x} \qquad\qquad \frac{\partial f}{\partial a_2} = -a_1 x e^{-a_2 x}$$

$$\frac{\partial f}{\partial a_3} = \frac{1}{(x-a_4)^2 + a_5} \qquad\qquad \frac{\partial f}{\partial a_4} = \frac{2a_3(x-a_4)}{\{(x-a_4)^2 + a_5\}^2}$$

$$\frac{\partial f}{\partial a_5} = \frac{-a_3}{\{(x-a_4)^2 + a_5\}^2}$$

and they are required to calculate the elements of $\boldsymbol{\beta}$ and $\boldsymbol{\alpha}$ defined in Eqs. (6-100) and (6-103).

As initial estimates of the parameters, we can use the value of y_i at $x_i \approx 0$ to give $a_1 \approx 10$ and the rate of the exponential decay to give $a_2 \approx 0.5$. From the location of the Lorentzian peak, we obtain $a_4 = 3$, the height $a_3 \approx 5$, and the width $a_5 = 1$. Together with the recommended value of $\lambda = 0.001$ as the starting point, the calculation converges in four iterations to an accuracy of

$$\frac{\Delta\chi^2}{\chi^2} \leq 0.01$$

where $\Delta\chi^2$ is the absolute value of the difference between the values of χ^2 in two successive iterations. The fitted results are plotted as a smooth curve together with the 50 input data points in Fig. 6-14.

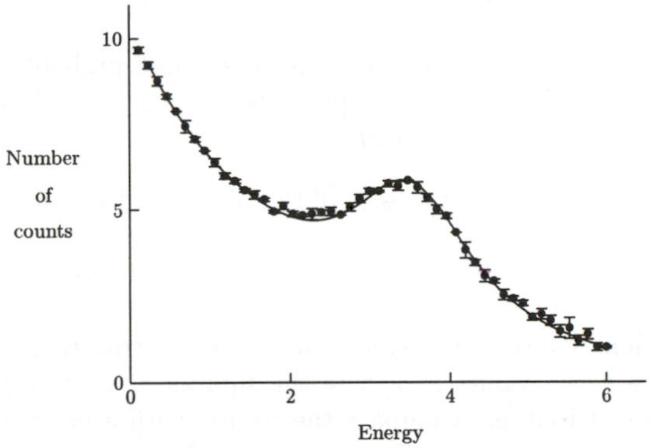

Figure 6-14: Nonlinear least-squares fit to a set of 50 hypothetical measurements in a radioactive decay experiment using Eq. (6-110). The χ^2 value is 67.6 with $a_1 = 10.1$, $a_2 = 0.51$, $a_3 = 3.7$, $a_4 = 3.5$, and $a_5 = 0.89$.

Uncertainties in the parameters In the case of a linear least-squares calculation, we have a direct relation between $y(x)$ and the parameters $a_1, a_2, \ldots, a_m$. As a result, there is a clear relation between the uncertainty in each of the parameters a_k and the uncertainties in the input quantities, such as that given by Eq. (6-24). For a nonlinear calculation, the final values are obtained as a result of an iterative search and, as a result, there is no longer any simple relation between the uncertainties in the input quantities and the final values of the parameters found. The definition of the uncertainties in the parameters obtained must, therefore, be assigned somewhat arbitrarily. A reasonable choice is to equate the square of the uncertainty in a_k with the kth diagonal element of the inverse of the curvature matrix:

$$\sigma_{a_k} = (\boldsymbol{\alpha}^{-1})_{k,k}.$$

It can be shown that, with this definition, the value of χ^2 increases approximately by unity if a_k is increased by σ_{a_k}. That is,

$$\chi^2(a_k \pm \sigma_{a_k}) \approx \chi^2(a_k) + 1.$$

The proof of this is given in Bevington.[4]

In carrying out a nonlinear least-squares fit, we need both the functional form of $y(x)$ and the partial derivative of $y(x)$ with respect to each parameter. For this reason, most computer codes for such calculations require, as input, the analytical forms for the partial derivatives. On the other hand, nonlinear least-squares calculations are not restricted to analytical functions. In such cases, numerical differentiation must be used.

Problems

6-1 Calculate the mean μ, medium $\mu_{1/2}$, and most probable value $\mu_{\max}$ for the distribution of Eq. (6-5). Compare the results with those obtained for the normal distribution of Eq. (6-9).

6-2 Show that the mean and variance of the binomial distribution of Eq. (6-12) are

$$\mu = Np \qquad\qquad \sigma^2 = Np(1-p).$$

6-3 Use a random number generator with an even distribution in the range $[-1, +1]$ to produce $n = 6$ values and store the sum as x. Collect 1000 such sums and plot their distribution. Compare the results with a normal distribution of the same mean and variance as the x collected. Calculate the χ^2-value. Repeat the calculations with $n = 50$. Compare the two χ^2-values obtained.

6-4 Show that, for a normal distribution given by Eq. (6-9), the central moments μ_r of Eq. (6-10) are given by

$$\mu_r = \frac{1}{\sigma\sqrt{2\pi}}\int_{-\infty}^{+\infty}(x-\mu)^r e^{-(x-\mu)^2/2\sigma^2}\,dx = \begin{cases} 0 & \text{for } r \text{ odd} \\ \left(\frac{\sigma^2}{2}\right)^m \frac{(2m)!}{m!} & \text{for } r = 2m \end{cases}$$

where $m = 1, 2, \ldots,$ is an integer. That is, all the odd central moments vanish and the even ones are related to the variance σ^2.

6-5 Find the relations between the uncertainties in x and y for the following functions: (*a*) $y = ax^b$, (*b*) $y = ae^{\pm x}$, (*c*) $y = a\ln(bx)$, and (*d*) $y = a/\sin x$.

6-6 Use the propagation of errors given by Eq. (6-24) to show that the uncertainty in the average of a quantity decreases in proportion to the inverse of the square root of the number of independent measurements taken.

6-7 Obtain the mean and variance for a χ^2-distribution of ν degrees of freedom given by Eq. (6-62).

6-8 A good way to test a computer program for linear least-squares calculations is to use a set of inputs produced with the relation

$$y_i = A + Bx_i + \delta y_i.$$

For the parameters A and B, adopt the values of 1.0 and 5.0, respectively. Use a random number generator to produce a set of 25 values of x_i in the range $[0, 1]$ and another 25 values for δy_i in the range $[-0.1, +0.1]$. Calculate the value of y_i from x_i and δy_i using the relation given. To each y_i, assign an uncertainty σ_i

using a random number generator in the range $[0, 0.1]$. Fit this set of 25 values of (x_i, y_i, σ_i) with the straight-line relation

$$y(x) = a_1 + a_2 x.$$

Compare the values of a_1 and a_2 obtained with those of A and B used to generate the input.

6-9 Repeat Problem 6-8 for 100 different sets of 25 input values generated with the same pair of values of A and B, but different random numbers for x_i, δy_i, and σ_i. Plot a histogram of the distribution of the values of a_1 and a_2. Are the results consistent with the uncertainties of a_1 and a_2 produced by the program?

6-10 Repeat Problem 6-8 except fit the input with the function

$$y(x) = a_1 + a_2 x + a_3 x^2.$$

Use the tests described in §6-5 and §6-6 to see if the additional quadratic term should be included.

6-11 The following ten sets of values are used as the input data used to produce Fig. 6-11:

$$(\cos\theta_i, W(\theta_i), \sigma_i) = \begin{array}{lll} (\ 0.000, 1148, 4) & (-0.174, 1152, 4) & (-0.342, 1166, 4) \\ (-0.500, 1184, 4) & (-0.643, 1208, 4) & (-0.766, 1234, 4) \\ (-0.866, 1274, 4) & (-0.940, 1297, 4) & (-0.985, 1316, 4) \\ (-1.000, 1324, 4) & & \end{array}$$

where σ_i is the estimated uncertainty in $W(\theta_i)$. Fit the values of $W(\theta)$ with a power series in $\cos^2\theta$

$$W(\theta) = b_0 + b_2 \cos^2\theta + b_4 \cos^4\theta.$$

What are the differences from a Legendre polynomial fit of the same order?

6-12 Check the significance of the $\cos^4\theta$ term in Problem 6-11. What about including a $\cos^6\theta$ term?

6-13 The following 20 sets of values

x_i	y_i	σ_i	x_i	y_i	σ_i	x_i	y_i	σ_i	x_i	y_i	σ_i
0.46	0.19	0.05	0.69	0.27	0.06	0.71	0.28	0.05	1.04	0.62	0.01
1.11	0.68	0.05	1.14	0.70	0.07	1.14	0.74	0.08	1.20	0.81	0.09
1.31	0.93	0.10	2.03	2.49	0.03	2.14	2.73	0.04	2.52	3.57	0.01
3.24	3.90	0.07	3.46	3.55	0.03	3.81	2.87	0.03	4.06	2.24	0.01
4.93	0.65	0.10	5.11	0.39	0.07	5.26	0.33	0.05	5.38	0.26	0.08

follow roughly a normal distribution

$$y(x) = a_1 \exp\left\{-\frac{1}{2}\left(\frac{x-a_2}{a_3}\right)^2\right\}$$

with σ_i as the uncertainty of y_i. Use a nonlinear least-squares method to find the values of a_1, a_2, and a_3.

6-14 As an alternative to using a nonlinear least-squares fit, the exercise given in Problem 6-13 can also be solved by defining $z = \ln y$ and using a linear least-squares fit for

$$z(x) = b_1 + b_2 x + b_3 x^2.$$

Compare the results obtained with the two methods.

Chapter 7

Monte Carlo Calculations

Monte Carlo techniques are useful in solving a variety of computational problems. Earlier, we made use of such an approach in §2-5 for an integral by evaluating the integrand at a randomly selected sample of points. Many problems, from simulation of experimental situations to complicated theoretical computations, can take advantage of such a sampling approach. We shall begin this chapter with a discussion on generating random numbers, the starting point of any Monte Carlo calculations.

7-1 Generation of random numbers

What is a random number? A simple answer is that it is a number chosen at random. More formally, we can say that a random number is a collection of random digits, drawn from 0, 1, 2, ..., 9, each with the probability of 1/10. Both definitions depend on the concept of randomness. Intuitively, all of us have a feeling of what randomness implies; however, it is not always easy to put it into words. From a practical point of view, it may be more appropriate to use the definition, given by Chaitin[10] in the May 1975 issue of *Scientific American*, which states that "a series of numbers is random if the smallest algorithm capable of specifying it to a computer has about the same number of bits of information as the series itself." There are several other ways to define randomness, but few of them are as able to capture the essence of the term without invoking a large amount of mathematics.

One way to obtain true random numbers is from natural sources. For example, we have seen earlier that the decay of radioactive nuclei is believed to be purely stochastic and, consequently, may be used as the source of random numbers. However, it is not a practical scheme, as the rate of most radioactive decays that can be used for such purposes are far too slow for any serious Monte Carlo calculations. It is also believed that for certain transcendental numbers, such as π, the higher digits are random and may be used as the source of random digits. Again, it is not a practical scheme in many applications.

For computational purposes, we seldom deal with true random numbers. In the first place the random numbers we use are usually generated by some algorithm. In general, we wish to use a simple one so as to produce the numbers quickly. As a result, the sequence obtained cannot be random by the definition of Chaitin. Furthermore, we may also wish our random numbers to be reproducible on demand, as this is essential, for example, in debugging a computer program that makes use of them. Technically, these are only *pseudo*-random numbers. Pseudo-random numbers that pass a set of randomness tests, such as typical statistical hypotheses of independence, uniformity, and goodness of fit, are often quite adequate for most practical applications that call for random numbers. Following general practice, we shall not make any distinction between pseudo-random numbers and *true* random numbers. It is, however, important to recognize that there are fundamental differences between them. More importantly, pseudo-random numbers that are adequate for one application may not be good enough for another. For this reason, extensive tests should be carried out before using any pseudo-random numbers for a new type of calculation. We shall return to the question of tests for randomness later in this section.

A good random number generator must be fast and simple to use. In addition, it should have the desired statistical properties. For example, because of the limited word length, the total number of different random numbers that can be generated on a computer is finite. In fact, for most generators, the actual range of numbers produced is much smaller than the largest integer that can be stored in a machine word. In addition to the range, the sequence of numbers produced will eventually repeat itself, as we are using an algorithm consisting of a limited number of instructions. The number of different results before a generator repeats itself is known as the period. Clearly, it is desirable to have as long a period as possible. Another important requirement is that the numbers generated should not be correlated among themselves. There are many different types of correlation that can exit among a sequence of numbers, and it may not be possible to find a generator that can satisfy all the requirements. Fortunately, most applications are sensitive to only a few of these properties and, as a result, generators that are, in principle, imperfect are often useful in practice.

Uniformly distributed random integers In Monte Carlo calculations, we usually have to take random samples of certain quantities with definite distributions. For this reason, the random numbers we use should follow the same distributions. For example, in a one-dimensional Monte Carlo integration, we wish to sample the variable of integration with uniform probability within the integration limits. For this purpose, we need random numbers with a uniform distribution. On the other hand, if we wish to simulate neutrons scattering off a nuclear target, the random numbers must have the angular distribution of neutrons scattering off the same nucleus. We shall see later that it is possible, in principle, to produce random numbers of any desired distribution starting with a set having a uniform distribution in the range $[0, 1]$. Such random numbers, or uniform deviates in the more formal language, are in turn generated from random integers distributed uniformly in the interval $[0, M]$. Here M is the largest random integer that can be produced by the algorithm. For this reason, we shall start our discussion with methods to generate random integers.

A simple way to obtain a random integer is to read the timer, or system clock, on the computer we are using. Let us assume, for the convenience of discussion here, that the system clock is measured in units of microseconds (μs) and it is set to zero at some arbitrary time, such as 00:00:00 Greenwich Mean Time (GMT) on January 1, 1970, used on many operating systems. A simple function can be written to extract the time t as integers in μs every time it is called. If t_1 and t_2 are the two values obtained in two successive calls separated by a few seconds, the most significant digits of the two integers will be the same, as they represent the time in units much larger than a second. On the other hand, the three least significant digits, for example, are likely to be random. This is especially true if the timing function is called by some sort of manual control, such as pushing a particular key on the keyboard. The source of randomness in this case comes from the fact that it is not possible for us to control our reaction time to be more accurate than the order of 0.1 s. If we obtain a sequence of such values, t_1, $t_2, \ldots$, then

$$r_i = t_i \bmod 1,000$$

for $i = 1, 2, \ldots$, is a sequence of random integers, as the mod, or *modulus*, $1,000$ operation eliminates all the leading significant figures except the last three.

There are several problems in such a generator. In the first place, there can only be, at most, 1,000 different random numbers in the series with the above method. (That is, our generator has a period of 1,000.) This comes from the simple fact that our numbers can only be integers in the interval 0 to 999. In the second place, we cannot use an automated way to read the clock, as any such method will likely produce a series of highly correlated numbers. Finally, we cannot repeat the sequence on those occasions when we need to reproduce the same set of random integers. For these reasons, the only practical use of the system clock as a random number generator is to set the "seed" for some other random number generator, as we shall see later.

Another interesting way to produce random integers is the middle-square method of von Neumann. If we square an integer consisting of several digits, both the most significant digits and the least significant digits are predictable. However, it is more difficult to predict the digits in the middle. For example, if we square an integer of three digits,

$$(123)^2 = 15,129$$

the result is a five-digit integer. By chopping off the leading and trailing digit, we obtain the number 512. If we square this number and retain only the third, fourth, and fifth digits, we obtain the number 214. In this way, a sequence of random integers can be obtained, each constructed from the square of the previous one with only the middle part of the digits retained. Although this method has been in use for a long time, it is no longer a preferred way to generate random integers. Tests have shown that there are many ways such a sequence can develop into a bad direction. For example, if for any reason a member of the sequence becomes zero, all the remaining members of the sequence will obviously be zero.

The linear congruence method The most popular way to generate random integers with a uniform distribution is the linear congruence method. In this approach, a random integer X_{n+1} is produced from another one X_n through the operation

$$X_{n+1} = (aX_n + c) \bmod m \tag{7-1}$$

where the modulus, m, is a positive integer. The multiplier a and increment c are also positive integers but their values must be less than m. To start off the sequence of random integers X_0, X_1, $X_2, \ldots$, we need to input an integer X_0, generally referred to as the *seed* of the random sequence.

The spirit of Eq. (7-1) is very similar to that behind the system-clock and middle-square methods discussed above. Instead of taking the square, the seed is multiplied by an integer a, and to prevent a bad sequence from developing, an increment c is added to the product. Similar to the system-clock approach, only the least significant part of the sum (less than m) is retained through the modulus operation.

The quality of random integers generated by the linear congruence method depends critically on the values of m, a, and c selected. Because of the modulus operation, all the random integers produced by this method must be in the range $[0, m-1]$. For this reason, we want m to be as large as possible. On the other hand, m^2 cannot be larger than the longest integer the computer can store in its memory. It is also clear that the choices of these three quantities are related with each other. For example, in order to have a long period, it is known that

(1) c must be relative prime to m (a and m share no common factors other than 1).

(2) $(a-1)$ is a multiple of p, if p is a prime factor of m, and

(3) $(a-1)$ is a multiple of 4, if m is a multiple of 4.

For example, a possible set of values for computers with word length of 32 bits is

$$m = 714,025 \qquad a = 1,366 \qquad c = 150,889.$$

The maximum number of different random numbers that can be generated is 714,205 in this case. Other possible combinations of three integers are given in Press and coauthors[44] and detailed discussions of the considerations that must go into the choices are given in Knuth.[31]

Conversion to random numbers If instead of random integers we wish to have random (floating) numbers, we can convert an integer X_n in the interval $[0, m]$ to a floating number R_n in interval $[a, b]$. For most generators, the interval is taken to be $[0, 1]$. If the largest integer produced by the linear congruence method of Eq. (7-1) is m, any random integer X_n may be converted to a random number in the interval $[0, 1]$ by dividing X_n with m. To carry out this procedure on a computer, we must first change both X_n and m into floating numbers and then take the quotient

$$R_n = X_n/m.$$

Box 7-1 Function RSHFL(ISEED)
Improved random number generator by shuffling

Argument list:
ISEED: Random number seed on input. Returns the seed for the next one.
Initialization:
(a) Select a stack $S(j)$ of length $L = 97$.
(b) Store $S(j)$ with random numbers produced by a linear congruence generator.
(c) Mark the stack as initialized.
1. Make sure that the stack is initialized.
2. Generate a new random number X.
3. Convert X to an integer j in the range $[1, L]$.
4. Return $S(j)$ as the random number required.
5. Replace $S(j)$ by X.

Note that, since X_n in Eq. (7-1) is needed to generate the next random integer, it must be preserved in a computer code.

Because of the wide range of applications of random numbers, computer operating systems and high-level languages are usually equipped with random number generators. A slight variant of Eq. (7-1) is often found on 32-bit machines,

$$X_{n+1} = \frac{aX_n + c}{d} \bmod m \tag{7-2}$$

with

$$m = 32,768 \qquad a = 1,103,515,245 \qquad c = 12,345 \qquad d = 65,536.$$

The range of random integers generated by this method is $0 \leq X_n < 2^{15}$. There is, however, a small problem if we wish to implement Eq. (7-2) using a language such as Fortran. Since the integer a is larger than 2^{30}, there is a high probability that an overflow will take place on a 32-bit word length computer when it is multiplied by a random integer X_n. To avoid the problem this portion of code can be written in a different language such as C in a way that it may be called by a Fortran program.

Improving randomness by shuffling One weak point of the linear congruence method is that every random integer is generated from the one before it. As a result, there is a strong possibility for the random numbers to be sequentially correlated. One consequence of such a correlation may be illustrated by the following example. If we use groups of three random numbers $(R_{3n}, R_{3n+1}, R_{3n+2})$ as the coordinates (x, y, z) of points inside a cube of length 1 in arbitrary units on each side, there is the tendency for the points to fall mainly on a small number of surfaces inside the cube rather than filling up the volume evenly. An illustration of this is provided by Problem 7-1.

In spite of such a shortcoming, the linear congruence method is widely used, as it is an efficient way to generate random numbers. Several remedies have been introduced to overcome its weakness. A simple method to alleviate the correlation tendency is to shuffle the sequence. That is, a large number of random numbers is generated by the linear congruence method and stored in an array. Let the size of the array be L, preferably a prime number. Anytime we need a random number, we take it from the array rather than having it generated directly. However, instead of taking them in sequence, a random integer j in the range $[1, L]$ is generated. The required random number is then taken from the jth location of the array. To prevent the same random number from being used again, the jth location in the array is replaced by a new random number produced by the generator. This is implemented in the algorithm of Box 7-1. For simplicity, we can use the linear congruence generator to fill the array, as the most serious problem of correlation is now corrected by the shuffling.

Alternate procedures Another way to improve the linear congruence method is to construct a new formula that makes use of more than one previous member of the sequence. Instead of X_n alone, as in Eq. (7-1), we can include an additional member on the right side of the equation,

$$X_{n+1} = (X_n + X_{n-k}) \bmod m$$

with k being a number larger than 15 or so to avoid possible correlations between X_n and X_{n-k}. For example, the following choice is given in Knuth:[31]

$$X_{n+1} = (X_{n-24} + X_{n-55}) \bmod m \qquad \text{for} \qquad n \geq 55. \tag{7-3}$$

This is known as an additive generator. A variant of this is often used in practice is the subtractive generator,

$$X_{n+1} = (X_{n-55} - X_{n-24}) \bmod m. \tag{7-4}$$

Among other advantages, this method can be made completely *portable*. That is, the computer program may be written in a high-level language, such as Fortran or C, and can be made to run on any machine that has a compiler for the language. The algorithm is summarized in Box 7-2.

It is also possible to design methods to generate random numbers using special features of the central processing unit of the computer. Two such methods are given by Marsaglia, Narasimhan, and Zaman;[37] and Chiu and Guu.[11] In general, it is useful to have more than one random number generator available. As we shall see below, it is not possible to have a complete set of tests for randomness. As a result, there is always a possibility that our Monte Carlo results are the artifact of the random numbers used. One way to ensure against such a pitfall is to run the calculations using different generators.

Tests of randomness How do we know that a given set of numbers is random? This is not an easy question to answer, especially for the *pseudo*-random numbers we

Box 7-2 Function RSUB(ISEED)
Subtraction method of random number generation

Argument list:
ISEED: Random number seed on input. Returns the seed for the next one.
Initialization:
(a) Define three constants $m = 10^9$, $i_s = 21$, and $i_r = 30$.
(b) Set up a stack $S(j)$ of length $L = 55$ for integers.
(c) Store the absolute value of the seed in $S(L)$.
(d) Let $k = 1$ and j =ISEED.
(e) For $\ell = 1$ to $(L - 1)$, carry out the following steps to store the stack:
  (i) Let $\ell_x = i_s * \ell \bmod L$.
  (ii) Store the value k in $S(\ell_x)$.
  (iii) Replace k by $(j - k)$. If $k < 0$, repeat with j increased by 10^9.
  (iv) Let $j = S(\ell_x)$
(f) Randomize the stack by carrying out the following steps three times:
  (i) For $\ell = 1$ to 24, replace $S(\ell)$ by $S(\ell) - S(\ell + 31)$.
  (ii) For $\ell = 25$ to L, replace $S(\ell)$ by $S(\ell) - S(\ell - 24)$.
  (iii) If $S(\ell)$ becomes negative, increase $S(\ell)$ by 10^9.
(g) Set stack counter $j_s = 0$.
(h) Mark the stack as initialized.
1. Re-initialize the stack if the input seed is negative.
2. Increase the stack counter j_s by 1.
3. If $j_s > L$,
  (a) Replace the stack by repeating the steps in (f) of the initialization step.
  (b) Set stack counter $j_s = 1$.
4. Return (float) $S(j_s) \times 10^{-9}$ as the uniform random number in $[0, 1]$.

are dealing with here. We know that it cannot fulfill all the fundamental criteria for randomness. For most practical needs, however, the requirement for randomness is not as stringent as the definitions imply. For example, in a one-dimensional Monte Carlo integration, we are essentially satisfied with a set of random numbers that distribute evenly in the integration interval. On the other hand, for a two-dimensional Monte Carlo integration, we have the additional concern of whether a pair of adjacent random numbers are correlated.

In general, there are many pitfalls in constructing a random number generator, as well as a number of ways to make bad use of a good generator. One such case of the latter category is given in Problem 7-3 as illustration. For this reason, testing random numbers is an important part of any Monte Carlo calculation. All the tests are based on statistical measures. Since there is no single measure that encompasses all the desirable features, it is difficult, if not impossible, to design a comprehensive test for randomness. A large number of measures are available in the literature, and a summary of some of the more important ones is given in Knuth.[31] Here we shall

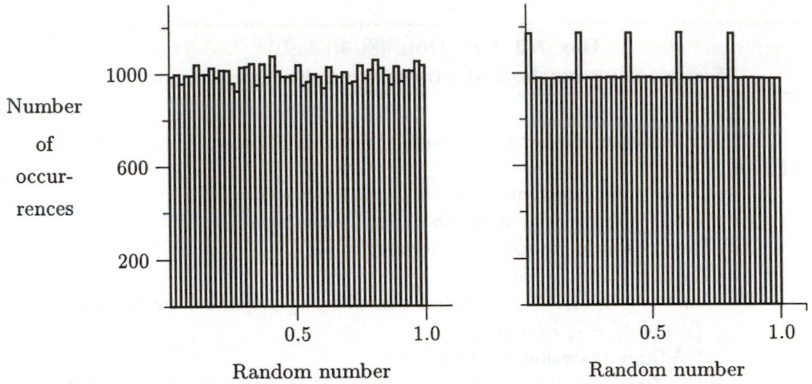

Figure 7-1: Distribution of 50,000 random numbers generated using the subtraction method (left) and a "bad" linear congruence method (right).

only describe three: distribution, correlation, and run tests.

Frequency test For random numbers evenly distributed in the interval $[0, 1]$, the first obvious test is to see whether the probability of finding it anywhere in the interval is indeed the same. As we did in §2-5, we can divide the range of possible values into a number of equal size bins and count the number of random numbers that fall into each bin. If the distribution is uniform in the interval, the number in each should be the same as in the others within statistical fluctuations.

As a practical procedure, let the total number of bins be N_{bin}. For uniform random numbers in the range $[0, 1]$, it is convenient to choose the width of each bin to be

$$w = 1/N_{\text{bin}}.$$

For a random number X_j, we can find the bin number k to which it belongs by multiplying X_j with N_{bin} and taking the integer part of the product,

$$k = \mathsf{INT}(X_j \times N_{\text{bin}}) + 1 \tag{7-5}$$

where we have added 1 to the result so that the bin number starts with 1. The function $\mathsf{INT}(x)$ takes the integer part of a floating number x and returns an integer that is less or equal to x, as we saw earlier in Table 1-2. There is a small probability that k will take on the value $(N_{\text{bin}}+1)$ in Eq. (7-5). This happens under the extremely unlikely occasion where $X_j = 1$ due to round-off errors. To make allowance for such a possibility an extra bin should be added.

If the total number of random numbers tested is n, the number expected in each bin is $(n \times w)$ on the average. Since the random numbers are assumed to be uncorrelated, the distribution in each bin follows a Poisson distribution, as we saw

Box 7-3 Subroutine FQNCY(ISEED,N_TTL,R_GEN)
Frequency test of random numbers

Argument list:
- ISEED: Random number seed.
- N_TTL: Total number of random numbers to be tested.
- R_GEN: Name of the external function that generates the random numbers.

Initialization:
- (a) Set up an array for 50+1 histogram bins.
- (b) Zero the bins.

1. Carry out the following steps for all the random numbers to be tested:
   - (a) Generate a random number X.
   - (b) Find the bin number X belongs to.
   - (c) Increase the count in that bin by 1.
2. Carry out the χ^2-test for an even distribution:
   - (a) Obtain the number of degrees of freedom.
   - (b) Calculate the contribution of each bin to the χ^2 using Eq. (7-6).
3. Output the χ^2-value, number of degrees of freedom, and count in each bin.

in §6-1. The standard deviation of such a distribution is $\sqrt{nw}$, and this provides us with a measure of the average fluctuation we can expect in each bin. A χ^2-test may be applied to the overall distribution in all the bins. If M_j is the actual number found in bin j, the χ^2 is then

$$\chi^2 = \sum_{j=1}^{N_{\text{bin}}} \frac{(M_j - nw)^2}{nw}. \tag{7-6}$$

The number of degrees of freedom is one less than the number of bins,

$$\nu = N_{\text{bin}} - 1$$

since we have a constraint that the sum of the occupancies in all the bins must equal to n. Obviously, if the χ^2 has a large value, the sequence of random numbers is not evenly distributed in the interval. On the other hand, a perfectly even distribution, one with $\chi^2 = 0$, is also an indication of departure from randomness, as fluctuation is an integral part. If the sequence is truly random, we expect that the departure for any one bin from the average value of nw is given by a normal distribution, as we saw in §6-2. For our present case, the normal distribution centers around zero with a variance of nw. This gives us an expected value for the reduced χ^2 to be $\chi^2_\nu \sim 1$. As a result, a small χ^2-value [$\chi^2 << \nu$ or $Q(\chi^2|\nu) \geq 0.99$] implies a lack of fluctuation and, hence, a sign of departure from randomness as well.

The results of applying a frequency test to random numbers generated from the linear congruence and subtractive methods are given in Table 7-1. To have an idea of the distribution on a finer scale, the two sequences were analyzed using 50 bins

($N_{\rm bin} = 50$). For good statistics, the total number of random numbers used in each case is $n = 50,000$. This gives us a value of $nw = 1,000$. The χ^2-values for both cases are around 50, showing that both sequences are evenly distributed.

For comparison, we include also the result obtained using a "bad" linear congruence generator:

$$X_{j+1} = (137X_j + 187) \bmod 256.$$

The χ^2-value in this case is 174, giving a probability $Q(\nu = 49|\chi^2 = 174) = 0.000$ for being an evenly distributed set of numbers. As expected, shuffling of the sequence of random number does not change the χ^2-values for the results of both the good and the bad linear congruence generators. This comes from the simple fact that the frequency test is not concerned with the order in which the random numbers are generated. The algorithm for the test is summarized in Box 7-3.

An interesting feature of the bad linear congruence generator is shown by comparing the two histograms in Fig. 7-1. On the left we find the histogram for 50,000 random numbers produced using a good generator. Since $nw = 1,000$ in each of the 50 bins, the fluctuations are expected to be $\sqrt{1,000} = 31.6$ on the average. On the whole, this is roughly what is observed. On the other hand, the histogram for random numbers produced by the bad generator shows regular peaks that cannot be statistical in nature, concurring with the large χ^2-value obtained in the test.

Serial correlation test There are many ways that members of a sequence of random numbers are correlated with each other. The easiest one to test for is linear correlation. The primary aim here is to find out whether there is a tendency for each random number to be followed by another one of a particular type. For example, it will be quite unacceptable if a large random number is always followed by a small one (or a large one). As a check, we can define a correlation coefficient between a sequence of n random numbers $X_1, X_2, \ldots, X_n$ as

$$C = \frac{n(\sum_{i=1}^{n} X_{i-1}X_i) - (\sum_{i=1}^{n} X_i)^2}{n\sum_{i=1}^{n} X_i^2 - (\sum_{i=1}^{n} X_i)^2} \tag{7-7}$$

where X_0 in the first term of the numerator may be taken to be equal to X_n.

The coefficient C defined above is slightly different from other such coefficients, for example, that given by Eq. (6-52). There, the correlation is between two different sets of variables $\{x_i\}$ and $\{y_i\}$. In contrast, we are taking here a sum over products of the type $X_{i-1}X_i$, $X_iX_{i+1}, \ldots$, instead of x_iy_i, $x_{i+1}y_{i+1}, \ldots$. As a result, the expected values of the two types of coefficients are different from each other. In the present case, the value of C for an uncorrelated sequence is expected to be in the range $[\mu_n - 2\sigma_n, \mu_n + 2\sigma_n]$ for 95% of the time, with

$$\mu_n = -\frac{1}{n-1} \qquad\qquad \sigma_n = \frac{1}{n-1}\sqrt{\frac{n+1}{n(n-3)}}.$$

In contrast, ordinary correlation coefficients, such as the one in Eq. (6-52), are bound in the range $[-1, +1]$.

Table 7-1: Tests of random number generators.

Test	Frequency χ^2	Serial correlation C	Run-up χ^2
Expected value	$\nu = 49$	$[-0.028, +0.028]$	$\nu = 6$
Linear congruence method			
Bad	174	−0.022	682
After shuffling	171	0.001	842
Good	41	−0.023	4.7
After shuffling	44	−0.003	2.8
Subtraction method	56	−0.003	4.9

The results of a correlation test for the three different random number generators, a bad linear congruence generator, a good linear congruence generator, and a subtraction generator — the same three used in the frequency test earlier — are also listed in Table 7-1. In each case, a set of 5,000 random numbers is used. The fact that results from linear congruence methods have a tendency to be correlated is reflected by the relatively large absolute values of both the good and the bad generator. Shuffling among the random numbers destroys such short-range correlations, as can be seen from the much smaller absolute values of C obtained. Furthermore, since the main difference between the good and the bad linear congruence generators is in the distribution of the random numbers produced, the results of the two methods are fairly similar as far as the correlation test is concerned. The value of C for the subtraction method is -0.003 in the test carried out, essentially a null value for a set of 5,000 numbers. In fact, repeated tests with different sets of 5,000 numbers, each set produced by a different seed, give values of C of roughly the same magnitude with both positive and negative signs.

Run test A more sensitive test of the correlation among random numbers is the length of a run of either increasing or decreasing size. Here, the aim is to check whether there is a tendency for consecutive random numbers in a sequence to be decreasing (run-down test) or increasing (run-up test) in value. To simplify the discussion, we shall only consider the latter.

By the length ℓ_{up} of a run, we mean the number of random numbers in a sequence that is increasing in value. A run starts at X_i if $X_i < X_{i-1}$ and the run continues till (but excluding) X_{i+j} if

$$X_i < X_{i+1} < X_{i+2} < \cdots < X_{i+j-1} > X_{i+j}.$$

The length of such a run is defined to be

$$\ell_{\text{up}} = j.$$

Box 7-4 Subroutine RUNUP(ISEED,N_TTL,R_GEN)
Run-up test of a sequence of random numbers

Argument list:
- ISEED: Random number seed.
- N_TTL: Total number of random numbers to be tested.
- R_GEN: Name of the external function to generate the random numbers.

Initialization:
- (a) Set up an array $K_{\text{up}}(i)$ of length $\ell_{\max} = 7$ for the histogram values.
- (b) Zero the array and the counter for the total number of runs.

1. Generate a random number and store it as X_{old}. Initialize $\ell_{\text{up}} = 1$.
2. Generate another random number and store it as X_{new}.
3. Compare X_{new} with X_{old}:
   - (a) If $X_{\text{new}} > X_{\text{old}}$,
     - (i) Increase the run length ℓ_{up} by 1.
     - (ii) Store X_{old} as X_{new}.
     - (iii) Go back to step 2.
   - (b) If $X_{\text{new}} \leq X_{\text{old}}$, the run ends:
     - (i) Increase run counter N_{ttl} by 1.
     - (ii) If $\ell_{\text{up}} \leq \ell_{\max}$, increment the histogram bin $K_{\text{up}}(\ell_{\text{up}})$ by 1.
     - (iii) If $\ell_{\text{up}} > \ell_{\max}$, increment the histogram bin $K_{\text{up}}(\ell_{\max})$ by 1.
     - (iv) Go back to step 1 if more runs are needed.
4. Calculate the χ^2-value using Eq. (7-9) and output the histogram.

Let us illustrate the point using random digits as an example. For the sequence

$$\begin{array}{ll} & \cdots 9|138|7|69|7|5|3\ |1\cdots \\ \ell_{\text{up}} = & \quad\ 3\quad 1\ \ 2\ \ 1\,1\,1 \end{array}$$

there are five complete runs and the length of each is as indicated.

For simplicity, we shall ignore the possibility of two consecutive random numbers being equal to each other in any of our discussions. However, in practical applications such a possibility, however remote, cannot be ignored. For this reason we shall make an approximation in an actual computer program to carry out the calculation and consider all the lesser signs ($<$) above to be lesser or equal signs ($\leq$), even though the statistics is defined in terms of the lesser sign alone.

In the definition above, the end of one run appears as the head of another run. For this reason, it is rather difficult to calculate the expected value for the distribution of ℓ_{up}. To simplify the situation, Knuth[31] suggested that one should disregard the comparison of the last random number of a run with the next one, resulting in runs that are uncorrelated with each other. For the random digit example above, this corresponds to the following method of counting:

$$\begin{array}{ll} & \cdots 9|138|\,7\,|69|\,7\,|5|\,3\ \ |1\cdots \\ \ell_{\text{up}} = & \quad\ 2\ \ \odot\ \ 2\ \odot 1 \odot \end{array}$$

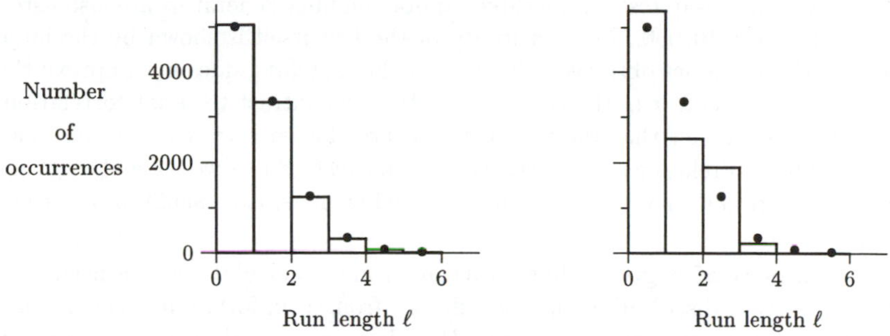

Figure 7-2: Distributions of 10,000 runs of length ℓ for random numbers generated using the subtraction method (left) and a "bad" linear congruence method (right). Expected values from Eq. (7-8) are indicated as dots.

where the symbol $\odot$ indicates run of length 1 being discarded. As a result, some of the runs have reduced lengths compared with those in the corresponding correlated case above, such as the first run (length 2 instead of 3). The probability of obtaining a run of length ℓ for a completely uncorrelated sequence is given by

$$p_\ell = \frac{1}{\ell!} - \frac{1}{(\ell+1)!}. \tag{7-8}$$

That is, in a sample of m runs constructed from n random numbers, we expect mp_ℓ runs with length ℓ. Since, in general, each run involves more than one random number, $m < n$. The algorithm to make a run-up test is summarized in Box 7-4.

We can apply a χ^2-test to see if the distribution of the length ℓ corresponds to our expectation of a random sequence. If the number of uncorrelated runs of length ℓ in a sample of random numbers is K_ℓ, the following quantity

$$\chi^2 = \sum_{\ell=1}^{L} \frac{1}{mp_\ell}(K_\ell - mp_\ell)^2 \tag{7-9}$$

follows a χ^2-distribution, as can be seen by comparing with Eq. (6-41). The weighting factor σ_i^2 for each point is $(mp_\ell)^{-1}$ here, based on the assumption of a Poisson distribution for the number of runs of a given length ℓ. From Eq. (7-8), we see that the probability of observing a long sequence decreases rapidly. Since $6! = 720$, there is very little point in extending L, the upper limit of the summation, to beyond 6. On the other hand, since there is no restriction for the upper limit of the length ℓ, the number of degrees of freedom for the χ^2-distribution is $\nu = L$ (rather than $L-1$). For a good sequence of random numbers, we expect a χ^2-value of approximately ν or a reduced χ^2-value of $\chi^2_\nu \approx 1$. The actual distributions for two sequences of random numbers produced by two different generators are plotted in Fig. 7-2 as illustration.

The results of a run-up test on the three random number generators are also listed in Table 7-1 for $m = 10,000$. The sensitivity of the test itself is shown by the large differences in the χ^2-values obtained. The reason that shuffling does not improve the results of the bad generator in the run-up statistics, in contrast to serial correlation, may be traced to the fact that the run test measures longer-term correlations than the next-neighbor correlations in a serial correlation test. Since we have chosen, for purposes of illustration, a very poor random number generator, shuffling does not improve the results.

Random numbers of a given distribution In many calculations, we need random numbers with a distribution that is different from a uniform one. The general approach to obtain a set with distribution $P(x)$ is to start with a uniform set and change it into the one required. There are several ways to carry out such a transformation and we shall describe here three representative ones. By transformation, we are not necessarily restricted here to those that can be carried out analytically. In fact, on many occasions the required distribution may not even be represented by an analytical function. A summary of the various methods commonly used is found in Abramowitz and Stegun.[1] Methods of obtaining many standard distributions can be found in Press and coauthors.[44] Fast algorithms for specific distributions can also be found in recent literature, such as those given for logarithmic and normal distributions by Fernandez and Rivero.[21]

Let us start the discussion by assuming that we have in hand a set of random numbers evenly distributed in the interval $[0, 1]$. The probability of finding a number in the interval dx around x is given by

$$p_u(x)\,dx = \begin{cases} dx & \text{for } 0 \le x \le 1 \\ 0 & \text{otherwise.} \end{cases} \tag{7-10}$$

The normalization condition here is

$$\int_{-\infty}^{+\infty} p_u(x)\,dx = 1. \tag{7-11}$$

The subscript u emphasizes the fact that the distribution is uniform. Our present interest is to find a transformation from x to y such that the random number y has the desired distribution $P(y)$. It is convenient for us to assume that $P(y)$ has the same normalization as that for $p_u(x)$ given in Eq. (7-11). From the transformation of probability, we have the equality

$$P(y)\,dy = p_u(x)\,dx. \tag{7-12}$$

This gives us

$$P(y) = \left|\frac{dx}{dy}\right| p_u(x)$$

where the Jacobian of the transformation is given by the absolute value of the derivative of x with respect to y, coming from the fact that probabilities cannot be negative.

For some $P(y)$, it is possible to find the transformation analytically. As an illustration of this approach, consider the exponential distribution

$$P(y)\,dy = e^{-y}dy. \tag{7-13}$$

To obtain the transformation function $y(x)$, we can start with Eq. (7-12). For $p_u(x)\,dx$ given by Eq. (7-10), we obtain

$$e^{-y} = x$$

by equating the absolute values of the indefinite integrals of both sides of Eq. (7-12). On inverting this relation, we obtain the transformation

$$y(x) = -\ln(x)$$

that changes a set of uniformly distributed random numbers $\{X_i\}$ to a set $\{Y_i\}$ having an exponential distribution. Such a method to carry out the transformation works well in cases where it is possible to carry out the indefinite integral

$$x = F(y) = \int P(y)\,dy$$

and the inverse of $F(y) = x$ can be found. In practice, the relation

$$y = F^{-1}(x) \tag{7-14}$$

can be complicated in form. In such cases it may be more efficient to adopt some other method, such as the ones discussed below.

Random number with a normal distribution Besides the uniform distribution, random numbers with a normal distribution are useful in many applications. For the convenience of discussion, let us consider a set centered around zero and having a variance of unity. That is,

$$P(y)\,dy = \frac{1}{\sqrt{2\pi}}e^{-y^2/2}dy. \tag{7-15}$$

In this case, a simple transformation from a uniform set cannot be worked out easily, as we have done above for an exponential distribution. On the other hand, transformation between two pairs of such random numbers, (x_1, x_2) and (y_1, y_2), of the following form

$$y_1 = \sqrt{-2\ln x_1}\,\cos(2\pi x_2) \qquad y_2 = \sqrt{-2\ln x_1}\,\sin(2\pi x_2) \tag{7-16}$$

produces y_1 and y_2 with a normal distribution. This may be seen from the relation between the two joint probabilities, $P(y_1, y_2)\,dy_1dy_2$ for y_1 and y_2 and $p_u(x_1, x_2)\,dx_1dx_2$ for x_1 and x_2, is given by a simple extension of Eq. (7-12):

$$P(y_1, y_2)\,dy_1dy_2 = p_u(x_1, x_2)\,dx_1dx_2.$$

The Jacobian of the transformation in this case is the determinant

$$\frac{D(x_1,x_2)}{D(y_1,y_2)} = \begin{vmatrix} \frac{\partial x_1}{\partial y_1} & \frac{\partial x_1}{\partial y_2} \\ \frac{\partial x_2}{\partial y_1} & \frac{\partial x_2}{\partial y_2} \end{vmatrix} = -\frac{x_1}{2\pi} = -\frac{1}{2\pi} e^{-(y_1^2+y_2^2)/2}$$

as can be shown by an explicit calculation. Since this is the product of two normal distributions, one for y_1 and the other for y_2, we see that the joint distribution function $P(y_1, y_2)$ is a product of two independent normal distributions. Note that the joint distribution function of x_1 and x_2 is the product of $p_u(x_1)$ and $p_u(x_2)$, each of which is given by Eq. (7-10).

In practical applications, the transformation given by Eq. (7-16) is too time consuming to carry out because of the trigonometry functions involved. To speed up the calculations, we can regard x_1 and x_2 as the coordinates of a point on a two-dimensional surface. Since x_1 and x_2 are individually distributed uniformly in the interval $[0, 1]$, the point (x_1, x_2) is uniformly distributed in a square of length 1 on each side. If we change (x_1, x_2) into (X_1, X_2) through the transformation

$$X_1 = 2x_1 - 1 \qquad\qquad X_2 = 2x_2 - 1$$

and reject those with

$$R^2 \equiv X_1^2 + X_2^2 \geq 1$$

we have a set of points (X_1, X_2) distributed evenly in a circle of radius 1 on a two-dimensional surface. Now if we use $(2\pi z_2)$ to represent the angle between the vector $\boldsymbol{R}$ from the origin at the center of the circle to the point (X_1, X_2) and the horizontal axis, we have the relations

$$\cos(2\pi z_2) = \frac{X_1}{R} \qquad\qquad \sin(2\pi z_2) = \frac{X_2}{R}.$$

If we further define another variable z_1 as

$$z_1 = R^2$$

we can write down a transformation similar to Eq. (7-16)

$$\begin{aligned} y_1 &= \sqrt{-2\ln z_1}\cos(2\pi z_2) = \sqrt{-2\ln R^2}\,\frac{X_1}{R} = X_1\sqrt{\frac{-2\ln R^2}{R^2}} \\ y_2 &= \sqrt{-2\ln z_1}\sin(2\pi z_2) = \sqrt{-2\ln R^2}\,\frac{X_2}{R} = X_2\sqrt{\frac{-2\ln R^2}{R^2}}. \end{aligned} \qquad (7\text{-}17)$$

The advantage here is that, instead of two trigonometry functions and a square root in Eq. (7-16), only a square root is needed.

In this approach, some loss of efficiency comes from pairs of random numbers with $R > 1$ being discarded. The fraction retained is $\pi/4 \sim 0.79$ on the average,

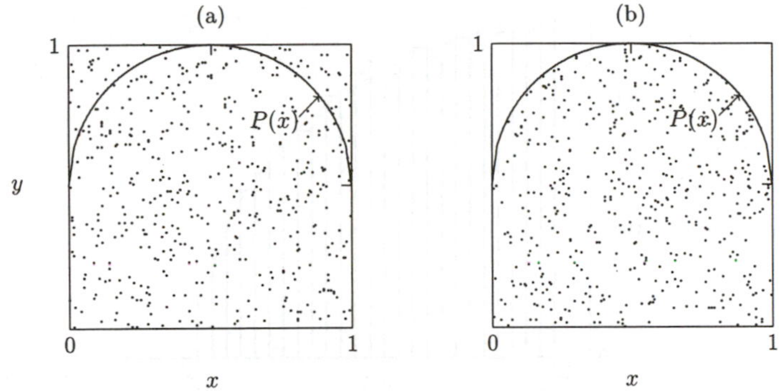

Figure 7-3: (a) Random points, each specified by two random numbers as its x- and y-coordinates, uniformly filling a square area. (b) When points with $y > P(x)$ are rejected, the x values of the remainder have distribution $P(x)$.

given by the ratio of the area of a unit circle to that of a square of length 2 on each side. Programming for the transformation is given as an exercise in Problem 7-5. An approximate scheme, based on the central limit theorem, is given in Problem 7-4 and is used in many applications.

Rejection method If the desired distribution of random numbers is not given by an analytical function, or if the inverse function for the indefinite integral of $P(x)$ does not exist, the transformation method of Eq. (7-14) fails. In such cases, the rejection method may be used. The basic idea here is quite simple. If we throw away those that are outside $P(x)$, the remainder must be distributed in the way we want. This type of approach was used, in part, in generating random numbers with a normal distribution above.

Consider a two-dimensional surface. As we have seen earlier, the location of each point on this surface may be specified by the values of its x- and y-coordinates with respect to some fixed system of reference. A random point on this surface may be chosen by selecting two random numbers X_1 and X_2, and let the coordinates $x = X_1$ and $y = X_2$. If both X_1 and X_2 are uniformly distributed in the interval $[0, 1]$, the random point has an equal probability of being anywhere in the square area, as shown in Fig. 7-3(a). If a large number of such random points are available, the square area will be uniformly covered. On the other hand, if we reject those points with $y > P(x)$, where $P(x)$ is the desired distribution, the remaining points fill only the area underneath the curve $P(x)$, as shown in Fig. 7-3(b). Consider now only the distribution of the values of the x-coordinates of the points remaining. Since the ratio of the numbers of points in two small intervals, one at $x = a$ and the other at $x = b$, is given by $P(x = a)/P(x = b)$, the distribution of x values is given by the probability $P(x)\,dx$. In more practical terms, if we have a collection of N points, represented

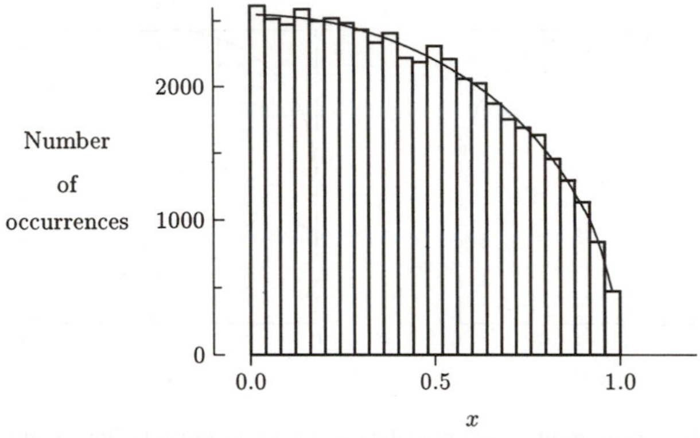

Figure 7-4: Distribution of 50,000 semicircular random numbers generated by the rejection method. The smooth curve is given by Eq. (7-18).

by their x- and y-coordinates $\{X_1(i), X_2(i)\}$, for $i = 1, 2, \ldots, N$, and we discard all those points [both $X_1(i)$ and $X_2(i)$] with $X_2(i) > P(x = X_1(i))$, the remaining random numbers $X_1(i)$ have the desired distribution $P(x)$.

A simple example may be helpful here to illustrate the principle behind the rejection method. Let us consider the semicircular distribution

$$P(x) = \frac{2}{\pi}\sqrt{1 - x^2} \qquad \text{for} \qquad |x| \leq 1. \tag{7-18}$$

For $x > 0$, the distribution covers only a quarter of a circle of unit radius with the center of the circle located at the origin. To produce random numbers having such a distribution with the rejection method, we start out with a uniform set in interval $[0, 1]$. For each pair of such random numbers, $X_1(i)$ and $X_2(i)$, we shall consider $X_1(i)$ as the x-coordinate and $X_2(i)$ as the y-coordinate of a random point in a square area of unit length on each side. If $\{X_1^2(i) + X_2^2(i)\} \geq 1$, the point is outside the quarter-circle and we reject the point. Another point is produced by taking a new pair of random numbers $X_1(i+1)$ and $X_2(i+1)$. If $\{X_1^2(i+1) + X_2^2(i+1)\} < 1$, the point is inside the unit circle and we retain $X_1(i+1)$ as one of the random numbers with a semicircular distribution. To show that the accepted sequence of X_1 follows the required distribution, a set of 50,000 such random numbers is plotted in the form of a histogram in Fig. 7-4. The fact that the distribution is semicircular may be seen by comparing the histogram with the quarter-circle superimposed in the form of a solid curve. The method is essentially the same as that used earlier to produce random numbers with a circular distribution that serve as the intermediate quantities to generate random numbers with a normal distribution.

Since the quarter-circle occupies most of the area of the square covered by our original random points, only a small fraction of the points is rejected. We can calcu-

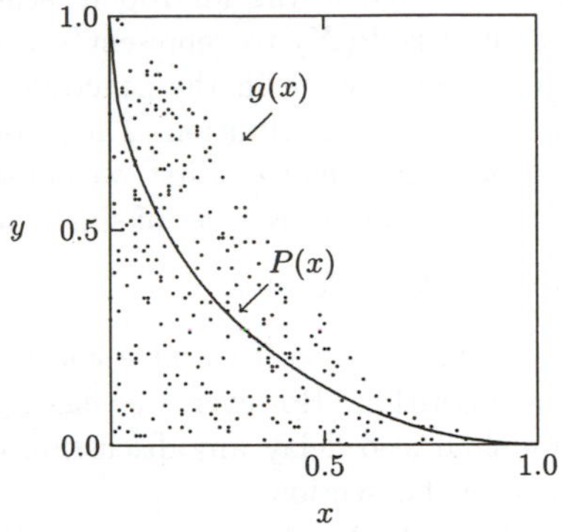

Figure 7-5: Improving the efficiency of the rejection method by transforming first to a normal distribution $g(x)$. Since $P(x)$ (solid curve) fills a much larger part of the area underneath $g(x)$, a smaller fraction has to be rejected.

late the ratio from the area of a quarter circle ($=\pi/4$) and that of a square (unity). However, since each point requires two random numbers, only $\pi/8$ of all those generated belong to the distribution $P(x)$. The additional factor of $\frac{1}{2}$ arises from the fact that only X_1 (but not X_2) can be used.

In general, if the area underneath the desired distribution $P(x)$ is only a small fraction of the rectangular area covered by pairs of uniform random numbers, it is more efficient to transform the uniform set first to a distribution $g(x)$ satisfying the condition $g(x) \geq P(x)$ before applying the rejection method. The function $g(x)$ must be chosen such that the transformation method can be applied with ease. Furthermore, $g(x)$ should be as close to $P(x)$ as possible in order to minimize the number of points to be rejected. For example, for the distribution $P(x)$ shown in Fig. 7-5, a normal distribution for $g(x)$ may be used. Only points between $P(x)$ and the normal distribution are discarded.

The Metropolis algorithm A method to produce random numbers of a given distribution that is especially convenient for problems in statistical mechanics and related applications was given by Metropolis and coauthors.[39] Instead of individual random numbers, the interest here is mainly in the distribution of random points in a multidimensional space, such as those representing members of a canonical ensemble in statistical mechanics. If a system consists of N particles, each of which is specified by its location (x, y, z) and momentum (p_x, p_y, p_z), the dimension of the phase space is $d = (3+3)N$. Each point in this N-particle space then requires d quantities to label it. To simplify the notation, we shall use a single bold faced letter,

$$\boldsymbol{X} \equiv (x_1, x_2, \ldots, x_d)$$

to represent the location of a point in such a space.

The way a random point is selected in the Metropolis scheme is closely related to the rejection method. Let us use $P(\boldsymbol{X})$ to represent the required distribution. Furthermore, we assume that at some stage in the generation there are already n points $\boldsymbol{X}_1, \boldsymbol{X}_2, \ldots, \boldsymbol{X}_n$ with the correct distribution. The algorithm to find the next point $\boldsymbol{X}_{n+1}$ is given by the following procedure. First, we construct a trial point $\boldsymbol{X}_t$ from the last accepted one $\boldsymbol{X}_n$ by adding to it a set of small changes $\boldsymbol{\delta}$. That is,

$$\boldsymbol{X}_t = \boldsymbol{X}_n + \boldsymbol{\delta}. \tag{7-19}$$

The size of $\boldsymbol{\delta}$ depends on the problem we wish to solve and is an important consideration in the efficiency of the algorithm. However, we shall ignore this question for the time being. Similarly, we shall also delay any discussion on how to obtain the starting point $\boldsymbol{X}_0$ until the end of this section.

The decision whether $\boldsymbol{X}_t$ should be accepted depends on the ratio

$$r = \frac{P(\boldsymbol{X}_t)}{P(\boldsymbol{X}_n)}. \tag{7-20}$$

If $r \geq 1$, the trial point $\boldsymbol{X}_t$ is accepted. This condition may be stated as

$$\boldsymbol{X}_{n+1} = \boldsymbol{X}_t \qquad \text{for} \qquad r \geq 1. \tag{7-21}$$

If $r < 1$, a decision has to be made whether to accept the trial point. In practical terms, a random number x, uniformly distributed in $[0, 1]$, is generated. If $x > r$, the trial point is rejected; otherwise it is accepted. In other words,

$$\boldsymbol{X}_{n+1} = \boldsymbol{X}_t \qquad \text{with probability } r \qquad \text{for} \qquad r < 1. \tag{7-22}$$

This way of deciding whether to accept $\boldsymbol{X}_t$ for $r < 1$ is very similar to the spirit of flipping a coin when we have to make an unbiased yes or no decision.

We shall now show that the sequence of random points $\boldsymbol{X}_1, \boldsymbol{X}_2, \ldots$, has the desired distribution $P(\boldsymbol{X})$. Let $N(\boldsymbol{X})$ be the density of points in the neighborhood of $\boldsymbol{X}$ and $K(\boldsymbol{X} \to \boldsymbol{Y})$ be the transition probability for the system to move from point $\boldsymbol{X}$ to $\boldsymbol{Y}$. At equilibrium, the density in the neighborhood of any point must be constant in time. This, however, does not mean the system is static — it simply implies that transitions out of the region are balanced by transitions into the same region. In other words, the equilibrium condition is given by

$$\sum_i N_e(\boldsymbol{X})K(\boldsymbol{X} \to \boldsymbol{Y}_i) - \sum_i N_e(\boldsymbol{Y}_i)K(\boldsymbol{Y}_i \to \boldsymbol{X}) = 0 \tag{7-23}$$

where $N(\boldsymbol{Z})$ is the density in the vicinity of point $\boldsymbol{Z}$ and the subscript e emphasize that the system is at equilibrium. This is, essentially, a statement of the principle of detailed balance. Away from equilibrium, the change in $N(\boldsymbol{X})$ is governed by

$$\begin{aligned}\Delta N(\boldsymbol{X}) &= \sum_i \{N(\boldsymbol{X})K(\boldsymbol{X} \to \boldsymbol{Y}_i) - N(\boldsymbol{Y}_i)K(\boldsymbol{Y}_i \to \boldsymbol{X})\} \\ &= \sum_i N(\boldsymbol{Y}_i)K(\boldsymbol{X} \to \boldsymbol{Y}_i)\left\{\frac{N(\boldsymbol{X})}{N(\boldsymbol{Y}_i)} - \frac{K(\boldsymbol{Y}_i \to \boldsymbol{X})}{K(\boldsymbol{X} \to \boldsymbol{Y}_i)}\right\}.\end{aligned} \tag{7-24}$$

An alternate way to state the condition of equilibrium of Eq. (7-23) is then

$$\frac{N_e(\boldsymbol{X})}{N_e(\boldsymbol{Y}_i)} = \frac{K(\boldsymbol{Y}_i \to \boldsymbol{X})}{K(\boldsymbol{X} \to \boldsymbol{Y}_i)}. \tag{7-25}$$

This condition implies also that the system is stable. For example, if the ratio of $N(\boldsymbol{X})/N(\boldsymbol{Y}_i)$ is larger than $K(\boldsymbol{Y}_i \to \boldsymbol{X})/K(\boldsymbol{X} \to \boldsymbol{Y}_i)$, we have $\Delta N(\boldsymbol{X}) > 0$ and the system moves away from the region around $\boldsymbol{X}$. The reverse becomes true if $N(\boldsymbol{X})/N(\boldsymbol{Y}_i) < K(\boldsymbol{Y}_i \to \boldsymbol{X})/K(\boldsymbol{X} \to \boldsymbol{Y}_i)$.

The equilibrium distribution of points can be shown to be equal to $P(\boldsymbol{X})$. To do this we note that, from the way the random points are generated, the transition probability $K(\boldsymbol{X} \to \boldsymbol{Y})$ is a product of two factors,

$$K(\boldsymbol{X} \to \boldsymbol{Y}) = T(\boldsymbol{X} \to \boldsymbol{Y})A(\boldsymbol{X} \to \boldsymbol{Y}) \tag{7-26}$$

where $T(\boldsymbol{X} \to \boldsymbol{Y})$ is the probability that, given that the system is at $\boldsymbol{X}$, the trial value $\boldsymbol{X}_t$ is $\boldsymbol{Y}$. In other words, $T(\boldsymbol{X} \to \boldsymbol{Y})$ is determined by the way we select $\boldsymbol{\delta}$ in Eq. (7-19). For this reason, it must be symmetric with respect to $\boldsymbol{X}$ and $\boldsymbol{Y}$:

$$T(\boldsymbol{X} \to \boldsymbol{Y}) = T(\boldsymbol{Y} \to \boldsymbol{X}). \tag{7-27}$$

The other factor, $A(\boldsymbol{X} \to \boldsymbol{Y})$, in Eq. (7-26) is the probability of accepting the trial point and is given by Eqs. (7-21) and (7-22). In terms of $T(\boldsymbol{X} \to \boldsymbol{Y})$ and $A(\boldsymbol{X} \to \boldsymbol{Y})$, the equilibrium condition Eq. (7-25) may be stated as

$$\frac{N_e(\boldsymbol{X})}{N_e(\boldsymbol{Y})} = \frac{T(\boldsymbol{Y}_i \to \boldsymbol{X})}{T(\boldsymbol{X} \to \boldsymbol{Y}_i)}\frac{A(\boldsymbol{Y}_i \to \boldsymbol{X})}{A(\boldsymbol{X} \to \boldsymbol{Y}_i)} = \frac{A(\boldsymbol{Y}_i \to \boldsymbol{X})}{A(\boldsymbol{X} \to \boldsymbol{Y}_i)} \tag{7-28}$$

where we made use of Eq. (7-27) to cancel out the dependence on $T(\boldsymbol{X} \to \boldsymbol{Y})$.

To make connection with Eqs. (7-21) and (7-22), let us consider the case of starting from $\boldsymbol{X}$ and using $\boldsymbol{Y}$ as the trial point. Here, the ratio

$$\frac{P(\boldsymbol{Y})}{P(\boldsymbol{X})} > 1 \qquad \text{for} \qquad P(\boldsymbol{Y}) > P(\boldsymbol{X}).$$

In this case, the trial point $\boldsymbol{Y}$ is accepted. In terms of $A(\boldsymbol{X} \to \boldsymbol{Y})$, we can state the same result by saying that

$$A(\boldsymbol{X} \to \boldsymbol{Y}) = 1 \qquad \text{for} \qquad P(\boldsymbol{Y}) > P(\boldsymbol{X}).$$

At the same time, for $P(\boldsymbol{Y}) > P(\boldsymbol{X})$, the ratio

$$r = \frac{P(\boldsymbol{X})}{P(\boldsymbol{Y})} < 1$$

and $A(\boldsymbol{Y} \to \boldsymbol{X}) = r$. The equilibrium condition Eq. (7-28) is then given by

$$\frac{N_e(\boldsymbol{X})}{N_e(\boldsymbol{Y})} = \frac{r}{1} = \frac{P(\boldsymbol{X})}{P(\boldsymbol{Y})}. \tag{7-29}$$

Conversely, the ratio

$$r = \frac{P(\boldsymbol{Y})}{P(\boldsymbol{X})} < 1 \qquad \text{for} \qquad P(\boldsymbol{Y}) < P(\boldsymbol{X})$$

and

$$A(\boldsymbol{X} \to \boldsymbol{Y}) = r = \frac{P(\boldsymbol{Y})}{P(\boldsymbol{X})}$$

or $A(\boldsymbol{Y} \to \boldsymbol{X}) = 1$ for $P(\boldsymbol{Y}) < P(\boldsymbol{X})$. Again, we find that the equilibrium condition is given by Eq. (7-29): The density at equilibrium at any point $\boldsymbol{X}$ is proportional to the probability $P(\boldsymbol{X})$ at the point. This, in turn, demonstrates that the Metropolis algorithm of accepting or rejecting a trial point by Eqs. (7-21) and (7-22) generates an equilibrium distribution $P(\boldsymbol{X})$.

The disadvantage of the Metropolis algorithm is that, from the way each trial point is generated by Eq. (7-19), the sequence of points $\boldsymbol{X}_i$, $\boldsymbol{X}_{i+1}, \ldots$, is strongly correlated. This is especially severe at the beginning of a sequence, as it is difficult to choose a starting point $\boldsymbol{X}_0$ that is a proper member of the set. As a result, a whole sequence of points that is outside the distribution we wish to have may be produced. One way to avoid the need to find a good starting point is to use an arbitrary one and discard it, together with the first few generated from it. Because of its statistical mechanics origin, the technique of discarding the initial part of a sequence is often called "thermalizing" the distribution. The same basic principle may also be used to reduce the strong correlation between successive random points by discarding a few points between any two that are adopted as members of the physical ensemble. An application of the Metropolis algorithm is given later in §7-5 for the Ising model.

7-2 Molecular diffusion and Brownian motion

As an elementary application of Monte Carlo methods, we shall examine molecular diffusion as a *random walk* problem. In its simplest form, it may be described in the following way. At some initial time t_0, we have a concentration of molecules at a point x_0 in space. If the temperature of the system is finite, each molecule possesses some nonzero kinetic energy. As a result, the molecules diffuse into the surroundings in the absence of barriers. The rate of diffusion is determined by two factors, the velocity of each molecule and the probability of collision with other molecules occupying the same space. The former is characterized by the temperature T and the latter by the mean free path $\overline{\ell}$.

A well-known application of the random walk approach to diffusion is Einstein's illustration of Brownian motion. The aim here is to simulate the movement of a particle in a fluid. To simplify the picture, we can view the motion of a Brownian particle as one that takes a step of constant size $\overline{\ell}$ in a time interval τ. If we further confine the space to one dimension, the displacement is limited either to taking a step forward $(+\overline{\ell})$ or a step backward $(-\overline{\ell})$ in each time step. For a given particle, its location at time t is then proportional to the excess of the number of positive steps over that of negative ones (or the other way around). In a time interval Δt, the total number of steps taken is

$$n = \frac{\Delta t}{\tau}.$$

If a particle starts from $x = 0$ at $t = 0$, its location $x(t)$ at time t is given by

$$x(t) = m\bar{\ell}$$

where m is the difference between the number of steps in the positive direction, $(n+m)/2$, and the number of steps in the negative direction, $(n-m)/2$. Since the direction taken in each step, + or −, is completely independent of any other steps, the probability for the difference to be m is given by

$$p(n,m) = P_B\Big(n, \frac{1}{2}\{n+m\}, \frac{1}{2}\Big) = \binom{n}{\{n+m\}/2}\Big(\frac{1}{2}\Big)^n = \Big(\frac{1}{2}\Big)^n \frac{n!}{(\frac{n+m}{2})!(\frac{n-m}{2})!}$$

where $P_B(n, \{n+m\}/2, \frac{1}{2})$ is the binomial distribution, defined earlier in Eq. (6-12), for choosing $(n+m)/2$ objects out of n, each with a probability $p = \frac{1}{2}$ to be selected.

The distribution $p(n,m)$ is symmetric with respect to $m = 0$, corresponding to the point $(n+m)/2 = n/2$ in $P_B(n, \frac{1}{2}\{n+m\}, \frac{1}{2})$. This means that it is equally likely to find a particle in the positive direction as in the negative direction and the mean position of a large number of particles $\overline{x(t)}$ is zero. However, the variance of the distribution increases with time:

$$\overline{x^2(t)} = \frac{\bar{\ell}^2}{\tau} t. \tag{7-30}$$

This comes from the fact that the variance of $P_B(n, \frac{1}{2}\{n+m\}, \frac{1}{2})$, given in Problem 6-2, is $n/4$, and $n = t/\tau$ increases linearly with t. The physical reason that the standard deviation of the distribution $\{\overline{x^2(t)}\}^{1/2}$ is proportional to $t^{1/2}$ rather than t comes from collisions with other particles. In the absence of such collisions, a particle will continue its motion along a straight line at constant velocity and the standard deviation of the distribution will increase linearly with time. In our one-dimensional model, each collision has a 50% probability of turning the Brownian particle backward (and the other 50% forward). As a result, growth in the standard deviation of the distribution is reduced from that in the absence of collisions.

Such a simple model of Brownian motion may be simulated by a Monte Carlo calculation. Let the location of a particle at time t be $x(t)$. At time $(t+\tau)$, its location is given by

$$x(t+\tau) = x(t) \pm \bar{\ell}. \tag{7-31}$$

The sign on the right side is completely random. To mimic such a probabilistic event, we can, in principle, pick a random number r uniformly distributed in the interval $[0,1]$. If $r < 0.5$, we take $x(t+\tau) = x(t) - \bar{\ell}$, and if $r \geq 0.5$, we take $x(t+\tau) = x(t) + \bar{\ell}$.

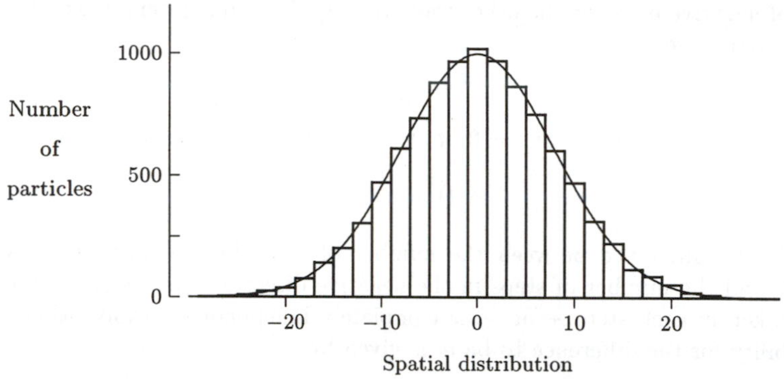

Figure 7-6: Spatial distribution of Brownian particles after 64 time steps compared with a normal distribution (solid curve).

In most practical calculations of this type, however, a much simpler and faster method of using a random bit generator, such as the one described in §A-4, is preferred.

Monte Carlo simulation The interest of our simulation is the distribution of Brownian particles along x at time t. Since the step size is fixed to be $\bar{\ell}$, we can measure all the distances using $\bar{\ell}$ as the unit. Similarly, since the time is always in steps of τ, we can use τ as the unit of time. The distribution at any time t is then given by the number of particles at integer number of length units from the origin. An array of length $(2n+1)$ with

$$n = t/\tau$$

may be used to store the number of particles at each one of the possible locations.

The actual calculation may be carried out in the following way. Each particle starts its random walk at $t = 0$ from its initial position at $x = 0$. The location at time $t = \tau$ is either $+1$ or -1 in units of $\bar{\ell}$, depending on whether the random bit generated is 1 or 0. Let us represent the result as $x(t = \tau)$. The location at $t = 2\tau$ is obtained using Eq. (7-31), with $x(t)$ on the right side given by the location at $t = \tau$. The resulting location gives us the result for $x(t = 2\tau)$. From this, we can obtain the location at the next time interval $t = 3\tau$, and so on. Each time, the random bit generator is used to determine whether the motion is to take a step forward $(+\bar{\ell})$ or backward $(-\bar{\ell})$. After n time steps the position is recorded by increasing the appropriate counter by 1. This completes the motion of one particle. All the steps are then repeated for another particle with a different set of random bits. This is repeated for as many particles as needed to produce good statistics for the final distribution.

As a practical procedure, we can carry out the calculation for some large number,

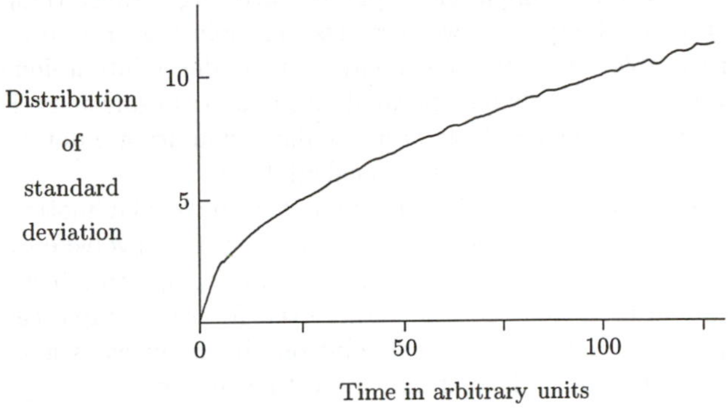

Figure 7-7: Standard deviation of the spatial distribution of Brownian particles as a function of time. Since the variance grows linearly with time, the standard deviation increases as the square root of time.

like a thousand, at a time. After each thousand particles have gone through n random steps, the contents of all the counters are checked to see if there are enough particles in them for a meaningful statistical analysis of the distribution. When this is true, we can proceed to the next step of calculating the mean and variance of the distribution. Otherwise, the calculation is repeated for another thousand particles. As mentioned earlier, the distribution gets broader as the number of time steps taken by each particle is increased. To maintain the same accuracy, larger numbers of particles are required if we wish to examine distributions with longer time spans.

The behavior of a collection of one-dimensional Brownian particles is well known. The sharp initial distribution of a delta function is quickly dispersed, as particles drift to both sides. After a few time steps, the distribution becomes an essentially normal one centered at the origin. A typical result after 64 time steps is shown in Fig. 7-6. The distribution becomes more diffused with time, as more and more particles move away from the origin. The shape of the distribution, however, remains normal. To demonstrate the fact that the standard deviation of the distribution increases with the square root of time, the results obtained in a Monte Carlo calculation are displayed in Fig. 7-7.

In general, the primary purpose of a Monte Carlo simulation is not to repeat calculations that can be done easily by analytical methods, such as the one-dimensional random walk here. In fact, we have tailored our example to mimic the conditions under which the analytical results are derived. As far as the numerical work is concerned, many of the simplifications are not essential. For this reason, it is possible for us to carry out much more realistic simulations of the true physical situation. For example, real molecular motions depend on the probability of colliding with other particles, instead of always moving a constant distance $\bar{\ell}$ in time τ. Furthermore, the collisions are not necessarily simple hard-sphere scattering. Many things can take place, including the possibility that two particles may join together to form a single molecule. Under suitable conditions, a gas may even condense into a liquid. Some of these considerations are not easily incorporated into analytical solutions. For this reason, many of the recent understandings of molecular dynamics are obtained through computer simulations involving Monte Carlo calculations.

There are also important limitations in simulating molecular motion on a computer. For example, any realistic collections of gas molecules involve on the order of 10^{23} (Avogadro number) particles. Obviously, we cannot hope to follow the motion of each one when the total number is so large, even for very simple cases. Nevertheless, simulations compliment analytical solutions in many ways and help us in understanding a variety of complicated but interesting problems.

7-3 Data simulation and hypothesis testing

In Chapter 6, we examined several methods of carrying out least-squares fits to data. Our primary aim there was to find the values of the parameters in an expression so as to obtain the best possible description of the data. From a slightly different point of view, the same calculations may also be regarded as a test of our idea, or hypothesis, of how the data should behave. If the expression used to fit the data is based on a sound understanding of the physical system, the uncertainties in the parameters obtained will be small, reflecting mainly the uncertainties in the measurements. On the other hand, if we made an incorrect interpretation of the system under investigation, our fit will be poor and the value of the χ^2 will be much larger than that resulting from the uncertainties in the measurements alone. In this way, the goodness of fit serves also as an indication of how well the expression used in the fit represents the true nature of the system we are studying. However, the more likely situation is one in which results are not clear-cut and a more quantitative criterion is required to base on our judgment.

Confidence region and confidence level A statistical measure of whether a hypothesis should be accepted or rejected is usually stated in terms of the *confidence region*, or confidence interval, at a given *confidence level.* Each of these terms has a specific meaning in statistics. In the place of formal definitions, it is perhaps more useful here to illustrate what they mean using an example.

Consider the case of measuring the mass of a neutrino. In the report of Lubimov and coauthors,[34] the mass μ_ν of an electron antineutrino $\overline{\nu}_e$ is given as 14 eV$\leq \mu_\nu c^2 \leq$ 46 eV at a confidence level of 99%. This means that, if the same experiment to measure the rest mass energy $\mu_\nu c^2$ is performed 100 times, we expect, on the average, the results of 99 measurements will be in the region between 14 and 46 eV.[1]

[1] The electron-volt, abbreviated eV, is a convenient unit of mass μ for subatomic particles in terms of their rest mass energy μc^2, where c is the speed of light. A rest mass energy of 1 eV is equivalent to a mass of 1.78×10^{-36} kg. The mass of an electron μ_e is 9.11×10^{-31} kg or $\mu_e c^2 = 0.511 \times 10^6$ eV.

Let us try to see the meaning of this statistical statement in terms of what one can actually do in an experiment. As emphasized several times earlier in Chapter 6, q_{exp}, the value obtained in a measurement, is not necessarily equal to the true value q_{true} for the quantity we are measuring. In fact, if we repeat the measurement many times, it is likely that the outcome will be slightly different each time. Let us use $p_{\text{exp}}(q)$ to represent the probability distribution of the measured values. Our interest here is to infer from the measured values q_{exp} and their distribution $p_{\text{exp}}(q)$ the true value q_{true} for the quantity q. From a statistical point of view, we can take q_{true} as the mean value of $p_{\text{exp}}(q)$, obtained after a very large number of measurements, as we did in Eq. (6-1). In practice, however, it is not possible to carry out the large number of measurements required to give a reasonable representation of $p_{\text{exp}}(q)$. Under such circumstances, one must try to do the best one can to arrive at an estimate of the true value by some other means.

Before we see how Monte Carlo calculations can help us to acquire some additional confidence in our measured results, let us see first the mathematical relation between confidence level and confidence region. Pretend for the moment that the probability distribution $p_{\text{exp}}(q)$ is known. The probability of obtaining a measured value q_{exp} in the region $[q^\ell, q^h]$ is given by the integral

$$P(q^\ell \leq q_{\text{exp}} \leq q^h) = \int_{q^\ell}^{q^h} p_{\text{exp}}(q)\, dq. \tag{7-32}$$

In practice, the problem one faces is the inverse. The probability distribution $p_{\text{exp}}(q)$ is unknown and cannot be established without carrying out a large number of measurements. If it is not possible to repeat the experiment many times, the best we can do is to have an estimate of q_{true} from the measured values we have in hand. For this purpose, we select first a confidence level, or *confidence coefficient* P, for example 99%, and ask for the (confidence) interval $[q^\ell, q^h]$ in which the integrated probability $P(q^\ell \leq q_{\text{exp}} \leq q^h)$ is equal to 0.99. For the probability distribution, we can use the measured distribution $p_{\text{exp}}(q)$, if it is available, or make the best estimate we can for it. Usually, a combination of the two approaches is used. For example, we know that the neutrino mass μ_ν cannot be a negative quantity. As a result, we can restrict the distribution $p(q)$ to the region with $q \geq 0$, as we shall see in an example later.

Alternatively, we can ask the same question in terms of the functional relation between the set of measured values $\{y_i\}$ and some independent variable $\{x_i\}$. Since we do not have a complete knowledge of the relation between y and x, the expression contains a number of adjustable parameters $\boldsymbol{a} \equiv (a_1, a_2, \ldots, a_m)$ in the form

$$y = f(a_1, a_2, \ldots, a_m; x).$$

This is the type of question we asked ourselves in Chapter 6 on least-squares calculations. If $f(\boldsymbol{a}; x)$ is used to fit to a number of measured values and the parameters $\boldsymbol{a}$ are determined up to some uncertainties $\delta\boldsymbol{a}$, we may wish to know the probability of getting a particular set of results for $\boldsymbol{a}$ in the interval $[\boldsymbol{a}^\ell, \boldsymbol{a}^h]$. Analogous to

Eq. (7-32), we have

$$P(a_1^\ell \leq a_1 \leq a_1^h, a_2^\ell \leq a_2 \leq a_2^h, \ldots, a_m^\ell \leq a_m \leq a_m^h; x)$$
$$= \int_{a_1^\ell}^{a_1^h} \int_{a_2^\ell}^{a_2^h} \cdots \int_{a_m^\ell}^{a_m^h} f(a_1, a_2, \ldots, a_m; x)\, da_1 da_2 \cdots da_m.$$

The confidence region in this case is a volume V in the m-dimensional parameter space, made of the product $\Pi_{i=1}^m (a_i^h - a_i^\ell)$. It is also possible in this case to define the confidence volume in terms of contours of constant χ^2-value. However, we shall not pursue this particular alternative here.

The relation between confidence limit and confidence region becomes somewhat complicated when the distribution of one or more of the parameters extends over into the unphysical region. For example, in an experiment, the measured value of the neutrino rest mass energy may turn out to be 30 eV with an uncertainty of 40 eV. If the probability distribution is a normal one, there is the probability of

$$\frac{1}{40\sqrt{2\pi}} \int_{-\infty}^{0} e^{-\{(x-30)/40\}^2/2} dx = 0.23$$

or 23% that the true value of $\mu_\nu c^2$ is less than zero. A straightforward interpretation of the results means that, if we repeat the experiment 100 times, it is likely that in 23 of the 100 measurements the value of μ_ν will turn out to be negative. In reality, mass is a positive quantity and a negative value should be rejected.

It is clear that a naive statistical approach is unsatisfactory here. Instead, it may be more correct to exclude the unphysical region from our probability distribution. There is no clear guidance in probability theory for the proper procedure to follow in this case. In physics, it is often recommended that, instead of Eq. (7-32), the relation between confidence limit and confidence region should be given by

$$P(q^\ell \leq q_{\rm exp} \leq q^h) = \frac{\int_{q^\ell}^{q^h} p_{\rm exp}(q)\, dq}{\int_{\rm phys.reg.} p_{\rm exp}(q)\, dq}.$$

In this case the denominator on the right side normalizes the probability distribution $q(p)$ to unity within the physical region (and it is implicit that both q^ℓ and q^h are within the physical region).

Example of neutrino mass measurement There are several ways that Monte Carlo calculations may be used to test our hypothesis about the distribution of measured values. Let us again use neutrino mass measurements as an example. Most of the experiments to date involve the β-decay of tritium (t), a nucleus made of two neutrons and one proton,

$$t \rightarrow {}^3\text{He} + e + \overline{\nu}_e.$$

The nucleus in the final state is ^{3}He, a light isotope of helium made of two protons and one neutron. In addition, an electron (e) as well as an electron antineutrino ($\overline{\nu}_e$)

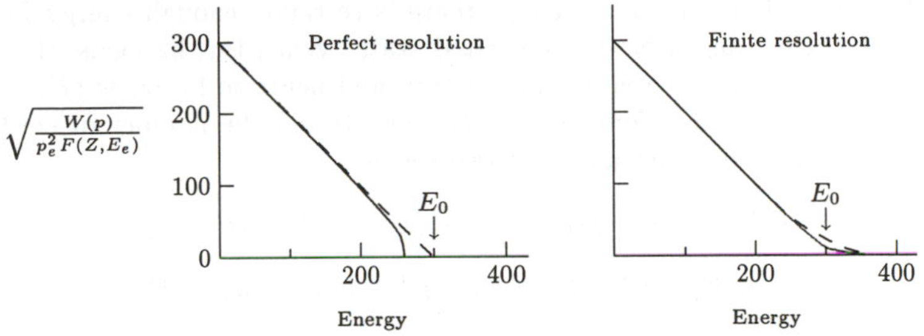

Figure 7-8: Schematic diagram of a Kurie plot given by Eq. (7-34) for neutrino mass $\mu_\nu \neq 0$ (solid curves) and $\mu_\nu = 0$ (dashed curves).

are emitted. The total amount of energy released in the decay, usually referred to as the end-point energy, is 18,578 eV, and is shared between the kinetic energies of the electron and the neutrino, and the excitation energy of the helium atom.

In the absence of uncertainties due to the apparatus used for the measurement, the number of electrons emitted with energy E_e is given by

$$\begin{aligned} \mathcal{W}(E_e) &= CF(Z,E_e)P_e^2(E_0-E_e)\{(E_0-E_e)^2-(\mu_\nu c^2)^2\}^{1/2} \\ &\xrightarrow[\mu_\nu\to 0]{} CF(Z,E_e)P_e^2(E_0-E_e)^2 \end{aligned} \qquad (7\text{-}33)$$

where C is a constant depending on the nuclear structure of the initial and final states involved in the decay and $F(Z,E_e)$ is the Fermi function that expresses the influence of the Coulomb field of the final nucleus on the electron emitted. The momentum P_e is that of the electron and is related to its kinetic energy E_e through the relation $P_e^2/2\mu_e \approx E_e$. If the decay goes to the ground state of the ^{3}He atom, the maximum energy an electron can have is E_0 if $\mu_\nu = 0$.

For all the known nuclear β-decays, E_0 is much greater than any reasonable value expected of the rest mass energy of the neutrino. As a result, the expression for $\mathcal{W}(E_e)$ is often approximated by the final form of Eq. (7-33), with the μ_ν term ignored. In this approximation, a plot of $\sqrt{\mathcal{W}(E_e)/(F(Z,E_e)P_e^2)}$ as a function of E_e gives a straight line that intercepts the horizontal axis at electron energy $E_e = E_0$, as shown in Fig. 7-8. This is known as a Kurie plot (for more details, see standard textbooks on nuclear physics, such as Wong,[59]). A finite neutrino mass is indicated by a slight departure from the straight line near the end-point energy, as can be seen from Eq. (7-33) and indicated by the solid curves in Fig. 7-8.

In nuclear reactions, the energies involved are usually much higher than the excitation energies of the neutral atoms involved. As a result, atomic effects do not enter into any of the considerations. This is not true for neutrino mass measurements using tritium β-decay. The value of $\mu_\nu c^2$ is expected to be on the order of 30 eV or less, and there is an excited state of the ^{3}He atom at energy $E^* = 43$ eV above the ground state. In the decay of the tritium nucleus, there is certainly enough energy for the final state of the ^{3}He atom to be in the excited state. When this happens, the total amount of kinetic energy available to the electron and neutrino becomes $(E_0 - E^*)$, instead of E_0. We must, therefore, modify Eq. (7-33) and the number of electrons emitted with energy E_e becomes a sum of two terms:

$$\begin{aligned}\mathcal{W}(E_e) &= CF(Z, E_e)P_e^2\big[p_1(E_0 - E_e)\{(E_0 - E_e)^2 - (\mu_\nu c^2)^2\}^{1/2} \\ &\quad + p_2(E_0 - E^* - E_e)\{(E_0 - E^* - E_e)^2 - (\mu_\nu c^2)^2\}^{1/2}\big] \qquad (7\text{-}34)\end{aligned}$$

where p_1 and p_2, with the normalization condition $(p_1 + p_2) = 1$, are the probabilities of having, respectively, the ground and excited state of the ^{3}He as the final state. Unfortunately, our understanding of atomic physics at the moment is not sufficient for us to calculate p_1 and p_2 reliably, and this becomes one of the sources of uncertainties in deducing the value of μ_ν.

Experimentally, one can only detect the electron emitted in the decay, as neutrinos interact only weakly with matter. If our interest is in the mass of the neutrino, it must be deduced indirectly from the shape of the distribution for the number of electrons emitted at a given energy E_e. Since the expected neutrino mass is small, its influence on the electron distribution is noticeable only where $(E_e - E_0) \approx 0$, as can be seen from Eq. (7-33). The most important region in a measurement is therefore around electron energy $E_e \sim E_0$. Unfortunately, this is also the place where the number of electrons emitted in the β-decay approaches zero, as can be see from the Kurie plot. For this reason, it is difficult to accumulate enough counts for a sound statistical analysis. Furthermore, any limitations in the measuring equipment tend also to skew the observed distribution in such a way that it can be easily confused with effects due to finite μ_ν. A typical set of measured results in the region around the end-point energy is shown schematically in Fig. 7-9.

Because of finite instrumental resolution, the number of electrons $N(P)$ observed with momentum P is modified from that given by Eq. (7-34). The actual form depends on the particular experimental setup and we shall not be concerned with this question here (for details, see Lubimov and coauthors,[34]). In the analysis of Lubimov and coauthors,[35] the following expression is used:

$$N(P_i) = A\sum_k \mathcal{W}(E_k)\{1 + \alpha(P_0 - P_k)\}R_{i,k} + \phi$$

where A is a normalization constant and $\mathcal{W}(E_e)$ is given by Eq. (7-34). The other factors are related to the detection equipment: α is a correction constant, $R_{i,k}$ is the resolution function or the probability for an electron emitted with momentum P_k and detected as an electron with momentum P_i, and ϕ is the background. Many of these factors cannot be determined exactly, either experimentally or from our theoretical knowledge. They, together with the value of μ_ν, are treated as parameters to be determined by a least-squares fit to the actual data obtained. Since these parameters

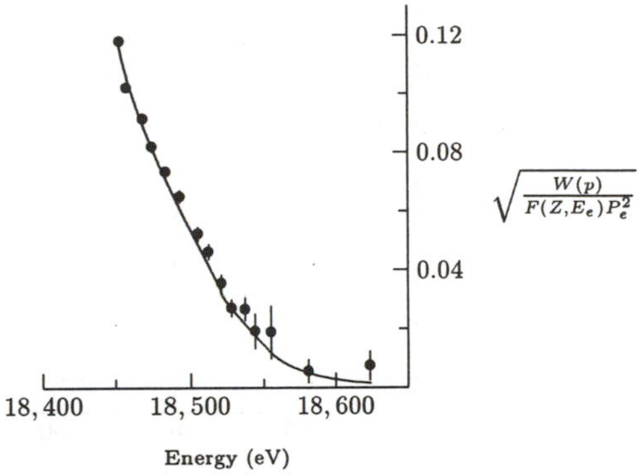

Figure 7-9: Schematic diagram of a Kurie plot for tritium β-decay. The solid line represents the case if the neutrino has a small mass of ~ 30 eV/c^2.

reflect only the equipment and the type of sample (including atomic effects) used, they are independent of the value of neutrino mass of interest to us. Except for statistical fluctuations, these parameters will remain constant if we repeat the experiment using the same type of sample and the same apparatus.

Estimate of confidence region using Monte Carlo calculation Since the experiment is very difficult to perform, it is not easy to accumulate a large number of measured values. Under such circumstances, Monte Carlo calculations may be used to test the reliability of the value of μ_ν obtained. Our interest here is to find the likelihood of obtaining the value of μ_ν deduced from the data. For this purpose, we can simulate on a computer the decay of the same number of tritium nuclei as found in the actual experiment. For each decay, we have no way of determining the energy E_k and the momentum P_k of the electron emitted other than that they must satisfy energy and momentum conservation laws. For the simulation, we can replace E_k and P_k with random numbers of the correct distributions and subject them to constraints imposed by the conservation laws. Similarly, the chance of reaching either one of the two possible final ^{3}He atomic states is also random, except that their probabilities are given by p_1 and p_2, with the values of p_1 and p_2 determined by the most reliable calculations available. The electron then goes through the detection apparatus and is recorded as having a momentum P_i, with a probability given by the resolution function $R_{i,k}$, corrected by the factor $\alpha(P_0 - P_k)$. After the right number of decays have been generated, we can add in the background in the form of random numbers having the distribution ϕ. The result is a set of Monte Carlo data. A set of such values may be treated as if it were obtained from an experiment, and should be subjected to the same analysis as the original data. When this is done, a value of μ_ν is obtained. In general, the result will be different from that deduced from the experiment, as the

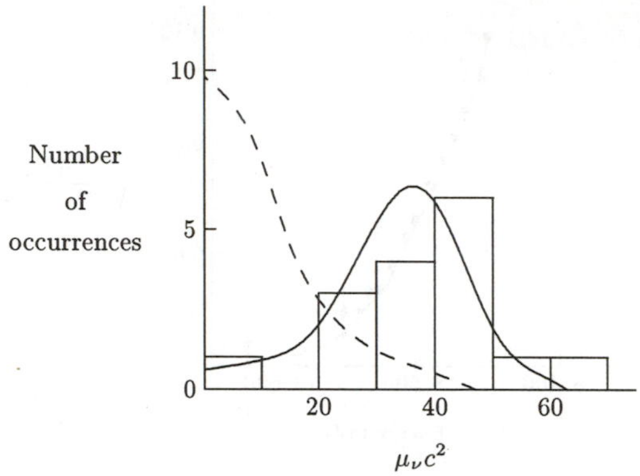

Figure 7-10: Schematic diagram showing the distribution of neutrino mass obtained from a Monte Carlo simulation of the experimental data for $\mu_\nu c^2 = 35$ eV (solid curve) and $\mu_\nu c^2 = 0$ (dashed curve). (Adapted from Lubimov and coauthors.[34])

individual pieces of data are not identical to those obtained in the actual experiment. If the results of several different sets of this type of data are accumulated, we have a distribution of the possible values of μ_ν. Figure 7-10 gives a plot of such a distribution made with the values given in the article of Lubimov and coauthors.[34] By setting a confidence level of 99%, the mass of electron antineutrino was determined to be in the interval 26 eV$\leq \mu_\nu c^2 \leq$ 46 eV in this way.

The procedure outlined above for testing our hypothesis about a given physical situation and to establish a confidence region is a general one and can be applied to many similar cases. The approach may be summarized in the following way:

(1) The physical system is described by a set of parameters $\boldsymbol{a}_{\text{true}}$ of unknown values.

(2) Through measurement we obtain a set of data points $\boldsymbol{D}_{\text{exp}}$.

(3) Using least-squares techniques, we obtain from $\boldsymbol{D}_{\text{exp}}$ a set of best-fit values for the parameters $\boldsymbol{a}_{\text{fit}}$.

(4) Simulate the experimental data with $\boldsymbol{a}_{\text{fit}}$ using Monte Carlo techniques. Each one of these simulations gives us a set of "data" $\boldsymbol{D}_{\text{sim}}(i)$.

(5) Least-squares analyses are performed on the simulated data $\boldsymbol{D}_{\text{sim}}(i)$. From each group of simulated results, a set of values $\boldsymbol{a}_{\text{fit}}(i)$ is obtained for the parameters.

(6) From the distribution of $\boldsymbol{a}_{\text{fit}}(i)$ for $i = 1, 2, \ldots$, the confidence interval of $\boldsymbol{a}$ is established for a given confidence limit.

In the above steps, it is understood that the fitting function and the least-squares procedure applied to the simulated data are the same as those used on the real data. Since Monte Carlo calculations are often easier to carry out than experiments, the method becomes a powerful way to provide a meaningful interpretation of the actual data. In particular, the technique is indispensable in situations where the experiment cannot be repeated.

In this section, we have essentially followed the traditional approach to statistical treatment of data. For the most part, these methods were developed before the birth of modern computers. As a result, there is a reluctance to involve extensive computation. One evidence is that statistical measures applied to the analyses are usually restricted to those that have analytical results. With the widespread use of computers, many other methods of statistical analyses are now possible. We shall not go into these new methods here. For an introduction, one can start with, for example, the *Scientific American* article on *Computer-Intensive Methods in Statistics* by Diaconis and Efron.[18]

7-4 Percolation and critical phenomena

An interesting feature in many physical systems is the long-range order induced by short-range interactions. For example, the typical range of forces acting between molecules is on the order of the diameter of a single molecule, and yet collective behavior can take place involving as many as millions of molecules. A good example is ferromagnetism. It is well known that many atoms and molecules possess weak magnetism resulting from intrinsic spin and orbital motion of the electrons. The magnetism observed in ferromagnetic materials, iron, nickel, and cobalt, is, however, much stronger and requires the alignment of a large number of atoms. In other words, many atoms must act in a coherent or collective manner in order to produce the observed strength. On the other hand, the interaction between the atoms in ferromagnetic materials has only a range on the order of 10 angstroms, not much larger than the diameter of each molecule. Similar phenomena are observed in many other types of collective behavior where the role played by the interaction is a minor one and the physics is determined primarily by statistical considerations. A class of such phenomena is *percolation*, a special type of phase transition that is observed in condensed matter and other many-body systems.

Phase transition In statistical mechanics, we are usually dealing with systems made of large numbers of individual atoms or molecules. The influence of the surroundings is characterized in terms of such parameters as temperature, pressure, and applied magnetic field. Because of the dominance of large numbers, typically on the order of Avogadro number, the properties of these systems usually vary smoothly as functions of the external parameters. The exceptions occur at phase transitions. Typical examples are water changing from liquid to gas (steam), a ferromagnetic material being raised to temperatures above its Curie point and losing its special magnetic property, and a superconductor turning into a normal one.

The changes in the properties occurring during a phase transition can often be characterized by a single parameter, known as the *order* parameter. For example, for ferromagnetic materials, we may use as the order parameter the magnetic susceptibility

$$\chi_m = \frac{\partial M}{\partial H} \tag{7-35}$$

where M is the magnetization and H is the magnetic field. Alternatively, we may use the specific heat

$$C_v = \left(\frac{\partial U}{\partial T}\right)_v \tag{7-36}$$

where U is the internal energy and T is the temperature. The term order parameter comes from the observation that quantities, such as χ_m and C_v, characterize the change from an ordered state, such as the ferromagnetic state or the superconducting state of a material, to a disordered one.

Critical phenomena For some materials, the order parameter is discontinuous across a phase transition. This is the case of a first-order phase transition. Our interest here is mainly in second-order phase transitions in which the order parameter itself is continuous at the phase transition but its first derivative is not.

A simple illustration, given by Schulman and Seiden,[48] is the spread of a hypothetic disease, "percolitis." Consider a very special group of N individuals living in a completely isolated community. For simplicity, we shall assume that each individual makes one and only one contact each day with every other individual in the community. Furthermore, we shall assume that percolitis has an incubation period of exactly one day and lasts also only for a day. The way the disease is spread is by direct contact between two individuals and the probability of catching the disease from an affected individual is p in each contact with such an individual. At a given time t, measured in units of days, let the number of individuals afflicted with percolitis be $n(t)$. The probability that any one individual in the community is suffering from the disease is given by

$$\rho(t) = \frac{n(t)}{N}.$$

The probability for any one not having percolitis at time t is therefore $\{1 - \rho(t)\}$.

Our immediate interest here is the probability for an individual to catch percolitis. In a contact with a diseased individual, the probability not to catch percolitis is $(1-p)$. Among the $(N-1)$ contacts one makes with different individuals in each day, we need only be concerned with those involving individuals afflicted with percolitis. Since there are $n(t)$ such individuals in the community, the fraction of the community not catching the disease is $(1-p)^{n(t)}$. This is then the fraction of the population not afflicted with percolitis on day $(t+1)$. By definition, this is equal to $\{1 - \rho(t+1)\}$. As a result, we have the equation

$$1 - \rho(t+1) = (1-p)^{n(t)}. \tag{7-37}$$

The relation may be used to solve for $\rho(t)$. For this purpose, let us define

$$x \equiv Np$$

and, on making use of the following infinite series expansion of the exponential function, as we did earlier in Eq. (6-15),

$$e^x = \lim_{s\to\infty}\left(1+\frac{x}{s}\right)^s$$

the right side of Eq. (7-37) may be put into the form

$$(1-p)^{n(t)} = \left(1-\frac{x}{N}\right)^{n(t)} = \left(1-\frac{x\rho(t)}{n(t)}\right)^{n(t)} \approx e^{-x\rho(t)}.$$

On substituting this result to the right side of Eq. (7-37), we obtain

$$\rho(t+1) = 1 - e^{-x\rho(t)}. \tag{7-38}$$

If a continuous process is used for contracting the disease rather than our simple step function in t, the equivalent relation will be in the form of a differential equation. For our purpose here, it is not worthwhile to bring in the additional complication in mathematics involved in deriving the more proper result.

We wish now to find out the fraction of the population that will ultimately be afflicted with percolitis at the same time. This is given by

$$\rho_{\max} = \lim_{t\to\infty}\rho(t).$$

At $t=\infty$, Eq. (7-38) takes on the form

$$\rho_{\max} = 1 - e^{-x\rho_{\max}}. \tag{7-39}$$

There are many ways to solve this equation. It is instructive for us to make use of a graphical approach. The behavior of the left side of Eq. (7-39) as a function of $\rho_{\max}$ is simple and given by

$$y_1 = \rho_{\max}.$$

The behavior of the right side of the equation has the form

$$y_2 = 1 - e^{-x\rho_{\max}}.$$

The value of $\rho_{\max}$ that satisfies both equations depends also on x, related to the probability of catching percolitis on each contact. The forms of y_2 for three typical values of x are shown in Fig. 7-11.

For a given x, the solution of Eq. (7-39) occurs at the intersection of $y_1(\rho_{\max})$ with $y_2(\rho_{\max})$. If $x \leq 1$, the value of y_2 is always less than y_1, as can be seen in Fig. 7-11. In this case, there is only one possible solution at $\rho_{\max}=0$. This means

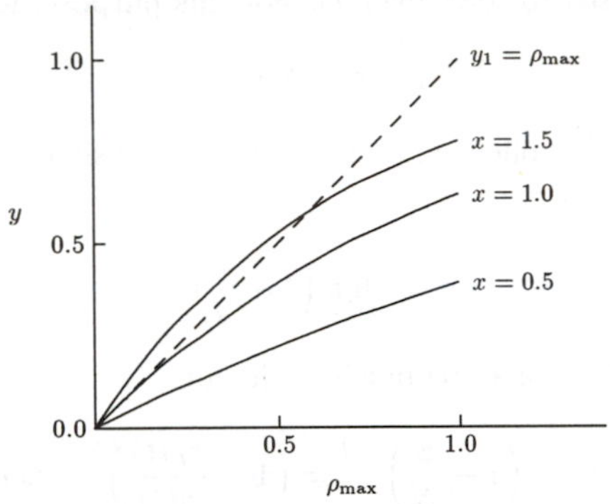

Figure 7-11: Illustration of a second-order phase transition given by Eq. (7-39). The solution is given by the intersection of $y_1(\rho_{\max}) = \rho_{\max}$ (dashed line) and $y_2(\rho_{\max}) = 1 - \exp(-x\rho_{\max})$ (solid curves). For $x \leq 1$, the only solution is $\rho_{\max} = 0$. For $x > 1$, there is a second solution that increases with x.

that percolitis cannot spread if p is less than $1/N$. That is, it will eventually die out in the community. However, for $x > 1$, the value of y_2 starts off larger than $y_1 = \rho_{\max}$ and falls eventually below the value of y_1 as $\rho_{\max}$ approaches the maximum value of 1. We now have a solution of Eq. (7-39) for $\rho_{\max} > 0$. The actual value depends on x, but we are not particularly concerned with the precise answer here. The important thing for us to realize is that we have a critical point at $x = 1$. Let us refer to this value of x as x_c. Below $x = x_c$, percolitis will eventually die out in the community, and above $x = x_c$, there will ultimately be a stable number in the community suffering from percolitis on any particular day. At $x = x_c$, we have a phase transition. The order parameter here is $\rho_{\max}$ and its value as a function of x is continuous at $x = x_c$. However, the first derivative $\partial\rho_{\max}/\partial x$ is discontinuous at $x = x_c$. The phase transition is, therefore, second order.

The example is a nice and effective way to illustrate the phenomenon of phase transition. In spite of its simplicity, the model is fairly close to what is observed in nature. In fact, Schulman and Seiden were able to demonstrate by computer simulation that the structure of spiral galaxies can come from a percolation phase transition if the stars are rotating with different velocities. This is a departure from the traditional points of view based on the dynamics of the system.

Critical exponents A critical phenomenon is usually associated with the divergence of some physical quantity. For our percolitis example, the quantity is the relaxation time

$$\Delta(t) \equiv \rho_{\max} - \rho(t)$$

the time for the disease to reach its equilibrium level in the community. It can be shown (L.S. Schulman and P.E. Seiden,[47]) that, as $\rho(t) \to \rho_{\max}$,

$$\Delta(t) \sim \text{constant} \times e^{-t/\xi}$$

where ξ is a time constant, generally known as the *correlation length.* Near the critical point, we have

$$\xi \sim |x - x_c|^{-1}.$$

In general, the exponent is not a simple integer as we have here. Instead, it often has the form

$$\xi \sim |x - x_c|^{-\nu}. \tag{7-40}$$

The quantity ν is known as the critical exponent.

For most critical phenomena studied, the phase transition takes place at a particular temperature, T_c, known as the critical temperature. For different systems, the value of T_c may be different, but the behaviors at temperatures near T_c are often very similar. For example, below T_c, the magnetization of a ferromagnetic material decreases with increasing temperature and vanishes at $T = T_c$. At $T \sim T_c$, the magnetization is given by a simple power law,

$$\mathcal{M} \propto (T_c - T)^{\beta}.$$

Similarly, at the critical temperature, the magnetization is proportional to the magnetic field following the power law

$$\mathcal{M} \propto H^{1/\delta}.$$

At zero magnetic field ($H = 0$), the specific heat C and magnetic susceptibility χ_m are given by

$$C \propto \begin{cases} (T - T_c)^{-\alpha} \\ (T_c - T)^{-\alpha'} \end{cases} \qquad \chi_m \propto \begin{cases} (T - T_c)^{-\gamma} & \text{for } T > T_c \\ (T_c - T)^{-\gamma'} & \text{for } T < T_c. \end{cases}$$

For iron, the values are found to be $\alpha = \alpha' = 0.12 \pm 0.01$, $\beta = 0.34 \pm 0.02$, and $\gamma' \approx \gamma = 1.333 \pm 0.015$ at the critical temperature of $T_c = 1044$ K, and, for nickel at $T_c = 631.58$ K, $\delta = 4.2 \pm 0.1$. For a more comprehensive introduction to the subject of critical phenomena, the interested reader should consult specialized books such as *Modern Theory of Critical Phenomena* by Ma,[36] *Introduction to the Renormalization Group and to Critical Phenomena* by Pfeuty and Toulouse,[40] and *Statistical Mechanics* by Huang.[28]

One reason for critical phenomena to be of interest is the fact that critical exponents are observed to be essentially the same for a variety of materials. In general, these materials may have quite different critical temperatures. However, their behaviors are very similar to each other near their critical points. This is known as *universality* and some quantities are more universal than others. The reason why there should be any difference is an intriguing question as well. Since the values of these exponents are not integers, they cannot come from some simple analytical relations. Furthermore, since the behavior near a critical point is singular in nature, standard perturbative methods are not applicable and new methods of investigation must be explored. For this reason, Monte Carlo calculations and computer simulations have become an important tool in the study of critical phenomena.

Percolation One way to make computer simulations of critical phenomena is through percolation studies. The usual meaning of percolation is the process by which water passes through a porous body, such as a filter. In physics, it is best described by the example of a collection of random, two-dimensional clusters of conductors. Consider two parallel conducting bars, each of length L and separated from each other by a distance, for simplicity, equal to their length. Let us divide the area between the top and bottom conductors into $(N \times N)$ small squares. Electric current can flow from one of these squares to its neighbor if both are occupied by conductors. For simplicity, we shall consider that any square not occupied by a conductor is filled with an insulator. A current can flow between two conductors that are located side by side, like this ■■, and on top of each other, like this █. However, diagonal passages, such as ▀▄ and ▄▀, are not allowed.

For our present purpose, a group of conductors that are joined together so that charge carriers can flow between them is called a *cluster*. In percolation studies we are interested in the formation and size of such clusters. In particular, we want to know whether a current can flow between the top and bottom conducting bars if the space in between is filled randomly with small square conductors.

The answer depends on the number of sites occupied by conductors. In other words, it is related to the probability p that a square is occupied by a conductor. An illustration is given in Figs. 7-12 and 7-13 for a case of $N = 60$. In Fig. 7-12*a*, the (60×60) squares in between the two bars are filled randomly with probability $p = 0.1$. That is, one-tenth of the lattice sites are occupied by conductors and the rest are filled with insulators. It is easy to see that most of the conductors are in isolated clusters and there is no possibility for a charge carrier to move from the top bar to the bottom bar. One way of classifying such a state is to say that the average size of conducting clusters is small. When p is increased to 0.3, that is, 30% of the sites are occupied, the cluster sizes become much larger. However, no continuous path can be found between the top and bottom, as can be seen by looking at Fig. 7-12*b*. On the other hand, a careful examination of the $p = 0.6$ result, shown in Fig. 7-13*a*, reveals that there are several such paths available. In this case, charge carriers can now "percolate" between the top and the bottom bars. When the value of p is increased to 0.9, the whole space is a single large cluster (with 10% holes), as can be seen in Fig. 7-13*b*.

If the size of our two-dimensional lattice is infinite instead of (60×60), the critical probability, or percolation threshold, is known to take place at $p_c = 0.59275$. For finite-size lattices, the transition is not a sharp one, and this is one of the limitations of computer simulation, as we shall see in more detail later. The onset of percolation

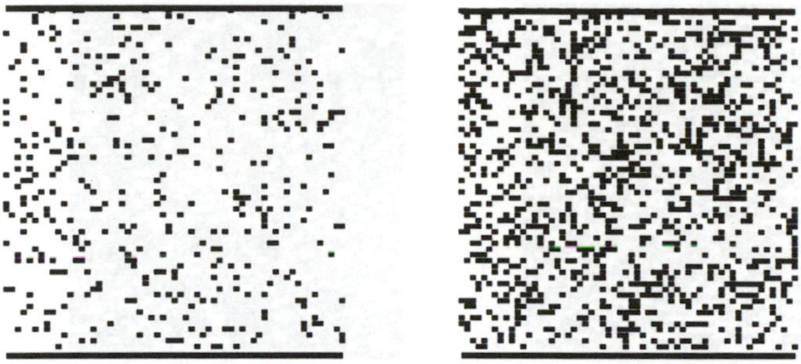

(a) (b)

Figure 7-12: Random conductors, indicated as black squares, filling up a square lattice with probability $p = 0.1$ in (a) and $p = 0.3$ in (b).

may be regarded as a phase transition. For our random conductor example, the space in between the two conducting bars changes from an insulator to a conductor when there is at least one cluster extending from one bar to the other. Instead of a random collection of conductors, we can imagine the same phenomenon taking place in a more realistic situation in the form of an idealized polymerization process, whereby (noninteracting) molecules coagulate into large molecules. The probability p in this case is analogous to temperature T in critical phenomena we encountered earlier. In this way, many behaviors of a system near the critical point may be examined through percolation studies.

In addition to our simple conducting square example, many other types of lattices can also be simulated on a computer. For example, instead of allowing conduction between only neighboring sites in a square lattice, we can have lattices of other shapes such as a triangle. In the place of two-dimensional lattices, we can construct three-dimensional ones. Our square lattice can be generalized into a cubic one if, in addition to conduction with sites to the immediate left, right, top, and bottom, we include also front and back. In fact, for computer simulation, we can work with arbitrary dimensions, a question of interest in many situations but difficult to carry out in controlled experiments.

There are some subtle differences between percolation and critical phenomena, but we shall not enter into a discussion of such finer points here. Our main purpose here is to give an introduction to the role of computer simulation in percolation studies, using a square lattice in two dimensions as the example. For a more comprehensive review the reader should consult monographs such as *Introduction to Percolation Theory* by Stauffer[52] and the series on *Phase Transitions and Critical Phenomena* edited by Domb and Lebowitz.[19]

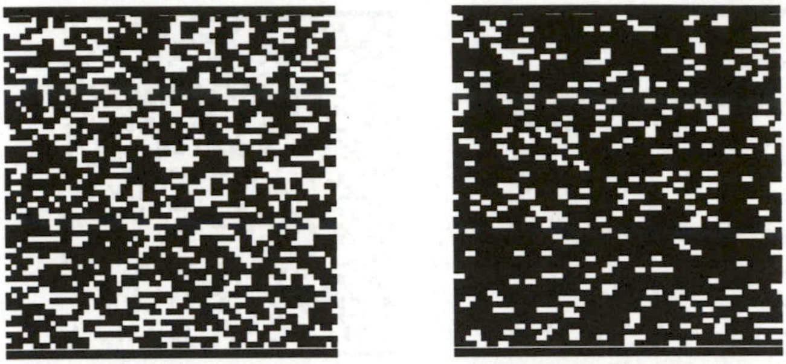

(a) (b)

Figure 7-13: Random conductors, indicated as black squares, filling up a square lattice with probability $p = 0.6$ in (a) and $p = 0.9$ in (b).

Simulation of a square lattice Let us examine some of the details in carrying out a computer simulation of percolation using a square lattice. There are three steps involved. The first is to divide the area into a lattice of small squares. For simplicity, we shall take the area to be square with each side divided into N equal parts. The size of N depends on the total number of little squares we wish to have. For our illustrative examples above, a value of $N = 60$ was used. This is sufficiently small so that a personal computer can handle the calculations involved with ease. On the other hand, we cannot hope to reproduce all the details of realistic situations involving the order of 10^{23} particles. We shall soon see that, even for a larger N of 500, finite size effects remain significant.

The second step in the simulation is to put small square conductors randomly with probability p into the $(N \times N)$ lattice sites. This may be done by starting from the top row and going through each of the N sites one by one, from the left to the right. For each square, a random number x with uniform distribution in the interval $[0, 1]$ is generated. If $x \leq p$, a conductor is put into the site, and if $x > p$, the site is filled with an insulator.

The third step is to analyze the result. We shall be concerned with only two questions here: the sizes of the clusters and whether there is a contagious path of conductors from the top to the bottom. If the lattice is small, such as the (60×60) example used in Figs. 7-12 and 7-13, it is perhaps possible to check the conduction path by hand. In more realistic applications, where the lattices are much larger, a better method is needed. In fact, this is the most involved part of the simulation.

We shall describe a method based on the technique published by Hoshen and Kopelman[27] whereby the analysis of the result is carried out along with the process of filling the lattice. To group occupied sites into clusters, we shall assign a label to each cluster. All sites belonging to the same cluster bear the same label. For example, the first cluster that appears as we go through the lattice sites is numbered 1, the second 2, and so on. To simplify the checking process, the insulator sites can be assigned the label either 0 or some number X that is larger than any possible label for an occupied site. We shall adopt the latter choice as it is slightly more efficient.

The assignment of a cluster label to a lattice site is carried out as the site is filled with either a conductor or an insulator. For a two-dimensional lattice, it is possible to find out whether a site newly occupied by a conductor is a member of an existing cluster by checking whether the site to its left and just above are conductors. There are three possibilities. The first is that both neighbors are unoccupied. Here, we have the beginning of a new cluster. The counter for the total number of clusters is increased by 1 and the value is assigned as the cluster number of the new site. The second possibility is that either the left or the top cell is occupied. In this case, the newly occupied site belongs to the same cluster as the occupied neighbor and, as a result, inherits the same cluster label.

The third possibility is that both neighbors are occupied. Two different situations can occur here. The first is that both neighbors belong to the same cluster. The newly occupied site joins them in the same cluster and takes on the same cluster label. The second situation is that the two neighbors belong to two different clusters. As a result of filling the site in question with a conductor, the two clusters that are separated until now are joined together by the newly occupied site to form a single cluster. We shall adopt the rule here that, when two different clusters are joined together to form a single cluster, the lower one of the two cluster labels becomes the label for the union.

At this stage, we can, in principle, go back and change the labels of all the members of the cluster to the smaller cluster label. This is very inefficient, as can be seen from the following considerations. To be able to change the cluster label of a site, the information must be stored in a large array and kept up to date. A more efficient way is to associate a counter $L_c(i)$ with each cluster. If cluster i is an isolated cluster, $L_c(i)$ keeps track of the number of members in the cluster. That is, each time a new member is added to cluster i, the counter $L_c(i)$ is increased by 1. When two clusters i and j merge, the value of $L_c(j)$ is added to that of $L_c(i)$ (assuming $j > i$), such that

$$L_c(i) + L_c(j) \to L_c(i).$$

As a result, $L_c(i)$ now reflects the size of the newly merged cluster. Instead of updating the labels of all the former members of cluster j, we shall change $L_c(j)$ to become a pointer, stating the fact that now cluster j no longer exists and all its members are merged with those of cluster i. This can be achieved by, for example, making the assignment

$$L_c(j) = -i.$$

In this way, whenever we encounter a cluster label j, we examine the value of its counter $L_c(j)$. If $L_c(j) > 0$, it is a good cluster with $L_c(j)$ members. On the other hand, if $L_c(j) < 0$, cluster j is delinquent and we shall refer, instead, to the counter of the cluster whose label is given by $|L_c(j)|$. It is possible that the counter cluster label $L_c(j)$ "points" to is also a pointer. In this case, we shall follow the new pointer to another label, and so on, until we come to a good cluster number. Since the number of members in each of the good clusters is stored in the counter, we have also the advantage of having the size of each cluster available to us when we finish with the process of filling the lattice.

This method also gives us a natural way to find out whether the top and bottom bars are connected so that charge carriers can percolate through the lattice. As we saw earlier, if a lattice site is occupied by a conductor, it bears the label of the cluster to which it belongs. If it is occupied by an insulator, the label is X. For each row, we can store this information in a one-dimensional array of N integers. For conduction to go from the top (first row) to the bottom (last row), these two rows must share at least one common cluster. Thus, by comparing the array of cluster labels for the first row with a similar array for the last row, taking care to change all the cluster labels in both rows first to those of "good" clusters, rather than pointers, we can tell at once whether there can be an electric current flowing between them.

Note that, besides the cluster counters $L_c(i)$, only two one-dimensional arrays are needed in this method, one to keep a copy of the occupancy pattern of the first row and one for the row just above the one we are in the process of filling. When a new site is filled with a conductor, we need the information on the cluster label of the site to the left and on top to determine the cluster to which the newly occupied site belongs. Let us assume that this information for row i is stored in the array ℓ_k for $k = 1, 2, \ldots, N$. Consider now site j of row $(i+1)$. The cluster label of the neighbor on top is in ℓ_j. This is all the information we need for site j, as far as the relation with the previous row is concerned. Once this is used, the information in location ℓ_j is no longer needed and we can use it to store the cluster label of site j for row $(i+1)$. If this rule is followed from the beginning of filling row $(i+1)$, the information concerning the site to the left, site $(j-1)$ in row $(i+1)$, is stored in location ℓ_{j-1}. In other words, as we fill the sites in row $(i+1)$, the information in array ℓ_j is replaced by the cluster label of the new row. In this way, the two pieces of information we need to assign the cluster label to a new site in row $(i+1)$ are both available in a single array ℓ_k.

For the convenience of programming, we shall add a dummy site to the left of the first one in each row and fill this site with an insulator. In this way, the starting cell is the second member of a one-dimensional array and comparisons of cluster labels can be carried out without having to pay special attention to whether a cell is the first one in the row and has no left neighbor. The algorithm is summarized in Box 7-5.

Finite size and other problems Using the algorithm described above, a percolation study is made on a (500×500) square lattice. Since the percolation threshold p_c for an infinite square lattice is known to occur at $p_c = 0.59275$, we restrict our study to occupation probability between 0.5 to 0.7. The calculated results show that the average cluster size increases from 7.5 to 95, as p goes from 0.5 to 0.7. However, there

Box 7-5 Program DM_PRCOL
Percolation on a Square Lattice

Definitions:

- N: Number of sites in a row and number of rows in the lattice.
- X: Label for an empty site.
- n_c: Number of different clusters.
- $\ell(i)$: Cluster number of site i of a row.
- $f(i)$: Cluster number of site i for the first row.
- $L_c(i)$: If > 0, the length of cluster i; if < 0, a pointer indicating that $|L_c(i)|$ is the cluster it is merged with.

Initialization:

(a) Input random number seed and occupation probability p.

(b) Initialize $n_c = 1$ and set the first element of a row $\ell(1) = X$.

Subprograms used:

- CLPS: Eliminate merged clusters from the array.
- ANALYSIS: Analysis of cluster size.
- RSUB: Uniform random number generator (Box 7-2).

1. First row, fill the site $i = 2$ to $(N + 1)$:
   (a) Generate a random number r.
   (b) If $r \leq p$, the site is to be occupied. Let $\ell(i) = c$, with c given by:
      (i) If left neighbor empty, increase n_c by 1, set $L_c(n_c) = 1$ and $c = n_c$.
      (ii) If occupied, set $c = \ell(i-1)$ and increase $L_c(\ell(i))$ by 1.
   (c) If $r > p$, mark the site as unoccupied by setting $\ell(i) = X$.
2. Keep a copy of $\ell(i)$ in $f(i)$.
3. For rows 2 to N, carry out steps 4 to 7 for $i = 2$ to N in each row:
4. Generate a random number r.
5. If $r > p$, mark the site as unoccupied by setting $\ell(i) = X$.
6. If $r \leq p$, the site is to be occupied. The cluster number c is given by:
   (a) If both the neighbor to the left and above are empty, increase the cluster number n_c by 1 and set $c = n_c$ and $L_c(n_c) = 1$.
   (b) If the neighbor on the left or above is occupied, let y equal the cluster number of the occupied site.
      (i) If $L_c(y) > 0$, y is a "good" cluster; set $c = y$ and add 1 to $L_c(y)$.
      (ii) If $L_c(y) < 0$, y is "delinquent;" let $y = -L_c(y)$ and return to (i).
   (c) If both the neighbor to the left and above are occupied:
      (i) Find the "good" cluster number y_1 of $\ell(i-1)$ and y_2 of $\ell(i)$ using steps (i) and (ii) of (b) above.
      (ii) If $y_1 = y_2$, let $y = y_1$.
      (iii) If $y_1 \neq y_2$, merge the two clusters by
         Setting $y = \min(y_1, y_2)$ and $z = \max(y_1, y_2)$.
         Letting $L_c(y) = L_c(y) + L_c(z)$ and $L_c(z) = -y$.
7. At any stage, if $n_c \geq X$, use CLPS to eliminate the merged clusters from the list and update the cluster numbers in $\ell(i)$ and $f(i)$.
8. When all N lines are filled, call ANALYSIS to carry out the following steps:
   (a) Group the clusters according to size and output the number in each.
   (b) Find the number of clusters common to the top and bottom.

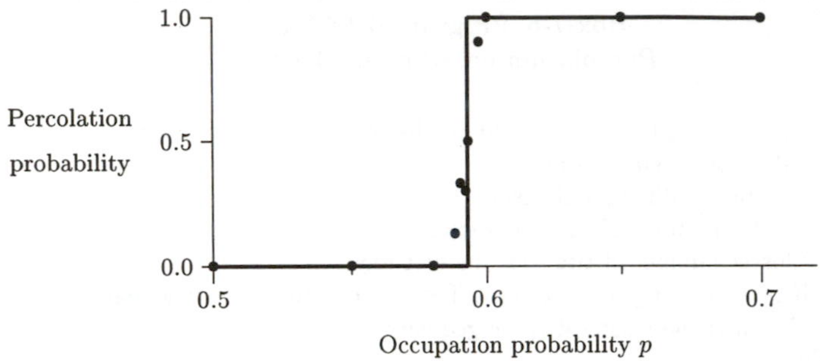

Figure 7-14: Probability of percolation as a function of occupation. The dots are the results for a lattice of (500 × 500) sites and the solid line is for an infinite lattice, with percolation threshold $p_c = 0.5927$.

are no dramatic changes around $p = 0.59$. It is perhaps more interesting to examine instead the changes in the size of the largest cluster. For example, at $p = 0.5$, clusters larger than 1,000 are very rare. At $p = 0.55$, the average cluster size is 12 and the size of the largest cluster is around 3,000. At $p = 0.58$, the corresponding numbers are 20 and 10,000. When we get to $p = 0.59$, large clusters consisting of over 50,000 members begin to appear, even though the average size remains around 23. At $p = 0.6$, the average size is 24, hardly different from that for $p = 0.59$. On the other hand, the largest clusters now have over 100,000 members. For our (500 × 500) square lattice with 60% occupation ($p = 0.6$), the total number of occupied sites is only 150,000. This means that almost 70% of all the conductors are connected as a single cluster. At $p = 0.7$, the same fraction increases to 98%.

The percolation phenomenon for which we are looking is connected to the question of whether there is a phase transition in the region between $p = 0.59$ and $p = 0.60$, accompanied by some sharp changes taking place. The results obtained above do not show any discontinuity in the properties of the system; however, they are sufficiently encouraging for us to take a closer look at our simulation. For this purpose, let us examine whether there is a sudden change as far as percolation is concerned. That is, whether there is some value of p below which it is extremely unlikely for a charge carrier to move from the first row of the lattice to the last one, and above which there is a good chance for conduction. Since we are doing a Monte Carlo simulation, a single calculation cannot determine the outcome. This point may be illustrated by the following consideration. At the low occupation probability of $p = 0.002$, there are, on the average, only 500 sites occupied in the lattice. On the other hand, even with such a small number, there is an extremely small, but nonzero, probability that all 500 occupied sites align on top of each other and form a conduction path. The same low probability occurrence can also take place for no percolation at high p values.

This can happen at $p = 0.998$, for example, if all 500 insulators are in the same row. For this reason, the simulation for given p values must be repeated many times, each starting with a different input random number seed. Any conclusion to be drawn from the simulation must be inferred from the distribution of many sets of calculated values.

For simplicity, we shall only carry out ten different runs for each p value. The probability of observing a percolation in the ten runs, the fraction of ten runs with percolation, for different values of p between 0.5 and 0.7 is shown in Fig. 7-14. It is quite clear from the figure that the simulated results do not indicate a sharp transition in percolation as a function of p, as required of a critical phenomenon. This is not a surprise as far as our calculations are concerned. We have seen in the previous paragraph how percolation can take place in samples of finite size even at very low values of p. However, the probability of observing percolation at low p values decreases with increasing size of the lattice. To illustrate this point, let us consider a square lattice of $(\ell \times \ell)$ sites. If the probability that any one site is occupied is p, the probability for all the sites in a column being occupied is then p^ℓ and goes to zero (since $p < 1$) in the limit that $\ell \to \infty$. By the same token, a sharp change in a percolation simulation can only appear for an infinite lattice.

In any computer simulation, it is only possible to work with a finite ℓ. One way to have some feeling for the size of a realistic lattice is to examine the number of atoms in a sample of reasonable size. For such an estimate, the order of magnitude is given by the Avogadro number, 10^{23}. If this number of atoms is arranged in the form of a cubic lattice, each side will have more than 10^7 atoms. A realistic simulation may not actually require ℓ of this size, but certainly a size much larger than anything we can attempt at this moment on a computer. As a result, finite-size effect becomes one of the problems in percolation studies using a computer. For a simple square lattice, a percolation threshold exists only in the limit of infinite lattice, and we must somehow infer the results for an "infinite" lattice from studies made with finite ones. The procedure to carry out this task is more complicated and specialized than we can go into here.

7-5 The Ising model

In the previous section, we considered the formation of clusters from particles that have no interaction with each other. For many problems in condensed matter physics and field theory, the interaction between neighbors is important, even for particles localized on a lattice. Historically, the best known example of the statistical properties of a system that includes interaction between neighboring particles is perhaps the Ising model.

The model was originally introduced to understand ferromagnetism. Ferromagnetic materials are characterized by the existence of a magnetic moment even in the absence of external fields. On a macroscopic scale, the strength of magnetism is measured in terms of the *spontaneous* magnetic moment, and it decreases smoothly with

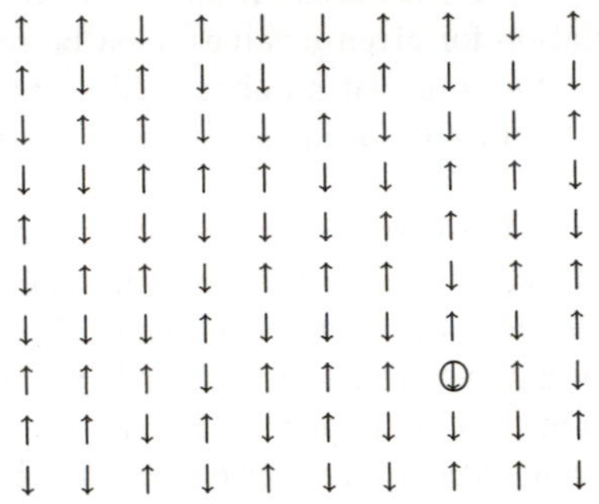

Figure 7-15: Schematic representation of a two-dimensional Ising model lattice with nearest-neighbor interaction $+\epsilon$ for two adjacent spins oriented in the same direction and $-\epsilon$ in opposite directions. For the site marked by a circle, they are $-\epsilon$, $-\epsilon$, $-\epsilon$, and $+\epsilon$, clockwise starting from the neighbor to the left.

increasing temperature until a critical value T_c, the Curie temperature, is reached. For iron, this occurs at $T_c = 770°$ C. For cobalt and nickel, the corresponding values are 1130° C and 358° C, respectively. A simple model to understand the magnetic property of these materials is to consider all the molecules to be localized in space. For convenience, we shall further constrain the model to be two-dimensional in the form of a square lattice. Generalizing to three-dimensional lattices is fairly straightforward, but we shall not do it here.

For our square lattice, the only degree of freedom at each site is in the orientation of the molecular spin $\boldsymbol{J}$. Let us use the symbol $J_{i,j}$ to represent the value of $\boldsymbol{J}$ at site (i, j). To simplify the discussion, we shall only allow two possible orientations, $J_{i,j} = +1$ (up) or -1 (down). The directions may be aligned with an external magnetic field if one is present. Otherwise, the spin orientations have meaning only in relation with each other. An example of such a lattice is shown in Fig. 7-15.

The model Hamiltonian In our model, the only "internal" interactions are those between the spins of two nearest neighbors $J_{i,j}$ and $J_{i',j'}$, with $(i', j') = (i \pm 1, j)$ or $(i, j \pm 1)$. The interaction energy involved may be parametrized in the following way. If the spins of both sites are oriented in the same direction, the contribution to the energy of the whole system is $+\epsilon$; and if they are in the opposite direction, the amount is $-\epsilon$. The Hamiltonian for such an interaction takes on the form

$$\hat{H}_{\text{int}} = -\epsilon \sum_{\langle {i,j \atop i',j'} \rangle} \hat{\sigma}_0(i,j)\,\hat{\sigma}_0(i',j') \tag{7-41}$$

where $\hat{\sigma}_0(i,j)$ is an operator that gives $+1$ if the spin of the molecule at site (i,j) is up and -1 if down. The summation over indices i and j are over all the lattice sites and the angle brackets imply that the possible values of (i', j') are restricted to

$(i \pm 1, j)$ and $(i, j \pm 1)$. We shall defer till later the introduction of a better indexing scheme for the lattice sites so that this restricted sum can be carried out conveniently in practice.

Although we have limited the orientations of the spins to two definite directions, the model is still a classical one. For a fully quantum mechanical system, the interaction Hamiltonian takes only the form

$$\hat{H}_{\text{int}} = -\epsilon \sum_{\langle {i,j \atop i',j'} \rangle} \hat{\boldsymbol{\sigma}}(i,j) \cdot \hat{\boldsymbol{\sigma}}(i',j')$$

where, for a spin-half system, $\hat{\boldsymbol{\sigma}}(i,j)$ is the Pauli spin matrix operator acting on the spin of the particle at site (i,j). This is known as the Heisenberg model. We shall, however, restrict ourselves to the classical case and consider only the Ising model.

In general, we can also include a term in the Hamiltonian that represents the interaction with an external magnetic field B supplied, for example, by a solenoid. The total Hamiltonian for the system then becomes

$$\hat{H} = -\epsilon \sum_{\langle {i,j \atop i',j'} \rangle} \hat{\sigma}_0(i,j)\hat{\sigma}_0(i',j') - \eta B \sum_{i,j} \hat{\sigma}_0(i,j) \tag{7-42}$$

where η is the strength of the interaction with the external B field. To simplify the notation, we shall absorb η into the definition of B and omit it from now on. Both ϵ and B are now in energy units and they may be regarded as the two parameters of the model. The energy of the system for a given configuration $\mathcal{C}$ of the assembly of $(N \times N)$ spins is then

$$E_{\mathcal{C}} = -\epsilon \sum_{\langle {i,j \atop i',j'} \rangle} J_{i,j} J_{i',j'} - B \sum_{i,j} J_{i,j}. \tag{7-43}$$

It is a function of the orientation of each one of the spins in the lattice.

Since there are $(N \times N)$ spins in our system and each can have two possible orientations, there is a total of $2^{N \times N}$ possible arrangements or *configurations*. The complete set of $2^{N \times N}$ configurations forms a canonical ensemble. If the system is ergodic, the probability $p(\mathcal{C})$ for the system to be in configuration $\mathcal{C}$ is given by

$$p(\mathcal{C}) = \frac{1}{Z(\epsilon, B)} e^{-E_{\mathcal{C}}/kT} \tag{7-44}$$

where $e^{-E_{\mathcal{C}}/kT}$ is the Boltzmann factor, with $E_{\mathcal{C}}$ being the energy of the configuration given by Eq. (7-43). The other factors in the argument of the exponential function are k, the Boltzmann constant, and T, the temperature.

The normalization of the probability distribution Eq. (7-44) is provided by the partition function

$$Z(\epsilon, B) = \sum_{\mathcal{C}} e^{-E_{\mathcal{C}}/kT}.$$

Following standard statistical mechanics procedures, we shall calculate this quantity first. From $Z(\epsilon, B)$, other quantities of interest to the thermodynamics of the system may be obtained, such as the internal energy

$$\mathcal{E} = kT^2 \frac{\partial}{\partial T} \ln Z(\epsilon, B) = \sum_{\mathcal{C}} p(\mathcal{C}) E_{\mathcal{C}} \tag{7-45}$$

and magnetization

$$\mathcal{M} = kT \frac{\partial}{\partial B} \ln Z(\epsilon, B) = \sum_{\mathcal{C}} p(\mathcal{C}) \Big(\sum_{i,j} J_{i,j} \Big). \tag{7-46}$$

From these, we obtain the specific heat

$$C_B = \frac{\partial \mathcal{E}}{\partial T} = \frac{1}{kT^2} \Big\{ \sum_{\mathcal{C}} p(\mathcal{C}) E_{\mathcal{C}}^2 - \mathcal{E}^2 \Big\} \tag{7-47}$$

and magnetic susceptibility

$$\chi_m = kT \frac{\partial \mathcal{M}}{\partial B} = \sum_{\mathcal{C}} p(\mathcal{C}) \Big(\sum_{i,j} J_{i,j} \Big)^2 - \mathcal{M}^2. \tag{7-48}$$

For our illustrative example, we shall be mainly concerned with $\mathcal{E}$ and $\mathcal{M}$, as it is far easier to obtain good numerical accuracies for these two quantities than for C_B and χ_m, which are derivatives of $\mathcal{E}$ and $\mathcal{M}$.

Infinite two-dimensional model For an infinite two-dimensional system, it is possible to obtain an analytical solution for the Ising model. The results show that there is a singularity in all the thermodynamic functions at critical temperature T_c given by

$$2 \tanh^2 \frac{2\epsilon}{kT_c} = 1.$$

The same result may also be written in the form $z \equiv e^{-2\epsilon/kT_c} = \sqrt{2} - 1$, corresponding to the value

$$\frac{\epsilon}{kT_c} = 0.44.$$

In the absence of an external field ($B = 0$), the internal energy has the form

$$\mathcal{E} = -\epsilon \coth \frac{2\epsilon}{kT} \Big\{ 1 + \frac{2}{\pi} K_1(\zeta) \Big(2 \tanh^2 \frac{2\epsilon}{kT} - 1 \Big) \Big\}$$

where

$$\zeta = \frac{2 \sinh \frac{2\epsilon}{kT}}{\cosh^2 \frac{2\epsilon}{kT}}$$

and $K_1(\zeta)$ is the complete elliptic integral of the first kind

$$K_1(\zeta) = \int_0^{\pi/2} \frac{d\phi}{1 - \zeta^2 \sin^2 \phi}.$$

At temperatures $T \approx T_c$, the specific heat has the form

$$C_B \approx \frac{2k}{\pi}\left(\frac{2\epsilon}{kT_c}\right)^2\left\{-\ln\left(\left|1-\frac{T}{T_c}\right|\right)+\ln\left(\frac{kT_c}{2\epsilon}\right)-\left(1+\frac{\pi}{4}\right)\right\}$$

and the spontaneous magnetization per spin takes on the value

$$\frac{1}{N^2}\mathcal{M} = \begin{cases} \dfrac{(1+z^2)^{1/4}(1-6z^2+z^4)^{1/8}}{\sqrt{1-z^2}} & \text{for } T < T_c \\ 0 & \text{for } T > T_c. \end{cases}$$

The calculations leading to these results are somewhat involved and can be found, for example, in Huang.[28]

Computer simulation of a finite square lattice The advantage of carrying out a computer simulation for the Ising model lies in the wider freedom in choosing the interactions and in the relative ease in extending to three dimensions. The disadvantage is that we have to be contented with a finite system. Even for a modest value of $N = 10$, the total number of possible configurations in a canonical ensemble for our two-dimensional system is $2^{N\times N} = 2^{100} \approx 10^{30}$. For such large numbers, it is not feasible to carry out the calculations by averaging over all the members, as implied by Eqs. (7-45) to (7-48). The alternative is to use a sampling approach. For this reason, Monte Carlo techniques enter into studies on Ising and related models.

For our square lattice, each configuration is characterized by the orientations, $J_{i,j} = \pm 1$, of the $(N \times N)$ spins. As a result, it takes N^2 random numbers (random bits is adequate here since each $J_{i,j}$ has only two possible orientations) to construct a member of the ensemble. The statistical weight of a particular configuration in the ensemble is given by Eq. (7-44). Since we can only take a relatively small sample of all the possible configurations, the Metropolis algorithm, discussed at the end of §7-1, becomes the method of choice. In this scheme, we start with a given configuration $\mathcal{C}$. From this, we generate a trial configuration $\mathcal{C}_t$ by going through each of the N^2 spins and check whether the orientation of any one should be flipped. Using Eq. (7-43), we find that the energy change connected with a reorientation of the spin at site (i, j), with original spin orientation $J_{i,j}$, is given by

$$\Delta E = -2J_{i,j}\{\epsilon(J_{i-1,j} + J_{i+1,j} + J_{i,j-1} + J_{i,j+1}) + B\} \tag{7-49}$$

where the first term, associated with the parameter ϵ, comes from the interaction with the four neighboring sites and the second term, associated with the parameter B, arises from the interaction with an external magnetic field.

In the Metropolis algorithm, the decision whether the trial configuration $\mathcal{C}_t$ should be accepted or rejected depends on the ratio r of Eq. (7-20) and is connected with the relative probabilities of having the two configurations $\mathcal{C}_t$ and $\mathcal{C}$. From Eqs. (7-44) and (7-49), we find that the value of r is given by

$$r = \frac{e^{-E_{\mathcal{C}_t}/kT}}{e^{-E_{\mathcal{C}}/kT}} = e^{-\Delta E/kT}. \tag{7-50}$$

For our square lattice, there are only $(2 \times 5) = 10$ possible values for r. The factor 2 comes from the two possible orientations of $J_{i,j}$, and the factor 5 corresponds to the five different possible values, 0, ± 2, and ± 4, for the sum of the relative orientations of the four neighboring spins, as can be seen in the first term of Eq. (7-49). The ten values for a given set of parameters (ϵ, B) can be stored as a (2×5) matrix and we do not have to recalculate any of them each time they are needed. We shall refer to this matrix as the *decision matrix*.

As discussed on page 314, the direction of the spin at site (i, j) is reversed if $r > 1$. For $r \leq 1$, a uniformly distributed random number x in the interval $[0, 1]$ is generated. If $x \leq r$, the spin orientation is changed. If $x > r$, the trial configuration is rejected and the spin at site (i, j) is unchanged. Thus, it is necessary to make a "sweep" over all $(N \times N)$ spins to find a new configuration $\mathcal{C}_{n+1}$ from $\mathcal{C}_n$ by making a choice whether to flip the spin direction at each site. This is the part of the calculation that takes most of the time, and any improvement in the speed, such as putting the decision matrix as a stored array, is important. The steps of the Metropolis scheme to sweep through the lattice are given in Box 7-6.

It is convenient to discuss here two other technical points in the calculation. The first is that each random configuration generated by the Metropolis algorithm is strongly correlated with the previous one from which it started. This is undesirable. One way to minimize the correlation is to discard a few configurations between any two that are accepted as members of our sample. This means that several sweeps through the lattice have to be made before we can select a configuration whose physical properties we shall examine and include in the ensemble averages we wish to study. For the same reason, it is also a good practice to discard the first few configurations generated from the one we use as the starting point. As we saw earlier, this may be regarded as a "thermalization" process, since the starting configuration may not be at equilibrium.

The second point concerns finite-size effects arising from the small value of N used in a computer calculation. To partially compensate for this limitation, we can impose a periodic boundary condition. In other words, we shall consider that our square lattice is only one of many identical ones in the space, with the top of our lattice bordering the bottom of another. Similarly, the left side is in contact with the right side of another. In practical terms, this means that we make the following replacements of the spins at the borders of our lattice:

$$J_{i-1,j} \to J_{N,j} \quad \text{for} \quad i = 1 \qquad\qquad J_{i+1,j} \to J_{1,j} \quad \text{for} \quad i = N$$
$$J_{i,j-1} \to J_{i,N} \quad \text{for} \quad j = 1 \qquad\qquad J_{i,j+1} \to J_{i,1} \quad \text{for} \quad j = N.$$

Again, this condition can be imposed by adding one more row to the top of our $(N \times N)$ array, with spin orientations identical to those of the last row, and by adding one more row to the bottom that has the same spin orientations as the first row. Similarly, an additional column is put on the left as well as on the right of our lattice.

Box 7-6 Subroutine SWEEP(MS,NP1,NDMN,DS_MTX)
One sweep of the lattice according to the Metropolis algorithm

Argument list:

- MS: Array for the spin orientations of the square lattice.
- NP1: Number of spins in a row (or column) plus 1.
- NDMN: Dimension of the MS array in the calling program.
- DS_MTX: Decision matrix.

Initialization:

(a) Construct in the calling program the decision matrix of Eq. (7-50).
(b) Set up the periodic boundary condition.

1. Check each site (i, j) of the lattice to see if the spin orientation should be reversed:
   (a) Impose the periodic condition:
      (i) If $i = 1$, use $J_{N,j}$ for $J_{i-1,j}$.
      (ii) If $i = N$, use $J_{1,j}$ for $J_{i+1,j}$.
      (iii) If $j = 1$, use $J_{i,N}$ for $J_{i,j-1}$.
      (iv) If $j = N$, use $J_{i,1}$ for $J_{i,j+1}$.
   (b) Set up the two indices for the decision matrix.
      (i) Find the value of the spin $J_{i,j}$ and use this to build the index i_s.
      (ii) Find the sum of four neighboring spins $J_{i-1,j} + J_{i+1,j} + J_{i,j-1} + J_{i,j+1}$ and use the value to build the index j_s.
   (c) Use i_s and j_s to find the ratio r of Eq. (7-50) from the decision matrix.
   (d) If $r > 1$, change the sign of the spin $J_{i,j}$.
   (e) If $r \leq 1$, generate a uniform random number x in $[0, 1]$.
      (i) If $x \leq r$, change the sign of spin $J_{i,j}$.
      (ii) If $x > r$, the sign of $J_{i,j}$ is not changed.
2. Return the new configuration in array MS.

Ensemble averages Since the configurations are selected according to probability $p(\mathcal{C}) \sim \exp -\{E_{\mathcal{C}}/kT\}$, an ensemble average of physical quantities is equivalent to averaging over the selected configurations. For example, the internal energy of Eq. (7-45) becomes

$$\mathcal{E} = \sum_{\mathcal{C}} p(\mathcal{C}) E_{\mathcal{C}} \longrightarrow \frac{1}{N_{\text{ens}}} {\sum_{\mathcal{C}}}' E_{\mathcal{C}}$$

where N_{ens} is the number of configurations sampled. The prime over the summation indicates the fact that the configurations included have been selected with the appropriate weighting factor $p(\mathcal{C})$. Furthermore, since $p(\mathcal{C})$ is already built into the sampling process, it is no longer necessary to include it explicitly in the summation.

For each member of the ensemble, all four thermodynamic quantities of interest to us, $\mathcal{E}$, C_B, $\mathcal{M}$, and χ_m, are derived from two terms, the spin-spin term and the spin term. The former depends on the factor

$$t_s(\mathcal{C}) = \sum_{i,j} J_{i,j}(J_{i-1,j} + J_{i+1,j} + J_{i,j-1} + J_{i,j+1}) \tag{7-51}$$

Box 7-7 Program DM_ISING
Two-dimensional Ising model
Energy, specific heat, magnetization, and susceptibility for a square lattice

Subprograms called:
- INI_LAT: Initializing the lattice.
- CALC_EM: Calculate the spin-spin and spin-B contributions for a configuration.
- OUT_GRP: Output the ensemble average.
- SWEEP: One Metropolis algorithm sweep through the lattice (Box 7-6).
- DISPLAY: Display the lattice.

Initial settings:
- N_{therm}: Number of thermalization sweeps (=20).
- N_{skip}: Number of sweeps to skip between ensembles (=5).
- N_{ens}: Number of members in an ensemble (= 100).
- N_{group}: Number of ensembles in the calculation (=20).

1. Input the random number seed, number of spins in each row N, energy parameter ϵ/kT, and external field B/kT.
2. Initialize the random number generator and the decision matrix Eq. (7-50).
3. Construct an initial configuration of the $(N \times N)$ lattice using INI_LAT:
   (a) For each lattice site, generate a uniform random number in $[0, 1]$.
   (b) If $x > 0.5$, the spin orientation is up ($J_{i,j} = +1$).
   (c) Otherwise, the spin orientation is down ($J_{i,j} = -1$).
4. Thermalize the lattice by making N_{therm} calls to SWEEP.
5. Construct an ensemble by carrying out the following steps N_{group} times:
   (a) Make a sweep through the lattice using SWEEP.
   (b) Use CALC_EM to calculate and store t_s of Eq. (7-51) and t_b of (7-52).
   (c) Call SWEEP N_{skip} times to reduce correlation between ensemble members.
6. Repeat the previous step N_{group} times to construct a collection of ensembles.
7. Use OUT_GRP to calculate and output the ensemble averages and standard deviations of $\mathcal{E}$, C_B, $\mathcal{M}$, and χ_m from t_s and t_b using Eq. (7-53).

and the latter,

$$t_b(\mathcal{C}) = \sum_{i,j} J_{i,j}. \tag{7-52}$$

In terms of the values of these two terms for each configuration, the ensemble averages may be expressed as

$$\mathcal{E} = \frac{1}{N_{\text{ens}}} \sum_{\mathcal{C}}{}' \left\{ -\epsilon\, t_s(\mathcal{C}) - B t_b(\mathcal{C}) \right\} \qquad \mathcal{M} = \frac{1}{N_{\text{ens}}} \sum_{\mathcal{C}}{}' t_b(\mathcal{C})$$

$$C_B = \frac{1}{kT^2 N_{\text{ens}}} \sum_{\mathcal{C}}{}' \left\{ -\epsilon\, t_s(\mathcal{C}) - B t_b(\mathcal{C}) - \mathcal{E} \right\}^2 \qquad \chi_m = \frac{1}{N_{\text{ens}}} \sum_{\mathcal{C}}{}' \left\{ t_b(\mathcal{C}) - \mathcal{M} \right\}^2. \tag{7-53}$$

The steps of the calculation are given in Box 7-7.

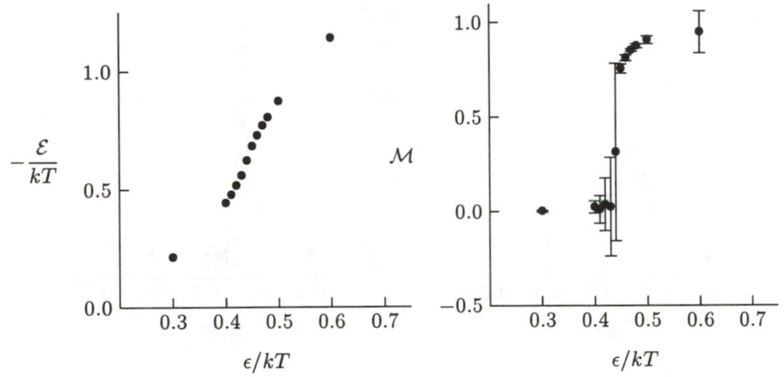

Figure 7-16: Internal energy $\mathcal{E}/kT$ and magnetization $\mathcal{M}$ per spin as functions of ϵ/kT for a (60×60) square Ising lattice with external field $B = 0$. Each point is the average of 20 ensembles consisting of 100 samples in each. Error bars indicate the sizes of one standard deviation in the distribution among the 20 ensembles. For $\mathcal{E}$, they are too small to be visible.

The results of a simulation for $\mathcal{E}/kT$ and $\mathcal{M}$ on a square lattice of $N = 60$ are given in Fig. 7-16. Each point in the plot is the average of 20 ensembles, with 100 members sampled in each. The external magnetic field is held at $B = 0$ so as to mimic the situation of spontaneous magnetization. In each case, the values of the internal energy $\mathcal{E}$ and magnetization $\mathcal{M}$ are plotted as functions of ϵ/kT. Since all the energies are measured in units of kT, the horizontal axes may be interpreted as variations of the temperature for a constant value of the spin-spin interaction strength ϵ (with temperature T decreasing for increasing ϵ/kT).

In the figure, we see that (the negative of) the internal energy varies smoothly with temperature, without anything resembling a discontinuity or shape change to indicate a phase transition. The magnetization, on other hand, is essentially zero below $\epsilon/kT \approx 0.45$ and increases with decreasing temperature (increasing ϵ/kT) beyond this point. The fluctuations in the value of $\mathcal{M}$ are large just below (higher temperatures) the expected place of a critical point. This makes it difficult to identify whether the results show a phase transition. Part of the reason for the large fluctuations comes from the fact that, in the absence of an external magnetic field, there is no reference direction with which the magnetic dipole moment associated with each spin can align. As a result, it is equally probable for the spins to be preferentially in the $J_{i,j} = +1$ direction in one member of the ensemble and the $J_{i,j} = -1$ direction in another. At temperatures below the critical point (large ϵ/kT), the spontaneous magnetic moment provides a reference direction and, as a result, fluctuations in the values are smaller. In spite of the ambiguity caused by fluctuations, the sharp rise in $\mathcal{M}$ at $\epsilon/kT \approx 0.45$

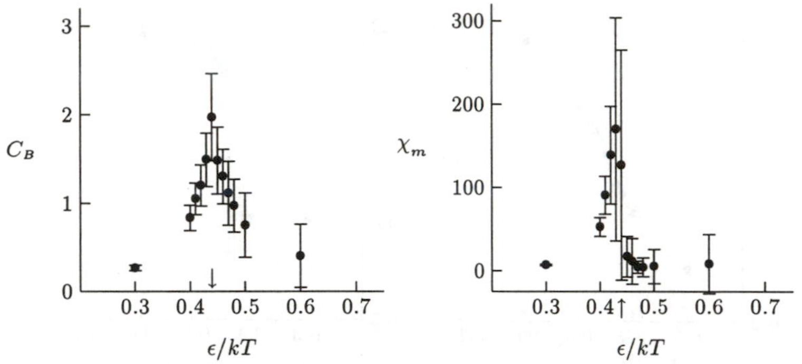

Figure 7-17: Specific heat per spin and magnetic susceptibility as functions of ϵ/kT for a (60×60) square Ising lattice with external field $B = 0$. Critical point for the corresponding infinite system occurs at $\epsilon/kT = 0.44$ marked by an arrow.

provides an indication for a second-order phase transition in the results shown in the figure.

The results for specific heat C_B and magnetic susceptibility χ_m are plotted in Fig. 7-17. As can be seen from Eq. (7-53), these two quantities are proportional to the variances of, respectively, $\mathcal{E}$ and $\mathcal{M}$. For this reason, we expect large variations in their values from one ensemble to another. Because of our finite lattice and sample sizes, the fluctuations in the simulated results can be significant. In spite of this, we see a cusp in each case at roughly the analytical result of $\epsilon/kT = 0.44$ for an infinite system, indicating the possible existence of a critical point.

The real interest in a computer simulation of the Ising model is not in the simple two-dimensional case we have shown here. With a little more effort, we can apply essentially the same method to carry out a model calculation in three dimensions. In this case, analytical solutions are difficult to obtain. Furthermore, it is much easier to introduce other physical effects in a computer modeling, such as interactions with the next-nearest neighbors. The main shortcoming we have is in interpreting the computer-generated results because of finite size effects, as we saw above. The limitations on the size of N we can use come from the amount of computation involved. For the three-dimensional case, we must sweep through all N^3 spins several times before we can produce a single member of the ensemble. As a result, the computational time increases as N^3.

7-6 Path integrals in quantum mechanics

In quantum mechanics, the usual way to find the time evolution of a state, characterized by the wave function $|\Psi(t)\rangle$, is to solve the time-dependent Schrödinger equation

$$i\hbar\frac{\partial}{\partial t}|\Psi(t)\rangle = \hat{H}|\Psi(t)\rangle \tag{7-54}$$

where, in the nonrelativistic limit, the Hamiltonian operator $\hat{H}$ is made of a sum of kinetic and potential energies. In terms of operators in coordinate space, the Hamiltonian for a particle of mass μ may be written as

$$\hat{H} = -\frac{\hbar^2}{2\mu}\nabla^2 + V(\boldsymbol{r}, t).$$

From the solution, we obtain the eigenvalues and wave functions. In general, it is difficult to solve Eq. (7-54) as a differential equation for any realistic interaction potential $V(\boldsymbol{r}, t)$ one encounters in physical systems. An alternative in such cases is to use the path integral method.

For a free particle, there is no potential acting on it and we have

$$V(\boldsymbol{r}, t) = 0.$$

If the wave function at $t = 0$ is a plane wave with momentum $\boldsymbol{p}$,

$$\psi(\boldsymbol{r}, 0) = \frac{1}{(2\pi\hbar)^{3/2}}\, e^{i(\boldsymbol{p}\cdot\boldsymbol{r})/\hbar} \tag{7-55}$$

the solution of Eq. (7-54) at any time t has the familiar form

$$\psi(\boldsymbol{r}, t) = \frac{1}{(2\pi\hbar)^{3/2}}\, e^{i(\boldsymbol{p}\cdot\boldsymbol{r} - Et)/\hbar} \tag{7-56}$$

where $E = p^2/2\mu$ (and $p = |\boldsymbol{p}|$) is the kinetic energy of the particle.

Bra-ket notation For the convenience of discussion and to make connection with standard references on the topic, we shall adopt the bra-ket notation here. Let $|\Psi\rangle$ represent a state vector. Similar to a vector $\boldsymbol{v}$ in three-dimensional space, a coordinate system must be set up to specify it further. For example, in a Cartesian system, $\boldsymbol{v}$ is given by the values of its projections on the x-, y-, and z-axes. Our state vector $|\Psi\rangle$ is a more general quantity and may require a continuum of values to identify it. For example, in coordinate representation, $|\Psi\rangle$ is specified by giving its "components" $\psi(\boldsymbol{r}_1)$, $\psi(\boldsymbol{r}_2), \ldots, \psi(\boldsymbol{r}_i), \ldots$, at a continuum of points $\boldsymbol{r}_1, \boldsymbol{r}_2, \ldots, \boldsymbol{r}_i, \ldots$ in coordinate space. Alternatively, we can realize the same state vector in momentum representation by giving its components $\psi(\boldsymbol{p}_1)$, $\psi(\boldsymbol{p}_2), \ldots, \psi(\boldsymbol{p}_i), \ldots$, at a continuum of points $\boldsymbol{p}_1$, $\boldsymbol{p}_2, \ldots, \boldsymbol{p}_i, \ldots$ in momentum space.

Formally, we can extract a particular component of $|\,\Psi\,\rangle$, for example $\psi(\boldsymbol{r})$, by defining a function $|\,\boldsymbol{r}\,\rangle$ as the probability amplitude of a particle at $\boldsymbol{r}$. Since $|\,\boldsymbol{r}\,\rangle$ forms a complete set of states, we have the orthonormal condition

$$\langle \boldsymbol{r}|\boldsymbol{r}'\rangle = \delta(\boldsymbol{r}-\boldsymbol{r}')$$

where $\delta(\boldsymbol{r}-\boldsymbol{r}')$ is the Dirac delta function given in Eq. (3-39). The overlap of $|\,\Psi\,\rangle$ with $|\,\boldsymbol{r}\,\rangle$, or the "projection" of $|\,\Psi\,\rangle$ on $|\,\boldsymbol{r}\,\rangle$, gives us the $\boldsymbol{r}$-component of $|\,\Psi\,\rangle$. That is,

$$\psi(\boldsymbol{r}) = \langle \boldsymbol{r}|\Psi\rangle.$$

Similarly, if $|\,\boldsymbol{p}\,\rangle$ represents the wave function of a particle with momentum $\boldsymbol{p}$, the components of $|\,\Psi\,\rangle$ in momentum representation are given by

$$\psi(\boldsymbol{p}) = \langle \boldsymbol{p}|\Psi\rangle.$$

Returning to our free particle example, we see that the right side of Eq. (7-55) is nothing but the coordinate representation of a particle with momentum $\boldsymbol{p}$. As a result, we have the relation

$$\langle \boldsymbol{r}|\boldsymbol{p}\rangle = \frac{1}{(2\pi\hbar)^{3/2}}\, e^{i(\boldsymbol{p}\cdot\boldsymbol{r})/\hbar}. \tag{7-57}$$

In this notation, the transformation between the momentum and coordinate representations of the state vector $|\,\Psi\,\rangle$ may be expressed as

$$\langle \boldsymbol{p}|\Psi\rangle = \iiint d^3r\, \langle \boldsymbol{p}|\boldsymbol{r}\rangle\langle \boldsymbol{r}|\Psi\rangle. \tag{7-58}$$

This is the same as inserting a complete set of intermediate states $|\,\boldsymbol{r}\,\rangle$ into the "matrix element" $\langle \boldsymbol{p}|\Psi\rangle$.

The free particle We shall now derive an expression for the wave function of a free particle at time $t \neq 0$ in terms of a propagator that takes it from time $t = 0$. In general, a free particle at location $\boldsymbol{r}$ is represented by a wave packet made of a linear combination of plane waves each with a different momentum. At $t = 0$, the wave function for a particle with momentum $\boldsymbol{p}$ is given by Eq. (7-55). For our wave packet, the wave function at time $t = 0$ is then a linear combination of such components weighted by the factor $\langle \boldsymbol{p}|\Psi(t=0)\rangle$:

$$\psi(\boldsymbol{r},0) = \langle \boldsymbol{r}|\Psi(t=0)\rangle = \frac{1}{(2\pi\hbar)^{3/2}} \iiint d^3p\, e^{i\boldsymbol{p}\cdot\boldsymbol{r}/\hbar}\langle \boldsymbol{p}|\Psi(t=0)\rangle. \tag{7-59}$$

This is, essentially, a Fourier transform of the wave function from coordinate space to momentum space. For our later needs, it is more convenient to put it into the following form:

$$\langle \boldsymbol{r}|\Psi(t=0)\rangle = \iiint d^3p\, \langle \boldsymbol{r}|\boldsymbol{p}\rangle\langle \boldsymbol{p}|\Psi(t=0)\rangle$$

where we have made use of Eq. (7-57) to change the integrand from that in Eq. (7-59).

At time t, the wave function $\psi(\boldsymbol{r}, t)$ for a free particle in coordinate space, following Eq. (7-56), has the form

$$\langle \boldsymbol{r}|\Psi(t)\rangle = \iiint d^3p\, \langle \boldsymbol{r}|\boldsymbol{p}\rangle e^{-iEt/\hbar} \langle \boldsymbol{p}|\Psi(t=0)\rangle.$$

This may be expressed as a relation between the wave function in coordinate space at time t and that at time $t = 0$ by inserting a complete set of states $|\,\boldsymbol{r}'\,\rangle$ on the right side,

$$\begin{aligned}\langle \boldsymbol{r}|\Psi(t)\rangle &= \iiint d^3r' \iiint d^3p \langle \boldsymbol{r}|\boldsymbol{p}\rangle e^{-iEt/\hbar} \langle \boldsymbol{p}|\boldsymbol{r}'\rangle\langle \boldsymbol{r}'|\Psi(t=0)\rangle \\ &= \iiint d^3r' K(\boldsymbol{r}, t; \boldsymbol{r}', 0)\langle \boldsymbol{r}'|\Psi(t=0)\rangle \qquad (7\text{-}60)\end{aligned}$$

the same as we have done in Eq. (7-58). For later convenience, we have rewritten the final result in terms of

$$K(\boldsymbol{r}, t; \boldsymbol{r}', 0) \equiv \iiint d^3p \langle \boldsymbol{r}|\boldsymbol{p}\rangle e^{-iEt/\hbar} \langle \boldsymbol{p}|\boldsymbol{r}'\rangle. \qquad (7\text{-}61)$$

Physically, we may regard $K(\boldsymbol{r}, t; \boldsymbol{r}', 0)$ as the transition probability amplitude for a particle at location $\boldsymbol{r}$ and time $t = 0$ to that at location $\boldsymbol{r}'$ and time t. Similarly, the wave function at some other time t' may be expressed in terms of that at $t = 0$,

$$\langle \boldsymbol{r}'|\Psi(t')\rangle = \iiint d^3r\, K(\boldsymbol{r}', t'; \boldsymbol{r}, 0)\langle \boldsymbol{r}|\Psi(t=0)\rangle \qquad (7\text{-}62)$$

as there is nothing special about time t or t'.

Comparing Eq. (7-62) with (7-60), we can relate the wave functions at two different times t_f and t_i in the following way:

$$\langle \boldsymbol{r}_f|\Psi(t_f)\rangle = \iiint d^3r_i\, K(\boldsymbol{r}_f, t_f; \boldsymbol{r}_i, t_i)\langle \boldsymbol{r}_i|\Psi(t_i)\rangle \qquad (7\text{-}63)$$

where

$$K(\boldsymbol{r}_f, t_f; \boldsymbol{r}_i, t_i) = \iiint d^3p \langle \boldsymbol{r}_f|\boldsymbol{p}\rangle e^{-iE(t_f-t_i)/\hbar} \langle \boldsymbol{p}|\boldsymbol{r}_i\rangle. \qquad (7\text{-}64)$$

The quantity $K(\boldsymbol{r}_f, t_f; \boldsymbol{r}_i, t_i)$ may also be interpreted as a "propagator" that connects the wave functions of a free particle at two different times. It is also possible to regard it as the overlap of the wave functions $\psi(\boldsymbol{r}_f, t_f)$ and $\psi(\boldsymbol{r}_i, t_i)$,

$$K(\boldsymbol{r}_f, t_f; \boldsymbol{r}_i, t_i) = \langle \boldsymbol{r}_f, t_f|\boldsymbol{r}_i, t_i\rangle \qquad (7\text{-}65)$$

since the right side of Eq. (7-63) may be regarded as an "expansion" of $\psi(\boldsymbol{r}_f, t_f)$ in terms of the complete set of states $\psi(\boldsymbol{r}_i, t_i)$.

For a free particle, it is possible to carry out the integral on the right side of Eq. (7-64) explicitly. The result,

$$K(\boldsymbol{r}_f, t_f; \boldsymbol{r}_i, t_i) = \left\{\frac{\mu}{2\pi i\hbar(t_f - t_i)}\right\}^{3/2} e^{i\frac{\mu}{2\hbar}\frac{(\boldsymbol{r}_f-\boldsymbol{r}_i)^2}{t_f-t_i}} \tag{7-66}$$

may be verified in the following way. First, we construct a differential equation satisfied by the propagator $K(\boldsymbol{r}, t; \boldsymbol{r}', t')$ for a free particle. This may be done by substituting Eq. (7-64) into the Schrödinger equation, Eq. (7-54), and using the condition that, at $t' = t$,

$$K(\boldsymbol{r}, t; \boldsymbol{r}', t) = \delta(\boldsymbol{r} - \boldsymbol{r}').$$

The resulting differential equation has the form

$$i\hbar\frac{\partial}{\partial t}K(\boldsymbol{r}, t; \boldsymbol{r}', t') = -\frac{\hbar^2}{2\mu}\nabla^2 K(\boldsymbol{r}, t; \boldsymbol{r}', t')$$

where ∇^2 operates only on $\boldsymbol{r}$ and not on $\boldsymbol{r}'$. It is obvious that the right side of Eq. (7-66) is a solution of such an equation.

Classical action The argument of the exponential function on the right side of Eq. (7-66) can be shown to be proportional to the classical action. In classical mechanics, the action for the path of a particle that moves from a point $\boldsymbol{r}_i$ at time t_i to another one $\boldsymbol{r}_f$ at time t_f is defined as the following integral over time:

$$\mathcal{S}(\boldsymbol{r}_f, t_f; \boldsymbol{r}_i, t_i) = \int_{t_i}^{t_f} \mathcal{L}\, dt$$

where $\mathcal{L}$ is the lagrangian. For a free particle, $\mathcal{L}$ is independent of time and is given by the kinetic energy

$$\mathcal{L} = \frac{1}{2}\mu \boldsymbol{v}^2.$$

For our example here, the velocity is a constant and may be found using the relation

$$\boldsymbol{v} = \frac{\boldsymbol{r}_f - \boldsymbol{r}_i}{t_f - t_i}.$$

As a result, the classical action has the explicit form

$$\mathcal{S}(\boldsymbol{r}_f, t_f; \boldsymbol{r}_i, t_i) = \frac{1}{2}\mu \boldsymbol{v}^2(t_f - t_i) = \frac{\mu}{2}\frac{(\boldsymbol{r}_f - \boldsymbol{r}_i)^2}{t_f - t_i}.$$

In terms of $\mathcal{S}(\boldsymbol{r}_f, t_f; \boldsymbol{r}_i, t_i)$, we can write the propagator of Eq. (7-66) as

$$K(\boldsymbol{r}_f, t_f; \boldsymbol{r}_i, t_i) = \left\{\frac{\mu}{2\pi i\hbar(t_f - t_i)}\right\}^{3/2} e^{i\mathcal{S}(\boldsymbol{r}_f,t_f;\boldsymbol{r}_i,t_i)/\hbar}. \tag{7-67}$$

Although the result is derived for the case of a free particle, the form is actually more general. If a particle moves in a potential $V(\boldsymbol{r})$ that depends only on the coordinates, the propagator is identical to that given in Eq. (7-67). The only difference is that we must replace the free-particle lagrangian in the action $\mathcal{S}(\boldsymbol{r}_f, t_f; \boldsymbol{r}_i, t_i)$ with the appropriate one that includes the interaction potential. A derivation of the more general case in terms of Hamiltonians can be found in Lee[33] and Ryder.[46]

We can put Eq. (7-65) in a more convenient form by rewriting it in the usual way for carrying out calculations in quantum mechanics. A solution of the Schrödinger equation may be obtained in the following manner. Ignoring for the moment the fact that $\hat{H}$ is an operator, we may rewrite Eq. (7-54) formally as

$$\frac{d\Psi(t)}{\Psi(t)} = -\frac{i}{\hbar}\hat{H}\,dt.$$

Since $\hat{H}$ is independent of time in the Schrödinger representation we are using here, it is possible to integrate both sides of the equation between times t_i and t_f and obtain the result

$$\ln\Psi(t_f) - \ln\Psi(t_i) = -\frac{i}{\hbar}\hat{H}(t_f - t_i).$$

This is equivalent to

$$\Psi(t_f) = e^{-i\hat{H}(t_f - t_i)/\hbar}\Psi(t_i).$$

The meaning of an operator in the argument of a mathematical function may be interpreted in terms of a series expansion of the function, as we shall see later in Eq. (8-96). In bra-ket notation, the relation above may be written as

$$|\Psi(t_f)\rangle = e^{-i\hat{H}(t_f - t_i)/\hbar}|\Psi(t_i)\rangle. \tag{7-68}$$

Using this result, we can express the right side of Eq. (7-65) in terms of a matrix element between states $|\,\boldsymbol{r}_i, t\,\rangle$ and $|\,\boldsymbol{r}_f, t\,\rangle$:

$$K(\boldsymbol{r}_f, t_f; \boldsymbol{r}_i, t_i) = \langle \boldsymbol{r}_f, t_f|e^{-i(t_f - t_i)\hat{H}/\hbar}|\boldsymbol{r}_i, t_i\rangle. \tag{7-69}$$

Let $|\,n\,\rangle$ be a member of a complete set of eigenstates of $\hat{H}$ with eigenvalue E_n,

$$\hat{H}|n\rangle = E_n|n\rangle$$

and

$$\psi_n(\boldsymbol{r}, t_f) = \langle n|\boldsymbol{r}, t_f\rangle.$$

By inserting a complete set of states $|\,n\,\rangle$ into the matrix element on the right side of Eq. (7-69), we obtain the result

$$K(\boldsymbol{r}_f, t_f; \boldsymbol{r}_i, t_i) = \sum_n \psi_n^*(\boldsymbol{r}_f, t_f)\psi_n(\boldsymbol{r}_i, t_f)\, e^{-i(t_f - t_i)E_n/\hbar}. \tag{7-70}$$

In this way, we see that the propagator $K(\boldsymbol{r}_f, t_f; \boldsymbol{r}_i, t_i)$ may be constructed once we have all the eigenfunctions of $\hat{H}$. However, this is not always feasible and the path integral method of Feynman provides us with an alternate method of solving the problem.

Path integral One way to evaluate $K(\boldsymbol{r}_f, t_f; \boldsymbol{r}_i, t_i)$ is to divide the time interval between t_i and t_f into two parts, one from t_i to t_k and the other from t_k to t_f. Using Eq. (7-63), we obtain the wave functions at times t_k and t_f in coordinate space as

$$\langle \boldsymbol{r}_k | \Psi(t_k) \rangle = \iiint d^3 r_i \, K(\boldsymbol{r}_k, t_k; \boldsymbol{r}_i, t_i) \langle \boldsymbol{r}_i | \Psi(t_i) \rangle \tag{7-71}$$

$$\langle \boldsymbol{r}_f | \Psi(t_f) \rangle = \iiint d^3 r_k \, K(\boldsymbol{r}_f, t_f; \boldsymbol{r}_k, t_k) \langle \boldsymbol{r}_k | \Psi(t_k) \rangle. \tag{7-72}$$

It is possible to combine these two results and obtain the wave function $\langle \boldsymbol{r}_f | \Psi(t_f) \rangle$ at time t_f in terms of $\langle \boldsymbol{r}_i | \Psi(t_i) \rangle$ at t_i by substituting the expression for $\langle \boldsymbol{r}_k | \Psi(t_k) \rangle$ given in Eq. (7-71) into the right side of Eq. (7-72). This gives us the relation

$$\langle \boldsymbol{r}_f | \Psi(t_f) \rangle = \iiint d^3 r_i \iiint d^3 r_k K(\boldsymbol{r}_f, t_f; \boldsymbol{r}_k, t_k) K(\boldsymbol{r}_k, t_k; \boldsymbol{r}_i, t_i) \langle \boldsymbol{r}_i | \Psi(t_i) \rangle.$$

Comparing with Eq. (7-63), we obtain the following relation between the propagators:

$$K(\boldsymbol{r}_f, t_f; \boldsymbol{r}_i, t_i) = \iiint d^3 r_k \, K(\boldsymbol{r}_f, t_f; \boldsymbol{r}_k, t_k) K(\boldsymbol{r}_k, t_k; \boldsymbol{r}_i, t_i).$$

Instead of propagators, it is perhaps more instructive to express the same relation in terms of the overlap between two wave functions in the form

$$\langle \boldsymbol{r}_f, t_f | \boldsymbol{r}_i, t_i \rangle = \iiint d^3 r_k \langle \boldsymbol{r}_f, t_f | \boldsymbol{r}_k, t_k \rangle \langle \boldsymbol{r}_k, t_k | \boldsymbol{r}_i, t_i \rangle. \tag{7-73}$$

The physical meaning of the last two equations is that the transition from $(\boldsymbol{r}_i, t_i)$ to $(\boldsymbol{r}_f, t_f)$ may be taken as the sum of transitions from $(\boldsymbol{r}_i, t_i)$ to all the possible intermediate points $\boldsymbol{r}_k$ at time t_k, followed by a transition from $\boldsymbol{r}_k$ to $(\boldsymbol{r}_f, t_f)$. This is similar in spirit to Huygens's principle in geometric optics.

By subdividing the time interval $[t_i, t_f]$ into more than two parts, we can split the path into even smaller segments. For latter convenience, we shall divide the time between t_i and t_f into n equal intervals, each of duration δ_t. Taking $t_0 = t_i$ and $t_n = t_f$, the value of t_k at the end of interval k is given by

$$t_k = t_0 + k\delta_t.$$

For a particular path, let the location of the particle at time t_k be $\boldsymbol{r}_k$. Analogous to Eq. (7-73), the propagator, expressed in the form of the overlap between two wave functions, is now the integral

$$\begin{aligned} \langle \boldsymbol{r}_f, t_f | \boldsymbol{r}_i, t_i \rangle = & \int \cdots \int d^3 r_1 d^3 r_2 \cdots d^3 r_{n-1} \\ & \times \langle \boldsymbol{r}_f, t_f | \boldsymbol{r}_{n-1}, t_{n-1} \rangle \langle \boldsymbol{r}_{n-1}, t_{n-1} | \boldsymbol{r}_{n-2}, t_{n-2} \rangle \cdots \langle \boldsymbol{r}_1, t_1 | \boldsymbol{r}_i, t_i \rangle. \end{aligned}$$

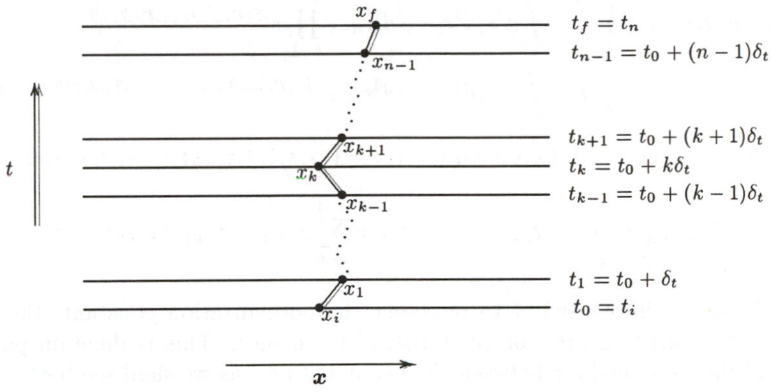

Figure 7-18: Example of a path from (x_i, t_i) to (x_f, t_f). It goes from (x_i, t_i) to (x_1, t_1), then from (x_1, t_1) to (x_2, t_2), and so on, with the final point at (x_f, t_f).

In terms of propagators, this may be put in the following way:

$$K(\boldsymbol{r}_f, t_f; \quad \boldsymbol{r}_i, t_i) = \int \cdots \int d^3r_1 d^3r_2 \cdots d^3r_{n-1}$$
$$\times K(\boldsymbol{r}_f, t_f; \boldsymbol{r}_{n-1}, t_{n-1}) K(\boldsymbol{r}_{n-1}, t_{n-1}; \boldsymbol{r}_{n-2}, t_{n-2}) \cdots K(\boldsymbol{r}_1, t_1; \boldsymbol{r}_i, t_i).$$

The integrations here are taken over all the possible intermediate points to which the various paths can lead. One such path for the simple case of a single spatial dimension is shown in Fig. 7-18 as an illustration. For simplicity, we have approximated the motion from $\boldsymbol{r}_k$ to $\boldsymbol{r}_{k+1}$ by a straight line, the shortest distance joining the two points. As a result, each path consists of a series of zigzags made of short straight segments. In the limit that the number of intervals $n \to \infty$, the entire path becomes a smooth curve.

For each segment of the path, Eq. (7-67) applies and the propagator is given in terms of the classical action $\mathcal{S}(\boldsymbol{r}_{k+1}, t_{k+1}; \boldsymbol{r}_k, t_k)$ for the interval from t_k to t_{k+1}:

$$K(\boldsymbol{r}_{k+1}, t_{k+1}; \boldsymbol{r}_k, t_k) = \left(\frac{\mu}{2\pi i\hbar\delta_t}\right)^{3/2} e^{i\mathcal{S}(\boldsymbol{r}_{k+1}, t_{k+1}; \boldsymbol{r}_k, t_k)/\hbar} \tag{7-74}$$

where

$$\mathcal{S}(\boldsymbol{r}_{k+1}, t_{k+1}; \boldsymbol{r}_k, t_k) = \int_{t_k}^{t_{k+1}} \mathcal{L}\, dt.$$

Substituting this result into Eq. (7-74), we obtain the transition probability amplitude from $(\boldsymbol{r}_i, t_i)$ to $(\boldsymbol{r}_f, t_f)$ in terms of the classical action

$$\begin{aligned} K(\boldsymbol{r}_f, t_f; \boldsymbol{r}_i, t_i) &= \frac{1}{Z}\int \cdots \int d^3r_1 d^3r_2 \cdots d^3r_{n-1} \prod_{k=1}^{n-1} e^{i\mathcal{S}(\boldsymbol{r}_{k+1}, t_{k+1}; \boldsymbol{r}_k, t_k)/\hbar} \\ &= \frac{1}{Z}\int \cdots \int d^3r_1 d^3r_2 \cdots d^3r_{n-1}\, e^{i\mathcal{S}(\boldsymbol{r}_n, t_n; \boldsymbol{r}_{n-1}, t_{n-1}; \cdots; \boldsymbol{r}_0, t_0)/\hbar} \end{aligned} \quad (7\text{-}75)$$

where, for convenience, we have used $(r_n, t_n) = (r_f, t_f)$, $(r_0, t_0) = (r_i, t_i)$, and

$$\mathcal{S}(\boldsymbol{r}_n, t_n; \boldsymbol{r}_{n-1}, t_{n-1}; \cdots; \boldsymbol{r}_0, t_0) = \sum_{k=0}^{n-1} \mathcal{S}(\boldsymbol{r}_{k+1}, t_{k+1}; \boldsymbol{r}_k, t_k).$$

In Eq. (7-75), we have used Z to represent the normalization constant, the same symbol as the partition function in statistical mechanics. This is done on purpose because of the close analogy between the two quantities, as we shall see better later.

There are two major advantages in using the path integral method for solving quantum mechanical problems and both are contained in Eq. (7-75). The first is that only the classical action enters into the calculation of the propagator. Since $\mathcal{S}(\boldsymbol{r}_n, t_n; \boldsymbol{r}_{n-1}, t_{n-1}; \cdots; \boldsymbol{r}_0, t_0)$ does not involve operators and wave functions, it is, in general, easier to evaluate. The usual quantum mechanical approach of solving the time-dependent Schrödinger equation, such as that used in arriving at Eq. (7-70), is applicable only for simple cases. For more realistic situations, approximations such as perturbation methods are often used. However, if the interaction involved is not weak, perturbation techniques do not apply and the path integral method becomes one of the few available alternatives. The second major advantage is that the action is evaluated in small time intervals. By making the interval δ_t sufficiently small, it is possible to approximate the integral by the product of the average value of the integrand and δ_t, similar to what we have done in the rectangular rule of numerical integration. This greatly simplifies the method and makes it possible to solve problems involving complicated interactions, such as those occurring between quarks in quantum chromodynamics.

The connection between the path integral approach and Monte Carlo calculations is also demonstrated by Eq. (7-75). For good accuracy, it is necessary to make the number of intervals n large. This, in turn, means that there are many possible paths. In fact, in the limit that $n \to \infty$, the number of possible paths is also infinite. In the same spirit as that used in the Ising model, it is not necessary to consider all the possible paths. A sample is taken and all the physically interesting quantities are evaluated as averages over the paths sampled. The analogy to statistical mechanics is sufficiently close that the application of the path integral method to quantum field theory has been given the name statistical field theory.

Application to a harmonic oscillator As an illustration, we shall carry out a trivial application of the path integral method using Monte Carlo techniques. For simplicity, consider only a single spatial dimension x. Mathematically, Eq. (7-70) holds also if we take $t_i = 0$ and $t_f = -it_{\text{total}}$. Furthermore, if the Hamiltonian is time independent, the basis states $\psi_n(x)$ can also be made to be time independent. Under these conditions, Eq. (7-70) reduces to the form

$$K(x_f, -it_{\text{total}}; x_i, 0) = \sum_n \psi_n^*(x_f)\psi_n(x_i)e^{-E_n t_{\text{total}}/\hbar}.$$

In the limit of large t_{total}, the term with the smallest energy E_n dominates the sum on the right side. As a result, the relation may be approximated as one for the absolute square of the ground state wave function:

$$|\psi_0(x)|^2 \approx e^{E_0 t_{\text{total}}/\hbar} K(x, -it_{\text{total}}; x, 0). \tag{7-76}$$

Thus, by evaluating the propagator $K(x, -it_{\text{total}}; x, 0)$ for some large value of t_{total}, we can obtain the absolute square of the ground state wave function of a system.

For our illustrative example, we shall use the potential

$$V(x) = \frac{1}{2}\mu\omega^2 x^2 \tag{7-77}$$

given earlier in Eq. (4-1) for a particle of mass μ in a one-dimensional harmonic oscillator well with frequency ω. To calculate the propagator using Eq. (7-75), we need to sum over the classical action $\mathcal{S}$ for a sequence of small time intervals, each of duration δ_t. In each interval, the action is given by

$$\mathcal{S}(x_{k+1}, t_{k+1}; x_k, t_k) = \int_{t_k}^{t_{k+1}} \mathcal{L}\, dt.$$

In terms of the complex time τ, given by $t = -i\tau$, this integral takes on the form

$$\mathcal{S}(x_{k+1}, -i\tau_{k+1}; x_k, -i\tau_k) = -i\int_{\tau_k}^{\tau_{k+1}} \mathcal{L}\, d\tau.$$

Since we have restricted ourselves to the nonrelativistic limit and our potential is time independent, the classical lagrangian is

$$\mathcal{L} = \frac{\mu}{2}\left(\frac{dx}{dt}\right)^2 - V(x).$$

On replacing t by $-i\tau$, we have

$$\mathcal{L} = -\frac{\mu}{2}\left(\frac{dx}{d\tau}\right)^2 - V(x).$$

The right side has the same form as the negative of the total energy, with τ taking over the role of time.

If the complex time interval $\delta_\tau = i\delta_t$ is sufficiently short, the classical action may be approximated by the average value in the interval. The result is

$$\mathcal{S}(x_{k+1}, -i\tau_{k+1}; x_k, -i\tau_k) = -i\int_{\tau_k}^{\tau_{k+1}} \{-E(x,\tau)\}\, d\tau \approx i\overline{E}(x_k, \tau_k)\delta_\tau \tag{7-78}$$

where $\overline{E}(x_k, \tau_k)$ is the average value of the total energy (in terms of complex time τ) in the interval τ_k to τ_{k+1} and δ_τ is the length of the interval ($\delta_\tau = \tau_{k+1} - \tau_k$). The sum of actions over a complete path is then

$$\begin{aligned}\mathcal{S}(x, \tau_n; x_{n-1}, \tau_{n-1}; \ldots; x, 0) &= \sum_{k=1}^{n-1} \mathcal{S}(x_{k+1}, -i\tau_{k+1}; x_k, -i\tau_k) \\ &\approx i\delta_\tau \overline{E}(x, x_1, x_2, \ldots, x_{n-1}, x) = i\delta_\tau \sum_{k=0}^{n-1} \overline{E}(x_k, \tau_k)\end{aligned}$$

where we have put $x = x_0$ in obtaining the final equality.

Using Eq. (7-75), the propagator becomes

$$\begin{aligned}K(x, -it_{\text{total}}; x, 0) &= \frac{1}{Z}\int \cdots \int dx_1 dx_2 \cdots dx_{n-1}\, e^{i\mathcal{S}(x,\tau_n;x_{n-1},\tau_{n-1};\cdots;x,0)/\hbar} \\ &= \frac{1}{Z}\int \cdots \int dx_1 dx_2 \cdots dx_{n-1}\, e^{-\frac{\delta_\tau}{\hbar}\overline{E}(x_1,x_2,\ldots,x_{n-1})}.\end{aligned}$$

By inserting this result into Eq. (7-76), we obtain the absolute square of the ground state wave function in the form

$$|\psi_0(x)|^2 = \frac{e^{E_0 t_{\text{total}}/\hbar}}{Z}\int \cdots \int dx_1 dx_2 \cdots dx_{n-1}\, e^{-\frac{\delta_\tau}{\hbar}\overline{E}(x,x_1,x_2,\ldots,x_{n-1},x)}. \qquad (7\text{-}79)$$

Since the wave function is normalized to unity,

$$\int_0^\infty |\psi_0(x)|^2 dx = 1$$

the constant Z here is given by integrating over x the integral on the right of Eq. (7-79),

$$Z = e^{E_0 t_{\text{total}}/\hbar}\int \cdots \int dx\, dx_1 dx_2 \cdots dx_{n-1}\, e^{-\frac{\delta_\tau}{\hbar}\overline{E}(x,x_1,x_2,\ldots,x_{n-1},x)}.$$

It has the same form as a partition function in statistical mechanics when we make the analogy of $\delta_\tau/\hbar$ with $1/kT$.

Again, it is convenient in our calculations to convert the Schrödinger equation of Eq. (7-54) into a simpler form by defining a dimensionless length η and a dimensionless time ξ using the following relations:

$$x = \eta\sqrt{\frac{\hbar}{\mu\omega}} \qquad\qquad t = \xi\frac{1}{\omega}.$$

On replacing (x, t) by (η, ξ) in Eq. (7-54), we obtain for the harmonic oscillator potential of Eq. (7-77),

$$i\frac{\partial\psi}{\partial\xi} = -\frac{1}{2}\left\{\frac{\partial^2\psi}{\partial\eta^2} - \eta^2\right\}. \qquad (7\text{-}80)$$

The Hamiltonian and all the energies are now in units of $\hbar\omega$.

In the time interval between τ_k and τ_{k+1}, the contribution to the classical action may be approximated by the average of the sum of kinetic and potential energies in the interval. Since x_k is the location of the particle at time τ_k, the kinetic energy during the small time interval $[\tau_k, \tau_{k+1}]$ is proportional to $(x_{k+1} - x_k)^2$. Furthermore, since we are measuring the time in units of ω^{-1}, the appropriate form of the kinetic energy contribution to the action for the time interval is

$$T_{[\tau_k,\tau_{k=1}]} = \frac{1}{2}\mu\Big(\frac{x_{k+1} - x_k}{\tau_{k+1} - \tau_k}\Big)^2 = \frac{1}{2}\mu\omega^2(x_{k+1} - x_k)^2.$$

The classical action of Eq. (7-78) now takes on the form

$$\begin{aligned}\mathcal{S}(x_{k+1}, -i\tau_{k+1}; x_k, -i\tau_k) \;&\approx\; -i\delta_\tau\Big\{\frac{\mu}{2}(x_{k+1} - x_k)^2 + \frac{\mu\omega^2}{2}\Big(\frac{x_{k+1} + x_k}{2}\Big)^2\Big\}\\ &=\; -i\frac{\hbar\delta_\xi}{2}\{(\eta_{k+1} - \eta_k)^2 + \tfrac{1}{4}(\eta_{k+1} + \eta_k)^2\}\end{aligned}$$

where

$$\delta_\tau = \delta_\xi\frac{1}{\omega}$$

and η_k is the location of the kth point of the path in units of $\sqrt{\hbar/(\mu\omega)}$. In time interval δ_τ, the particle moves from η_k to η_{k+1}. The contribution to the kinetic energy from the kth time interval is then $(\eta_{k+1} - \eta_k)^2$ in units of $\hbar\omega$. For the contribution from the potential energy term, we can take the average value in the interval by assuming the particle to be located at the middle point between η_k, its position at the beginning of the interval, and η_{k+1}, its position at the end of the interval.

The square of the ground state wave function given in Eq. (7-79) is then

$$|\psi_0(\eta)|^2 = \frac{e^{E_0 t_{\text{total}}/\hbar}}{Z}\int\cdots\int d\eta_1 d\eta_2\cdots d\eta_{n-1}\, e^{-\frac{1}{2}\delta_\xi\overline{E}(\eta,\eta_1,\eta_2,\ldots,\eta_{n-1},\eta)} \tag{7-81}$$

with

$$\overline{E}(\eta, \eta_1, \eta_2, \ldots, \eta_{n-1}, \eta) = \sum_{k=0}^{n-1}\{(\eta_{k+1} - \eta_k)^2 + \tfrac{1}{4}(\eta_{k+1} + \eta_k)^2\}. \tag{7-82}$$

It is advantageous to establish an evenly spaced grid system for η in the numerical calculation by choosing a constant step size δ_η. All the values of our dimensionless length η can now be specified as integer multiples of the step size

$$\eta_k = m_k\delta_\eta.$$

In terms of δ_η and δ_ξ, the energy for a path is given as

$$\overline{E}(\eta_1, \eta_2, \ldots, \eta_{n-1}) = \delta_\eta^2\sum_{k=0}^{n-1}\{(m_{k+1} - m_k)^2 + \tfrac{1}{4}(m_{k+1} + m_k)^2\}.$$

As we shall see later, this choice depends also on the step size δ_ξ for time as well.

We are now ready to calculate the value of $|\psi_0(\eta)|^2$ using Eq. (7-81). The Metropolis algorithm is used to sample possible paths given by η, η_1, $\eta_2, \ldots$, η_n. Since we are calculating the absolute square of a wave function, rather than some sort of transition amplitude that takes the system from one location to another, the last point of the path η_n must coincide with the starting point η. For a given path, the weighting factor of its contributions to the ensemble average is $\exp\{-\frac{1}{2}\delta_\eta \overline{E}(\eta, \eta_1, \eta_2, \ldots, \eta_{n-1}, \eta)\}$. Since this factor is included in selecting the path, each one chosen by the algorithm contributes unity to the (unnormalized) value of $|\psi_0(\eta)|^2$. After a large number of samples are taken, the values of $|\psi_0(\eta)|^2$ for all η are normalized by the total number of paths sampled.

The efficiency of the calculation may be improved by making the following adjustments. Since the last point in the path must be located at the same point in space as the starting point, a selected path $\{\eta_0, \eta_1, \ldots, \eta_{n-1}, \eta_n = \eta_0\}$ that begins and ends at η_0 may also be regarded as a path that starts at location η_1. This can be done by moving the first point η_0 to position η_n. The result is a path going through the points $\{\eta_1, \ldots, \eta_{n-1}, \eta_0, \eta_{n+1} = \eta_1\}$. The energy is not changed by this move, as can be seen from Eq. (7-82). However, we have a "new" path that begins and ends at $\eta = \eta_1$ and contributes to $|\psi_0(\eta)|^2$ at $\eta = \eta_1$. This process may be repeated by moving η_2 to the point η_{n+2} and we have a path for $\eta = \eta_2$. This argument can be continued until we come to η_{n-1}. As a result, each selected path may be counted as n different paths, each contributing to $|\psi_0(\eta)|^2$ at $\eta = \eta_0, \eta_1, \ldots, \eta_{n-1}$ (and adding n to the normalization).

In principle, we can apply the Metropolis algorithm in the same way as we did earlier in the Ising model by going through each point $k = 0, 1, \ldots, (n-1)$, one by one and selecting a new path by varying the value of η_k. Because of the constraint that the path must terminate at the same spatial point as the starting point of the path, a path has to be rejected if $\eta_n \neq \eta$. Since this occurs at the end of constructing a new path, it becomes rather wasteful, as a lot of work has already gone into constructing the partial path. A more effective alternative is to select a point k randomly along the path and see if its location should be varied according to the rules of the Metropolis algorithm. The trial paths are constructed by varying only η_k for amounts that are δ times the step size. Using Eq. (7-82), the change in energy because of $\eta_k \to \eta_k + \delta\delta_\eta$ or $m_k \to m_k + \delta$ is given by

$$\begin{aligned}\Delta_k\overline{E} &= \frac{\delta_\eta^2}{2}\Big\{(m_{k+1} - m_k - \delta)^2 - (m_{k+1} - m_k)^2 + (m_k + \delta - m_{k-1})^2 - (m_k - m_{k-1})^2 \\ &\quad + \tfrac{1}{4}\Big[(m_{k+1} + m_k + \delta)^2 - (m_{k+1} + m_k)^2 + (m_k + \delta + m_{k-1})^2 - (m_k + m_{k-1})^2\Big]\Big\} \\ &= \delta_\eta^2\{\delta^2 + \delta(2m_k - m_{k-1} - m_{k-1}) + \tfrac{1}{4}[\delta^2 + \delta(2m_k + m_{k-1} + m_{k+1})]\}.\end{aligned}$$

Similar to Eq. (7-50), the trial path is accepted if

$$r = \exp -\left\{\frac{\delta_\xi}{2}\Delta_k\overline{E}\right\} \tag{7-83}$$

Box 7-8 Program DM_PTHIT
Path integral calculation of harmonic oscillator wave function

Subprograms called:
- THE_VAL: Compares the results with theoretical values.
- PROB_NRML: Normal probability function by rational polynomial approximation.
- M_RATIO: Generate the Metropolis probability ratios of Eq. (7-84).
- V_PATH: Sampling the path.
- THERM: Thermalization sweeps.
- PATH_PRINT: Output the path.
- RSUB: Uniform random number generator (Box 7-2).

Parameters:
- t_{total}: Total time (=30).
- n: Number of intervals (=128).
- δ_η: Spatial step size (input as $\alpha\delta_\xi = \alpha t_{\text{total}}/n$).
- n_{therm}: Number of thermalization steps (=300).
- n_{var}: Number of random points to select a new path (=300).

Initialization:
- (a) Zero the array for $|\psi(\eta_k)|^2$.
- (b) Generate a set of n random numbers $\{x_k\}$ and use $5x_k$ as the initial path.
- (c) Use M_RATIO to construct the decision table $r_{\text{k.e.}}$ and $r_{\text{p.e.}}$ of Eq. (7-84).

1. Use THERM to carry out the following steps n_{therm} times for thermalization:
   - (a) Generate a random number x and use $k = (nx + 1)$ for path variation.
   - (b) Generate a random number x', trial variation is $+1$ for $x' > 0.5$, -1 otherwise.
   - (c) Determine the change in energies for a move of η_k by ± 1.
   - (d) Calculate r of Eq. (7-83) using (7-84) if not in the stored decision tables.
   - (e) If $r_k \geq 1$, the trial variation is accepted. If $r_k < 1$,
     - (i) Generate another uniform random number x'' in $[0, 1]$.
     - (ii) If $x'' \leq r_k$, the trial variation is accepted; otherwise it is rejected.
2. Use V_PATH to vary the path — same as THERM except $|\psi(\eta_k)|^2$ is increased by unity each time.
3. Use THE_VAL to normalize and output the value of $|\psi(\eta_k)|^2$.

is greater than 1. For $r < 1$, it is accepted with a probability r (that is, a uniform random number x in the interval $[0, 1]$ is generated and the trial path is accepted only if $x \leq r$). Since every point along an accepted path contributes to $|\psi_0(\eta)|^2$, we add 1 to $|\psi_0(\eta)|^2$ at $\eta = \eta_k + \delta$ if the new path is accepted and to $|\psi_0(\eta)|^2$ at $\eta = \eta_k$ if the new path is rejected.

For the convenience of calculation, we shall take the absolute value of δ to be a constant; only the sign is arbitrary. Thus, in addition to using a random number to determine the point k, we need another random number (a random bit will be adequate) to determine the sign of δ. The obvious choice for the value of $|\delta|$ is the step size for our dimensionless length η. As a result, we have $\delta = \pm 1$, with the sign determined by the random number generated.

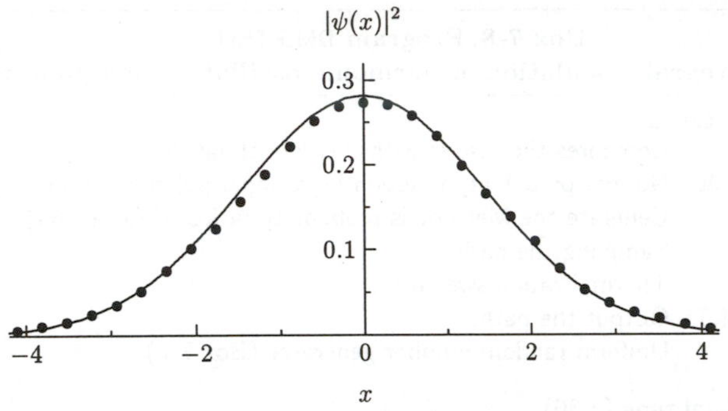

Figure 7-19: Absolute square of the ground state wave function of a harmonic oscillator calculated using the path integral method (dots) compared with the exact results (solid curve).

The efficiency of the Metropolis algorithm may be further improved by decomposing r of Eq. (7-83) into a product of kinetic and potential energy parts:

$$r_{\text{k.e.}} = (b_{\text{k.e.}})^{i_k} \qquad r_{\text{p.e.}} = (b_{\text{p.e.}})^{i_p} \tag{7-84}$$

where

$$b_{\text{k.e.}} = \exp -\{\delta_\eta^2 \delta_\xi\} \qquad i_k = 1 \pm (2m_k - m_{k+1} - m_{k-1})$$
$$b_{\text{p.e.}} = \exp -\{\tfrac{1}{4}\delta_\eta^2 \delta_\xi\} \qquad i_p = 1 \pm (2m_k + m_{k+1} + m_{k-1}).$$

In this way, the values of $b_{\text{k.e.}}$ and $b_{\text{p.e.}}$ most often encountered may be kept as two one-dimensional arrays. For these cases, the calculation of r may be carried out as a product of the stored values of $b_{\text{k.e.}}$ and $b_{\text{p.e.}}$ recalled from the arrays. In this way, explicit calculations of the corresponding factors are needed only for values of i_k or i_p that are outside the range of the stored values.

We need a starting path for the Metropolis algorithm. This may be chosen arbitrarily. However, before we can begin taking samples, it is necessary to "thermalize" the initial choice, just as we did earlier for the Ising model. The steps involved in the thermalization process are essentially the same as in selecting a path.

In addition to the initial path, three other quantities must also be chosen at the start of the calculation, the total time t_{total}, the number of time intervals n, and the step size δ_η in x. To make the ground state the dominant component in Eq. (7-76), we have assumed that $t_{\text{total}} \to \infty$. In practice, a finite value must be adopted for t_{total}. As a result, we always have small admixtures of excited states in the value of $|\phi(x)|^2$ obtained. A comparison of the results calculated using different values of t_{total} was given by Lawande, Jensen, and Sahlin.[32] In general, values of $t_{\text{total}} > 20$

(in units of ω^{-1}) were found to be necessary for the harmonic oscillator potential to obtain good agreements with the analytical form of

$$\psi_0(\eta) = \pi^{-1/4} e^{-\eta^2/2} \tag{7-85}$$

given earlier in Eq. (4-9). For the results shown in Fig. 7-19, a value of $t_{\text{total}} = 30$ was used.

The number of steps for the path to develop from $\tau = 0$ to $\tau = t_{\text{total}}$ depends on the numerical accuracy we wish to achieve in the calculation. For the illustration, a value of 128 was used. The choice depends in part on the step size δ_η. Since the ground state wave function for a harmonic oscillator is essentially confined to the region $-3 \leq \eta \leq 3$, a sufficient number of η_k must lie within this interval. A value a few times t_{total}/n seems to be satisfactory. The actual steps of the calculations are summarized in Box 7-8.

Lattice gauge calculation The path integral method is used extensively in field theoretical calculations where perturbation techniques are inappropriate. One of the more interesting applications involves strong interactions between quarks in quantum chromodynamics (QCD). Here, the strength of the interaction at energies accessible to laboratory experiments is too strong for methods based on perturbation expansions. In 1974, Wilson[58] suggested the use of a numerical approach that is now known as lattice gauge calculations. The word lattice comes from the fact that the space is divided into grids similar to what we have done in the example above. However, the problem is much more complicated here since we must take the coordinate space to be three dimensional. A modest grid that divides each spatial direction into 16 parts gives us $(16)^3 = 4096$ points. In contrast to the single degree of freedom of $\delta = \pm 1$ at each site in our harmonic oscillator example, there are now many more, given by the product of the different types of quarks and the degrees of freedom each quark can assume, such as color and spin. It is obvious that an exact calculation is impossible and sampling of the multidimensional space becomes the only viable alternative. Although the basic calculations involved are not too different from our simple example above, the time development of such a complex system becomes far more complicated. In fact, the calculations have become one of the challenges of computational physics and computers.

7-7 Fractals

Among the new studies initiated by computers, fractals are perhaps the one area that received the widest attention. From a relatively obscure mathematical curiosity, which may be traced back to its origins in the early 1960s, fractals have grown to be a popular form of computer recreation as well a branch of serious studies. What are fractals? Why are they of interest to physicists? It is perhaps more useful to begin the discussion by examining how the two diagrams shown in Figs. 7-20 and 7-21 are made.

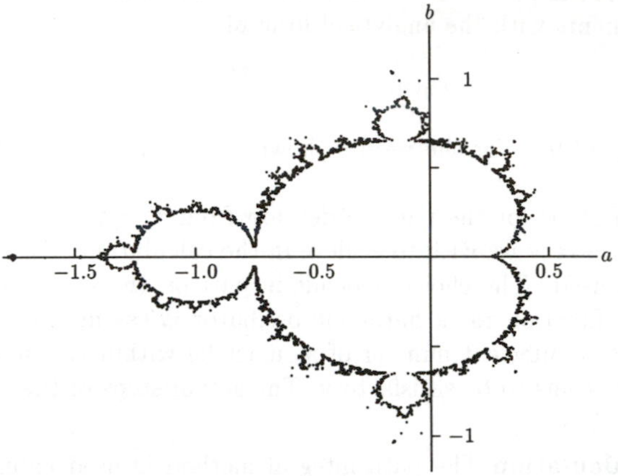

Figure 7-20: A Mandelbrot set made of complex numbers $c = a + ib$. The ones in the area bound by the curve shown have the property that the magnitude of $z_n = z_{n-1}^2 + c$ remains finite when $n \to \infty$.

Mandelbrot set and Julia set Let $z_n = x_n + iy_n$ be a complex number defined by the recursive relation

$$z_n = z_{n-1}^2 + c \tag{7-86}$$

where $c = a + ib$ is another complex number. If we start the recursion from some value, such as $z_0 = 0$, and repeat several times the process of generating z_n from z_{n-1} in the way specified by Eq. (7-86), we find that there are two possible results. The first is that the magnitude of z_n becomes unbound as n increases. The second is that the magnitude of z_n remains finite regardless of how many times we iterate the relation. The outcome clearly depends on the value of c used. It is therefore possible to divide the set of all complex numbers $\{c\}$ into two groups according to the behavior after iterating Eq. (7-86) a large number of times. A Mandelbrot set is a group that contains all c such that

$$\lim_{n\to\infty} |z_n| = \text{finite}.$$

A graphical display of the set may be carried out by plotting c in the complex plan. In Fig. 7-20, complex numbers $c = a + ib$ belonging to a Mandelbrot set are represented by points within the area enclosed by the curve. Points outside the area belong to the other group, and they produce values of z_n whose magnitudes become infinite eventually as we apply Eq. (7-86) more and more times.

Instead of c, we can also separate the starting complex numbers z_0 into two groups according to the criterion of whether the magnitude of z_n stays finite or not after carrying out the iteration Eq. (7-86) an infinite number of times. A collection

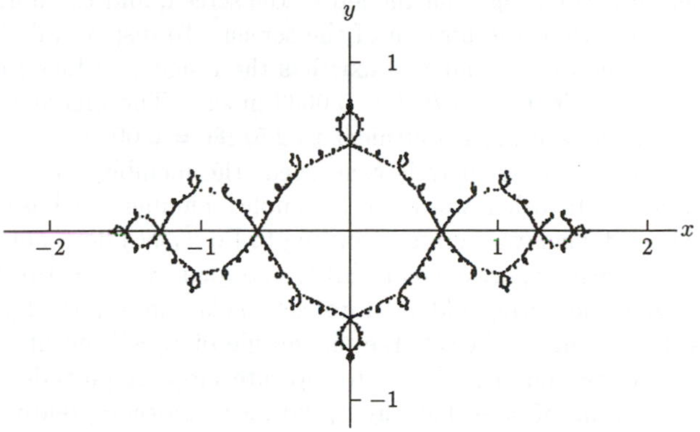

Figure 7-21: A Julia set containing all the complex numbers $z_0 = x + iy$ for which $|z_n|$, with $z_n = z_{n-1}^2 + c$, remains finite as $n \to \infty$. ($c = -1$ here.)

of z_0 that leads to finite z_n is know as a Julia set. Since the result of iterating Eq. (7-86) now depends on the value of c used, there are many possible Julia sets, each corresponding to a different choice of the complex number c. The set enclosed by the curve in Fig. 7-21 corresponds to the choice $c = -1$. Other choices are also possible and many of them display interesting structures.

Coloring a Mandelbrot plot The usual pictures that one sees of Mandelbrot and Julia plots are very colorful and fascinating, such as those that appeared in the *Scientific American* articles by A.K. Dewdney[16] and references therein. The rendition of color for these plots is done in the following way. It is sufficient for us to consider here only Mandelbrot sets, as it is quite easy to adapt the same method to Julia sets. For simplicity, we shall treat only the situation of displaying the results on the color monitor of a computer. The screen of a computer monitor is divided into a number of pixels. We shall take the total number of available pixels to be $(N_x \times N_y)$. That is, there are N_x dots in a horizontal line and N_y vertical lines on a screen. For a color monitor, each pixel can appear in a number of different colors and we shall take the number to be N_c. Typical values of N_x, N_y, and N_c are, respectively, 640, 480, and 16, or higher.

To display a Mandelbrot set, we shall consider the screen as the complex plane and each pixel on the screen is a point in this plane. Following the usual convention in displaying complex numbers, we shall use the horizontal direction to represent the real part and the vertical direction, the imaginary part. To put a complex number $c = a + ib$ on the screen, we need to select an origin and a scale factor. For Fig. 7-20, we have assumed that there are 640 horizontal pixels on a line and 400 lines on the screen. The origin of our coordinate system is at pixel (461,236). The vertical axis is then along pixels 461, counting from the left of the screen, and the horizontal axis along line 236, counting from the bottom of the screen. To display all the numbers belonging to the Mandelbrot set, our real axis has the range of values from -1.8 to 0.7, with each pixel equivalent to $2.5/640 = 0.0039$ in size. The imaginary axis goes from -1.25 to 1.25, with each pixel equivalent to $2.5/480 = 0.0052$.

As far as the color of the display is concerned, the members of a Mandelbrot set are not of interest. In other words, any complex number c whose magnitude never becomes unbound when we iterate according to Eq. (7-86) does not produce an interesting effect. The rich display of color that is so impressive in pictures of fractals is obtained from regions bordering a Mandelbrot set. This comes in the following way. If a point c does not belong to the set, the magnitude of z_n will eventually become very large. The criterion commonly used to indicate large magnitude is $|z_n| > 2$. We shall not attempt a proof here that any z_n with a magnitude greater than 2 will eventually become unbound if we iterate further with Eq. (7-86), and we shall simply take it as the limit for z_n to be unbound. For a Mandelbrot set, different c outside the set reach this limit at different rates. For example, some c may only take $n = 5$ iterations to reach the limit of $|z_n| > 2$, whereas others may take $n = 100$. If we associate a color with each value of n for a complex number c to reach the level of $|z_n| > 2$, a colorful picture will be produced.

In an actual computation, we must choose some value n_{max} as the largest number of iterations we wish to carry out. For all practical purposes, n_{max} is the "infinity" for our iteration process. To generate a plot, each complex number c is now subject to the recursive process given by Eq. (7-86), starting with $z_0 = 0$. At the end of each iteration, the magnitude of z_n is checked. If $|z_n| > 2$, the complex number c is not a member of the Mandelbrot set. The iteration process stops and the pixel corresponding to this value of c is given a color by the following procedure. Let us use the symbol n_{esc} to represent the number of iterations taken. Since the number of available colors N_c is usually smaller than n_{max}, we cannot assign a unique color to every possible value of n_{esc}. Instead, we shall associate the pixel with a color I_c, given by the relation

$$I_c = n_{\text{esc}} \bmod N_c. \tag{7-87}$$

In this way, I_c is a number in the range $[0, N_c]$, and each integer value, 0, 1, 2, ..., corresponds to a particular color, such as black, blue, green, ... If the value of $|z_n|$ after an iteration is less than or equal to 2, the iteration process is repeated. If after n_{max} iterations, $|z_n|$ is still less than or equal to 2, we have a member of the Mandelbrot set. The pixel associated with this value of c is colored black. Obviously, our picture is more colorful if we have larger values of N_c at our disposal. Similarly, our picture will be more interesting, and more accurate, if we are willing to tolerate larger values of n_{max}.

For better efficiency, it is advantageous to carry out the complex arithmetics involved in terms of the real numbers, as we saw earlier in discussions related to complex matrices in §5-7. This can be done by handling the real and imaginary parts

Box 7-9 Program DM_MNDLB
Make a Mandelbrot plot

Subprograms used (Computer system dependent):
SET_PXL: Turn the pixel at (IX,IY) to color KOL.
ST_LINE: Bring the present pixel position to (IX,IY) with color KOL.
Parameters in the calculation:
N_x: Number of pixels in a horizontal line (= 640).
N_y: Number vertical lines on the screen (= 480).

1. Input:
   (a) The range of complex number c.
   x_{start}: Minimum value of the real part.
   x_{range}: Range of the real part.
   y_{start}: Minimum value of the imaginary part.
   y_{range}: Range of the imaginary part.
   (b) $n_{\max}$: The maximum number of iterations allowed.
2. Find the correspondence between c and pixel position.
3. Initialize the graphics screen (system dependent).
4. Iterate each value of c in the range using Eq. (7-86):
   (a) Imaginary part (vertical direction): For $i_y = 1$ to N_y:
   (i) Start with the minimum value specified in the input.
   (ii) Increase by a pixel each time i_y increases by 1.
   (b) Real part (horizontal direction): For $i_x = 1$ to N_x:
   (i) Start with minimum value specified in the input.
   (ii) Increase by a pixel each time i_x increases by 1.
   (c) Iterate z_n according to Eq. (7-86) up to a maximum of $n_{\max}$ times:
   (i) Start with $x_0 = y_0 = 0$.
   (ii) Calculate x_n and y_n using Eq. (7-88) and let $r^2 = x_n^2 + y_n^2$.
   (iii) If $r^2 > 4$, color the pixel with color index I_c given by Eq. (7-87).
   (iv) If $r^2 \leq 4$ after $n_{\max}$ iterations, color the pixel black.

of the complex number calculation in Eq. (7-86) separately:

$$x_n = x_{n-1}^2 - y_{n-1}^2 + a \qquad\qquad y_n = 2x_{n-1}y_{n-1} + b \tag{7-88}$$

where we have used x_n and y_n to be, respectively, the real and imaginary parts of z_n. Similarly, a is the real part of c and b, the imaginary part.

In making an actual Mandelbrot plot, it is more convenient to think in terms of each individual pixel on the screen and translate it to the corresponding value of complex number c, rather than the other way around. The calculations involved may be thought of in terms of three loops. The first scans the N_y rows in the entire screen one at a time, starting from the bottom. For each row, the second loop examines the N_x pixels one by one. For each c associated with a particular pixel, a third loop handles the iteration given by Eq. (7-86), using $z_0 = 0$ as the starting point.

Anytime the magnitude of z_n becomes greater than 2, it is output to the screen with a color index I_c given by Eq. (7-87). If after $n_{\max}$ iterations the magnitude of z_n remains less than or equal to 2, the complex number is considered to be a member of the Mandelbrot set and the pixel is given the color black (blank in Fig. 7-20). The algorithm outlined in Box 7-9 is quite simple, but it may take a long time to run if $n_{\max}$ is chosen to be a large number. For the plot in Fig. 7-20, a modest value of $n_{\max} = 16$ is used. However, for the more colorful plots of a small part of the border area, much larger numbers of iterations are required. Values of $n_{\max}$ in excess of 100 are quite common and this can be very time consuming on a computer. Some of the interesting ranges of c examined are given in the article by Dewdney cited above.

The construction of a Julia plot is very similar to that of a Mandelbrot plot. The pixels on a computer screen now represent points in the complex plane for the possible values of z_0, and each plot is for a fixed value of c. The color of each pixel is assigned according to the number of iterations it takes to start with z_0 and end up with a magnitude of z_n that is greater than 2. Instead of the value $c = -1$ used in Fig. 7-21, different Julia plots can be obtained with other values.

Fractals and fractal dimension The beauty of fractal pictures, however, does not provide us with much insight into the question of what fractals are and the ways in which they are useful in physics. The usual definition of fractals is that they are objects with fractional dimension. To understand this definition, we must first describe what is the meaning of the term "dimension."

Consider an ordinary object, for example, a cube with each side of length ℓ. The volume is equal to ℓ^3. For this reason, it is called a *three-dimensional* object. More formally, we can say the object has dimension $d = 3$. A printed page, on the other hand, is a two-dimensional object ($d = 2$), as the thickness of the paper on which it is printed is immaterial. A line is a one-dimensional object ($d = 1$), since it has no width or thickness. For our discussion here, it is convenient to define dimension in a slightly different way. For most objects, the density is a constant. As a result, the mass of a three-dimensional object is proportional to the third power of its linear size. Similarly, the mass of a sheet of paper (of constant thickness and density) is proportional to the square of its linear dimension, and the mass of a piece of wire (of constant cross section) is proportional to the length. In other words, the mass $\mu(\ell)$ of an object, with linear dimension ℓ, is given by the relation

$$\mu(\ell) \sim \ell^d \tag{7-89}$$

where d is the dimension of the object.

More generally, the same relation can be expressed in terms of a scale change. If we double the linear dimension of a three-dimensional object, the mass is increased by 2^3. In general, the scaling relation may be written as

$$\mu(\lambda\ell) = \lambda^d \mu(\ell). \tag{7-90}$$

This is a better definition for dimension d than Eq. (7-89), since it is in the form of a mathematical identity.

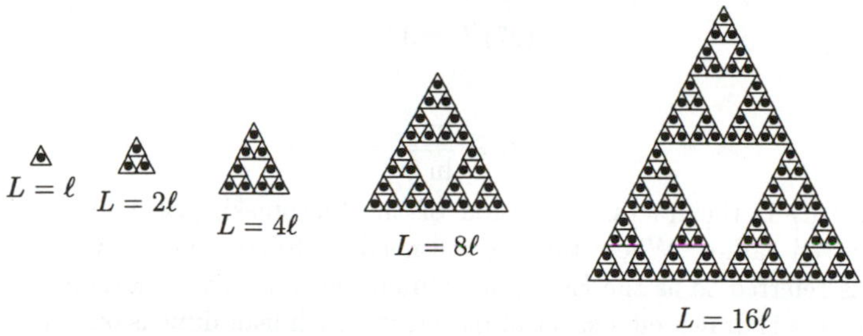

Figure 7-22: The Sierpinski gasket as an example of self-similar objects, by repeatedly putting three identical equilateral triangles on top of each other to form one with twice the length of each side.

Almost all the objects we encounter in daily life have integer values for their dimensions. However, there is also a class of objects we can construct that has non-integer dimensions, or fractals. Before we go into the subject on the significance of fractals, we shall first give the example of a Sierpinski gasket to see how objects with fractional dimensions can occur.

Consider an equilateral triangle with all three sides having the same length ℓ. For the ease of argument, we shall take the mass of each such triangle to be unity. Next, we put three such triangles together in such a way that the top corner of triangle 2 touches the lower-left corner of triangle 1, the top corner of triangle 3 touches the lower-right corner of triangle 1, and the lower-right corner of triangle 2 touches the lower-left corner of triangle 3. The result is a triangle with the length of each side $L = 2\ell$, as shown in Fig. 7-22. Since the $L = 2\ell$ triangle is made of three $L = \ell$ triangles, its mass is three units, or

$$\mu(2\ell) = 3\mu(\ell).$$

We can now repeat the process by putting three $L = 2\ell$ triangles together in the same way as we did in forming the $L = 2\ell$ triangles themselves. This results in a triangle with the length of each side $L = 4\ell$. The mass of this new triangle is

$$\mu(L = 4\ell) = 9\mu(\ell)$$

as can be seen by looking at Fig. 7-22. By the same token, the mass of an $L = 8\ell$ triangle is

$$\mu(L = 8\ell) = 27\mu(\ell).$$

The process can be repeated as many times as we wish. Each time the length of the resulting triangle is doubled and the mass is tripled. The scaling relation is then

$$\mu(2^n\ell) = 3^n\mu(\ell).$$

Comparing with Eq. (7-90), we obtain the result

$$(2^n)^{d_f} = 3^n$$

with

$$d_f = \frac{\ln 3}{\ln 2}.$$

The quantity d_f is the fractal dimension of the Sierpinski gasket and is different from d discussed earlier. Where there is the need to differentiate between d and d_f, the former is referred to as the Euclidean dimension, for obvious reasons. Since the Sierpinski gasket is drawn on a sheet of paper, its Euclidean dimension is $d = 2$. The fractal dimension is $d_f = 1.5849\ldots$, and it is not an integer.

Self-similarity of fractals Another property that characterizes fractals is the existence of self-similarity over many different length scales. Take the Sierpinski gasket as an example. From a distance, a large gasket, for example, with $L = n\ell$ for some large n like 2^{100}, will look like an equilateral triangle made of three other similar triangles, each with sides half the length of the large one. On a slightly closer examination, each of the three constituent triangles is again made of three similar triangles. Such a similarity in form persists for many length scales (100 in this case). Usually, the self-similarity observed in fractals is not exact, unlike our Sierpinski gasket example. In general, it is to be understood in terms of statistical measures, as in the case of one handful of fine sand being similar to another handful from the same pile.

In this sense, the borders of Mandelbrot and Julia plots are fractals. As we zoom in closer and closer to examine a smaller and smaller area of the border, more and more details are revealed. Each magnification displays finer details of the area, and the pictures consist of different shapes and are made of different combinations of colors. However, no reasonable statistical measures can determine that they are different in any significant way from the pictures of a different magnification.

Many physical objects possess this property of self-similarity. The most obvious example is a crystal. If we break up a large crystal into small pieces, each one will have essentially the same shape as the large crystal. The small crystals may be broken up further and even smaller crystals of similar shapes are obtained. For this reason, the study of fractal growth has become a topic of broad interest in statistical physics. As an application of Monte Carlo techniques, we shall give below one such example in the form of diffusion-limited aggregation.

Example of diffusion-limited aggregation Aggregation is the process by which particles are collected together to form larger entities, or *aggregates.* Consider a volume of space that is empty to start with. Slowly, we can release into it particles that have some form of short-range attraction among themselves. As the particles diffuse throughout the volume because of random thermal motion, there is a finite chance that two of them will come into contact with each other. If the force acting between the two particles is attractive, there is a finite probability that they will stick together and form a larger unit. Since the new unit is larger, there is an even

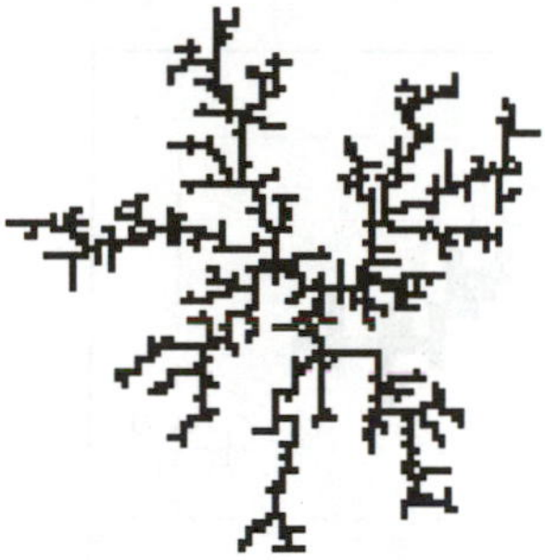

Figure 7-23: Pattern generated by a diffusion-limited aggregation algorithm. The result is a fractal, resembling a large variety of physical processes.

higher probability for another particle to collide and adhere to it. The additional particle makes the aggregate even larger and increases the rate of growth further. This process continues as long as the supply of free particles is not exhausted and no instabilities develop as the size gets larger. In the absence of any preferences as to where a new particle should join the aggregate, the resulting object for a two-dimensional system will look similar to that shown in Fig. 7-23, a pattern observed in a variety of situations, such as the path of an electric discharge and the growth of lichen on rocks. On the other hand, if there are preferences where the particles should stick, the results may become objects such as snowflakes. Things formed in this way are fractals because of the self-similarity property if any part of it is examined in more detail. It is also possible to define a fractional dimension for such objects, but we shall not be going into the question here.

The diffusion-limited aggregation (DLA) model is designed to carry out computer simulations of such processes. For simplicity, let us consider a case in two (Euclidean) dimensions. Such a surface may be divided into a lattice consisting of $(N \times N)$ sites. At the start, our "aggregate" consists of a single particle in the center of the grid. The diffusion of other particles is simulated by allowing one particle to enter into the space at a time and wander around the lattice in the form of a random walk. If the particle comes within the interaction distance of those in the aggregate, there is some finite probability for it to become a member by sticking to an existing particle.

The simplest DLA model we can construct is to assign unity to the sticking probability if the wandering particle comes into an empty site next to an occupied one. Once this takes place, the wandering particle is stuck and becomes a member of the aggregate (and the site in which it stops becomes an occupied site). The process is now repeated by letting another particle into the space. Since the motion of the wandering particle is random, it is possible that it will move off the area. We shall ignore such "lost" particles. In this way, each particle we introduce into the lattice

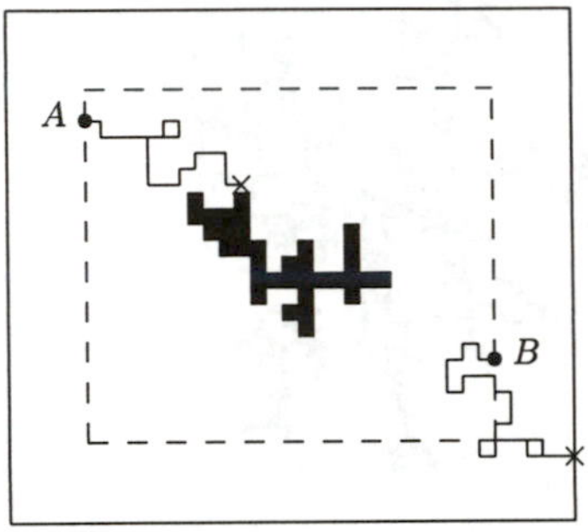

Figure 7-24: Generating a diffusion-limited aggregation pattern. Each particle starts its wandering through the lattice by random walk from the launching zone (dashed border). If a particle gets into a site next to an occupied one, it becomes a member, as shown by the one started from point A. Particle B, on the other hand, wanders off in the wrong direction. Once it reaches the "kill" zone (solid line border), it is considered to be permanently lost.

either becomes a part of the aggregate or wanders off the lattice.

The algorithm is a very simple one. To simulate random walk in two-dimensional space, we can extend the basic idea of the one-dimensional random walk in §7-2. One way to do this is to use a random number x that is uniformly distributed in the interval $[0,1]$. If $x < \frac{1}{4}$, the particle moves up one lattice site; if $\frac{1}{4} \leq x < \frac{1}{2}$, it moves down one lattice site; if $\frac{1}{2} \leq x < \frac{3}{4}$, it moves to the left by one lattice site; and if $x \geq \frac{3}{4}$, it moves to the right by one lattice site. Every time the particle arrives at a new location, we check whether there is a nearest-neighbor site that is occupied; i.e., whether there is a particle above, below, to the left, or to the right. If so, the particle becomes a part of the aggregate. Otherwise, the random walk resumes again. The process for each particle stops either when the particle moves off the area or becomes a part of the aggregate.

In principle, the growth of the aggregate should be independent of the point of launch where new particles are introduced into the lattice. This is true as long as the launching point is sufficiently far away from any part of the aggregate so that the particle has an equal probability of sticking to any available site on the perimeter. On the other hand, if a particle is launched from long distances away, it takes a large number of random walk steps for it to reach the aggregate. From a computational point of view, this is extremely time consuming. For this reason, a *launch zone* that is closer to the existing aggregate is used instead of the border of the lattice. Usually, this consists of a circle around the center of the lattice with a radius R_{launch} that is much larger than the maximum extent of the aggregate. Instead of the border of the

Box 7-10 Program DM_DLA
Simulation of diffusion-limited aggregation

Parameters:
- ℓ_{dist}: Distance of the launch zone from the nearest member.
- k_{dist}: Distance of the kill zone from the launch zone.
- m_{dist}: Size of an artificial occupation zone at the start.
- $n_{\max}$: Maximum size of the aggregate.

Subprogram called:
- INIT: Initialize the lattice, launch and kill zones, and so on.
- LNCH: Set up a random point on the launch zone to launch a particle.
- SET_PXL: Turn the pixel at (IX,IY) to color KOL (computer system dependent).
- RSUB: Uniform random number generator in the interval $[0,1]$ (Box 7-2).

Initialization:
- (a) In INIT:
  - (i) Zero the entire lattice and set the central site to 1.
  - (ii) Set up the launch and kill zones and an artificial occupation boundary.
- (b) Initialize the graphics screen and input a random number seed.

1. Launch a particle using LNCH:
   - (a) Generate a random number r_1 in $[0,4]$ and take the integer part to be i_{xy}.
   - (b) Launch along the top, bottom, right, or left for $i_{xy}=0$, 1, 2, or 3, respectively.
   - (c) Generate a random number r_2 to determine where the particle starts.
2. Random walk of the particle through the lattice:
   - (a) Generate a random number r_3 in $[0,4]$, and take the integer part to be j_{xy}.
   - (i) Move up ($j_{xy}=0$) down (1), left (2), or right (3) by one lattice site.
   - (c) If the particle is outside the kill zone, go back to step 1.
   - (d) Check for sticking by finding out if a neighboring site is occupied. If so, mark the site of the particle as occupied, and
     - (i) Increase the counter for the size of the aggregate by 1.
     - (ii) Plot the location of the particle using SET_PXL.
     - (iii) Check if the launch and kill zones need to be enlarged.
     - (iv) If the aggregate is smaller than $n_{\max}$, go back to step 1.

     Otherwise, go back to step (a) and carry out another random walk step.
3. Stop if a sufficient number of particles have accumulated at the aggregate.

lattice, each new particle starts its random walk journey from a random point on this circle. As the aggregate grows in size, the launch radius is increased. For the sake of computational efficiency, it is better to use a square launch zone rather than a circular one. By avoiding any calculations of the circumference and angles associated with a circle, it is possible to confine the operations related to the launching process (other than generating random numbers) to integers. In Fig. 7-24, the launch zone is shown as dashed lines.

A large amount of computational time can also be saved if we impose a kill zone around the aggregate. Instead of waiting for the wandering particle to go off the lattice before declaring it to be lost, any particle that goes beyond the kill zone is discarded. Again, the kill zone is increased in size as the aggregate grows. Obviously, the kill zone should be at some distance outside the launch zone to allow the possibility for a particle that wanders outside the launch zone to move back inside. In Fig. 7-24, the kill zone is indicate by the solid square box. The steps of the DLA algorithm are outlined in Box 7-10.

The calculation stops when enough particles are accumulated in the aggregate. This limit may be set at the beginning of the calculation to be some number $N_{\max}$. In general, it is necessary to choose $N_{\max}$ to be much smaller than $(N \times N)$, the total number of sites in the lattice. As in other problems with a finite lattice, the border, or finite size, effect can be important. For example, for the algorithm to work, it is necessary that the kill zone lies within the lattice, the launch zone well within the kill zone, and any occupied lattice well within the launch zone. As a result, a fairly large lattice is required to build an aggregate of reasonable size.

It is possible to change the sticking probability from unity for a neighboring site to one having a preference in direction. In this way, different patterns can be generated to simulate different physical situations. It is also of interest to color the particles according to the order of arrival so that one can see how the aggregate grows with time. One can also add to the visual effect by displaying the location of a particle as it wanders throughout the lattice. These are left as exercises, as it has more to do with the graphical presentation of results than with Monte Carlo calculations.

Problems

7-1 As an example of bad choices for the parameters of the linear congruence method of Eq. (7-1),

$$X_{n+1} = (aX_n + c) \bmod m$$

Knuth[31] gives the values of $m = 256$, $a = 137$, and $c = 187$. Use this generator to select random points in a square surface area of 10 cm on a side. Plot the distribution of points obtained.

7-2 Apply frequency, correlation, and run-up tests to the random numbers generated in Problem 7-1. Which of the three tests provides a quantitative statement for the fact that the random points do not fill the square evenly?

7-3 As an example of bad ways of using random generators, consider a Monte Carlo calculation that requires 250 random numbers as the input. To find the statistical distribution of the results, we shall run each calculation 100 times. To make the 100 calculations independent of each other we need 100 different seeds for the random number generator, one for each set of 250 random numbers required for each calculation. There are two ways we can carry out this part of the calculation: (*a*) Use the random number to generate the 100 seeds first and then use each seed to generate 250 random numbers. (*b*) Starting with one seed, generate 250 random numbers for the first calculation, and then 250 more random numbers for the next calculation, and so on. What is wrong with approach (a)?

7-4 A popular way to generate random numbers with a normal distribution is to sum n random numbers with a uniform distribution in the interval $[0, 1]$. Apply a frequency test to the results carried out with $n = 3$, 4, and 5. Find out how good a normal distribution can be obtained in this way.

7-5 Start with a uniform random number generator and obtain a sequence of 1,000 random numbers with a normal distribution using the transformation technique of Eq. (7-16). Repeat the process except now use the transformation of Eq. (7-17) instead. Compare the computer time taken to obtain each set.

7-6 Extend the Ising model simulation of §7-5 to a three-dimensional, cubic lattice with $n = 8$ divisions along each side.

7-7 A simple two-dimensional random walk problem may be stated in the following way. Consider a vertical area with infinite array of pegs placed at equal intervals in the form

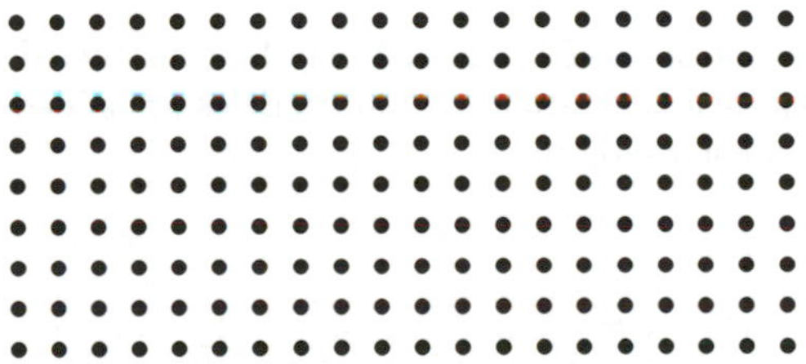

A small ball dropping down from the top may hit one of the pegs and, as a result, scatter randomly either to the left or the right. The ball may, in turn, strike another peg one row down, and so on. Such an arrangement is sometimes referred to as a "probability machine." To simplify the problem, we can assume that the scattering always leads to another peg to the left or to the right one row down. Construct a program to simulate the movement of a ball dropping down the array. Demonstrate through computer simulation that, if all the balls originate from the same point at the top, the resulting distribution of balls at the bottom is a normal one.

7-8 A plausible reason for the spread of the Indo-European languages is due to the introduction of a new method of agriculture (see C. Renfrew[45]). Use a two-dimensional random walk to simulate the spread of the farming technique by assuming that the method passes only from parents to children. If each farmer on coming of age moves 18 km in a random direction from the parents' farm to establish a new one and the interval between two generations is 25 years, find the average rate for the spread of the technique and, hence, the language spoken by the people.

7-9 A simple study of self-similar diagrams may be carried out with the Von Koch snowflake curve. Take a straight line and divide it into three equal segments. Replace the middle segment with two straight lines of the same length as the segment and use them to form the two sides of an equilateral triangle. The complete curve now takes on the form . Repeat the process for each one of the four segments and form a curve with sixteen segments, and so on. Design an algorithm to carry out this construction on a computer and try it out.

7-10 Modify the algorithm in Box 7-9 to display a Julia set. Construct the plot with $c = (-1 + 0i)$ and $c = (-0.9 + 0.12i)$.

7-11 Instead of mapping out a complete Mandelbrot set as done in Fig. 7-20, concentrate the plot on a small area near the border. As the starting point, some of the ranges suggested in the *Scientific American* articles of Dewdney[16, 17] may be used.

7-12 A colorful plot of the DLA model may be constructed by changing the color of the pixels each time after, say, 50 particles are added to the aggregate. Try this out on a color monitor.

Chapter 8

Finite Difference Solution of Differential Equations

Solving differential equations by numerical methods is one of the well developed areas of computational techniques. This arises in part from the needs in a broad range of fields, from engineering to weather forecasting. Besides the method of finite difference — our primary concern in this chapter — finite element methods are also used extensively and they are introduced in the next chapter. The subject of numerical solution to differential equations is vast and we can give here only a brief account of some of the more common approaches used in problems of interest. There are large numbers of excellent specialized books on both numerical and analytical solutions to various types of differential equations and we shall make reference to them at the appropriate places. Software packages and subroutine libraries are also available to handle certain standard forms of differential equations. Although we shall not make any direct reference to them in our introduction to the subject, they are useful in practical applications. There are also advanced finite difference techniques, such as multigrid, that are more appropriate for specialized treatment and we shall not touch upon them here.

8-1 Types of differential equations

A differential equation expresses the relation between a function, its derivatives, and the independent variables. The order of the equation is given by the highest-order derivative involved. For example, the equation for simple harmonic motion,

$$\mu \frac{d^2\phi}{dt^2} = -k\phi(t) \tag{8-1}$$

is a second-order differential equation, as the highest-order derivative is $d^2\phi/dt^2$. It is a *linear* differential equation, since ϕ and its derivatives appear in the first order and there are no products between them. As we shall see later, nonlinear differential

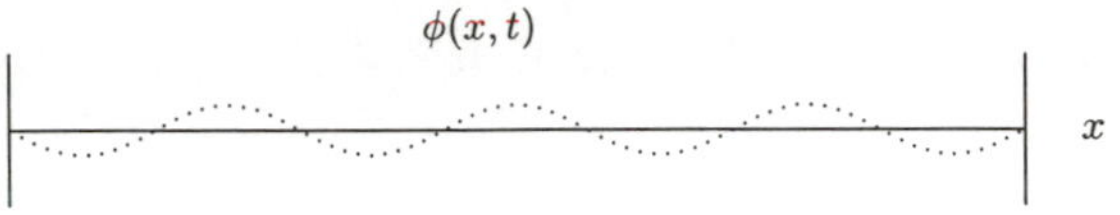

Figure 8-1: Schematic diagram of a vibrating string. The displacement $\phi(x,t)$ is a function of both x, the location along the string, and t, the time.

equations are more complicated to solve, similar to their counterparts in algebraic equations.

If ϕ has more than a single independent variable, partial derivatives enter into the equation as, for example, in the case of a vibrating string shown schematically in Fig. 8-1. Here, the vertical displacement $\phi(x,t)$ is a function of time t as well as the location x along the string. The equation of motion is a partial differential equation (PDE)

$$\frac{\partial^2\phi}{\partial x^2} - \frac{1}{v^2}\frac{\partial^2\phi}{\partial t^2} = 0 \tag{8-2}$$

where v is the velocity for the vibration to travel along the string. In contrast, Eq. (8-1) is an ordinary differential equation (ODE).

In general, we can write a two-dimensional (that is, two independent variables) second-order PDE in the form

$$p\frac{\partial^2\phi}{\partial x^2} + q\frac{\partial^2\phi}{\partial x\partial y} + r\frac{\partial^2\phi}{\partial y^2} + s\frac{\partial\phi}{\partial x} + t\frac{\partial\phi}{\partial y} + u\phi + v = 0 \tag{8-3}$$

where p, q, r, s, t, u, and v may be functions of the independent variables x and y, as well as the dependent variable ϕ and its derivatives. If $q^2 < 4pr$, it is called an *elliptic* equation; if $q^2 = 4pr$, it is a *parabolic* equation; and if $q^2 > 4pr$, it is a *hyperbolic* equation. The classification comes originally from the shape of the curve when a plane intersects a cone. In numerical work, the differences are not as important as some of the considerations below.

In Eq. (8-3), we have assumed implicitly that factors p, q, r, s, t, u, and v do not involve derivatives of ϕ higher than second order, otherwise it will be a PDE of higher order. If none of the factors involve ϕ or any of its derivatives, Eq. (8-3) is a linear PDE. Otherwise, it is a nonlinear PDE. As we shall see later, the method of solution depends critically on the order of the equation and whether it is linear or not.

The wave equation we saw earlier in Eq. (8-2) is an example of hyperbolic PDE. A good example of elliptic equations is the two-dimensional Poisson equation,

$$\frac{\partial^2\phi}{\partial x^2} + \frac{\partial^2\phi}{\partial y^2} = -\rho(x,y).$$

In electrostatics, it describes the field $\phi(x,y)$ of a charge distribution $\rho(x,y)$ in two dimensions. The standard example of a parabolic equation is the diffusion equation,

$$\frac{\partial\phi}{\partial t} = -\frac{\partial}{\partial x}\Big(D\frac{\partial\phi}{\partial x}\Big)$$

where ϕ may be the concentration of a certain kind of particle and D is its diffusion coefficient.

Initial and boundary value problems Differential equations can also be classified into initial value and boundary value problems. For numerical solutions, the differences between these two categories can often be more important than some of the considerations above. This comes from the fact that an initial value problem propagates the solution forward from the values given at the starting point. In contrast, a boundary value problem has constraints that must be satisfied at both the start and the end of the interval. The distinction between temporal and spatial variables is mainly in their physical significance. As far as numerical solutions are concerned, there are very few differences between them except, perhaps, in the manner they appear in the equation under certain circumstances. Our primary concern in the differences between initial value and boundary value problems is in the way the constraints are placed on the solution. If all the conditions that must be satisfied by the solution are given at one end of the interval, we have an initial value problem. On the other hand, if the constraints on the solution are applied to both ends of the interval, we have a boundary value problem. For partial differential equations, it may also happen that the conditions for some independent variables are given in the form of initial values and others as boundary conditions. If this happens, we have an initial value boundary problem, a mixture of both types in one.

As an example, let us consider the second-order differential equation given by Eq. (8-1). The only independent variable is t and let us assume that we are interested in the solution in the interval $t = [t_0, t_N]$. Since it is a second-order differential equation, two pieces of information must be supplied before we can solve it. If we specify them, for example, in terms of the value of $\phi(t)$ and its first-order derivative at $t = t_0$, we have an initial value problem, independent of whether the physical meaning of t is time or distance. On the other hand, if the two conditions are given as the values of $\phi(t)$ at $t = t_0$ and $t = t_N$, we have a boundary value problem. Needless to say, the two types call for quite different numerical methods.

Euler's method for initial value problems The general philosophy behind numerical solutions to differential equations can be illustrated using Euler's method to solve Eq. (8-1) as an initial value problem. The algorithm is simple and straightforward.

The equation may be put into the following form

$$\frac{d^2\phi}{dt^2} + \omega_0^2\phi(t) = 0 \tag{8-4}$$

by expressing the spring constant k and mass μ in terms of the angular frequency for simple harmonic motion

$$\omega_0 = \sqrt{\frac{k}{\mu}}.$$

Similar to what we did in numerical integrations, the first step in finding a numerical solution is to "discretize" the interval of interest into a grid or a mesh consisting of a finite number of points in the interval $t = [t_0, t_N]$. For a numerical solution, our aim is to find the values of $\phi(t)$ at a discrete number of points at $t = t_0, t_1, t_2, \ldots, t_N$ that satisfy Eq. (8-4). The distance between two consecutive points, or step size, is

$$h_i = t_{i+1} - t_i$$

and there are N such subintervals. In the limit of all $h_i \to 0$, we recover the continuous function in an analytical solution.

To solve for $\{\phi(t_1), \phi(t_2), \ldots, \phi(t_{N-1})\}$, we can convert Eq. (8-4) into a set of algebraic equations by rewriting the second-order derivative using finite difference. In terms of the central difference introduced in Eq. (2-52), the second-order derivative with respect to time at $t = t_i$ may be written in the form

$$\left.\frac{d^2\phi(t)}{dt^2}\right|_{t=t_i} \longrightarrow \frac{1}{h^2}\{\phi(t_{i+1}) - 2\phi(t_i) + \phi(t_{i-1})\} \tag{8-5}$$

as we have done in Eq. (2-55). For simplicity, we shall take the step size

$$h = t_{i+1} - t_i$$

to be a constant, independent of time and make use of the shorthand notation

$$\phi_i \equiv \phi(t_i).$$

The differential equation Eq. (8-4) may now be expressed as the relation between the value of $\phi(t)$ at three different times $t = t_{i-1}$, t_i, and t_{i+1}:

$$\phi_{i+1} - (2 - h^2\omega_0^2)\phi_i + \phi_{i-1} = 0. \tag{8-6}$$

This is called a finite difference equation (FDE), as it relates the differences in the values of $\phi(t)$ at nearby points.

As an initial value problem, we need two independent pieces of input on $\phi(t)$ at the starting time. Physically, these initial conditions may be supplied in terms of the location and velocity of the mass at $t = t_0$. For example, the mass may be displaced at the start by one unit in distance in the positive direction. That is,

$$\phi_0 \equiv \phi(t = t_0) = 1.$$

It is then released from this arrangement at $t = t_0$ without being given any initial velocity. This may be expressed by the condition

$$\left.\frac{d\phi(t)}{dt}\right|_{t=t_0} = 0.$$

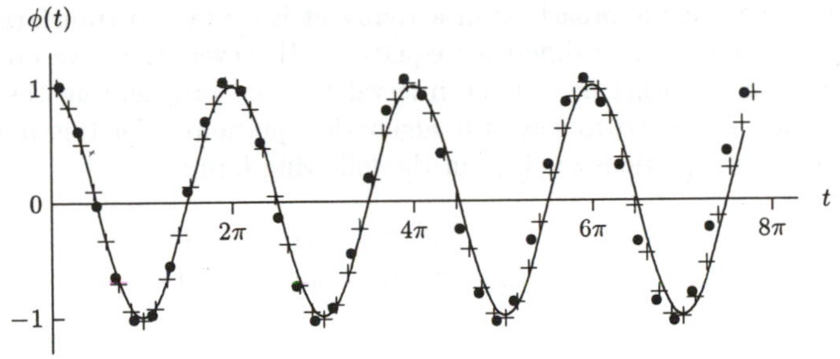

Figure 8-2: Solutions for the simple harmonic oscillation of Eq. (8-4): dots (•), numerical solution using Euler's method with step size $h = 0.63$; plus symbols (+), with $h = 0.4$; and solid curve, analytical solution.

Using finite differences, these two initial conditions may be approximated by the following assignments:

$$\phi_1 \approx \phi_0 = 1. \tag{8-7}$$

This provides us two of three ϕ values in Eq. (8-6) for $i = 1$. As a result, we can calculate ϕ_2. Once ϕ_2 is found, we can make use of its value, together with that of ϕ_1 to find ϕ_3 using Eq. (8-6) with $i = 2$. This process continues until we come to $i = N - 1$. In this way, the solution for $\phi(t)$ is propagated forward one step at a time and all the ϕ_i's are found one after another. The results are shown in Fig. 8-2. In more realistic applications, several improvements can be made on both the accuracy and efficiency of the calculation and we shall discuss three ways later in §8-2 and §8-3.

Example of a boundary value problem As a comparison, we shall solve the same differential equation as a boundary value problem. Instead of an oscillator, we may be interested in the sound produced in a hollow, cylindrical pipe. In this case, the dependent variable $\phi(x)$ represents the air pressure along the length x of the pipe. The differential equation for $\phi(x)$ has the same form as Eq. (8-4),

$$\frac{d^2\phi(x)}{dx^2} + \kappa^2\phi(x) = 0 \tag{8-8}$$

except that, physically, the factor κ^2 here is related to the compressibility of air. The constraints on the system in this case occur at the two ends of the pipe. For simplicity, we shall take the pressure to have some constant value p at $x = 0$ and $x = \ell$. In other words, we have a boundary value problem, with

$$\phi_0 = p \qquad\qquad \phi_N = p$$

as the boundary conditions.

We can use the same approach as in arriving at Eq. (8-6) to transform the differential equation into a finite difference equation. However, since we do not have enough information at either end of the interval to start propagating the solution, a different method is needed to solve the algebraic equations. For this purpose, we shall write the set of equations for $\{\phi_i\}$ in the following form:

$$\begin{aligned} -(2-h^2\kappa^2)\phi_1 + \phi_2 &= -p \\ \phi_1 - (2-h^2\kappa^2)\phi_2 + \phi_3 &= 0 \\ \cdots \quad \cdots \quad \cdots \quad & \quad \cdots \\ \phi_{N-3} - (2-h^2\kappa^2)\phi_{N-2} + \phi_{N-1} &= 0 \\ \phi_{N-2} - (2-h^2\kappa^2)\phi_{N-1} &= -p \end{aligned}$$

where $h = x_{i+1} - x_i$ and we have included the boundary conditions by putting the values of ϕ_0 and ϕ_N into the first and last one.

In matrix notation, this set of equations may be written as

$$\begin{pmatrix} d & 1 & & & & \\ 1 & d & 1 & & & \\ & 1 & d & 1 & & \\ & & \ddots & \ddots & \ddots & \\ & & & 1 & d & 1 \\ & & & & 1 & d \end{pmatrix} \begin{pmatrix} \phi_1 \\ \phi_2 \\ \phi_3 \\ \vdots \\ \phi_{N-2} \\ \phi_{N-1} \end{pmatrix} = \begin{pmatrix} -p \\ 0 \\ 0 \\ \vdots \\ 0 \\ -p \end{pmatrix}$$

where, for simplicity, we have used

$$d \equiv h^2\kappa^2 - 2.$$

Furthermore, we have omitted from the square matrix on the left side all the elements that are zero. The problem is now equivalent to one of finding the roots of a set of $(N-1)$ linear algebraic equations. In principle, we can use the technique discussed in §A-3 to find the solution. To achieve good accuracy, the step size must be small. Consequently, N may be quite large.

Stability and convergence Besides the obvious demand of efficiency, a numerical method has to be stable and the solution converges to that of the differential equation. The importance of these two additional requirements can be seen from the oscillator example we encountered earlier. Since the full solution is obtained by propagating from two initial values, the truncation error in all the steps can be cumulative and may eventually make the numerical results meaningless.

The main source of truncation errors comes from approximations introduced in replacing the differential equation by a finite difference equation. This may be illustrated using a generic first-order ordinary differential equation of the form

$$\frac{d\phi}{dt} = f(\phi, t). \tag{8-9}$$

In the method of Euler used above, a forward difference approximation is used to replace the derivative. As a result, our finite difference equation takes on the form

$$\frac{1}{h}\{\phi_{i+1} - \phi_i\} = f(\phi_i, t_i). \tag{8-10}$$

We can see the errors introduced in this approximation by making a Taylor series expansion of $\phi(t)$ at $t = t_{i+1}$ in terms of $\phi(t)$ and its derivatives at $t = t_i$

$$\phi_{i+1} = \phi_i + h\frac{d\phi}{dt}\bigg|_{t=t_i} + \frac{h^2}{2!}\frac{d^2\phi}{dt^2}\bigg|_{t=t_i} + \cdots. \tag{8-11}$$

If we replace $d\phi/dt$ at $t = t_i$ on the right side by $f(\phi_i, t_i)$ and compare the resulting expression with Eq. (8-10), we find that the leading order difference is proportional to $h^2 d^2\phi(t)/dt^2$ evaluated at $t = t_i$. Since errors of this type are often the most important ones in the numerical solution of differential equations, we shall ignore in this chapter other errors, such as those arising from finite length of a computer word. As a result, truncation errors may be defined here as the differences between the value $f(\phi_i, t_i)$ given by Eq. (8-10) and the exact value of $f(\phi, t)$ at $t = t_i$.

If the function $\phi(t)$ is well behaved, we do not expect $d^2\phi/dt^2$ in Eq. (8-11) to become very large anywhere in the range of t of interest to us. As a result, the truncation error of the method goes to zero as $h \to 0$. This is a necessary requirement for any numerical method and is sometimes called the *consistency* condition. However, consistency does not guarantee *convergence*, the requirement that the numerical results approach the true solution for the differential equation. There are methods, whose errors (differences from the true solution) actually increase as the step size h is decreased, even though it satisfies the consistency condition. Such methods are called *unstable*, and we shall not elaborate on these methods here. On many occasions, it may not be easy to demonstrate that a solution is stable and convergent without actually carrying out the calculations. This is especially true for nonlinear differential equations where, often, the condition for consistency can be demonstrated only after drastic approximations.

The Taylor series expansion of Eq. (8-11) provides us also with a way to improve the accuracy of our numerical solutions. For example, if we wish to reduce the truncation error in our finite difference equation to order h^3, we must somehow include contributions from the h^2 term into our approximation. There are many ingenious ways to do this, as we can see from the following illustration. Again, let us use the first-order ODE of Eq. (8-9) as an example, and rewrite Eq. (8-10) in the form

$$\phi_{i+1} = \phi_i + hf_i \tag{8-12}$$

where $f_i \equiv f(\phi_i, t_i)$. Only the values at mesh point i are used in this approximation to find the solution at $(i + 1)$. This is known as a one-step process. To reduce the truncation errors, we must incorporate additional information on the right side. This can be done by making use of the values at points other than $(i + 1)$. In general, we can express the solution at a point $(i+k)$ as a function of the values at the previous k points in the following way:

$$\phi_{i+k} = a_0\phi_i + a_1\phi_{i+1} + \cdots + a_{k-1}\phi_{i+k-1} + hF(\phi_i, t_i; \phi_{i+1}, t_{i+1}; \ldots; \phi_{i+k}, t_{i+k}) \tag{8-13}$$

where $a_0, a_1, \ldots, a_{k-1}$ are coefficients and F is a known function of both the dependent variable ϕ and independent variable t at mesh points $i, i+1, \ldots, i+k$. For $k=1$ and $F = f(\phi_i, t_i)$, we recover Eq. (8-12). For $k > 1$, the method represented by Eq. (8-13) is known as a multistep process. In practice, the function F is often replaced by a linear combination of the value of $f(\phi, t)$ at the previous k mesh points, such as

$$F(\phi_i, t_i; \phi_{i+1}, t_{i+1}; \ldots; \phi_{i+k}, t_{i+k}) = \sum_{j=0}^{k} \beta_j f_{i+j}.$$

For $k > 0$, this is known as a linear multistep process. If the coefficient β_k does not vanish, the value of ϕ_{i+k} appears also on the right side of the finite difference equation. In this case, the method is an implicit one. For our simple example of Eq. (8-12), and its equivalent form for multistep processes in which ϕ_{i+k} appear only on the left side of the equation, the method is known as an explicit one.

8-2 Runge-Kutta methods

One way to reduce truncation errors in initial value problems is through Runge-Kutta methods. As shown in §A-5, ordinary differential equation of order n can always be reduced into a set of n first-order ones in the form

$$\frac{d\boldsymbol{y}}{dt} = \boldsymbol{f}(\boldsymbol{y}(t), t) \tag{8-14}$$

where $\boldsymbol{y} \equiv \{y_1, y_2, \ldots, y_n\}$ and $\boldsymbol{f}(\boldsymbol{y}(t), t) = \{f_1(\boldsymbol{y}(t), t), f_2(\boldsymbol{y}(t), t), \ldots, f_n(\boldsymbol{y}(t), t)\}$. Because of this, we shall only consider first-order ODE here.

Higher-order approximations To the same level of accuracy as Eq. (8-6), the finite difference approximation to Eq. (8-14) may be expressed in the following form:

$$\frac{1}{h}\{\boldsymbol{y}(t_{k+1}) - \boldsymbol{y}(t_k)\} = \boldsymbol{f}(\boldsymbol{y}(t_k), t_k). \tag{8-15}$$

To make any improvement, we must include additional "input" information on the right side. For example, we can make use of

$$\boldsymbol{p} \equiv \boldsymbol{f}(\boldsymbol{y}(t_k), t_k) \qquad \boldsymbol{q} = \boldsymbol{f}(\boldsymbol{y}(t_k) + \alpha h\boldsymbol{p}, t_k + \alpha h)$$

where α is an adjustable parameter to be determined later. Eq. (8-15) may now be put in the form

$$\frac{1}{h}\{\boldsymbol{y}(t_{k+1}) - \boldsymbol{y}(t_k)\} = \beta_1\boldsymbol{p} + \beta_2\boldsymbol{q} \tag{8-16}$$

and we can make use of the two additional parameters, β_1 and β_2, together with α to minimize the truncation error.

Using a Taylor series expansion, we find that

$$q \approx \boldsymbol{y}_t(t_k) + \alpha h \boldsymbol{y}_{tt}(t_k) + \frac{1}{2}(\alpha h)^2\{\boldsymbol{y}_{ttt}(t_k) - \boldsymbol{f}_y \boldsymbol{y}_{tt}(t_k)\}$$

where, to simplify the notation, we have used

$$\boldsymbol{f}_y \equiv \frac{\partial \boldsymbol{f}}{\partial y} \qquad \boldsymbol{y}_t(t) \equiv \frac{d\boldsymbol{y}}{dt} \qquad \boldsymbol{y}_{tt}(t) \equiv \frac{d^2\boldsymbol{y}}{dt^2} \qquad \boldsymbol{y}_{ttt}(t) \equiv \frac{d^3\boldsymbol{y}}{dt^3}.$$

The truncation error in Eq. (8-16) is then

$$\begin{aligned} \mathcal{E} &= (1-\beta_1-\beta_2)\boldsymbol{y}_t(t_k) + h(\frac{1}{2} - \alpha\beta_2)\boldsymbol{y}_{tt}(t_k) \\ &\quad + h^2\{\tfrac{1}{6}\boldsymbol{y}_{ttt}(t_k) + \frac{1}{2}\alpha^2\beta_2[\boldsymbol{y}_{ttt}(t_k) - \boldsymbol{f}_y\boldsymbol{y}_{tt}(t_k)]\} + O(h^3). \end{aligned}$$

Because of the $\boldsymbol{f}_y\boldsymbol{y}_{tt}(t_k)$ term, it is not possible to eliminate all the h^2 dependence for an arbitrary function $\boldsymbol{y}(t)$. There are, however, several choices available to us to make $\mathcal{E}$ as small as possible. One possibility is to take $\alpha = 1$ and $\beta_1 = \beta_2 = \frac{1}{2}$. Alternatively, we can use the midpoint method by taking $\alpha = \frac{1}{2}$, $\beta_1 = 0$, and $\beta_2 = 1$. We shall not elaborate further on either of these two choices here.

Our interest lies in extending the above idea to construct a set of finite difference equations that is accurate to order h^4. One result commonly used in practical applications is the fourth-order Runge-Kutta formula

$$\boldsymbol{y}(t_{k+1}) = \boldsymbol{y}(t_k) + \tfrac{1}{6}h(\boldsymbol{p} + 2\boldsymbol{q} + 2\boldsymbol{r} + \boldsymbol{s}) \tag{8-17}$$

with

$$\begin{aligned} \boldsymbol{p} &= \boldsymbol{f}(\boldsymbol{y}(t_k), t_k) & \boldsymbol{q} &= \boldsymbol{f}(\boldsymbol{y}(t_k) + \tfrac{1}{2}h\boldsymbol{p}, t_k + \tfrac{1}{2}h) \\ \boldsymbol{r} &= \boldsymbol{f}(\boldsymbol{y}(t_k) + \tfrac{1}{2}h\boldsymbol{q}, t_k + \tfrac{1}{2}h) & \boldsymbol{s} &= \boldsymbol{f}(\boldsymbol{y}(t_k) + h\boldsymbol{r}, t_k + h). \end{aligned} \tag{8-18}$$

We shall not try to reproduce here the lengthy proof that truncation errors up to and including order h^4 vanish in Eq. (8-17). This is reminiscent of Simpson's rule for an integral of the form

$$y(t_{k+1}) - y(t_k) = \int_{t_k}^{t_{k+1}} y'(t)\,dt$$

which, we recall from Eq. (2-15), is accurate up to fourth order by taking weighted averages of neighboring points. Note that, to evaluate $\boldsymbol{p}$, $\boldsymbol{q}$, $\boldsymbol{r}$, and $\boldsymbol{s}$ of Eq. (8-18), the only values of $\boldsymbol{y}(t)$ needed are those at $t = t_k$. The algorithm for solving a system of n-coupled first-order ODEs in this way using a fixed step size h is outlined in Box 8-1.

Box 8-1 Subroutine RGKT4(H,BFY,ND_EQN,ND_PTS)
Fourth-order Runge-Kutta method
for a system of n-coupled first-order, initial value ODE

Argument list:
- H: Input step size h.
- BFY: Two-dimensional array for the solution $\boldsymbol{y}$.
- ND_EQN: Number of equations.
- ND_PTS: Total number of steps in t to be taken.

Subprogram to be supplied:
- FVAL: Calculate the value of $\boldsymbol{f}(\boldsymbol{y}, t)$ with input $\boldsymbol{y}(t)$ and t.

Initialization:
In the calling program, store the initial conditions in BFY.

1. Carry out the following calculations for step size h for each of the n ODEs:
   (a) For each t_k, calculate $\boldsymbol{p}$, $\boldsymbol{q}$, $\boldsymbol{r}$, and $\boldsymbol{s}$ of Eq. (8-18) by the following steps:
      (i) Call FVAL to calculate $\boldsymbol{p}$ using $\boldsymbol{y}_k$ and t_k.
      (ii) Call FVAL to calculate $\boldsymbol{q}$ using $(\boldsymbol{y}_k + \frac{1}{2}h\boldsymbol{p})$ and $t = (t_k + \frac{1}{2}h)$.
      (iii) Call FVAL to calculate $\boldsymbol{r}$ using $(\boldsymbol{y}_k + \frac{1}{2}h\boldsymbol{q})$ and $t = (t_k + \frac{1}{2}h)$.
      (iv) Call FVAL to calculate $\boldsymbol{s}$ using $(\boldsymbol{y}_k + h\boldsymbol{r})$ and $t = (t_k + h)$.
   (b) Calculate $\boldsymbol{y}_{k+1}$ from the values of $\boldsymbol{y}_k$, $\boldsymbol{p}$, $\boldsymbol{q}$, $\boldsymbol{r}$, and $\boldsymbol{s}$ using Eq. (8-17).
2. Return the values of $\boldsymbol{y}(t)$ in the array BFY.

Application to damped oscillation If we include a friction term into Eq. (8-4), we obtain the following second-order differential equation for a damped oscillator:

$$\mu\frac{d^2\phi}{dt^2} + b\frac{d\phi}{dt} + k\phi = 0 \tag{8-19}$$

where b is the constant of proportionality for the damping term due to friction and, as before, $\phi(t)$ is the displacement at time t, with μ as the mass and k, the spring constant. This type of second-order ordinary differential equation appears in a variety of other problems, including alternating-current circuits.

We can put Eq. (8-19) into the general form of

$$\frac{d^2\phi}{dt^2} + 2\gamma\frac{d\phi}{dt} + \omega_0^2\phi = 0 \tag{8-20}$$

by defining

$$2\gamma = \frac{b}{\mu} \qquad\qquad \omega_0^2 = \frac{k}{\mu}.$$

To solve the equation numerically as two coupled first-order ordinary differential equations, we shall follow the method used in arriving at Eq. (A-10) and define two new independent variables:

$$y_1(t) = \phi(t) \qquad\qquad y_2(t) = \frac{d\phi}{dt}.$$

In terms of $y_1(t)$ and $y_2(t)$, Eq. (8-19) is transformed into

$$\frac{d\boldsymbol{f}}{dt} = \boldsymbol{f}(\boldsymbol{y}(t), t)$$

with

$$f_1(t) = y_2 \qquad\qquad f_2(t) = -2\gamma y_2 - \omega_0^2 y_1.$$

At $t = 0$, we shall assume that the oscillator is displaced by an amount ϕ_0 and released with some velocity $(d\phi/dt)_{t=0}$. This gives us the initial conditions

$$y_1(t=0) = \phi_0 \qquad\qquad y_2(t=0) = \frac{d\phi}{dt}\bigg|_{t=0}.$$

To obtain the numerical solution, we shall, for simplicity, construct a mesh of points dividing the time interval into a number of constant size steps. The distance between two adjacent mesh points depends on the accuracy required as well as the total number of steps we wish to take and the time span we are interested in.

It is well known that there are three possible types of solution for Eq. (8-20), depending on the size of γ relative to that of ω_0. In the case of heavy damping with $\gamma^2 > \omega_0^2$, the system does not oscillate. The amplitude decays with two different time constants, $(\gamma - \zeta)$ and $(\gamma + \zeta)$, where

$$\zeta = \sqrt{\gamma^2 - \omega_0^2}$$

is a real and positive quantity. The solution of the differential equation takes on the form

$$\phi(t) = \frac{\phi_0 + \xi}{2}\, e^{-(\gamma-\zeta)t} + \frac{\phi_0 - \xi}{2}\, e^{-(\gamma+\zeta)t}$$

with the factor ξ given by

$$\xi = \frac{1}{\zeta}\left\{\frac{d\phi}{dt}\bigg|_{t=0} + \gamma\phi_0\right\}.$$

We shall not be concerned with this case any further, as there is nothing of interest to us here in terms of numerical solutions.

The decay is most rapid when $\gamma = \omega_0$. This is known as the case of critical damping and the results are shown in Fig. 8-3(a). The analytical solution is given by

$$\phi(t) = e^{-\gamma t}\left\{\phi_0 + \left(\frac{d\phi}{dt}\bigg|_{t=0} + \gamma\phi_0\right)t\right\}. \tag{8-21}$$

We see that the numerical solution, represented by the crosses in the figure, is in good agreement with Eq. (8-21), shown as a solid curve.

If we let $b = 0$, there is no damping and we return to the case of simple harmonic oscillation. The analytical solution for the initial condition of $d\phi/dt = 0$ at $t = 0$ has the form

$$\phi(t) = \phi_0 \cos\omega_0 t.$$

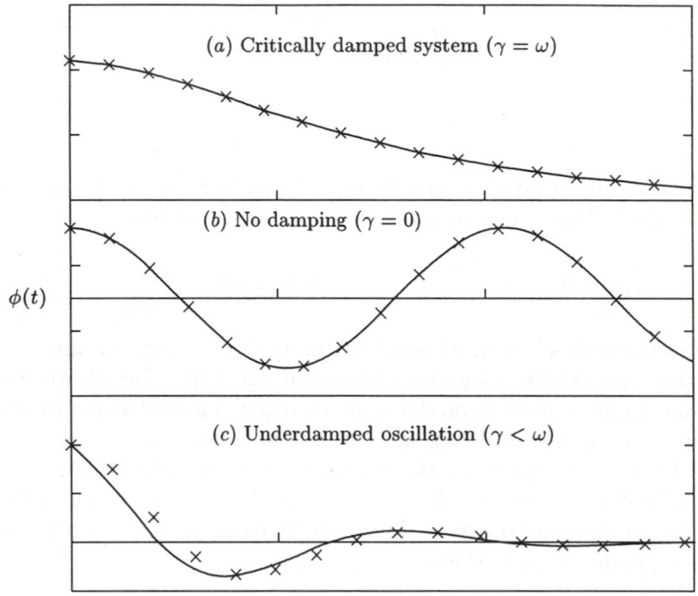

Figure 8-3: Solutions for a damped oscillator: (*a*) Critical damping ($\gamma = \omega_0$), (*b*) no damping ($\gamma = 0$), and (*c*) underdamping ($\gamma^2 < \omega_0^2$). In each case, the solid curve comes from an analytical solution and the crosses indicate the results of a fourth-order Runge-Kutta method. Differences in (*c*) arises from the fact that the equation for the underdamped case is a "stiff" one.

This is indicated by the solid curve in Fig. 8-3(*b*). Again, the numerical solution of Eq. (8-19), shown as crosses, is in good agreement with the analytic results.

The case of underdamping ($\gamma^2 < \omega_0^2$) is shown in Fig. 8-3(*c*). The analytical solution in this case is given by

$$\phi(t) = e^{-\gamma t}\{\phi_0 \cos \omega t + \xi \sin \omega t\}$$

where

$$\omega = \sqrt{\omega_0^2 - \gamma^2} \qquad\qquad \xi = \frac{1}{\omega}\left\{\frac{d\phi}{dt}\bigg|_{t=0} + \gamma\phi_0\right\}.$$

Here, we see that the accuracy of the numerical solution is relatively poor, as can be seen by the departures from the analytical values. This is an example of a "stiff" differential equation, arising from the fact that there are two time constants in the solution, ω and γ, and their values are rather different. As a result, the numerical solution becomes unstable, as if it does not know which of the two time scales to follow. We shall return to the question of stiff equation in §8-10.

Variable step size The efficiency of the Runge-Kutta method outlined above may be improved if we can use different step sizes in different regions. Similar to methods of numerical integration discussed in Chapter 2, it is possible to take much larger steps in a region where $\boldsymbol{f}(\boldsymbol{y}, t)$ is smooth. On the other hand, if the function varies rapidly in some intervals, smaller step sizes must be used to achieve the same accuracy. In a constant step size approach, the size of h is usually selected in such a way that no region has an error larger than some given amount. To achieve this goal, h must be taken to be the smallest value required by accuracy considerations among all the subintervals. In contrast, in a variable step size method, h needs only to be small enough to achieve the desired accuracy in each region. Obviously, this is profitable only if it is easy to find the optimum step size for a given region.

In order to make use of a variable step size approach, we must be able to estimate the truncation error. One way to do this is to start with some reasonable trial value for h and calculate $\boldsymbol{y}(t)$ in two different ways, one using two steps of size h each and one using one step of size $2h$. The difference between the results in these two approaches provides us with an indication of the truncation error associated with the trial value h. If acceptable, it can also give us a way to arrive at the size for the next step. On the other hand, if the truncation error is too large, the trial step size must be reduced and the calculations repeated using a smaller value.

We shall not be concerned here with the question of how to arrive at a first estimate of h in a calculation, as any reasonable value of h is adequate for the purpose. Let us assume that we have just completed step k that ends at $t = t_k$. For the next step, we shall adopt as trial step size the value h_t that was used as the step size for the previous step. Let $\boldsymbol{Y}_{2h}(t_k + 2h_t)$ represent the value of $\boldsymbol{y}(t + 2h_t)$ calculated using Eq. (8-17) with a step size of $2h_t$. That is,

$$\boldsymbol{Y}_{2h}(t_k + 2h_t) = \boldsymbol{y}(t_k) + \tfrac{1}{3}h_t(\boldsymbol{p}_{2h} + 2\boldsymbol{q}_{2h} + 2\boldsymbol{r}_{2h} + \boldsymbol{s}_{2h}) \tag{8-22}$$

where

$$\boldsymbol{p}_{2h} = \boldsymbol{f}(\boldsymbol{y}(t_k), t_k) \qquad \boldsymbol{q}_{2h} = \boldsymbol{f}(\boldsymbol{y}(t_k) + h_t\boldsymbol{p}_{2h}, t_k + h_t)$$

$$\boldsymbol{r}_{2h} = \boldsymbol{f}(\boldsymbol{y}(t_k) + h_t\boldsymbol{q}_{2h}, t_k + h_t) \qquad \boldsymbol{s}_{2h} = \boldsymbol{f}(\boldsymbol{y}(t_k) + 2h_t\boldsymbol{r}_{2h}, t_k + 2h_t).$$

The value of $\boldsymbol{y}(t)$ at $t = (t_k + 2h_t)$ can also be obtained in two separate steps. First, we take a step of h_t. This gives us the result

$$\boldsymbol{z}(t_k + h_t) = \boldsymbol{y}(t_k) + \tfrac{1}{6}h_t(\boldsymbol{p}_{h1} + 2\boldsymbol{q}_{h1} + 2\boldsymbol{r}_{h1} + \boldsymbol{s}_{h1}) \tag{8-23}$$

with

$$\boldsymbol{p}_{h1} = \boldsymbol{f}(\boldsymbol{y}(t_k), t_k) = \boldsymbol{p}_{2h} \qquad \boldsymbol{q}_{h1} = \boldsymbol{f}(\boldsymbol{y}(t_k) + \tfrac{1}{2}h_t\boldsymbol{p}_{h1}, t_k + \tfrac{1}{2}h_t)$$

$$\boldsymbol{r}_{h1} = \boldsymbol{f}(\boldsymbol{y}(t_k) + \tfrac{1}{2}h_t\boldsymbol{q}_{h1}, t_k + \tfrac{1}{2}h_t) \qquad \boldsymbol{s}_{h1} = \boldsymbol{f}(\boldsymbol{y}(t_k) + h_t\boldsymbol{r}_{h1}, t_k + h_t).$$

Note that $\boldsymbol{p}_{h1}$ is the same as $\boldsymbol{p}_{2h}$ of Eq. (8-22) and does not have to be evaluated again here. Starting from $\boldsymbol{z}(t_k + h_t)$, we take another step of size h_t,

$$\boldsymbol{Y}_{1h}(t_k + 2h_t) = \boldsymbol{z}(t_k + h_t) + \tfrac{1}{6}h_t(\boldsymbol{p}_{h2} + 2\boldsymbol{q}_{h2} + 2\boldsymbol{r}_{h2} + \boldsymbol{s}_{h2}) \tag{8-24}$$

this time with

$$\boldsymbol{p}_{h2} = \boldsymbol{f}(\boldsymbol{z}(t_k + h_t), t_k + h_t) \qquad \boldsymbol{q}_{h2} = \boldsymbol{f}(\boldsymbol{z}(t_k + h_t) + \tfrac{1}{2}h_t\boldsymbol{p}_{h2}, t_k + \tfrac{3}{2}h_t)$$
$$\boldsymbol{r}_{h2} = \boldsymbol{f}(\boldsymbol{z}(t_k + h_t) + \tfrac{1}{2}h_t\boldsymbol{q}_{h2}, t_k + \tfrac{3}{2}h_t) \qquad \boldsymbol{s}_{h2} = \boldsymbol{f}(\boldsymbol{z}(t_k + h_t) + h_t\boldsymbol{r}_{h2}, t_k + 2h_t).$$

Since both $\boldsymbol{Y}_{2h}(t_k + 2h_t)$ and $\boldsymbol{Y}_{1h}(t_k + 2h_t)$ are obtained using a fourth-order Runge-Kutta method, we expect the difference between them to be on the order h_t^5:

$$\boldsymbol{\Delta}(t_k, h_t) = \boldsymbol{Y}_{1h}(t_k + 2h_t) - \boldsymbol{Y}_{2h}(t_k + 2h_t) \sim O(h_t^5).$$

That is, if we carry out a Taylor series expansion of $\boldsymbol{y}(t)$ at $t = t_k$, we will find that $\boldsymbol{\Delta}(t_k, h_t)$ are proportional to $h_t^5\boldsymbol{y}^{(5)}(t_k)$, where $\boldsymbol{y}^5(t_k)$ are the fifth-order derivatives of $\boldsymbol{y}(t)$ at $t = t_k$.

The value of $\boldsymbol{\Delta}(t_k, h_t)$ is useful for two related purposes. The first is that we can compare $\boldsymbol{\Delta}(t_k, h_t)$ with $\mathcal{E}_{\max}$, the maximum truncation error we can tolerate. If $|\boldsymbol{\Delta}(t_k, h_t)| \leq \mathcal{E}_{\max}$, the step size of h_t is acceptable and the value of $\boldsymbol{Y}_{1h}(t_k + 2h_t)$ may be taken as the value of $\boldsymbol{y}(t_{k+1})$ at $t_{k+1} = (t_k + 2h_t)$. A second use of $\boldsymbol{\Delta}(t_k, h_t)$ is to estimate the optimum trial step size h_0 for the next subinterval $[t_{k+1}, t_{k+2}]$. Ideally, the value of h_0 should be such that

$$\boldsymbol{\Delta}(t_{k+1}, h_0) = \mathcal{E}_{\max}.$$

Since all we have is the value of $\boldsymbol{\Delta}(t_k, h_t)$ for the interval $[t_k, t_{k+1}]$, we can only find the optimum step size h_0 for the present step. If the solutions do not contain rapid fluctuations, we expect that the change in the size is small from one step to another. As a result, it is possible to make use of the value of h_0 in an estimate for the next step.

We saw earlier that $\boldsymbol{\Delta}(t_k, h_t) \sim h_t^5$ in our present approach. As a result, we have the approximate relation

$$\left|\frac{\mathcal{E}_{\max}}{\boldsymbol{\Delta}(t_k, h_t)}\right| \approx \left(\frac{h_0}{h_t}\right)^5.$$

From this, we obtain

$$h_0 \approx h_t\left|\frac{\mathcal{E}_{\max}}{\boldsymbol{\Delta}(t_k, h_t)}\right|^{1/5}. \tag{8-25}$$

This is only an approximate result; however, it is adequate to be used as a guide for estimating the optimum step size.

If, on the other hand, the trial step size h_t produces a truncation error that is larger than $\mathcal{E}_{\max}$, we must go back to Eq. (8-22) with a smaller h_t. We can also make use of Eq. (8-25) to find the optimum value for h_t. Let us represent the value as h_0.

Box 8-2 Subroutine RKODE(NPTS,T_BEG,T_END,T_ARY,BFY,ND_EQN,ND_PTS)
Runge-Kutta solution of n initial-value, first-order ODEs with step size control

Argument list:
- NPTS: Number of steps taken.
- T_BEG: Beginning of the interval.
- T_END: End of the interval.
- T_ARY: One-dimensional array for t_k.
- BFY: Two-dimensional array for $\boldsymbol{y}$.
- ND_EQN: Number of differential equations.
- ND_PTS: Maximum number of points.

Subprograms used:
- RK_STP: Fourth-order Runge-Kutta for step sizes h and $2h$ (cf. Box 8-1).

Initialization:
- (a) Define the maximum tolerable truncation error $\mathcal{E}_{\max}$.
- (b) Make an estimate for the initial step size h_t.

1. Check if the step size is below numerical accuracy.
2. Calculate the results of one step of $2h$ and two steps of h each using RK_STEP:
   - (a) Calculate $\boldsymbol{Y}_{2h}(t_k + 2h_t)$ of Eq. (8-22):
     - (i) From the values of t_k and $\boldsymbol{y}_k$, evaluate $\boldsymbol{p}_{2h}$, $\boldsymbol{q}_{2h}$, $\boldsymbol{r}_{2h}$, and $\boldsymbol{s}_{2h}$.
     - (ii) Find $\boldsymbol{Y}_{2h}(t_k + 2h_t)$ from $\boldsymbol{p}_{2h}$, $\boldsymbol{q}_{2h}$, $\boldsymbol{r}_{2h}$, and $\boldsymbol{s}_{2h}$.
   - (b) Calculate $\boldsymbol{Y}_{1h}(t_k + 2h_t)$ with Eqs. (8-23) and (8-24):
     - (i) From the value of $t = t_k$ and $\boldsymbol{y}_k$, evaluate $\boldsymbol{q}_{h1}$, $\boldsymbol{r}_{h1}$, and $\boldsymbol{s}_{h1}$.
     - (ii) Evaluate $\boldsymbol{z}(t_k + h_t)$ with Eq. (8-23) using $\boldsymbol{p}_{2h}$, $\boldsymbol{q}_{h1}$, $\boldsymbol{r}_{h1}$, and $\boldsymbol{s}_{h1}$.
     - (iii) From $t = (t_k + h_t)$ and $\boldsymbol{z}(t_k + h_t)$, calculate $\boldsymbol{p}_{h2}$, $\boldsymbol{q}_{h2}$, $\boldsymbol{r}_{h2}$, and $\boldsymbol{s}_{h2}$.
     - (iv) Evaluate $\boldsymbol{Y}_{1h}(t_k + 2h_t)$ from $\boldsymbol{p}_{h2}$, $\boldsymbol{q}_{h2}$, $\boldsymbol{r}_{h2}$, and $\boldsymbol{s}_{h2}$.
3. Find the maximum truncation error R of Eq. (8-27):
   - (a) If R is less than $\mathcal{E}_{\max}$, accept the step.
     - (i) Check if the maximum number of points is exceeded.
     - (ii) Store the results in the arrays for t and $\boldsymbol{y}$.
     - (iii) Find the next optimum step size using Eq. (8-26).
   - (b) If the maximum error is too large, decrease h and try again.
4. Return t_k in T_ARY and $\boldsymbol{y}_k$ in BFY.

Since $\boldsymbol{\Delta}(t_k, h_t)$ is larger than $\mathcal{E}_{\max}$, h_0 is expected to be smaller than h_t. We can go back to Eqs. (8-22) to (8-24) to calculate $\boldsymbol{Y}_{2h}$, $\boldsymbol{Y}_{2h}$, and $\boldsymbol{\Delta}$ with h_0 as the step size. This process may have to be repeated until the condition $\boldsymbol{\Delta}(t_k, h_t) \leq \mathcal{E}_{\max}$ is satisfied. As a practical matter, one must guard against the situation that, as we reduce the step size, the value of h_0 becomes smaller than the least significant figure of t (due to the limited word length of the computer memory) and is essentially zero as far as the numerical calculation is concerned. If this happens, we have a situation equivalent to having $h_t = 0$ and the calculation becomes meaningless.

Because the power on the right side of Eq. (8-25) is small, the value of h_0 obtained is not very sensitive to those of $\boldsymbol{\Delta}(t_k, h_t)$ and $\mathcal{E}_{\max}$. This is a good thing, as the relation is only an approximate one, coming from estimates of the truncation errors. It is recommended in Press et al., [44] to use a slightly modified form

$$h_0 = R_f h_t \left| \frac{\mathcal{E}_{\max}}{\boldsymbol{\Delta}(t_k, h_t)} \right|^{\eta} \tag{8-26}$$

where R_f is a factor on the order of 0.9 and the power is $\eta = 1/4$ if $\boldsymbol{\Delta}(t_k, h_t)$ is greater than $\mathcal{E}_{\max}$ and $\eta = 1/5$ otherwise. This is implemented in the algorithm outlined in Box 8-2.

The estimate of h_0 is slightly more complicated in practice for a system of n-coupled equations. Since there are n different dependent variables, $y_1(t)$, $y_2(t), \ldots,$ $y_n(t)$, the tolerance for error may be different for each of them. As a result, we have n different values for $\mathcal{E}_{\max}$ and there is a different $\boldsymbol{\Delta}(t_k, h_t)$ for each of the n-coupled first-order ODEs. To determine the optimum step size for the independent variable t, therefore, requires a choice among the n different estimates. There are no unique answers to this question. One possibility is to measure $\boldsymbol{\Delta}(t_k, h_t)$ by the average of $\boldsymbol{Y}_{1h}(t_k, 2h)$ and $\boldsymbol{Y}_{2h}(t_k, 2h)$. In this case, we may define the error in a step as

$$R = 2 \max\left\{ \left| \frac{\boldsymbol{Y}_{1h}(t_k + 2h_t) - \boldsymbol{Y}_{2h}(t_k + 2h_t)}{\boldsymbol{Y}_{1h}(t_k + 2h_t) + \boldsymbol{Y}_{2h}(t_k + 2h_t)} \right| \right\} \tag{8-27}$$

where the numerator is nothing but $\boldsymbol{\Delta}(t_k, h_t)$. Since R defined in this way is "dimensionless," we can specify the tolerance for error $\mathcal{E}_{\max}$ in the same way, such as a value of 10^{-4} in single precision calculations.

8-3 Solution of initial value problems by extrapolation

In the Romberg method of integration described in §3-5, we saw an example of using extrapolation techniques to reduce truncation errors due to finite step size. Here we shall apply a similar approach to solve a system of differential equations. The main advantage over the Runge-Kutta methods of §8-2 is the higher degree of accuracy that can be achieved with a relatively small amount of additional calculations. On the other hand, since extrapolation is safe only with smooth functions, the method is applicable only in regions where there are no singularities and the functions $\boldsymbol{f}(\boldsymbol{y}, t)$ do not have rapid variations.

Our ultimate goal here is the same as that of the previous section: to find the solution of a system of n-coupled first-order ordinary differential equations of the form

$$\frac{d\boldsymbol{y}}{dt} = \boldsymbol{f}(\boldsymbol{y}(t), t)$$

subject to a set of initial values at time $t = t_0$. The solutions are propagated to the end of the interval in two different stages. Conceptually, we can think of first dividing the interval $[t_0, t_d]$ into a number of big steps each one of size H that can vary from

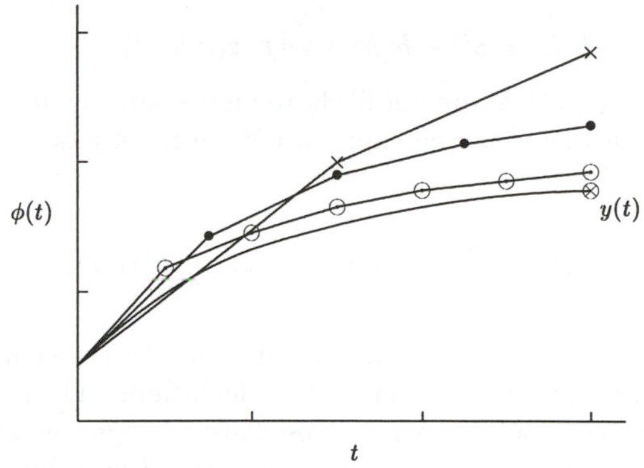

Figure 8-4: Schematic illustration of successive approximations to solve an ODE. The accuracy improves as the number of divisions increases from $\eta = 2$ (×), 4 (•), to 6 (⊙), leading eventually to the the exact value [smooth curve labeled $y(t)$] when extrapolated to an infinite number of intervals.

one region to another. Within each one of these big steps, the calculation is carried out using a number of smaller steps with size h. The extrapolation procedure is used only at the second stage, as it may be unreasonable to expect such a procedure to be valid in a large interval. To simplify the calculation, h is taken to be constant within a given big step by dividing H into η equal size subintervals.

The accuracy depends on the size of h, or the value of η, used. Clearly, the results improve with increasingly larger η, as illustrated schematically in Fig. 8-4. The advantage provided by the extrapolation technique is that we can find the value in the limit of $h = 0$ without actually having to carry out the calculations to $\eta \to \infty$. This is the principle behind Richardson's deferred approach to the limit we saw earlier in §3-5.

The strategy of the extrapolation method is quite simple. For each interval H, we start with an h obtained with some small η, like 2, and gradually decrease h by increasing η. When there are several values of $\boldsymbol{y}(t + H)$ obtained with different step size h, an extrapolation is carried out to find the value in the limit of $h = 0$. For a given h we can, in principle, calculate the value of $\boldsymbol{y}(t+H)$ using any of the methods described earlier. As a practical matter, it is more efficient to use one adapted for the purpose.

Modified midpoint method Consider a function $\boldsymbol{z}(t, h)$ defined on a discrete set of points, $t_k = t_0 + kh$ for $k = 0, 1, \ldots, \eta$. Starting from $t = t_0$, the values of $\boldsymbol{z}(t, h)$ for the first two points are given by

$$\boldsymbol{z}(t_0; h) = \boldsymbol{y}(t_0) \qquad\qquad \boldsymbol{z}(t_0 + h; h) = \boldsymbol{y}(t_0) + h\boldsymbol{f}(\boldsymbol{y}(t_0), t_0).$$

With these two as the start, the values of $\boldsymbol{z}(t, h)$ at the other points are defined by the relation

$$z(t+h;h) = z(t-h;h) + 2hf(z(t;h),t). \tag{8-28}$$

Because of truncation errors, $z(t;h)$ are not likely to be the same as $y(t)$ for any finite h. The difference between them may be expressed in terms of a series consisting of only even powers of h,

$$z(t;h) = y(t) + \sum_{k=1}^{\infty} h^{2k}\{u_k(t) + (-1)^{(t-t_0)/h} v_k(t)\} \tag{8-29}$$

where the functions $u_k(t)$ and $v_k(t)$ are independent of h. As we shall soon see, the forms of these functions are not important in actual calculations. The $(-1)^{(t-t_0)/h}v_k(t)$ term in the sum is, however, troublesome, as it oscillates in sign. As shown in Stoer and Bulirsch,[54] it can be eliminated in the leading order h^2 by taking the following linear combination

$$S(t;h) = \frac{1}{2}\Big\{z(t;h) + z(t-h;h) + hf(z(t;h),t)\Big\}.$$

The result is a well-behaved series at least up to order h^2:

$$S(t;h) = y(t) + h^2\Big\{u_1(t) + \frac{1}{4}\frac{d^2y}{dt^2}\Big\} + \sum_{k=2}^{\infty} h^{2k}\Big\{u_k(t) + (-1)^{(t-t_0)/h}v_k(t)\Big\}.$$

In the limit $h \to 0$, we have $S(t;h) = y(t)$. If several $S(t+H,h)$ are available, each one for a different h, we are in a position to extrapolate its value at $h = 0$ and, hence, the value of $y(t+H)$.

To implement the algorithm, we start by dividing the interval H into η equal steps. To make clear the dependence of the step size on η, we shall add a subscript to h in the form

$$h_\eta = \frac{H}{\eta}.$$

For a given h_η, the value of $S(t+H;h_\eta)$ is given by

$$S(T;h_\eta) = \frac{1}{2}\Big\{z(T;h_\eta) + z(T-h_\eta;h_\eta) + hf(z(T;h_\eta),T)\Big\} \tag{8-30}$$

where we have used the symbol $T = (t+H)$ to simplify the notation. To obtain the values of $z(T-h_\eta;h_\eta)$ and $z(T;h_\eta)$, we follow Eq. (8-28) and use the value of $y(t)$ already in our possession to define

$$z(t,h_\eta) = y(t) \qquad\qquad z(t+h_\eta;h_\eta) = z(t;h_\eta) + h_\eta f(z(t,h_\eta),t).$$

All the later values of $z(t+mh_\eta;h_\eta)$ with $m > 1$ are produced by the recursion relation

$$z(t+(m+1)h_\eta;h_\eta) = z(t+(m-1)h_\eta;h_\eta) + 2h_\eta f(z(t+mh_\eta;h_\eta),t+mh_\eta) \tag{8-31}$$

Box 8-3 Subroutine MIDPT(Y_IN,Y_OUT,T,H_BIG,NSTPS,NEQS,FVAL)
Modified midpoint method for a system of n first-order ODEs

Argument list:

Y_IN:	Input value of $\boldsymbol{y}(t)$.
Y_OUT:	Output value of $\boldsymbol{y}(t+H)$.
T:	Value of t.
H_BIG:	Value of H.
NSTPS:	Number of steps η.
NEQS:	Number of equations.
FVAL:	External subprogram to calculate $\boldsymbol{f}(\boldsymbol{y},t)$.

1. Define the step size h_η to be H/η.
2. Define ZM1 as $z(t;h_\eta)$ and ZM as $z(t+h_\eta;h_\eta)$.
3. Propagate using Eq. (8-31) starting from $m=1$ until $m=\eta$:
   (a) Calculate $\boldsymbol{y}(t+(m+1)h_\eta)$ using Eq. (8-31).
   (b) Store the results temporarily as TEMP.
   (c) Store the value of ZM in ZM1.
   (d) Move the value of TEMP into ZM.
4. Return the value of $\frac{1}{2}\{ZM+ZM1+h_\eta\boldsymbol{f}(ZM,t+H)\}$ as $\boldsymbol{y}(t+H)$.

obtained using Eq. (8-28). The necessary input to the right side of Eq. (8-30) is obtained from the calculated results for $m=\eta$ and $m=(\eta-1)$.

It is basically a midpoint method, since the solutions to the differential equations for most of the steps from t to $(t+H)$ are propagated from $t=(m-1)h$ to $t=(m+1)h$ using the value of $\boldsymbol{f}(\boldsymbol{y},t)$ at a point half way between the beginning and the end of the interval for the step. However, the first and last steps are different. For this reason, it is known as the *modified midpoint* method. A summary of the algorithm is given in Box 8-3. In principle, the method can be used to find the solution for a set of ODEs in general. However, it is not efficient by itself and is used here only as the inner core for the extrapolation method.

To extrapolate, we need a sequence of $\boldsymbol{S}(T;h_1)$, $\boldsymbol{S}(T;h_2),\ldots$ calculated with different values of h_k and η_k, for $k=1,2,\ldots$. The obvious choice of $\eta_k=2^k$ for the series is not practical here, as the size of h_k diminishes too quickly to be useful as a sequence for the purpose of extrapolation. An alternative is the sequence $\eta_k=\{2,\ 4,\ 6,\ 8,\ 12,\ 16,\ldots,\}$, which may be written as

$$\eta_k=2\eta_{k-2} \qquad \text{for} \qquad k>3.$$

For reasonable values of H, it is seldom necessary to go to very large k values.

Once $\boldsymbol{y}(t+H)$ are available for step sizes $h_1,\ h_2\ldots$, we can extrapolate the corresponding values for $h\to 0$. For the convenience of discussion, let us represent the values of $\boldsymbol{y}(t+H)$ calculated with step size $h_k=H/\eta_k$ as $\boldsymbol{F}(x_k)$. Because of Eq. (8-29), the independent variable x_k for $\boldsymbol{F}(x_k)$, as far as the extrapolation is concerned, is h_k^2. In other words, we have the relation

$$\lim_{h\to 0} \boldsymbol{F}(h^2) = \boldsymbol{y}(t+H). \tag{8-32}$$

Similar to extrapolations discussed in §3-5, we can regard $\boldsymbol{F}(x)$ as a function of x, with values known to us only at $x = x_1, x_2, \ldots$, corresponding to $h^2 = h_1^2, h_2^2, \ldots$. The polynomial extrapolation method of §3-5 may be used if we express $\boldsymbol{F}(x)$ as a polynomial in x. This is left as an exercise.

Rational polynomial extrapolation For better accuracy, we shall express $\boldsymbol{F}(x)$ as a ratio of two polynomials and apply the technique of rational polynomial interpolation to find $\boldsymbol{y}(t+H)$ in the limit of $h = 0$. The extrapolation may be carried out using essentially the Neville's algorithm outlined in Box 3-2.

We should not forget the fact that we are dealing with a set of n dependent variables here, $\boldsymbol{y}(t+H) = \{y_1(t+H), y_2(t+H), \ldots, y_n(t+H)\}$. As far as extrapolation is concerned, a calculation must be carried out for each of these n quantities. There is no difficulty in doing this, as each $y_k(t+H)$, for $k = 1$ to n, may be regarded as unrelated to the others as far as extrapolations are concerned. For each one, we shall assume that there are ξ different values $F(x)$, calculated with successive smaller step sizes h_k. Similar to Eqs. (3-18) and (3-19), we can define two sets of auxiliary quantities Θ_{mk} and Δ_{mk}, where the subscript $m = 0, 1, 2, \ldots, (\xi - 1)$ indicates the order of extrapolation, and the subscript $k = 1, 2, \ldots, (\xi - m)$ distinguishes the different elements in each order.

Because we are interested only in extrapolating to $x \equiv h^2 \to 0$, the calculations are somewhat simpler than those for the general case in §3-2. The recursion relations remain to be the same as given in Eq. (3-18)

$$\begin{aligned} \Theta_{(m+1)k} &= \frac{(\Delta_{m(k+1)} - \Theta_{mk})\Delta_{m(k+1)}}{\frac{x_k}{x_{m+k+1}}\Theta_{mk} - \Delta_{m(k+1)}} \\ \Delta_{(m+1)k} &= \frac{\frac{x_k}{x_{m+k+1}}\Theta_{mk}(\Delta_{m(k+1)} - \Theta_{mk})}{\frac{x_k}{x_{m+k+1}}\Theta_{mk} - \Delta_{m(k+1)}}. \end{aligned} \tag{8-33}$$

The starting values of Θ and Δ are given by $F(x_k)$

$$\Theta_{0k} = \Delta_{0k} = F(x_k)$$

the values of $\boldsymbol{y}(t+H)$, more precisely, those of $S(T, h_k)$ in Eq. (8-30), calculated with $h_k = H/\eta_k$. Let us represent the best estimate of the extrapolated value for order m as $F_m(0)$. For $m = 0$, this is given by the one calculated with the smallest step size. As a result, we have

$$F_0(0) = F(x_\xi)$$

where $x_\xi = h_\xi^2$ and h_ξ is the smallest step size used to calculate $\boldsymbol{y}(t+H)$. Since we are carrying out an extrapolation, the correction for order $m > 0$ is given by $\Theta_{m(\xi-m)}$,

Box 8-4 Subroutine EXTZS(T_B,T_D,T_ARY,BFY,EPS,N_BSTP,ND_PTS,ND_EQN)
Solution of first-order ODEs by extrapolating to zero step size

Argument list:
- T_B: Beginning of the interval for t.
- T_D: End of the interval for t.
- T_ARY: One-dimensional array for t.
- BFY: Two-dimensional array for $\boldsymbol{y}$.
- EPS: Maximum error allowed.
- N_BSTP: Number of big steps H taken.
- ND_PTS: Number of big steps dimensioned in the calling program.
- ND_EQN: Number of first-order ODEs in the system.

Subprogram used:
- MIDPT: Modified midpoint method of Box 8-3.
- EXTZO: Extrapolate to zero step size using Neville's algorithm (cf. Box 3-2).

Initialization:
- (a) Store the initial conditions as the first points of the solution.
- (b) Set the maximum extrapolation order to be 11, maximum tries $N_{\rm try}$ and extrapolations $N_{\rm ext}$, and big step number 10.
- (c) Use $\{2, 4, 6, 8, 12, 16, \ldots\}$ as the sequence for the values of η_k.

1. Set the size of big step H.
2. Use MIDPT to calculate $\boldsymbol{y}(t + H)$:
   Divide H into η_k steps of size h_η each. Store h_η^2 and $\boldsymbol{y}(t + H)$ obtained.
3. Call EXTZO to extrapolate the values of $\boldsymbol{y}(t + H)$ for $h^2 = 0$.
4. Find the maximum error $\mathcal{E}_{\max}$ among the ND_EQN extrapolated values.
5. If $\mathcal{E}_{\max} \leq$ EPS, accept the extrapolated result:
   - (a) Store the extrapolated values as $\boldsymbol{y}(t + H)$.
   - (b) If t is less than T_D:
     Go back to step 2 for the next time interval. Adjust H if needed.
   - (c) If $t \geq$ T_D, return the values of t and $\boldsymbol{y}$.
6. If $\mathcal{E}_{\max} >$ EPS, carry out further extrapolations:
   - (a) If no convergence after $N_{\rm ext}$ extrapolations. Reduce H and try again.
   - (b) If no convergence after $N_{\rm try}$ reductions, abort the calculation.

the last element of Θ for that order. As a result, the best estimate of the value for $h^2 \equiv x = 0$ for order m is given by

$$F_m(0) = F_{m-1}(0) + \Theta_{m(\xi-m)}.$$

Similar to what we did in §3-2, an estimate of the error of the process is provided by the correction term used for the highest order. In the present case, it is $\Theta_{\xi-1,1}$.

The algorithm outlined in Box 8-4 concentrates mainly on getting the value of $\boldsymbol{y}(t+H)$ starting with the known $\boldsymbol{y}(t)$. To start with, the modified midpoint method is used to obtain $\boldsymbol{y}(t + H)$ for the first two values of η_k. Once these are available, an attempt is made to extrapolate to zero step size. If the estimated error in the extrapolation is larger than the tolerance, one more set of $\boldsymbol{y}(t+H)$ will be calculated using the next smaller size for h_η and the extrapolation is carried out again. This is repeated until the errors in the extrapolated results are sufficient small. The values are returned as the solution for $\boldsymbol{y}(t+H)$ and the calculation proceeds to the next subinterval of size H.

The number of extrapolation steps actually taken at a given t gives us an estimate of the size of H to be used for the next subinterval. If the desired accuracy is reached with a small number of η_k, it is possible to make the next H larger than the present one. On the other hand, if we cannot reach the required numerical accuracy within a reasonable number of subdivisions into h_η, it is wise to use a smaller value for H for the next step. From the point of view of computational efficiency, we may want to carry out the calculations with a minimum number of steps in terms of H. This may not always be desirable from other considerations. On many occasions, we may wish to have a feeling for how the solution behaves as a function of the independent variable. In this case a minimum size must be imposed on H.

Another practical consideration is that there is no point in taking the extrapolation process to too high an order, as we are interested only in the value at $h = 0$. For this reason, the algorithm puts an upper limit ξ for the order of extrapolation. That is, if there are more $(\xi + 1)$ different sets of values available, the extrapolation makes use only of the last $(\xi + 1)$.

8-4 Boundary value problems by shooting methods

We saw earlier in §8-1 that, in general, it is more complicated to find the numerical solution of a boundary value problem than that of an initial value problem. The difference comes primarily from the fact that, in boundary value problems, not all the constraints are given at one location. As a result, we do not have enough information to propagate the solution. On the other hand, if we are able to make some estimates of the missing information, the calculations can be reduced to the same as an initial value problem. Whether the estimates are correct can be checked by comparing the solution with the values of the unused boundary conditions at the other end. If, in addition, we have a way to improve our estimates step by step towards satisfying all the boundary conditions, we have a way to solve a boundary value problem using techniques designed for initial value problems. This is the philosophy of shooting methods.

The projectile problem posed in Problems 8-1 and 8-2 gives a good illustration of method. If our objective is to find out the location where a projectile will land for some given initial velocity and angle of elevation, we have an initial value problem. On the other hand, if our interest is to send the projectile with a given initial velocity to a fixed point some distance away, we have a boundary value problem. The analogy to the shooting method here is that we adjust the angle of elevation by trial and error until the projectile lands at correct point.

Quantum mechanical scattering from a square-well potential Let us use quantum mechanical scattering as an example to illustrate the method. To start with, we shall apply it first to the simple case of a square well where analytical solution is available. Once the principles are established, we shall apply the method to scattering from a more realistic potential for which numerical solution is the only way.

For scattering, we need to solve the three-dimensional Schrödinger equation

$$-\frac{\hbar^2}{2\mu}\nabla^2\psi(\boldsymbol{r}) + V(r)\psi(\boldsymbol{r}) = E\psi(\boldsymbol{r}) \tag{8-34}$$

where μ is the reduced mass of the scattered particle and $V(r)$ is the potential. The energy of the particle outside the potential well is given by E and is a positive quantity. For spherically symmetric potentials, we can make use of procedures similar to §4-4 to separate out the angular dependence

$$\psi(\boldsymbol{r}) = \sum_{\ell=0}^{\infty}\frac{1}{r}u_\ell(r)Y_{\ell,m}(\theta,\phi) \tag{8-35}$$

where $Y_{\ell,m}(\theta,\phi)$ are the spherical harmonics. The only differential equation to be solved is that for the modified radial wave function $u_\ell(r)$,

$$-\frac{\hbar^2}{2\mu}\left\{\frac{d^2u_\ell}{dr^2} - \frac{\ell(\ell+1)}{r^2}u_\ell(r)\right\} + V(r)\,u_\ell(r) = E\,u_\ell(r). \tag{8-36}$$

For our purpose here, we shall be interested only in the angular momentum $\ell = 0$ case. The radial equation simplifies to

$$-\frac{\hbar^2}{2\mu}\frac{d^2u_0}{dr^2} + V(r)\,u_0(r) = E\,u_0(r). \tag{8-37}$$

For $\ell > 0$, we can replace $V(r)$ by the following effective potential:

$$\tilde{V}(r) = V(r) + \frac{\hbar^2}{2\mu}\frac{\ell(\ell+1)}{r^2}.$$

In this way, the differential equation is formally identical to Eq. (8-37).

For a square-well potential of height V_0,

$$V(r) = \begin{cases} V_0 & \text{for } r \le r_g \\ = 0 & \text{for } r > r_g \end{cases} \tag{8-38}$$

where r_g is the range of the potential. If V_0 is positive, the potential is repulsive, as shown in Fig. 8-5(a), and, if V_0 is negative, the potential is attractive, as shown of Fig. 8-5(b).

Since we are dealing with a second-order differential equation, two boundary conditions are required. The first comes from the fact that the wave function $\psi(\boldsymbol{r})$ must be finite at the origin. As a result, $u_0(r)$ vanishes at $r = 0$

$$u_0(r = 0) = 0. \tag{8-39}$$

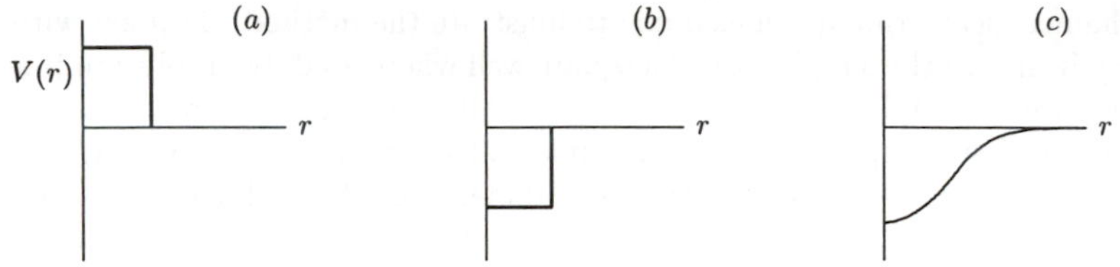

Figure 8-5: Scattering potential: (*a*) repulsive square well, (*b*) attractive square well, and (*c*) Woods-Saxon well.

The second boundary condition comes from the radial wave function at large r. Outside the potential well, $V(r) = 0$, the particle is a free one and the modified radial wave function is given by the equation

$$-\frac{\hbar^2}{2\mu}\frac{d^2u_0(r)}{dr^2} = Eu_0 \qquad \text{for} \qquad r \gg r_g.$$

This can be put into a form similar to Eq. (8-1):

$$\frac{d^2u_0(r)}{dr^2} + k^2u_0 = 0$$

where k is real and has the value

$$k = \frac{1}{\hbar}\sqrt{2\mu E}. \tag{8-40}$$

The solution of this equation is familiar, and we shall write it in the form

$$u_0(r) = A\sin(kr + \delta_0) \qquad \text{for} \qquad r \gg r_g. \tag{8-41}$$

Here A is the amplitude, related to the normalization of the wave function. The value of A is of no consequence to our problem here and, as a result, we can take

$$u_0(r_d) = 1 \tag{8-42}$$

for some large, fixed value of r_d as our second boundary condition. All the information of physical importance is contained in the phase factor δ_0, generally known as the *phase shift*. A schematic illustration of the role of δ_0 is given in Fig. 8-6.

The solution of the scattering problem for a square-well potential is well known in quantum mechanics. In the case of an attractive potential, the phase shift is given by

$$\delta_0 = m\pi - kr_g + \tan^{-1}\left(\frac{k}{\kappa}\tan\kappa r_g\right) \tag{8-43}$$

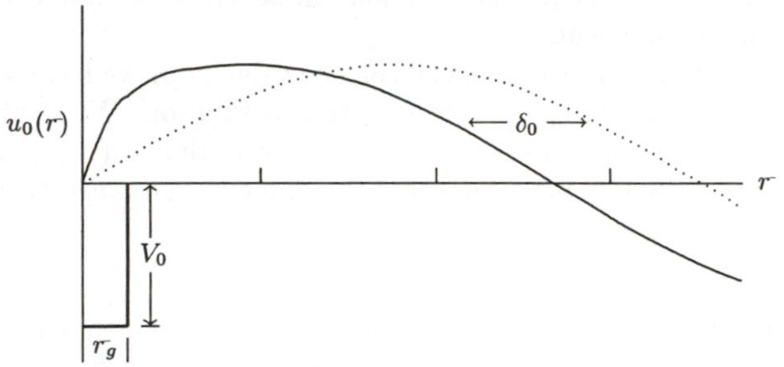

Figure 8-6: Phase shift δ_0. The scattered wave function (solid curve) is shifted in phase from that of a free particle (dotted curve) as a result of the interaction with a potential well.

where m is an integer and

$$\kappa = \frac{1}{\hbar}\sqrt{2\mu(E - V_0)}. \tag{8-44}$$

Since $E > 0$ in a scattering problem and $V_0 < 0$ for an attractive potential, κ is real. We shall confine our interest to low energies where $m = 0$.

The inside region As the first step in a numerical solution, we shall try to solve Eq. (8-37) only for the region inside the potential well, $r = [0, r_g]$. For the purpose of illustration, we shall use the value of $u_0(r_g)$ given by Eq. (8-41)

$$u_0(r_g) = \sin(kr_g + \delta_0) \tag{8-45}$$

with δ_0 obtained from Eq. (8-43) for $m = 0$ as the boundary condition at $r = r_g$.

The differential equation Eq. (8-37) may be simplified to the form

$$\frac{d^2u_0}{dr^2} + \frac{\hbar^2}{2\mu}\{E - V(r)\}u_0(r) = 0. \tag{8-46}$$

However, instead of solving it as a two-point boundary value problem with one boundary condition specified at each end of the interval, we wish to treat it as an initial value problem. Before we can do this, we need a second "initial" condition at $r = 0$, in addition to the one supplied by Eq. (8-39). This may be provided for us in the form of an estimate for du_0/dr at $r = 0$. Let us assume that the value is

$$\left.\frac{du_0}{dr}\right|_{r=0} = \beta_1 \tag{8-47}$$

where β_1 is some reasonable value based on physical grounds. The differential equation can now be solved using any one of the methods for initial value problems, such as the extrapolation method of the previous section. However, we shall not be concerned with such details at the moment.

The solution we obtain for $u_0(r)$ depends on the value of β_1 we have selected. Let us use the symbol $u_0(\beta_1; r)$ to represent this particular solution. When we propagate it from $r = 0$ to $r = r_g$, we obtain $u_0(\beta_1; r_g)$. It is unlikely that $u_0(\beta_1; r_g)$ satisfies the condition given by Eq. (8-45). Let us indicate the difference between them as

$$R(\beta_1) = u_0(\beta_1; r_g) - u_0(r_g). \tag{8-48}$$

Our aim now is to find a new value of $\beta = du_0/dr$ at $r = 0$ such that $R(\beta)$ vanishes.

For this purpose, we shall repeat the calculation with a different estimate β_2 that is also a reasonable choice. On solving the differential equation again as an initial value problem, we obtain

$$R(\beta_2) = u_0(\beta_2; r_g) - u_0(r_g).$$

In general, the functional dependence of R on β is complicated and nonlinear. However, if both our estimates are sufficiently close to the true solution, we can approximate $R(\beta)$ to be linear in β. This is reasonable at least in a small region where $R(\beta) \sim 0$. In this limit,

$$\beta = \beta_2 - R(\beta_2)\frac{\beta_2 - \beta_1}{R(\beta_2) - R(\beta_1)} \tag{8-49}$$

satisfies the requirement

$$R(\beta) = 0.$$

If this is true, we have solved our boundary value problem using techniques designed for initial value problems.

In general, our linear extrapolation for β used above in deriving Eq. (8-49) is not sufficiently accurate. As a remedy, we can iterate the process by defining a new value

$$\beta_{i+1} = \beta_i - \Delta\beta_i \tag{8-50}$$

where the necessary change from β_i, obtained in iteration i, is given by

$$\Delta\beta_i = R(\beta_i)\frac{\beta_i - \beta_{i-1}}{R(\beta_i) - R(\beta_{i-1})}. \tag{8-51}$$

In general, if we start from a value that is close to the solution, we have the assurance that

$$|R(\beta_{i+1})| < |R(\beta_i)|.$$

As a result, repeated applications of Eq. (8-50) will eventually lead to a $R(\beta)$ that is less than the error that can be tolerated. The algorithm is outlined in Box 8-5.

For ODEs with order higher than two, the more common situation is one of having n_1 boundary conditions given at one end of the interval and n_2 boundary

Box 8-5 Program DM_SHTSQ
Shooting method to obtain the scattering solution
For a one-dimensional square-well potential

Subprograms used:
- EXTZS: Solution of n first-order ODEs by extrapolation (Box 8-4).
- EXTZO: Extrapolation to zero step size using Neville's algorithm (cf Box 3-2).
- MIDPT: Modified midpoint method for n first-order ODEs (Box 8-3).
- FVAL: Value of $f(y,t)$ for a square well.

Initialization:
- (a) Map the interval to $x = [0, 1]$.
- (b) Let the two initial estimates be $\beta_1 = 1$ and $\beta_2 = 2$.
- (c) Reduce the second-order ODE to two first-order ODEs using Eq. (A-10).

1. Input energy E and well depth V_0.
2. Calculate the constant factors k of Eq. (8-40) and κ of (8-44).
3. Evaluate δ_0 of Eq. (8-43) and define the boundary condition at $x = 1$.
4. Solve the ODE by extrapolation using EXTZS:
   - (a) Set up the boundary value at $x = 0$.
   - (b) Make a first estimate β_1 as the value of $y_2(x_b)$.
   - (c) Use EXTZO to carry out the extrapolation to zero step size.
   - (d) Calculate the difference $R(\beta_1)$ of Eq. (8-48).
5. Solve the ODE again with a second estimate β_2.
6. Improve the estimate:
   - (a) Obtain a new value of β using Eq. (8-50).
   - (b) Check for convergence by comparing $|\Delta\beta/\beta|$ with the tolerance for error:
     - (i) If larger, solve the ODE again with the new value of β.
     - (ii) Otherwise, output the solution.

conditions given at the other end, with $(n_1 + n_2)$ equal to the number of equivalent first-order ODEs that must be solved. In this case, we must make n_2 estimates of the values at the start of the interval before we can apply the shooting method. A more sophisticated form to improve upon the estimates by iterative calculations can be found in Press and coauthors.[44]

For our illustrative example of the scattering wave function inside a square-well potential, the procedure given by Eq. (8-50) is more than adequate. Since $u_0(r) = 0$ at $r = 0$, our estimate β for the value of du_0/dr at $r = 0$ is equivalent to specifying the normalization of the wave function. At the other end of the interval, only the logarithmic derivative

$$L(r_g) \equiv \frac{1}{u_0(r_g)} \frac{du_0}{dr}\bigg|_{r=r_g} \tag{8-52}$$

is of physical significance. Since $L(r_g)$ is independent of the normalization, the solution, except the normalization, is independent of the value of β used. The reason for the simplicity comes from the fact that we have made use of the value of δ_0 given by the analytic solution of Eq. (8-43). In more realistic situations, several iterations are usually needed, as we shall see later.

Solution for the outside region Instead of using the analytic result for δ_0, we can try to find its value by solving the differential equation for the region $r = [r_g, \infty)$ as well. The two boundary conditions in this case are the value of $u_0(r_d)$ given in Eq. (8-42) and the logarithmic derivative $L(r_g)$ from the solution obtained for inside region.

Before we can start, we need to decide on what is a reasonable value for r_d. In principle, $r_d = \infty$. However, in practice, it has to be taken to be some large but finite value. For our present case, we can take r to be effectively infinite when we are far away from the potential well, for example, $r_d = 10r_g$. In principle, we can take even larger values; the price one pays is a loss of efficiency, as we have to cover a larger range of the value of r.

To make use of a shooting method for the outside region, we start from $r = r_d$ and work backwards toward $r = r_g$. For this purpose, we need to have, in addition to the condition given by Eq. (8-42), the value of a second one at $r = r_d$. Similar to Eq. (8-47) for the inside region, we can make an estimate for it and take

$$\left.\frac{du_0}{dr}\right|_{r=r_d} = \gamma.$$

We shall vary the value of γ until the logarithmic derivative at $r = r_g$ matches the one from the inside region.

We can use essentially the same procedure as that for the inside region to solve for $u_0(r)$ for $r = [r_g, r_d]$. Since the initial value problem must now be solved backwards with the dependent variable decreasing rather than increasing, some care is needed in the actual coding. For a given value of γ, the solution is propagated from $r = r_d$ to $r = r_g$. At $r = r_g$, the result is compared with $L(r_g)$ from the inside region and a new estimate of γ is obtained to improve the match. It is a relatively simple matter to replace u_0 by $L(r_g)$ in arriving at a similar expression as Eq. (8-51). This is left as an exercise.

Bidirectional shooting We have now a method to solve Eq. (8-37) for both the inside and the outside regions. In principle, we can use the shooting method to cover the entire range $[0, r_d]$. That is, we start the solution at one end, for example $r = 0$, with some estimate of the necessary second "initial" condition and propagate the solution forward all the way. At the other end, the solution is compared with the boundary condition given at $r = r_d$ and the difference is used to adjust the estimate, in the same way as we did above for each of the two subintervals. However, there are several advantages to dividing the region $[0, r_d]$ into two parts and using the bidirectional shooting method. In the first place, the whole interval may be too large for us to propagate the solution all the way and still expect good numerical accuracy. In the second place, the small and large r subintervals may have quite different properties. For example, in our scattering example, the step size can be

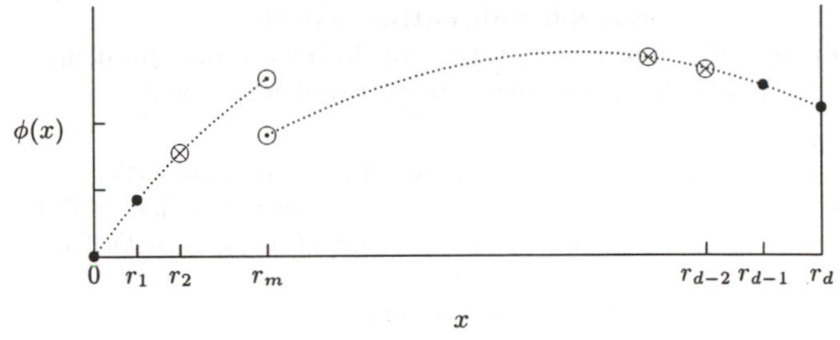

Figure 8-7: Solution of a boundary value problem by bidirectional shooting. The differential equation is treated as two initial value problems, one from each end of the intervals at $r = r_0$ and $r = r_d$. Estimates for the missing "initial" conditions are improved, step by step, by requiring the two solutions to match at r_m.

taken to be much larger in the outside region, where there is no interaction, and the division provides us with a natural way to do so. Finally, it may happen that, in certain problems, the truncation errors may accumulate and, as a result, it is not possible to propagate the solution all the way from one end of the interval to the other. In this case, it is helpful to start from both ends so that no single solution is propagated beyond where truncation errors become a serious problem.

We shall now solve the scattering problem for complete interval $[0, r_d]$ by starting from both ends and match the solution at some point $r = r_m$, in a way more or less the same as we have done earlier for the inside and outside separately. The principle behind the method of bidirectional shooting is quite straightforward. In each subinterval, the differential equation is solved as an initial value problem. In addition to the boundary condition provided by the physical problem, we introduce a second one based on estimates. Instead of the interval $[0, r_d]$ for our scattering problem, we can think more generally of an interval $r = [r_b, r_d]$ and divide it into two parts $[r_b, r_m]$ and $[r_m, r_d]$, with r_m as the matching radius. Let us use the symbol β to represent the supplemental condition we must have in the subinterval $[r_b, r_m]$ and the symbol γ for the corresponding quantity in the other subinterval $[r_m, r_d]$. At $r = r_m$, we have the logarithmic derivative $L^\ell(\beta)$ from the trial solution in the subinterval $[r_b, r_m]$ on the "left." Similarly, from the trial solution in the subinterval $[r_m, r_d]$ on the "right," we have the logarithmic derivative $L^r(\gamma)$ at the same point. If

$$L^\ell(\beta) = L^r(\gamma) \tag{8-53}$$

we have a solution for the whole region. The only question remaining now is to find a way to improve upon the estimated values of both β and γ so that the two logarithmic derivatives match at $r = r_m$.

Matching the logarithmic derivatives The actual implementation of the matching condition at $r = r_m$ requires some care. We shall adopt a simple algorithm here,

Box 8-6 Subroutine MATCH
Solution of second-order ODEs by bidirectional shooting
Example of scattering from a potential well

Subprograms used:
- EXTZS: Solution of n first-order ODEs by extrapolation (Box 8-4).
- EXTZO: Extrapolation to zero step size, using Neville's algorithm (cf Box 3-2).
- MIDPT: Modified midpoint method for a system of n first-order ODEs.
- FOUT: Value of $\boldsymbol{f}(\boldsymbol{y}, t)$ for the outside region.
- FINS: Value of $\boldsymbol{f}(\boldsymbol{y}, t)$ for the inside region.

Initialization:
- (a) Divide the region into two, $[r_b, r_m]$ and $[r_m, r_d]$, at matching radius r_m.
- (b) Set the boundary conditions as $y_0 = 0$ and $y_d = 1$.
- (c) Make estimates for the values of β_1, β_2, γ_1, and γ_2.

1. Two initial solutions for the inside region $r = [r_b, r_m]$ using EXTZS:
   - (a) Obtain a solution using y_0 and β_1 as the initial conditions.
   - (b) Obtain another solution with y_0 and β_2 as the initial conditions.
   - (c) Calculate the logarithmic derivatives for both cases at r_m.
   - (d) Calculate the value of B_m of Eq. (8-54).
2. Two solutions for the outside region $r = [r_m, r_d]$ using EXTZS:
   - (a) Obtain a solution using y_d and γ_1 as the initial conditions.
   - (b) Obtain another solution using y_d and γ_2 as the initial conditions.
   - (c) Calculate the logarithmic derivatives for both cases at r_m.
   - (d) Find the difference $R(\gamma_i)$ of Eq. (8-55).
   - (e) Use a modified form of Eq. (8-51) to calculate $\Delta\gamma$ so as to improve the match.
3. If $\Delta\gamma$ is larger than the tolerance for error, obtain a solution for the outside region with the new γ and calculate the logarithmic derivative $L^r(\gamma)$ again.
4. Improve the estimate for $[r_b, r_m]$ using $B_m = L^r(\gamma)$ at $r = r_m$.
   - (a) Find $\Delta\beta$ required to match B_m.
   - (b) If $\Delta\beta$ is larger than error tolerance, solve again with the new β value.
   - (c) Calculate $L^\ell(\beta)$ and use it as the condition at $r = r_m$ for $[r_m, r_d]$.
5. Check if $L^\ell(\beta_i)$ and $L^r(\gamma_j)$ are sufficiently close to each other.
   - (a) If so, normalize the solutions in two subintervals and return the results.
   - (b) If not, repeat steps 3 and 4 with the new values of B_m.

relying on the fact that, for most physical problems, we have the freedom of choosing the matching point at a place where the solution is well behaved. Furthermore, we can also depend on the problem to provide us with good initial estimates of β and γ. Let us begin with two different estimates, β_1 and β_2, of the values of du_0/dr at $r = r_b$ and, similarly, two different estimates, γ_1 and γ_2, at $r = r_d$. Since it is unlikely that we have chanced upon a set of choices that satisfy Eq. (8-53), we need a recipe to find successively better estimates for the values of β and γ.

The simple method described in Box 8-6 is constructed in such a way that, for each of the two subintervals $[r_b, r_m]$ and $[r_m, r_d]$, we can use the method of Box 8-5 to arrive

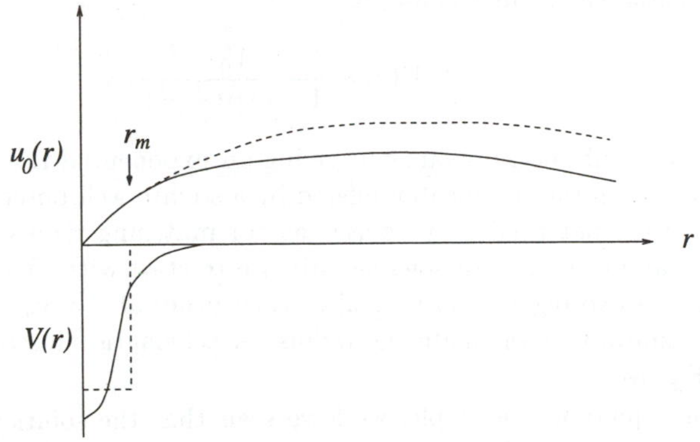

Figure 8-8: Radial wave function for a Woods-Saxon potential (solid curve) compared with that of a square-well potential (dotted curve). Differences outside are exaggerated so as to match the two solutions in the inside region.

at the next set of values for β and γ. For this purpose, each region requires a value at $r = r_m$ for the solution to "shoot" for. As far as each subinterval is concerned, this is the boundary condition at the other end that the solution must attempt to meet. There are several ways to provide such an artificial boundary condition. A simple approach is to take, as a start, the average of the logarithmic derivatives calculated in the subinterval $[r_b, r_m]$ with estimates β_1 and β_2. This gives us the value

$$B_m = \frac{1}{2}\{L^\ell(\beta_1) + L^\ell(\beta_2)\} \tag{8-54}$$

as the "boundary condition" for the solution from the other subinterval, $r = [r_m, r_d]$, to shoot for. Similarly, from the values of $L^r(\gamma_1)$ and $L^r(\gamma_2)$ we can obtain the value $\Delta\gamma$ from the differences

$$R(\gamma_j) = L^r(\gamma_j) - B_m \tag{8-55}$$

in an analogous manner as in Eq. (8-51). This gives us a new estimate γ_{j+1} for the value of du_0/dr at $r = r_d$.

We can now use the value of $L^r(\gamma_{j+1})$, obtained from solving the equation in the subinterval $[r_m, r_d]$, as the value B_m for the solution of the $[r_b, r_m]$ region to shoot for. With the logarithmic derivatives $L^\ell(\beta_i)$ for $i = 1$ and 2 we already have, a new value of β_{i+1} may be obtained. This can be used to obtain a new $L^\ell(\beta_{i+1})$ to serve as the value of B_m for the other subinterval. The process of iterating between the two subintervals $[r_b, r_m]$ and $[r_d, r_m]$ continues until we achieve convergence as defined by Eq. (8-53).

Our method is not restricted to the simple case of a square well potential given by Eq. (8-38). For example, we can use a more realistic potential, such as the Woods-

Saxon potential shown in Fig. 8-5(c):

$$V(r) = \frac{V_0}{1 + \exp\{\frac{r-r_0}{a}\}} \tag{8-56}$$

where a is the diffuseness parameter, giving an exponential rise to the radial dependence rather than the sharp cutoff offered by a square-well potential. In this case, r_0, the range of the potential, may be used as our matching radius r_m. For the value of $L(r_m)$, we can make some reasonable estimate to start with. The logarithmic derivatives from the two regions must equal to each other at $r = r_m$ for the two solutions to merge smoothly at the matching radius. A schematic illustration of this point is given in Fig. 8-7.

For our square-well example, we have seen that the solution in the subinterval $[r_b, r_m]$ is very stable. Consequently, one can arrive at the "final" value of the logarithmic derivative L^ℓ without much effort. The main part of the work involved is to find a value of γ such that $L^r(\gamma) = L^\ell$. The calculation depends very much on the question of how well we can estimate the values of γ_1 and γ_2 to start with. If the values are too far off from those required for convergence, the method of finding the next γ using Eq. (8-50) may fail and the calculation diverges. Fortunately, for such problems, we can usually rely on the physics to guide us to good estimates. For example, if we use the Woods-Saxon well of Eq. (8-38) for our scattering problem, we can make use of the known analytical solution of an equivalent square well as a guide. From Fig. 8-8, we see that the two solutions can be made similar to each other in the region of small r.

The idea of subdividing a region into two or more parts is, in general, a useful one for solving differential equations with shooting methods. We do not have to stop at two subdivisions, except that the physical problem we used as an example here does not really require it. A description of "multiple" shooting methods can be found in Sewell.[49]

8-5 Relaxation methods

Another approach to boundary value problems is to solve the finite difference equations using one of the numerical methods discussed in Chapter 5. To be more specific, we shall use, as the prototype for our discussion, the second-order differential equation

$$\frac{d^2\phi}{dx^2} + g(x)\phi(x) = 0 \tag{8-57}$$

in the interval $[x_0, x_N]$. The two boundary conditions required for this problem may be taken as

$$\phi(x_0) = \alpha_0 \qquad\qquad \phi(x_N) = \alpha_N. \tag{8-58}$$

The factor $g(x)$ can be either a constant or a function of x. The former corresponds to the case of Eq. (8-8), for which the solution is related to a sinusoidal wave form, and the latter is represented by Eq. (8-37) for the modified radial wave function of scattering from a Woods-Saxon potential. If $g(x)$ is a function of $\phi(x)$ or its derivatives, or both, it is a nonlinear differential equation, and we shall delay any discussion of such cases till §8-9.

To convert the second-order derivative in Eq. (8-57) to finite difference, we can follow the procedure used in Eq. (8-5) and make use of the central difference formula Eq. (2-52). For a constant step size, $h = (x_{k+1} - x_k)$, the result is

$$\left.\frac{d^2\phi}{dx^2}\right|_{x=x_k} \longrightarrow \frac{1}{h^2}\{\phi(x_{k+1}) - 2\phi(x_k) + \phi(x_{k-1})\}. \tag{8-59}$$

For unequal step sizes, the form is slightly more complicated:

$$\left.\frac{d^2\phi}{dx^2}\right|_{x=x_k} \longrightarrow \frac{2}{h_{k+1}h_k(h_{k+1}+h_k)}\{h_k\phi(x_{k+1}) - (h_{k+1}+h_k)\phi(x_k) + h_{k+1}\phi(x_{k-1})\} \tag{8-60}$$

where $h_k = (x_k - x_{k-1})$ and we have approximated $(x_{k+1/2} - x_{k-1/2})$ with $\frac{1}{2}(h_{k+1}+h_k)$. The $(N+1)$ mesh points are labeled from x_0 to x_N. Since there are two boundary conditions, the number of unknowns is $(N-1)$.

To simplify the discussion, we shall write most of the formulas in this section for the case of constant step size and they can be adapted, in principle, to the case of variable step size. In terms of finite difference equations, Eq. (8-57) takes on the form

$$\phi(x_{k+1}) - 2\phi(x_k) + \phi(x_{k-1}) + h^2 g(x_k)\phi(x_k) = 0. \tag{8-61}$$

Altogether there are $(N-1)$ such equations, as k takes on values from 1 to $(N-1)$. The total number of different $\phi(x_k)$'s is $(N+1)$ as $k = 0, 1, \ldots, N$. Instead of using the two boundary conditions to reduce the number of unknown quantities to $(N-1)$, it is more convenient for us to regard Eq. (8-58) as two additional equations to Eq. (8-61). In this way the set of algebraic equations we must solve now is

$$\begin{aligned} \phi(x_0) &= \alpha_0 \\ -\phi(x_{k-1}) + \{2 - h^2 g(x_k)\}\phi(x_k) - \phi(x_{k+1}) &= 0 \qquad \text{for} \quad k = 1, 2, \ldots, (N-1) \\ \phi(x_N) &= \alpha_N. \end{aligned}$$

In terms of matrices, this may be written as

$$\boldsymbol{AY} = \boldsymbol{b}. \tag{8-62}$$

Here, $\boldsymbol{A}$ is a square matrix of dimension $(N+1)$,

$$\boldsymbol{A} = \begin{pmatrix} 1 & 0 & 0 & 0 & \cdots & 0 \\ -1 & \eta_1 & -1 & 0 & \cdots & 0 \\ 0 & -1 & \eta_2 & -1 & \cdots & 0 \\ \vdots & \vdots & \vdots & \vdots & \ddots & \vdots \\ 0 & \cdots & \cdots & -1 & \eta_{N-1} & -1 \\ 0 & \cdots & \cdots & \cdots & 0 & 1 \end{pmatrix}$$

and $\eta_k \equiv \{2 - h^2 g(x_k)\}$. Both $\boldsymbol{Y}$ and $\boldsymbol{b}$ are column matrices with $(N+1)$ elements,

$$\boldsymbol{Y} = \begin{pmatrix} \phi(x_0) \\ \phi(x_1) \\ \phi(x_2) \\ \vdots \\ \phi(x_{N-1}) \\ \phi(x_N) \end{pmatrix} \qquad \boldsymbol{b} = \begin{pmatrix} \alpha_0 \\ 0 \\ 0 \\ \vdots \\ 0 \\ \alpha_N \end{pmatrix}.$$

The main feature to note here is that, although N may be a large number, the fraction of nonvanishing elements in each row of $\boldsymbol{A}$ is small, no more than three in the particular example we have here. The matrix $\boldsymbol{A}$ is, therefore, sparse. In general, the structure of $\boldsymbol{A}$ may be more complicated than our prototype here, but the fact that the matrix is sparse remains. Relaxation methods, and others based on them, take advantage of this feature of the matrix equation to solve boundary value problems.

Iterative method for solving a system of linear algebraic equations An efficient way to solve a system containing a large number of linear algebraic equations is to use an iterative approach. The basic principles involved may be described using a system of only four equations. In terms of numerical solutions to differential equations, this corresponds to the unrealistic situation of having only four points on the mesh. As we shall see below, the ideas apply to arbitrary number of equations.

The general form of a system of four equations for Eq. (8-62) may be written as

$$\begin{aligned} A_{1,1}Y_1 + A_{1,2}Y_2 + A_{1,3}Y_3 + A_{1,4}Y_4 &= b_1 \\ A_{2,1}Y_1 + A_{2,2}Y_2 + A_{2,3}Y_3 + A_{2,4}Y_4 &= b_2 \\ A_{3,1}Y_1 + A_{3,2}Y_2 + A_{3,3}Y_3 + A_{3,4}Y_4 &= b_3 \\ A_{4,1}Y_1 + A_{4,2}Y_2 + A_{4,3}Y_3 + A_{4,4}Y_4 &= b_4. \end{aligned} \tag{8-63}$$

It is always possible to arrange the equations in such a way that the diagonal elements of $\boldsymbol{A}$ are nonvanishing. We shall proceed with such an assumption and rewrite Eq. (8-63) as

$$\begin{aligned} Y_1 &= \frac{1}{A_{1,1}}\{b_1 - A_{1,2}Y_2 - A_{1,3}Y_3 - A_{1,4}Y_4\} \\ Y_2 &= \frac{1}{A_{2,2}}\{b_2 - A_{2,1}Y_1 - A_{2,3}Y_3 - A_{2,4}Y_4\} \\ Y_3 &= \frac{1}{A_{3,3}}\{b_3 - A_{3,1}Y_1 - A_{3,2}Y_2 - A_{3,4}Y_4\} \\ Y_4 &= \frac{1}{A_{4,4}}\{b_4 - A_{4,1}Y_1 - A_{4,2}Y_2 - A_{4,3}Y_3\}. \end{aligned} \tag{8-64}$$

This forms the starting point in our discussion on the various methods to solve Eq. (8-62).

An iterative solution begins with a set of initial estimates $\boldsymbol{Y}^{(0)}$. In actual problems, this set can come from, among other possibilities, the analytical solution of a similar situation. For example, if we are interested in obtaining the modified radial wave function for a particle scattered off a Woods-Saxon potential, as we did in the previous section, we can use the solution for an equivalent square-well potential as $\boldsymbol{Y}^{(0)}$. Unless we are dealing with an ill-conditioned case, the final solution is, in principle, independent of the initial estimate. On the other hand, a good starting point is essential in minimizing the number of iterations required.

The core of an iterative method is the way by which, given an approximate solution $\boldsymbol{Y}^{(i)}$, we can construct an improved one $\boldsymbol{Y}^{(i+1)}$. In this way, we can start from the trial solution $\boldsymbol{Y}^{(0)}$ and obtain $\boldsymbol{Y}^{(1)}$ that is a step closer to the final solution. In turn, $\boldsymbol{Y}^{(1)}$ may be used to obtain $\boldsymbol{Y}^{(2)}$, which is a further improvement, and so on. Convergence is achieved when the difference between $\boldsymbol{Y}^{(i+1)}$ and $\boldsymbol{Y}^{(i)}$ is less than some small, positive number ϵ, determined by the requirement of the problem and the accuracy that can be achieved with a particular type of computer. Obviously, efficiency, or the speed to reach convergence, is an important concern.

There are several ways to characterize the differences between two successive iterations and, hence, the manner in which we want the calculation to converge. For example, we can require that the average absolute values of the differences in each term be less than ϵ. Alternatively, we can make use of the root-mean-square (rms) difference instead. As a third possibility, we can demand that the difference at any point in the interval of interest be less than ϵ. In the last case, the maximum difference among all the points becomes our criterion for convergence. The choice between these three methods, and other variants of them, is determined by the problem, and we shall not elaborate on the question here. For our example in this section we shall adopt the first one: the average of the absolute difference between $\boldsymbol{Y}^{(i+1)}(x_j)$ and $\boldsymbol{Y}^{(i)}(x_j)$ for all x_j should be less than ϵ.

The simplest type of iterative formula we can construct based on Eq. (8-64) is

$$\begin{aligned}
Y_1^{(i+1)} &= \frac{1}{A_{1,1}}\left\{b_1 - A_{1,2}Y_2^{(i)} - A_{1,3}Y_3^{(i)} - A_{1,4}Y_4^{(i)}\right\} \\
Y_2^{(i+1)} &= \frac{1}{A_{2,2}}\left\{b_2 - A_{2,1}Y_1^{(i)} - A_{2,3}Y_3^{(i)} - A_{2,4}Y_4^{(i)}\right\} \\
Y_3^{(i+1)} &= \frac{1}{A_{3,3}}\left\{b_3 - A_{3,1}Y_1^{(i)} - A_{3,2}Y_2^{(i)} - A_{3,4}Y_4^{(i)}\right\} \\
Y_4^{(i+1)} &= \frac{1}{A_{4,4}}\left\{b_4 - A_{4,1}Y_1^{(i)} - A_{4,2}Y_2^{(i)} - A_{4,3}Y_3^{(i)}\right\}.
\end{aligned} \tag{8-65}$$

That is, all the input for calculating the next set of approximate results $\{\boldsymbol{Y}_j^{(i+1)}\}$ comes from those of the previous iteration $\{\boldsymbol{Y}_j^{(i)}\}$. This is known as the Jacobi method. For the more general case of N (instead of 4) algebraic equations, this may be written as

$$Y_j^{(i+1)} = \frac{1}{A_{j,j}}\left\{b_j - \sum_{k=1}^{j-1} A_{j,k}Y_k^{(i)} - \sum_{k=j+1}^{N} A_{j,k}Y_k^{(i)}\right\}. \tag{8-66}$$

A proof of the convergence of this method can be found in Dahlquist and Björck.[15]

We can make a slight improvement on the Jacobi method in the following way. Once $Y_1^{(i+1)}$ is found when we finish the calculations for the first one of the four equations of Eq. (8-65), the result may be used in solving the other three equations. Similarly, $Y_2^{(i+1)}$ may be used for solving the last two equations, and so on. The results for our $N = 4$ example are

$$\begin{aligned}
Y_1^{(i+1)} &= \frac{1}{A_{1,1}}\left\{b_1 - A_{1,2}Y_2^{(i)} - A_{1,3}Y_3^{(i)} - A_{1,4}Y_4^{(i)}\right\} \\
Y_2^{(i+1)} &= \frac{1}{A_{2,2}}\left\{b_2 - A_{2,1}Y_1^{(i+1)} - A_{2,3}Y_3^{(i)} - A_{2,4}Y_4^{(i)}\right\} \\
Y_3^{(i+1)} &= \frac{1}{A_{3,3}}\left\{b_3 - A_{3,1}Y_1^{(i+1)} - A_{3,2}Y_2^{(i+1)} - A_{3,4}Y_4^{(i)}\right\} \\
Y_4^{(i+1)} &= \frac{1}{A_{4,4}}\left\{b_4 - A_{4,1}Y_1^{(i+1)} - A_{4,2}Y_2^{(i+1)} - A_{4,3}Y_3^{(i+1)}\right\}.
\end{aligned}$$

This is known as the Gauss-Seidel method. The form for arbitrary N is

$$Y_j^{(i+1)} = \frac{1}{A_{j,j}}\left\{b_j - \sum_{k=1}^{j-1} A_{j,k}Y_k^{(i+1)} - \sum_{k=j+1}^{N} A_{j,k}Y_k^{(i)}\right\}. \tag{8-67}$$

Since $Y_j^{(i+1)}$ are closer to the final values than $Y_j^{(i)}$, we expect better convergence properties than the Jacobi method as a result. Implicit in the equations is that $Y_j^{(i+1)}$ are calculated one after another in the order $j = 1, 2, 3, \cdots, N$.

A more practical way is to use one of the relaxation methods. This may be derived by adding $Y_j^{(i)}$ to the right side of Eq. (8-67) outside the braces and subtracting the same amount from inside. The results for our $N = 4$ example take on the form

$$\begin{aligned}
Y_1^{(i+1)} &= Y_1^{(i)} + \frac{1}{A_{1,1}}\left\{b_1 - A_{1,1}Y_1^{(i)} - A_{1,2}Y_2^{(i)} - A_{1,3}Y_3^{(i)} - A_{1,4}Y_4^{(i)}\right\} \\
Y_2^{(i+1)} &= Y_2^{(i)} + \frac{1}{A_{2,2}}\left\{b_2 - A_{2,1}Y_1^{(i+1)} - A_{2,2}Y_2^{(i)} - A_{2,3}Y_3^{(i)} - A_{2,4}Y_4^{(i)}\right\} \\
Y_3^{(i+1)} &= Y_3^{(i)} + \frac{1}{A_{3,3}}\left\{b_3 - A_{3,1}Y_1^{(i+1)} - A_{3,2}Y_2^{(i+1)} - A_{3,3}Y_3^{(i)} - A_{3,4}Y_4^{(i)}\right\} \\
Y_4^{(i+1)} &= Y_4^{(i)} + \frac{1}{A_{4,4}}\left\{b_4 - A_{4,1}Y_1^{(i+1)} - A_{4,2}Y_2^{(i+1)} - A_{4,3}Y_3^{(i+1)} - A_{4,4}Y_4^{(i)}\right\}.
\end{aligned}$$

For an arbitrary N, the corresponding equations may be written as

$$Y_j^{(i+1)} = Y_j^{(i)} + \frac{1}{A_{j,j}}\left\{b_j - \sum_{k=1}^{j-1} A_{j,k}Y_k^{(i+1)} - \sum_{k=j}^{N} A_{j,k}Y_k^{(i)}\right\}. \tag{8-68}$$

Box 8-7 Program DM_SORWS
Successive overrelaxation method for boundary value problems
Example of scattering from a Woods-Saxon potential

Subprograms used:
- MESH: Construct a mesh of constant step size.
- INI_ARY: Initialize the array η of Eq. (8-62) and construct a trial solution.
- RELAX: Solve the FDE for a Woods-Saxon potential using SOR.

Initialization:
Use MESH to construct a fixed mesh of points $x_0, x_1, x_2, \ldots, x_N$.

1. Input energy E and parameters V_0, r_g, and a for the potential of Eq. (8-56).
2. Use INI_ARY to:
   (a) Calculate the diagonal elements η_k of $\boldsymbol{A}$ in Eq. (8-62).
   (b) Construct a trial solution $\boldsymbol{Y}^{(0)}(x_i)$.
3. Input the relaxation parameter ω of Eq. (8-69).
4. Use RELAX to carry out the relaxation calculations:
   (a) Set up the tolerance for error and the maximum number of iterations.
   (b) Update $\boldsymbol{Y}$ according Eq. (8-69) and calculate the estimated error.
   (c) Check the error,
      (i) If larger than the tolerance, iterate the solution.
      (ii) Otherwise, return the solution.

The second of the two terms on the right side may be taken as the correction to $Y_j^{(i)}$ produced in the iteration. As a result, it may by possible to modify the rate of convergence by including a *relaxation* parameter ω as the weighting factor

$$Y_j^{(i+1)} = Y_j^{(i)} + \frac{\omega}{A_{j,j}}\Big\{b_j - \sum_{k=1}^{j-1} A_{j,k}Y_k^{(i+1)} - \sum_{k=j}^{N} A_{j,k}Y_k^{(i)}\Big\}. \tag{8-69}$$

For $\omega < 1$, it is known as the underrelaxation method. For $\omega = 1$, we return to the Gauss-Seidel method. The main interest here is in $\omega > 1$, the overrelaxation method. Since it often happens that successive corrections to Y_j are of the same sign, it is quite likely that overrelaxation, with $1 < \omega < 2$, provides a faster way toward convergence than the Gauss-Seidel method. For this reason, successive overrelaxation (SOR) is a method of choice for boundary value problems and the factor ω is sometimes called the *acceleration* parameter. The algorithm, as applied to the example of scattering from a Woods-Saxon potential of the previous section, is given in Box 8-7.

Convergence For a discussion on the convergence criteria, it is convenient to write all three iterative equations, Eqs. (8-66), (8-67), and (8-69), formally into the form

$$\boldsymbol{Y}^{(i+1)} = \boldsymbol{P}\boldsymbol{Y}^{(i)} + \boldsymbol{q} \tag{8-70}$$

where $\boldsymbol{P}$ is the iteration matrix for each one of the three cases. The matrix $\boldsymbol{q}$, obtained from $\boldsymbol{b}$, represents the part that does not change from one iteration to another. Let us define the difference between the exact solution $\boldsymbol{Y}$ and $\boldsymbol{Y}^{(i)}$ as

$$\boldsymbol{e}^{(i)} \equiv \boldsymbol{Y} - \boldsymbol{Y}^{(i)}.$$

In terms of $\boldsymbol{e}^{(i)}$, we may write Eq. (8-70) in the form

$$\boldsymbol{e}^{(i+1)} = \boldsymbol{P}\boldsymbol{e}^{(i)} \tag{8-71}$$

as $\boldsymbol{q}$ is unchanged from one iteration to another.

The relation given by Eq. (8-71) may be interpreted in the following way. Starting with an initial estimate of the solution $\boldsymbol{Y}^{(0)}$ and the associated difference $\boldsymbol{e}^{(0)}$, we can generate $\boldsymbol{Y}^{(1)}$ and, hence, $\boldsymbol{e}^{(1)}$ using Eq. (8-71). From $\boldsymbol{Y}^{(1)}$ and $\boldsymbol{e}^{(1)}$, we obtain $\boldsymbol{Y}^{(2)}$ and $\boldsymbol{e}^{(2)}$, and so on. When this is repeated k times, we obtain the result

$$\boldsymbol{e}^{(k)} = \boldsymbol{P}\boldsymbol{e}^{(k-1)} = \boldsymbol{P}^2\boldsymbol{e}^{(k-2)} = \cdots = \boldsymbol{P}^k\boldsymbol{e}^{(0)}.$$

The process is convergent if we can show that

$$\lim_{k\to\infty} \boldsymbol{e}^{(k)} = 0. \tag{8-72}$$

Since $\boldsymbol{Y}^{(0)}$, and hence $\boldsymbol{e}^{(0)}$, is arbitrary, the only way that this limiting relation can be satisfied is that

$$\lim_{k\to\infty} \boldsymbol{P}^k = 0.$$

This is, then, the convergence criterion we are seeking for the iterative methods to solve boundary value problems in this section.

It is easier to restate the same criterion in terms of the eigenvalues of the matrix $\boldsymbol{P}$. Let λ_j be an eigenvalue of $\boldsymbol{P}$ and $\boldsymbol{v}_j$, the corresponding eigenvector. That is,

$$\boldsymbol{P}\boldsymbol{v}_j = \lambda_j \boldsymbol{v}_j \tag{8-73}$$

for $j = 1, 2, \ldots, N$. Since $\boldsymbol{v}_0, \boldsymbol{v}_1, \boldsymbol{v}_2, \ldots, \boldsymbol{v}_N$ form a complete set of states in the $(N+1)$-dimensional space, we can express the difference between the exact solution and our initial estimate in terms of them,

$$\boldsymbol{e}^{(0)} = \sum_{\ell=1}^{N} c_\ell^{(0)} \boldsymbol{v}_\ell$$

where the coefficients $c_\ell^{(0)}$ are constants whose actual values are not of immediate concern to us. Using this expansion, together with the relation of Eq. (8-73), we have

$$\boldsymbol{e}^{(1)} = \boldsymbol{P}\boldsymbol{e}^{(0)} = \sum_{\ell=1}^{N} c_\ell^{(0)} \boldsymbol{P}\boldsymbol{v}_\ell = \sum_{\ell=1}^{N} c_\ell^{(0)} \lambda_\ell \boldsymbol{v}_\ell.$$

By repeating the step k times, we find that

$$\boldsymbol{e}^{(k)} = \sum_{\ell=0}^{N} c_\ell^{(0)} (\lambda_\ell)^k \boldsymbol{v}_\ell.$$

The only way for this equation to satisfy the requirement of Eq. (8-72) is

$$|\lambda_\ell| < 1$$

for all ℓ. An alternative statement to the same effect is to define a quantity for the matrix $\boldsymbol{P}$ called *spectral radius*, given by the maximum absolute value of the eigenvalues:

$$\rho(\boldsymbol{P}) = \max_{\ell=1,2,\ldots,N} |\lambda_\ell|. \tag{8-74}$$

The condition of convergence for our iterative methods may be stated now as the requirement for the spectral radius of the corresponding point iteration matrix to be less than unity.

Optimum relaxation parameter The condition of convergence, however, does not give us any indication of the rate for a solution to reach convergence. For SOR, it is known that the optimum value of the relaxation parameter is

$$\omega_x = \frac{2}{1+\sqrt{1-\rho^2(\boldsymbol{P})}} \tag{8-75}$$

where $\rho(\boldsymbol{P})$ is the spectral radius of the iteration matrix. A discussion on how to arrive at this result is given in Smith.[51]

For our prototype equation Eq. (8-57), the matrix $\boldsymbol{A}$ is tridiagonal as can be seen from Eq. (8-62). The iteration matrix $\boldsymbol{P}$ is also tridiagonal with the diagonal elements equal to zero. However, the matrix is not symmetric and has the form

$$\boldsymbol{P} = -\begin{pmatrix} 0 & 0 & \cdots & & & & & \\ \frac{1}{\eta_1} & 0 & \frac{1}{\eta_1} & 0 & \cdots & & & \\ 0 & \frac{1}{\eta_2} & 0 & \frac{1}{\eta_2} & \cdots & & & \\ \vdots & \vdots & \vdots & \ddots & \vdots & \vdots & \vdots & \vdots \\ & & & & \cdots & \frac{1}{\eta_{N-1}} & 0 & \frac{1}{\eta_{N-1}} \\ & & & & & \cdots & 0 & 0 \end{pmatrix}.$$

A slight modification of the bisection technique described in §3-6 and §5-5 may be used to find the eigenvalues of this matrix. If the step size is constant for the entire interval, the tridiagonal matrix may be mapped to the form

$$\boldsymbol{T} = \begin{pmatrix} a & b & 0 & \cdots & \cdots & 0 \\ c & a & b & 0 & \cdots & 0 \\ 0 & c & a & b & \cdots & 0 \\ \vdots & \ddots & \ddots & \ddots & \ddots & \vdots \\ 0 & \cdots & 0 & c & a & b \\ 0 & \cdots & \cdots & 0 & c & a \end{pmatrix}.$$

The eigenvalues of such a matrix are given by

$$\lambda_k = a + 2\sqrt{bc}\cos\frac{k\pi}{N+1} \tag{8-76}$$

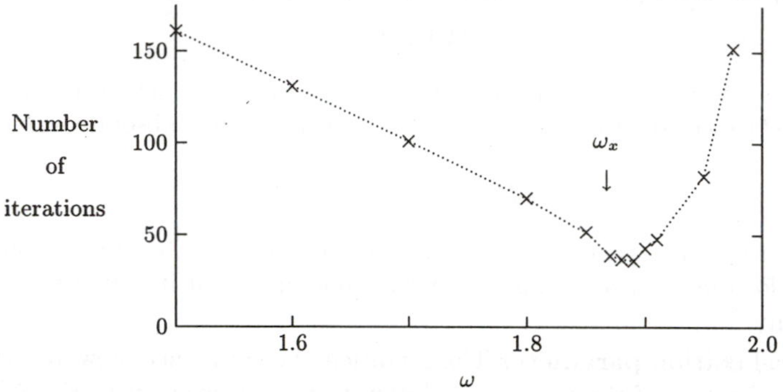

Figure 8-9: Number of iterations for different relaxation parameter ω to reach an accuracy of $\epsilon = 9 \times 10^{-5}$ in Eq. (8-37). An evenly spaced grid of 40 points in $[0, 3]$ is used for the Woods-Saxon well of Eq. (8-56). Dotted curve is drawn to guide the eye and the arrow indicates the value of $\omega_x = 1.87$ given by Eq. (8-75).

where $k = 1, 2, \ldots, N$ and N is the dimension of the matrix. For our matrix $\boldsymbol{P}$, we have $a = 0$ and $b = c \approx \frac{1}{2}$. If there are 40 equal-size steps in the interval, the largest eigenvalue is $\lambda_1 = \cos(\pi/41) = 0.997$. This gives a value of $\omega_x \approx 1.9$ for the optimum relaxation parameter. A more precise calculation using the exact values of b and c gives the value 1.87. The proof of Eq. (8-76) is also given in Smith.[51] A plot of the actual number of iterations in a calculation is given in Fig. 8-9.

In general, the matrix $\boldsymbol{P}$ remains sparse but may be more complicated in form than the tridiagonal one in our example here. Nevertheless, it is still quite possible to find the maximum absolute value of the eigenvalues relatively easily. Furthermore, since we are only using the spectral radius as a way to locate the optimum value of ω in the range $1 \leq \omega < 2$, we do not need highly accurate results.

In practice, the value of ω_x given above provides the fastest convergence only toward the end of a calculation using SOR. At the beginning of the iterative process, a smaller value of $\omega \approx 1$ is actually more efficient. A method, given in Press and coauthors,[44] of varying the value of ω for each iteration in the following manner,

$$\omega^{(i)} = \begin{cases} \dfrac{1}{1 - \frac{1}{2}\rho^2(\boldsymbol{P})} & \text{for } i = 1 \\[2ex] \dfrac{1}{1 - \frac{\omega^{(i-1)}}{4}\rho^2(\boldsymbol{P})} & \text{for } i > 1 \end{cases} \tag{8-77}$$

is also found to be very efficient in minimizing the number of iterations.

8-6 Boundary value problems in partial differential equations

The relaxation methods developed in the last section were aimed at ordinary differential equations. With only minor modifications, they may also be used to solve partial differential equations (PDEs) where we have more than one independent variable. As our prototype in this section, we shall use the Poisson equation

$$\nabla^2 V(\boldsymbol{r}) = -\rho(\boldsymbol{r}).$$

Such equations arise, for example, in electrostatics when we wish to find the potential $V(\boldsymbol{r})$ due to a charge distribution $\rho(\boldsymbol{r})$. For the convenience of discussion, we have absorbed all overall constants in the equation into the definition of the source term $\rho(\boldsymbol{r})$. In terms of the classifications of partial differential equations on page 378, it is an elliptic equation.

For simplicity we shall restrict ourselves to two spatial dimensions. In the Cartesian coordinate system, the PDE takes on the form

$$\frac{\partial^2 V}{dx^2} + \frac{\partial^2 V}{dy^2} = -\rho(x, y). \tag{8-78}$$

It describes the situation where the charge distribution is independent of the z-direction as, for example, in the case of an infinite straight line of charge that extends from $-\infty$ to $+\infty$ along the z-axis. To simplify the boundary conditions, we shall place the source in the middle of a hollow tube made of conducting material, also infinite in length along the z-direction. For convenience, we shall consider the cross-sectional area of the tube to be a square of length 2ℓ on each side. If we put the origin of our two-dimensional space at the center of the square, the boundary condition for this problem may be specified as

$$V(x = \pm\ell, y) = V(x, y = \pm\ell) = V_0$$

where V_0 is the potential of the conducting tube. If it is grounded, we have

$$V_0 = 0.$$

We shall adopt this value as the boundary condition in our example. It is a Dirichlet type of boundary condition, as the values of $V(x, y)$ are given on the boundary. Alternatively, as we have seen earlier in the discussions on ODEs, we can specify the boundary condition in terms of the derivatives at the boundary. Other forms are also possible, including linear combinations of $V(x, y)$ and its derivatives.

Using the same procedure as Eq. (8-59), we can replace both second-order derivatives in Eq. (8-78) with their equivalent finite differences. For this purpose, we must first adopt a two-dimensional mesh of points, such as the one shown schematically in Fig. 8-10. Our discussion is much simpler if we take an evenly spaced grid of size Δx

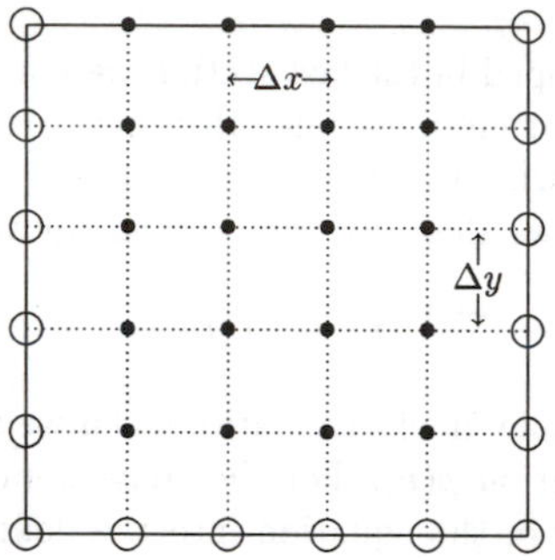

Figure 8-10: A two-dimensional mesh of points to transform a PDE of two independent variables into a set of FDEs. In general, $\Delta x \neq \Delta y$. Circles represent points where the values are given by boundary conditions and dots, points where the values must be obtained by solving the equations.

along the x-direction and Δy along the y-direction. The finite difference form of the Poisson equation of Eq. (8-78) is then

$$\frac{1}{\Delta x^2}\big\{V(x_{i+1},y_j) - 2V(x_i,y_j) + V(x_{i-1},y_j)\big\} + \frac{1}{\Delta y^2}\big\{V(x_i,y_{j+1}) - 2V(x_i,y_j) + V(x_i,y_{j-1})\big\} = -\rho(x_i,y_j). \tag{8-79}$$

If we follow the procedure of Eq. (8-60) instead, we can also obtain a set of FDEs with variable step sizes. The form of the equation in this case is more complicated and may not be easy to solved.

If Δy is taken to be equal to Δx, with $\Delta x = \Delta y \equiv h$, Eq. (8-79) simplifies to

$$V(x_{i+1},y_j) + V(x_{i-1},y_j) + V(x_i,y_{j+1}) + V(x_i,y_{j-1}) - 4V(x_i,y_j) = -h^2\rho(x_i,y_j). \tag{8-80}$$

This is the two-dimensional analog of Eq. (8-61). The difference between them may be seen by comparing the two diagrams in Fig. 8-11. In the case of a single independent variable, each of the equations represented by Eq. (8-61) involves the values of the dependent variable at three adjacent points on the mesh, $\phi(x_{k-1})$, $\phi(x_k)$, and $\phi(x_{k+1})$. We shall call such a group of points a computational "stencil," as shown schematically in Fig. 8-11(a). For the two-dimensional Poisson equation, each finite difference equation in the set represented by Eq. (8-80) involves five points, (i,j), $(i-1,j)$, $(i+1,j)$, $(i,j-1)$, and $(i,j+1)$, as shown in Fig. 8-11(b). This is a five-point stencil. We shall see later that different forms of PDE, as well as different methods to solve the differential equations, result in different stencils.

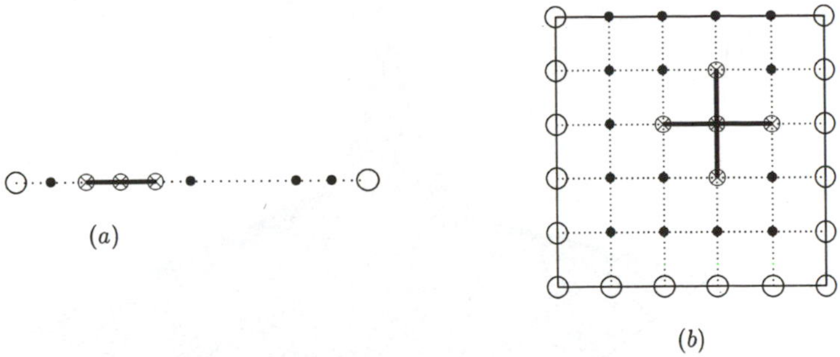

Figure 8-11: Computational stencils for boundary value problems. For the ODE of Eq. (8-61), each FDE involves three adjacent points, as shown in (*a*). For the PDE of Eq. (8-80), the relation is between the five points shown in (*b*). Boundary conditions are given at points shown as circles.

It is fairly straightforward to rewrite Eq. (8-80) as a recursive equation in the same way as we did in the previous section. The analog of Eq. (8-66) for the Jacobi method is

$$\begin{aligned} V^{(k+1)}(x_i, y_j) &= \frac{1}{4}\Big\{V^{(k)}(x_{i+1}, y_j) + V^{(k)}(x_{i-1}, y_j) + V^{(k)}(x_i, y_{j+1}) + V^{(k)}(x_i, y_{j-1}) \\ &\quad + h^2\rho(x_i, y_j)\Big\}. \end{aligned} \qquad (8\text{-}81)$$

The relation implies that the value of $V(x, y)$ at (x_i, y_j) in the next iteration is given by averaging over its four nearest neighbors in the present iteration, together with the value of $\rho(x, y)$ at (x_i, y_i).

To find the recursion equations for Gauss-Seidel and successive overrelaxation methods, we have a choice of several slightly different approaches, depending on the order in which we wish to "update" the elements in the calculations. The basic philosophy is to calculate $V^{(k+1)}(x, y)$ in iteration $k + 1$ using the updated values (those obtained in iteration $k + 1$), for two of the five elements in our five-point stencil and the other three from iteration k. This is identical in spirit to Eq. (8-69) where, for the three-point stencil of our second-order ODE, the value of one of the three elements involved is of the present iteration and the other two belong to the previous iteration. For our purpose here, we shall take the simplest approach by working on all the elements of one row at a time, from left to right in the order of increasing index i, before proceeding to the next row (increasing j). The resulting equation has the form

$$\begin{aligned} V^{(k+1)}(x_i, y_j) &= V^{(k)}(x_i, y_j) - \frac{\omega}{4}\Big\{h^2\rho(x_i, y_j) + V^{(k+1)}(x_{i-1}, y_j) + V^{(k+1)}(x_i, y_{j-1}) \\ &\quad -4V^{(k)}(x_i, y_j) + V^{(k)}(x_{i+1}, y_j) + V^{(k)}(x_i, y_{j+1})\Big\}. \end{aligned} \qquad (8\text{-}82)$$

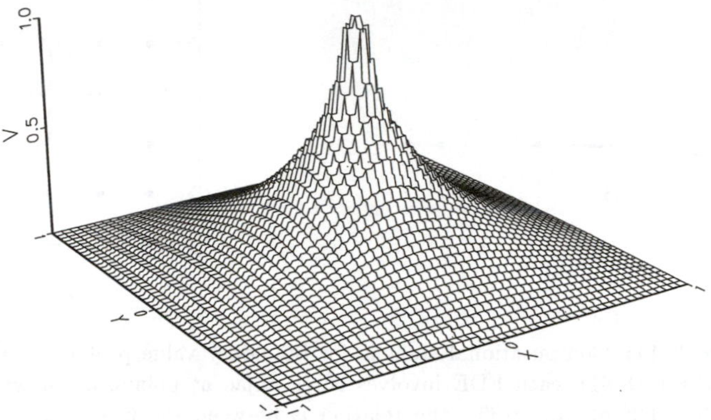

Figure 8-12: Potential distribution $V(x, y)$ in two dimensions for a point charge in the center of a square, subject to the boundary condition $V = 0$ at the edges. The PDE of Eq. (8-78) is solved using SOR with 31 mesh points on each side. With $\omega = 1.816$, an average difference of 10^{-5} between successive iterations is reached in 52 iterations and, with a variable ω given by Eq. (8-77), 46 iterations.

We recover the Gauss-Seidel method from this equation by putting the relaxation parameter ω to unity.

An alternate procedure may be found by noting that the "corrections" to $V(x_i, y_j)$ for even i or j come, respectively, from the neighboring points with odd i or j, and vice versa. This gives us the possibility of separating each iteration into two parts, one for the odd values of the indices and the other for the even values. There are certain advantages in taking such an approach but we shall not be going into this subject here.

The only thing remaining before embarking on the actual numerical calculations is the trial solution used as the starting point for the iterative calculations. The Poisson equation example here is sufficiently stable that we do not need to be overly concerned with this question. In fact, even the convenient choice of

$$V^{(0)}(x_i, y_j) = 0$$

for all i and j, can be adequate. The resulting potential surface for a point charge of magnitude q located at $x = y = 0$,

$$\rho(x, y) = \begin{cases} q & \text{for } x = y = 0 \\ 0 & \text{otherwise} \end{cases}$$

is shown in Fig. 8-12.

Box 8-8 Program DM_SORPS
Solution of partial differential equation using SOR
Example of a two-dimensional Poisson equation

Subprograms used:
INI_POS: Initializations for solving the Poisson equation using SOR.
SOR_ITR: SOR iterations for two-dimensional PDE.
Initialization:
(a) Choose a mesh of 31 points on each side and an error tolerance of $\epsilon = 10^{-5}$.
(b) Use INI_POS to:
(i) Set the initial trial solution as zero everywhere.
(ii) Define the charge density to be zero everywhere except the middle.
1. Input the relaxation parameter ω.
2. Use SOR_ITR to carry out the following calculations:
(a) Use Eq. (8-82) to find the difference of $V(x_i, y_j)$ between two iterations.
(b) Improve the solution by including the difference.
(c) Store the largest absolute value of the difference as the estimated error.
(d) If the error is larger than ϵ, iterate the solution again.
3. Return $V(x, y)$ as the solution.

For SOR, we also need to determine the optimum value of the relaxation parameter ω so as to "accelerate" the convergence as much as possible. The equivalent matrix to $\boldsymbol{A}$ of Eq. (8-62) has, in general, five nonvanishing elements in each row. It is still a sparse matrix, but its actual form depends on the order we wish to carry out the iterative calculation. It is possible to use a matrix diagonalization to find out the spectral radius of the point Jacobi iteration matrix and thence the value of ω using Eq. (8-75). Alternatively, we can adopt a variable value for ω that changes from one iteration to the next as in Eq. (8-77). For this purpose, we need the spectral radius of the point Jacobi matrix for the problem. In the case of the Poisson equation on a rectangular mesh of sides with lengths $(ph \times qh)$, the spectral radius for a set of Dirichlet boundary conditions is shown in Smith[51] to be

$$\rho(\boldsymbol{P}) = \frac{1}{2}\Big(\cos\frac{\pi}{p} + \cos\frac{\pi}{q}\Big). \tag{8-83}$$

For other cases, it may be possible to approximate the problem to a known one and use the spectral radius of its Jacobi matrix in place of the real one. Such topics belong to more advanced treatments on PDEs and we shall not be concerned with them here. The steps used in carrying out the calculations for our example are summarized in Box 8-8.

8-7 Parabolic partial differential equations

A good example of parabolic partial differential equations is the diffusion equation

$$\frac{\partial \phi}{\partial t} = \nabla \cdot D \nabla \phi(\boldsymbol{r}, t). \tag{8-84}$$

It arises in transport phenomena, for instance, in describing the concentration $\phi(\boldsymbol{r}, t)$ of gas molecules as a function of location and time. From the continuity equation,

$$\nabla \cdot \boldsymbol{j}(\boldsymbol{r}, t) + \frac{\partial \phi}{\partial t} = 0 \tag{8-85}$$

we find that the changes are related to the divergence of the current density $\boldsymbol{j}(\boldsymbol{r}, t)$. On the other hand, the current density itself is proportional to the gradient of the concentration:

$$\boldsymbol{j}(\boldsymbol{r}, t) = -D \nabla \phi(\boldsymbol{r}, t). \tag{8-86}$$

We recover Eq. (8-84) on substituting Eq. (8-86) into (8-85). In general, the diffusion coefficient D can also depend on coordinate $\boldsymbol{r}$ and time t. However, for a uniform medium, it may be taken as a constant.

In addition to equations related to the diffusion equation, such as heat conduction and fluid flow through porous medium, the time-dependent Schrödinger equation

$$i\hbar \frac{\partial \Psi}{\partial t} = -\frac{\hbar^2}{2\mu} \nabla^2 \Psi(\boldsymbol{r}, t) + V(\boldsymbol{r}) \Psi(\boldsymbol{r}, t) \tag{8-87}$$

is also an example of parabolic differential equations. From a physical point of view, parabolic (as well as hyperbolic) equations describe the time development of a system under a given set of boundary conditions. Since it involves both initial values and boundary conditions, it is an initial value boundary problem.

For a discussion of the numerical methods to solve such equations, it is more instructive to use Eq. (8-84) and simplify it to an equation involving only one of the three spatial coordinates. Furthermore, if D is a constant, we have a parabolic equation in one of its most elementary forms:

$$\frac{\partial \phi(x, t)}{\partial t} = D \frac{\partial^2 \phi(x, t)}{\partial x^2}. \tag{8-88}$$

To put the differential equation into a finite difference equation, we need to set up a fixed two-dimensional mesh of points, $j = 0, 1, 2, \ldots, N_x$ for x, and $k = 0, 1, 2, \ldots, N_t$ for t.

Although it makes very little difference to a computer program that the two coordinates in our mesh are referring to two different physical quantities, space and time, the problems we face are quite different for the two. In the present case, this comes mainly from the fact that the derivatives with respect to the two independent variables are of different order. In addition, the time dependence in the problem

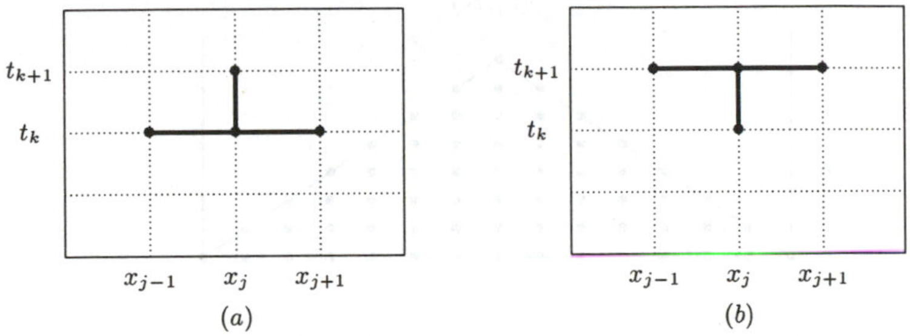

Figure 8-13: (*a*) Implicit and (*b*) explicit methods to find the solution of a PDE that is first order in t and second order in x.

requires an input set of initial conditions and the spatial dependence requires a set of boundary conditions. For example, the initial conditions may be given in terms of the values of $\phi(t_0, x_j)$ at x_0, $x_1, \ldots$, x_{N_x} for $t = t_0$. For the x-dependence, the two boundary conditions may be given in terms of the values of $\phi(t, x)$ at $x = x_0$ and x_{N_x}, one pair for each of the discrete set of values t_0, $t_1, \ldots$, t_{N_t}.

Implicit and explicit methods To construct the finite difference equations, we can make use of central difference for the second-order derivative

$$\left.\frac{\partial^2\phi}{\partial x^2}\right|_{x=x_j} \longrightarrow \frac{1}{(\Delta x)^2}\Big\{\phi(x_{j+1}, t) - 2\phi(x_j, t) + \phi(x_{j-1}, t)\Big\}. \qquad (8\text{-}89)$$

Similarly, we can use forward difference to change the time derivative to

$$\left.\frac{\partial\phi}{\partial t}\right|_{t=t_k} \longrightarrow \frac{1}{\Delta t}\Big\{\phi(x, t_{k+1}) - \phi(x, t_k)\Big\}. \qquad (8\text{-}90)$$

To put the two relations together to form the equivalent FDE for Eq. (8-88), we need to specify t in Eq. (8-89) and x in (8-90). Since we have used a central difference centered around x_j, we have good reasons to evaluate all the quantities on the right side of Eq. (8-90) at $x = x_j$.

On the other hand, there are two different ways we can associate the value of t in Eq. (8-89) with those on the mesh. The most obvious one is to evaluate all the quantities at $t = t_k$. The advantage of this scheme is that the FDE for Eq. (8-88) takes on the form

$$\frac{\phi(x_j, t_{k+1}) - \phi(x_j, t_k)}{\Delta t} = D\frac{\phi(x_{j+1}, t_k) - 2\phi(x_j, t_k) + \phi(x_{j-1}, t_k)}{(\Delta x)^2}. \qquad (8\text{-}91)$$

As a result, it is possible to calculate $\phi(x_j, t_{k+1})$ directly from the three input quantities $\phi(x_{j+1}, t_k)$, $\phi(x_j, t_k)$, and $\phi(x_{j-1}, t_k)$, as shown schematically in Fig. 8-13(*a*).

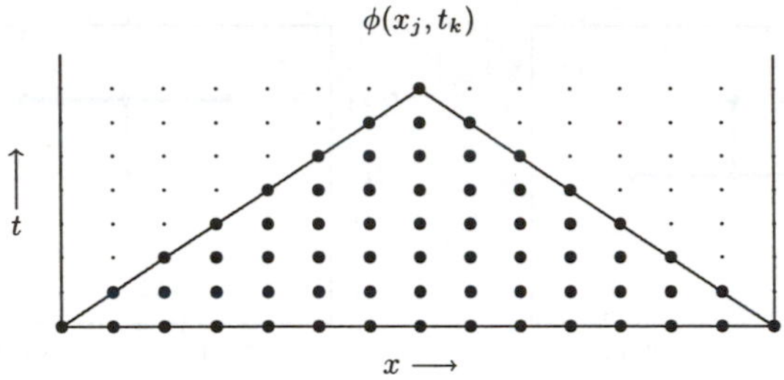

Figure 8-14: Schematic illustration of the limitations of explicit methods. Only points in the triangular domain of dependence can influence the value at (x_j, t_k).

This is an explicit method, as all the input quantities belong to an earlier time step and are available when they are needed. By starting from a set of initial conditions at $t = t_0$, all $\phi(x_j, t_k)$ for $k = 1$ may be calculated. Once this is done, we can proceed to $k = 2$, and so on. The disadvantage of this scheme is that the stability of the FDE is restricted to

$$\frac{2D\Delta t}{(\Delta x)^2} \leq 1.$$

The derivation of this condition is given in standard references on numerical solutions of partial differential equations. Here, we shall only try to provide a feeling how such a restriction arises.

It is quite easy to see the reason why one cannot take big steps in t in an explicit method. For the example above, the numerical solution at a point (x_j, t_k) depends only on those for the three points x_{j-1}, x_j, and x_{j+1} at the previous time step $t = t_{k-1}$. The values at these three points are, in turn, obtained from those at points, x_{j-2}, x_{j-1}, x_j, x_{j+1}, and x_{j+2}, at time $t = t_{k-2}$. When we continue to trace back in time in this way, we find that all the earlier points on which the value at (x_j, t_k) depends fall within the pyramid shown in Fig. 8-14. If the step size in time is too large compared with that in the spatial coordinate, the pyramid is narrow at the base and points at the end of the time interval depend only on a subset of the points at $t = t_0$. From a practical point of view, it is difficult to think of many situations in which the solution at any time is only a function of a part of the initial conditions. It is far more likely that every point in a proper solution at the end of the time interval involves all the points at $t = t_0$. As a result, numerical solutions obtained with large step sizes may not correspond to the true solution.

An alternative to the method described above is to evaluate all the quantities on the right side of Eq. (8-89) at $t = t_{k+1}$. The resulting FDE for the differential equation of Eq. (8-88) takes on the form

$$\frac{\phi(x_j,t_{k+1})-\phi(x_j,t_k)}{\Delta t}=D\frac{\phi(x_{j+1},t_{k+1})-2\phi(x_j,t_{k+1})+\phi(x_{j-1},t_{k+1})}{(\Delta x)^2}. \tag{8-92}$$

The advantage of this approach is that the FDE is unconditionally stable. Again, we shall not give the proof here. The method is, however, an *implicit* one. As shown schematically in Fig. 8-13(b), the stencil for this method involves four quantities, $\phi(x_j,t_k)$, $\phi(x_{j+1},t_{k+1})$, $\phi(x_j,t_{k+1})$, and $\phi(x_{j-1},t_{k+1})$. We shall see later that, instead of simply propagating the solution forward in time, the finite difference equation must be solved as a set of linear algebraic equations, the same as we did earlier in Eq. (8-62).

Crank-Nicholson method It is also possible to use a linear combination of explicit and implicit methods by taking a weighted average of the right sides of Eqs. (8-91) and (8-92). The result is known as the *Crank-Nicholson method.* For example, by taking equal weights, we obtain

$$\begin{aligned}\frac{\phi(x_j,t_{k+1})-\phi(x_j,t_k)}{\Delta t} &= \frac{D}{2}\Big\{\frac{\phi(x_{j+1},t_k)-2\phi(x_j,t_k)+\phi(x_{j-1},t_k)}{(\Delta x)^2}\\ &\qquad+\frac{\phi(x_{j+1},t_{k+1})-2\phi(x_j,t_{k+1})+\phi(x_{j-1},t_{k+1})}{(\Delta x)^2}\Big\}.\end{aligned} \tag{8-93}$$

This is the method we shall adopt for numerical calculations in all the examples in the rest of this section.

A good reason to use the Crank-Nicholson method for parabolic equations may be seen by looking at the time-dependent Schrödinger equation of Eq. (8-87) as an example. It is convenient to transform first the differential equation into a dimensionless form first. That is, we shall express energy, length, and time in our equation, respectively, in units of MeV (million electron volts), $\hbar/\{2\mu\times\text{MeV}\}^{1/2}$, and $\hbar$/MeV. In such a system, one unit of time is approximately 6.6×10^{-22} s, and one unit of energy is 1 MeV (1.6×10^{-13} J). The value of one unit of length depends on the mass μ of the particle involved. For a nucleon ($\mu c^2\approx10^3$ MeV), it is 4.5×10^{-15} m, and for an electron ($\mu c^2\approx0.5$ MeV), it is 2×10^{-13} m.

The resulting Schrödinger equation remains the same if we put $\hbar=1$ and $2\mu=1$ in Eq. (8-87), a practice used often in the literature to simplify the form of such equations. The basic points we wish to illustrate here may be carried out by restricting ourselves to one spatial dimension. In this limit, our partial differential equation takes on the form

$$i\frac{\partial\Psi}{\partial t}=\hat{H}\Psi(x,t) \tag{8-94}$$

where the Hamiltonian operator is given by

$$\hat{H}=-\frac{\partial^2}{\partial x^2}+V(x).$$

As shown earlier in Eq. (7-68), a formal solution to Eq. (8-94) may be written as

$$\Psi(x,t) = e^{-i(t-t_0)\hat{H}}\Psi(x,t_0). \tag{8-95}$$

In other words, we can regard $\exp\{-i(t-t_0)\hat{H}\}$ as the operator that "propagates" the wave function $\Psi(x,t_0)$ at time t_0 to $\Psi(x,t)$ at time t.

The meaning of having an operator $\hat{H}$ in the argument of an exponential function may be understood in terms of an infinite series expansion

$$\begin{aligned} e^{-i(t-t_0)\hat{H}} &= 1 + \{-i(t-t_0)\hat{H}\} + \frac{1}{2!}\{-i(t-t_0)\hat{H}\}^2 + \frac{1}{3!}\{-i(t-t_0)\hat{H}\}^3 + \cdots \\ &= \sum_{\ell=0}^{\infty}\frac{1}{\ell!}\{-i(t-t_0)\hat{H}\}^{\ell}. \end{aligned} \tag{8-96}$$

If we divide the time interval $[t_0, t]$ into N_t equal parts, it is easy to verify that the formal solution of Eq. (8-95), with the exponential function expressed in the series form, is identical to our FDE in the limit $N_t \to \infty$.

For an increment of Δt in time from t to $t+\Delta t$, Eq. (8-95) takes on the form

$$\Psi(x, t+\Delta t) = e^{-i(\Delta t)\hat{H}}\Psi(x,t). \tag{8-97}$$

On expressing all the quantities in terms of their values on the mesh points, we have the result

$$\begin{aligned} \Psi(x_j, t_{k+1}) &= e^{-i(\Delta t)\hat{H}}\Psi(x_j, t_k) \\ &= \Big[1 + \{-i(\Delta t)\hat{H}\} + \frac{1}{2!}\{-i(\Delta t)\hat{H}\}^2 + \cdots\Big]\Psi(x_j, t_k) \end{aligned} \tag{8-98}$$

for $j = 0, 1, 2, \ldots, N_x$. In terms of this expansion, the explicit method corresponds to truncating the series after the linear term in $\hat{H}$:

$$\begin{aligned} &\Psi(x_j, t_{k+1}) \approx [1 + \{-i(\Delta t)\hat{H}\}]\Psi(x_j, t_k) \\ &\quad = \Psi(x_j, t_k) - i\Delta t\Big\{\frac{\Psi(x_{j+1}, t_k) - 2\Psi(x_j, t_k) + \Psi(x_{j-1}, t_k)}{(\Delta x)^2} - V(x_j)\Psi(x_j, t_k)\Big\}. \end{aligned}$$

The implicit method corresponds to the same approximation except that, instead of Eq. (8-97), we start with the equation

$$\Psi(x_j, t_k) = e^{+i(\Delta t)\hat{H}}\Psi(x_j, t_{k+1})$$

that propagates $\Psi(x,t)$ backwards in time by Δt from $t = t_{k+1}$ to t_k.

Since $\hat{H}$ is Hermitian, the operators $\exp\{\pm i(\Delta t)\hat{H}\}$ are unitary. However, any approximations of the operator by truncating the series are not unitary in general. This causes several difficulties, the most serious being that, if the wave function $\Psi(x,t)$ is normalized at a given t, the normalization may not be preserved when it is propagated forward in time. This is true with both explicit and implicit methods. Unitarity must therefore be imposed as an additional condition, for example, by normalizing the wave function after each time step. In terms of the diffusion problem, the same difficulties appear in the form that the flux is not conserved.

We shall see that the Crank-Nicholson approach is an approximation that preserves unitarity. For this purpose, we shall replace the exponential function by the Cayley form, as suggested in Goldberg, Schey, and Schwartz:[24]

$$\begin{aligned} e^{-i(\Delta t)\hat{H}} &\approx \frac{1-\frac{1}{2}i(\Delta t)\hat{H}}{1+\frac{1}{2}i(\Delta t)\hat{H}} \\ &= \left\{1-\frac{1}{2}i(\Delta t)\hat{H}\right\}\left\{1-\frac{1}{2}i(\Delta t)\hat{H}+[\frac{1}{2}i(\Delta t)\hat{H}]^2+\cdots\right\}. \end{aligned} \tag{8-99}$$

Comparing this result with the series expansion of the exponential function given by Eq. (8-96), we see that the Cayley form is identical to the exact result up to the second power in the argument.

Since our finite difference equation corresponds to retaining only terms up to the first power, there is no additional loss in the accuracy by starting with the Cayley form instead of the series expansion. In place of Eq. (8-97), the propagation equation becomes

$$\Psi(x_j, t_{k+1}) = \frac{1-\frac{1}{2}i(\Delta t)\hat{H}}{1+\frac{1}{2}i(\Delta t)\hat{H}}\,\Psi(x_j, t_k).$$

When both sides of this relation are multiplied by the denominator on the right, we obtain the result

$$\left\{1+\frac{1}{2}i(\Delta t)\hat{H}\right\}\Psi(x_j, t_{k+1}) = \left\{1-\frac{1}{2}i(\Delta t)\hat{H}\right\}\Psi(x_j, t_k). \tag{8-100}$$

On rearranging the terms, we obtain a form that is exactly the same as given by the Crank-Nicholson method. Since the Cayley form of the operator is unitary, we have no difficulty in preserving the normalization of the wave function in this case. For this reason, it is used widely for problems of this type.

Solving the FDE by Gaussian elimination Let us address the problem of finding the roots of the FDE for both implicit and Crank-Nicholson methods at the same time. Returning to the diffusion equation in one spatial dimension, the FDE of Eq. (8-92) for the implicit method may be written in the form

$$\phi(x_{j-1}, t_{k+1}) - (2+\eta)\phi(x_j, t_{k+1}) + \phi(x_{j+1}, t_{k+1}) = -\eta\phi(x_j, t_k) \tag{8-101}$$

where

$$\eta = \frac{(\Delta x)^2}{D\Delta t}.$$

For the Crank-Nicholson method of Eq. (8-93), we have

$$\begin{aligned} &\phi(x_{j-1}, t_{k+1}) - (2+\eta')\phi(x_j, t_{k+1}) + \phi(x_{j+1}, t_{k+1}) \\ &\qquad = -\phi(x_{j-1}, t_k) + (2-\eta')\phi(x_j, t_k) - \phi(x_{j+1}, t_k) \end{aligned} \tag{8-102}$$

with

$$\eta' = \frac{2(\Delta x)^2}{D\Delta t}.$$

The only differences between the two forms are on the right side. Since all the quantities pertaining to $t = t_k$ are known by the time we come to solve the FDE at $t = t_{k+1}$, the differences are not significant for the discussion below.

Because the boundary conditions are given at $x = x_0$ and $x = x_{N_x}$, the equations for $j = 1$ and $j = (N_x - 1)$ are slightly different in form from those away from the boundaries. Consider first Eq. (8-101). Since $\phi(x, t)$ is defined only in the interval $[x_0, x_{N_x}]$, the equation has no meaning for $j < 1$. For $j = 1$, we have

$$\phi(x_0, t_{k+1}) - (2 + \eta)\phi(x_1, t_{k+1}) + \phi(x_2, t_{k+1}) = -\eta\phi(x_1, t_k).$$

However, $\phi(x_0, t_{k+1})$ is given by the boundary condition. By putting all the terms involving unknown quantities on the left side, we have

$$-(2 + \eta)\phi(x_1, t_{k+1}) + \phi(x_2, t_{k+1}) = -\eta\phi(x_1, t_k) - \phi(x_0, t_{k+1}).$$

Similarly, for $j = (N_x - 1)$, we have

$$\phi(x_{N_x-2}, t_{k+1}) - (2 + \eta)\phi(x_{N_x-1}, t_{k+1}) = -\eta\phi(x_{N_x-1}, t_k) - \phi(x_{N_x}, t_{k+1}). \qquad \text{(8-103)}$$

The same differences in form exist also for the Crank-Nicholson method. Again, only the right sides are different from what we have above for the implicit method and we shall not give them here. Note that, since the values of $\phi(x, t_{k+1})$ at $x = x_0$ and $x = x_{N_x}$ are given by the boundary conditions, there are a total of $(N_x - 1)$ unknown quantities $\phi(x_1, t_{k+1})$, $\phi(x_2, t_{k+1}), \ldots,$ $\phi(x_{N_x-1}, t_{k+1})$ for each t. With an equal number of equations given by either Eq. (8-101) or (8-102), there is no basic problem in obtaining the solution.

Both Eqs. (8-101) and (8-102) may be represented by the following:

$$\begin{array}{llllll} & b_1 y_1 & + \; c_1 y_2 & = d_1 & & \\ a_j y_{j-1} & + \; b_j y_j & + \; c_j y_{j+1} & = d_j & \text{for} \quad j = 2, 3, \ldots, (N-1) & \text{(8-104)} \\ a_N y_{N-1} & + \; b_N y_N & & = d_N & & \end{array}$$

where $N \equiv (N_x - 1)$. The values of coefficients a_j, b_j, c_j, and d_j are slightly different depending on whether we are using the implicit or the Crank-Nicholson method. In each case, they can be obtained by comparing with Eq. (8-101) or (8-102), as the case may be.

In matrix notation, these equations may be written as

$$\boldsymbol{A}\boldsymbol{y} = \boldsymbol{d}.$$

Box 8-9 Program DM_DFFUS
Crank-Nicholson solution of parabolic PDE Diffusion equation

Subprograms used:
- INIT: Initialize the function and other arrays.
- CALC_D: Calculate the matrix elements of $\boldsymbol{d}$ from $\boldsymbol{y}$ and η.
- PROP: Propagate the parabolic FDE by one time step (cf. Box 3-9).
- OUTPUT: Output the solution for a given time i_t.

Initialization:
- (a) Set the ranges $x = [0, x_d]$ and $t = [0, t_d]$, and the number of steps in each.
- (b) Construct a two-dimensional mesh with step sizes Δx and Δt.

1. Input the diffusion coefficient D.
2. Call INIT to:
   - (a) Zero the array $\{y_i\}$ as the initial condition.
   - (b) Define the arrays $\{a_i\}$, $\{b_i\}$, and $\{c_i\}$ of Eq. (8-104) using (8-103).
   - (c) Specify the boundary conditions.
3. Propagate the solution forward by one time step:
   - (a) Use CALC_D to calculate the values of d_j of Eq. (8-104) from $\boldsymbol{y}(x,t)$.
   - (b) Use PROP to:
     - (i) Forward eliminate the subdiagonal elements according to Eq. (A-7).
     - (ii) Find y_j at $(t + \Delta t)$ by back substitution using Eq. (A-8).
4. Increase time from t to $(t + \Delta t)$ and output the solution $\boldsymbol{y}(t)$ at regular intervals.

The square matrix $\boldsymbol{A}$, with dimension $(N \times N)$, is tridiagonal for our example. That is,

$$\boldsymbol{A} = \begin{pmatrix} b_1 & c_1 & 0 & \cdots & \cdots & 0 \\ a_2 & b_2 & c_2 & 0 & \cdots & 0 \\ 0 & a_3 & b_3 & c_3 & \cdots & 0 \\ \vdots & \vdots & \vdots & \ddots & \vdots & \vdots \\ 0 & \cdots & 0 & a_{N-1} & b_{N-1} & c_{N-1} \\ 0 & \cdots & 0 & 0 & a_N & b_N \end{pmatrix}. \tag{8-105}$$

In general, the matrix may be more complicated but remains sparse. The form of Eq. (8-104) is identical to that of Eq. (3-93) and can be solved using the Gaussian elimination method described in §A-3. Technically, we can improve the numerical accuracy by including pivoting, as we did in §5-1. This is left as an exercise.

As an illustration, we shall solve Eq. (8-88) for the temperature distribution along a one-dimensional conducting rod. The algorithm used is summarized in Box 8-9. As boundary conditions, we shall maintain one end of the rod at temperature 100°C and the other end at 0°C. For the initial condition, we can put the entire rod at temperature 0°C. The solution gives us the temperature at any point along the rod as a function of time. It may be obtained using the Crank-Nicholson method and the results are shown in Fig. 8-15 for the time $t = 1$, 10, 50, and infinity. Both the spatial coordinate and time are measured in arbitrary units, and, in these units, the

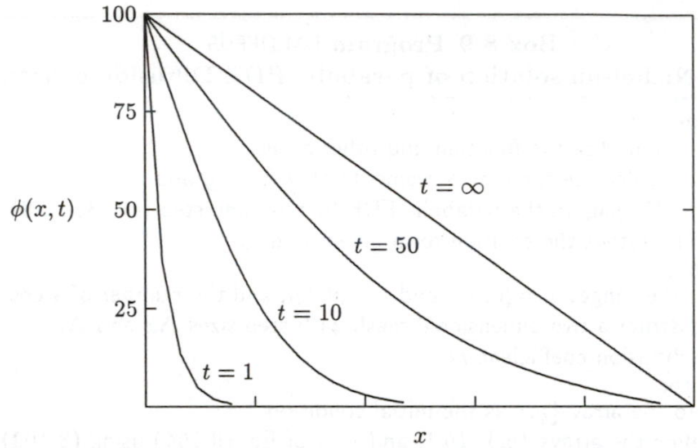

Figure 8-15: Temperature distribution along a one-dimensional conducting rod as a function of the distance from one end at different time t. The whole rod starts at 0°C at $t = 0$. The diffusion coefficient is $D = 0.2$ and one end of the rod is kept at 100°C while the other end at 0°C.

diffusion coefficient is taken to be $D = 0.2$.

Scattering of a wave packet As a second example, we shall apply the method developed to solve the scattering problem from a square-well potential. Instead of calculating the phase shift, as we did earlier in §8-4, we are now interested in the time development of the wave packet as it moves toward the barrier from some large distance away. When the packet gets close to the potential, it is influenced by the interaction and the wave function is changed as a result. In one spatial dimension, this is the problem discussed in Goldberg, Schey and Schwartz,[24] and the results appear in several textbooks on quantum mechanics as an illustration for the behavior of a wave packet in scattering.

The wave function of a particle traveling along the x-direction is governed by the Schrödinger equation Eq. (8-94). For the scattering problem, we shall assume that the particle is initially at $x = -x_c$, far away from where the potential $V(x)$ is nonzero. The square-well potential is centered at $x = 0$ and has the form

$$V(x) = \begin{cases} V_0 & \text{for } |x| \leq \frac{1}{2}r_g \\ 0 & \text{otherwise.} \end{cases}$$

For $|x_c| \gg r_g$, the particle is a free one. At this stage, the wave function, as well as the velocity of the particle, is completely given by the initial conditions we impose on the problem. The boundary conditions come from the requirement of quantum mechanics that the wave function must vanish at $x = \pm\infty$.

There are several possibilities in selecting the initial form of the wave function. The recommendation of Goldberg, Schey, and Schwartz is to use a Gaussian wave

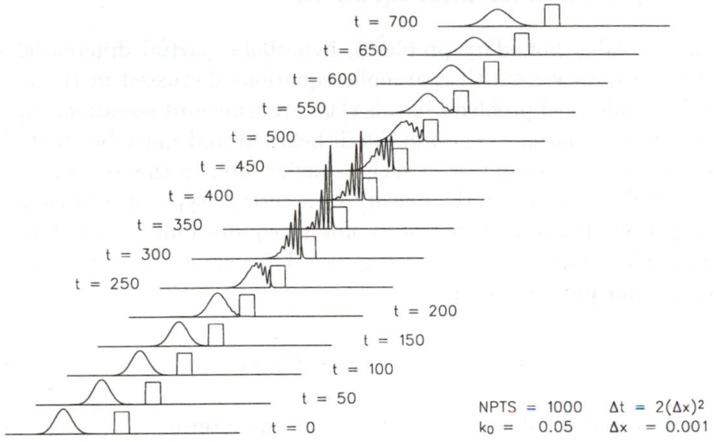

Figure 8-16: Scattering of a quantum mechanical particle from a one-dimensional repulsive square-well barrier centered at $x = L/2$. The interval $x = [0, L]$ is divided into 1,000 equal parts and the time step is taken to be $\Delta t = 2(\Delta x)^2$. The particle starts as a Gaussian wave packet centered at $x_c = L/4$ with $k_0 = 50\pi$ and is reflected back because of the high barrier ($V_0 = 2k_0^2$ and $r_g = 0.064L$).

packet:

$$\Psi(x, t = 0) = e^{ik_0 x - (x - x_c)^2/2\sigma_0^2} \tag{8-106}$$

where x_c is the location of the center and σ_0 is the width at $t = 0$. The wave number k_0 is related to the kinetic energy of the wave packet and, hence, to the velocity it travels along the x-direction. The physics of the problem is completely determined if we are given the values of V_0 and r_g for the potential well and k_0, x_c, and σ_0 for the initial wave packet.

To solve the resulting differential equation by Crank-Nicholson method, we start by constructing a fixed mesh of points to convert the partial differential equation into a set of finite difference equations. The method follows closely that used for the diffusion problem above and we shall not reproduce it again here. For a repulsive well ($V_0 > 0$), the wave functions obtained from the numerical solution at different times are shown in Fig. 8-16. To see the rapid fluctuations when the center of the packet is near the potential well, the step size in x must be taken to be fairly small. For this reason, the amount of computation is nontrivial but can still be carried out on a fast personal computer. Extension of the scattering problem to two spatial dimensions is given in Galbraith, Ching, and Abraham.[23]

8-8 Hyperbolic partial differential equations

As a class of initial value boundary problems, hyperbolic partial differential equations are different in nature from the parabolic equations discussed in the previous section. Examples of physical problems in this class are transport equations and wave equations. Hyperbolic equations can often be ill-behaved and must be treated with caution. We shall encounter examples of such behavior later in this section.

Let us use $\phi(\boldsymbol{r}, t)$ to represent the density of a certain type of particle at time t and spatial point $\boldsymbol{r}$. If the distribution is not in equilibrium, a net flow of the particles can take place between different regions. The changes in $\phi(\boldsymbol{r}, t)$ are given by the Boltzmann transport equation,

$$\frac{\partial \phi}{\partial t} = -\nabla \cdot \{\boldsymbol{v}\phi(\boldsymbol{r}, t)\} + f(\boldsymbol{r}, t) \tag{8-107}$$

where $\boldsymbol{v}$ is the velocity of the flow and $f(\boldsymbol{r}, t)$ is the source term.

Wave equations, on the other hand, describe phenomena such as the propagation of electromagnetic waves in space. In one spatial dimension, they take on the form

$$\frac{\partial^2 \phi}{\partial t^2} - \frac{1}{v^2}\frac{\partial^2 \phi}{\partial x^2} = 0. \tag{8-108}$$

Equations of this type occur also, for example, in describing the small amplitude vibrations of a stretched string, as we saw earlier in Eq. (8-2). Here, $\phi(x, t)$ is the displacement of the string at location x and time t, and v is the propagation velocity of the vibration along the string.

The method of characteristics One traditional way to solve hyperbolic equations is to use the method of characteristics. Let us briefly illustrate the ideas behind the method using as example the transport equation of Eq. (8-107). In the limit of one spatial dimension and constant velocity, the equation may be written as

$$a\frac{\partial \phi}{\partial t} + b\frac{\partial \phi}{\partial x} = c \tag{8-109}$$

where, for our transport equation example, a, b, and c are known quantities. By defining

$$p = \frac{\partial \phi}{\partial t} \qquad\qquad q = \frac{\partial \phi}{\partial x}$$

the equation takes on the form

$$ap + bq = c. \tag{8-110}$$

Along a curve C in the x-t plane, small changes in $\phi(x, t)$ may be expressed in terms of p and q:

$$d\phi = \frac{\partial \phi}{\partial t}dt + \frac{\partial \phi}{\partial x}dx = p\,dt + q\,dx \tag{8-111}$$

where dt/dx is the tangent to $\mathcal{C}$. We can eliminate p between Eqs. (8-110) and (8-111) and obtain

$$a\,d\phi - c\,dt = q(a\,dx - b\,dt).$$

If we choose the curve $\mathcal{C}$ such that it satisfies the condition

$$a\,dx - b\,dt = 0 \tag{8-112}$$

we have $\phi(x,t)$ as the solution of the differential equation

$$a\,d\phi - c\,dt = 0 \tag{8-113}$$

along $\mathcal{C}$. The curve is known as a characteristic curve. Physically, it may be interpreted as the trajectory followed by a particle in the convectional flow given by the transport equation.

We can regard Eqs. (8-112) and (8-113) as a pair of first-order ordinary differential equations,

$$\frac{dx}{dt} = \frac{b}{a} \qquad\qquad \frac{d\phi}{dt} = \frac{c}{a}$$

that replace the original first-order PDE of Eq. (8-109). The equation for the characteristic curve is given by the first of these two equations. Along such a curve, the solution may be propagated forward starting from initial conditions. This is straightforward if a, b, and c are known constants. In general, they can be functions of the independent variables as well as ϕ and its derivatives. In such cases, approximations are necessary and the propagation can only be carried out in small steps. The method is quite interesting, in particular, because of its close relation with the actual physical flow; however, we shall not make any further reference of the method here.

Finite difference method for transport equations It is also possible to propagate the solution of, for example, a transport equation forward in time using the method of finite difference. For the convenience of discussion, we shall divide the x-t plane of our one-spatial-dimension example into an evenly spaced, rectangular mesh of points with Δt as the step size in time and Δx as the step size along x. Since the derivative with respect to x is first order, we need one boundary condition for each t value on the mesh. Let us assume that this is given to us in terms of the values of $\phi(x,t)$ at $x = x_0$ and the domain of interest is in $x \geq x_0$. In this case, the most logical difference scheme is to take a forward one in t and a backward one in x. The resulting FDE for Eq. (8-109) may be written in the form

$$\frac{\phi(x_i,t_{k+1}) - \phi(x_i,t_k)}{\Delta t} + v\frac{\phi(x_i,t_k) - \phi(x_{i-1},t_k)}{\Delta x} = f(x_i,t_k)$$

where $v = b/a$ and $f = c/a$. The three-point stencil used in such an approach is shown in Fig. 8-17(a).

This is an explicit method, as the unknown quantities at $t = t_{k+1}$ are given directly in terms of those at $t = t_k$. We can see this more directly by rewriting Eq. (8-108) in the form

$$\phi(x_i,t_{k+1}) = (1-\eta)\phi(x_i,t_k) + \eta\phi(x_{i-1},t_k) + (\Delta t)f(x_i,t_k)$$

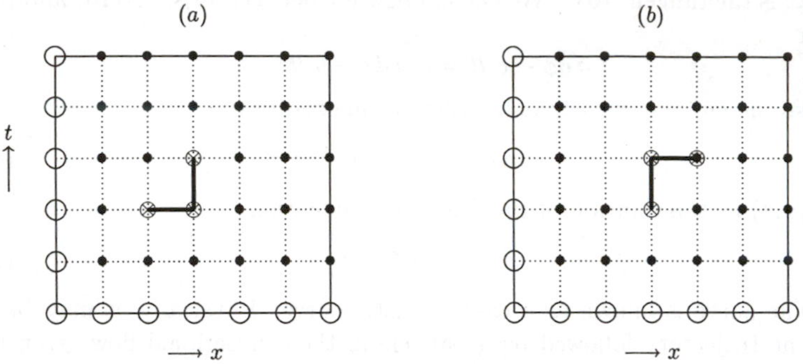

Figure 8-17: Stencils of (*a*) explicit and (*b*) implicit finite difference methods to solve transport equations of one spatial dimension. The initial and boundary conditions are assumed to be supplied at the points indicated by circles.

where

$$\eta = \frac{v(\Delta t)}{\Delta x}. \tag{8-114}$$

For the solution to be stable, it is necessary that the maximum value of η is less than or equal to unity. This is quite easy to understand if we examine the domain of dependence, in the same way we did in the previous section. Here, if $v\Delta t$ is larger than Δx, the solution at a point (x_i, t_k) is calculated from those of a small number of points at earlier times. On the other hand, physical information propagates at the velocity v. For $\eta > 1$, the numerical calculation is, therefore, moving forward at a rate faster than v. As a result, a smaller number of earlier points are used than required by the transport phenomenon itself. It is obvious that, in this case, the solution cannot be physically meaningful. This shows up in our calculation as an unstable solution. A mathematical derivation of this conclusion may be found in standard references on partial differential equations.

We can also design an implicit scheme to solve Eq. (8-109). If the initial and boundary conditions are supplied at mesh points indicated by circles in Fig. 8-17, it is possible to use a forward difference for both time and space to obtain the finite difference equation

$$\frac{\phi(x_i, t_{k+1}) - \phi(x_i, t_k)}{\Delta t} + v\frac{\phi(x_{i+1}, t_{k+1}) - \phi(x_i, t_{k+1})}{\Delta x} = f(x_i, t_k).$$

In terms of the η defined in Eq. (8-114), this may be put into the form

$$\eta\phi(x_{i+1}, t_{k+1}) + (1 - \eta)\phi(x_i, t_{k+1}) = \phi(x_i, t_k) + (\Delta t) f(x_i, t_k).$$

The stencil for this case is shown in Fig. 8-17(*b*). Since the calculation, in general, requires solving a set of simultaneous equations, it is not as simple to implement as the explicit methods. On the other hand, the method is stable.

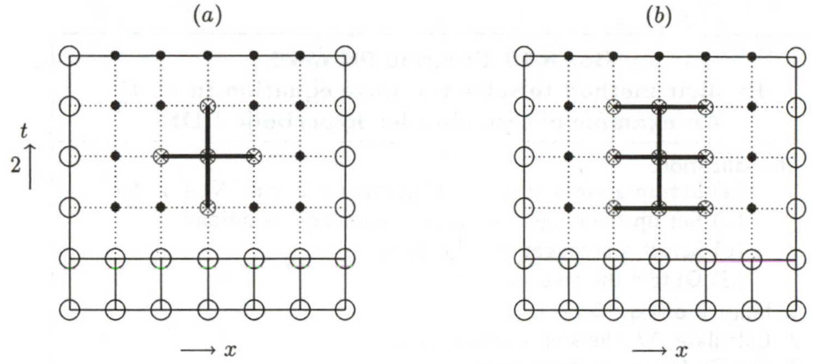

Figure 8-18: Stencils of (a) explicit and (b) implicit methods for wave equations. Circles indicate values supplied by initial and boundary conditions.

Wave equations For wave equations, both the time and spatial derivatives are second order. Each set of points along the x-axis requires two initial values. We shall assume that these are given to us in terms of the values of $\phi(x,t)$ at $t = t_0$ and $t = (t_0 + \Delta t)$, as indicated schematically in Fig. 8-18. For each set of points along the t-axis, two boundary values must be provided and we shall take that they are the values of $\phi(x,t)$ at $x = x_b$ and $x = x_d$, the two ends of the interval for x. Using central difference for both derivatives in Eq. (8-108), we obtain a FDE of the form

$$\frac{\phi(x_i,t_{k+1}) - 2\phi(x_i,t_k) + \phi(x_i,t_{k-1})}{(\Delta t)^2} - \frac{1}{v^2}\frac{\phi(x_{i+1},t_k) - 2\phi(x_i,t_k) + \phi(x_{i-1},t_k)}{(\Delta x)^2} = 0.$$

The five-point stencil is shown in Fig. 8-18(a). Since this is an explicit scheme, the finite difference equation can be put into a form such that we can solve directly for the only unknown quantity in it,

$$\phi(x_i,t_{k+1}) = -\phi(x_i,t_{k-1}) + \eta\phi(x_{i+1},t_k) + 2(1-\eta)\phi(x_i,t_k) + \eta\phi(x_{i-1},t_k). \tag{8-115}$$

Here $\eta = (\Delta t)^2/(v\Delta x)^2$. The condition of stability, $\eta \leq 1$, may be obtained from arguments based on the domain of dependence, as we did earlier for transport equations.

An implicit FDE equation for the wave equation may be constructed using the Crank-Nicholson scheme:

$$\frac{1}{(\Delta t)^2}\delta_t^2\phi(x_i,t_k) = \frac{1}{4v^2(\Delta x)^2}\{\delta_x^2\phi(x_i,t_{k+1}) + 2\delta_x^2\phi(x_i,t_k) + \delta_x^2\phi(x_i,t_{k-1})\} \tag{8-116}$$

where, to simplify the notation, we have used

$$\begin{aligned}\delta_t^2\phi(x_i,t_k) &\equiv \phi(x_i,t_{k+1}) - 2\phi(x_i,t_k) + \phi(x_i,t_{k-1}) \\ \delta_x^2\phi(x_i,t_k) &\equiv \phi(x_{i+1},t_k) - 2\phi(x_i,t_k) + \phi(x_{i-1},t_k).\end{aligned} \tag{8-117}$$

Box 8-10 Program DM_WAVE
Explicit method to solve the wave equation in (x, t)
An example of second-order hyperbolic PDE

Initialization:
- (a) Set up a mesh of $(N_x + 1)$ points for x and $(N_t + 1)$ for t.
- (b) Set up the range $[x_b, x_d]$ and boundary conditions.
- (c) Select a propagation velocity v.
- (d) Obtain the step size Δx.

1. Input η of Eq. (8-114).
2. Calculate Δt, the step size in time t.
3. Specify the initial conditions.
4. Propagate forward in time using Eq. (8-115).
5. Output $\phi(x_i, t_k)$ once every few time steps.

We can regard the right side of Eq. (8-116) as taking a weighted "average" of the three finite differences in space at times $t = t_{k-1}$, t_k, and t_{k+1}. Here, we have chosen to put twice the weight on the middle point compared with each of the two on the side. Obviously, other averaging methods can also be used. It is also possible to regard Eq. (8-116) as an expression that gives the three quantities $\phi(x_{i+1}, t_{k+1})$, $\phi(x_i, t_{k+1})$, and $\phi(x_{i-1}, t_{k+1})$ at $t = t_{k+1}$ in terms of the six at $t = t_k$ and t_{k-1}. The form is very similar to that in the parabolic example of the previous section, and the same method of solution may be used.

Vibrating string As an example, let us solve the wave equation of Eq. (8-108) for a stretched string. To fix the boundary conditions, we shall anchor the two ends of the string at $x = x_b$ and x_d to the walls of the laboratory. In other words,

$$\phi(x_b, t) = \phi(x_d, t) = 0.$$

Without losing any generality, we can take $x_b = -1$ and $x_d = 1$. Initially, at $t = 0$, the string is assumed to be at rest,

$$\left.\frac{d\phi}{dt}\right|_{t=0} = 0.$$

To set the string into vibration, we introduce a sinusoidal pulse near the end at $x = 1$ in the form

$$\phi(x, t = 0) = \begin{cases} A\sin(2\pi x) & \text{for } 0.5 \le x \le 1 \\ 0 & \text{otherwise.} \end{cases}$$

The subsequent behavior of the string is described by Eq. (8-108) and may be calculated by solving the FDE of Eq. (8-115). The algorithm is outlined in Box 8-10.

Since no damping is assumed in the problem, the energy represented by the initial pulse is conserved. For the particular initial conditions we have selected, the pulse

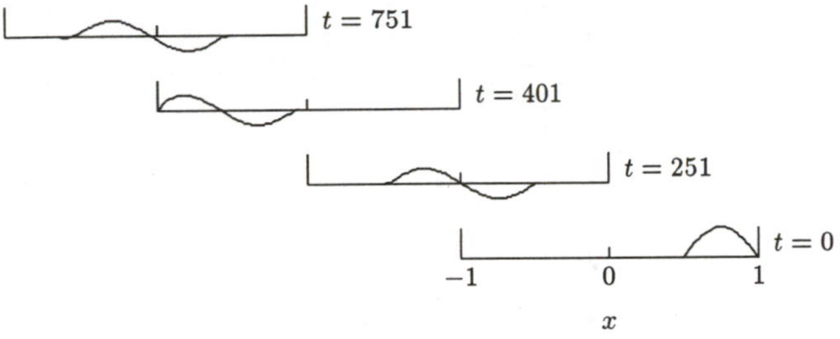

Figure 8-19: Propagation of a wave with $\eta = v(\Delta t)/(\Delta x) = 0.25$ along a stretched string fixed at both ends. In the absence of damping, the oscillation starts initially at one end is reflected back after $t = 401$ without any energy loss.

travels back and forth along the string between the two fixed ends with constant amplitude, as shown schematically in Fig. 8-19. In our representation of the string by a finite number of points in the FDE, only a limited number of normal modes can be present in the problem. If a sharp pulse is introduced into the system, instead of the smooth sine wave used here, the pulse shape is likely to be modified, as it may require more normal modes to represent a sharp pulse than our mesh can support. A finite number of points along x is equivalent to replacing the string by N_x discrete "masses" coupled together by tension in the connecting (massless) string. This is more or less opposite to the usual way of deriving the wave equation. For example, in standard texts for wave phenomena, it is common to obtain Eq. (8-108) by starting with N-coupled oscillators and the wave equation is obtained on invoking the limit of $N \to \infty$.

8-9 Nonlinear differential equations

So far, we have been dealing mainly with linear equations. In terms of the standard form of a second-order PDE given in Eq. (8-3),

$$p\frac{\partial^2\phi}{\partial x^2} + q\frac{\partial^2\phi}{\partial x\partial y} + r\frac{\partial^2\phi}{\partial y^2} + s\frac{\partial\phi}{\partial x} + t\frac{\partial\phi}{\partial y} + u\phi + v = 0$$

this means that the coefficients p, q, r, s, t, and u are functions of the independent variables x and y only. If any of them involve $\phi(x, y)$ or its derivatives, the equation becomes *nonlinear*. There are many possible forms of nonlinear differential equations and, in general, they are more difficult to solve. We are, however, more likely to encounter a special class, called *quasilinear equations*, where the highest-order derivatives are linear in the equation. For example, in the second-order PDE shown above, the coefficients p, q, r, s, t, u, and v do not involve any of the three second-order

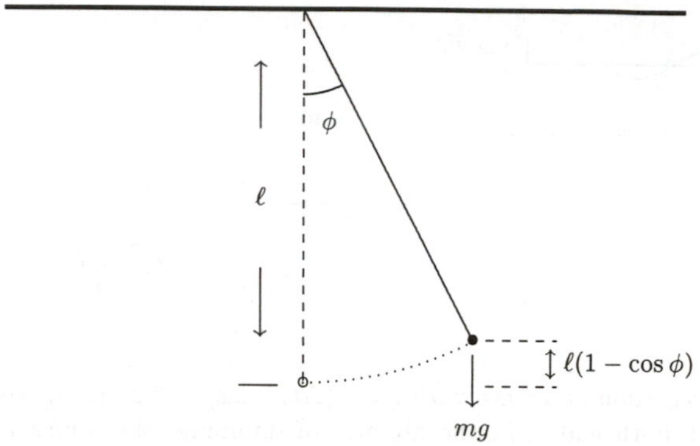

Figure 8-20: Pendulum with maximum angular displacement ϕ. If ϕ is large, it is no longer a simple harmonic oscillator and the motion is governed by the nonlinear differential equation of Eq. (8-118).

derivatives, $\partial^2\phi/\partial x^2$, $\partial^2\phi/\partial x\partial y$, and $\partial^2\phi/\partial y^2$, but they may be functions of $\partial\phi/\partial x$, $\partial\phi/\partial x$, and $\phi(x,y)$, as well as x and y. We shall limit ourselves to this special class of nonlinear differential equations.

As an example, consider a pendulum with angular displacement $\phi(t)$, as shown in Fig. 8-20. The oscillation is governed by the second-order ODE

$$\frac{d^2\phi}{dt^2} + \omega_0^2 \sin\phi(t) = 0 \tag{8-118}$$

where $\omega_0 = \sqrt{g/\ell}$, with g being the acceleration due to gravity and ℓ the length of the pendulum. It is a quasilinear equation, since $\phi(t)$ enters the equations as the argument of a transcendental function. If the displacement is sufficiently small, we can approximate $\sin\phi(t)$ by its first term in the series expansion

$$\sin\phi = \phi - \frac{1}{3!}\phi^3 + \cdots.$$

In this limit, we obtain the usual linear form for simple harmonic motion,

$$\frac{d^2\phi}{dt^2} + \omega_0^2\phi(t) = 0.$$

However, in general, the amplitude may not be small and we must go back to the nonlinear equation given by Eq. (8-118). As we shall soon see, the presence of nonlinear terms complicates solution to the corresponding finite difference equation.

A finite difference approximation to Eq. (8-118) may be constructed using a central difference scheme,

$$\phi(t_{k+1}) - 2\phi(t_k) + \phi(t_{k-1}) + (\Delta t)^2\omega_0^2\sin\phi(t_k) = 0. \tag{8-119}$$

Because of the $(\Delta t)^2$ factor, the influence of the nonlinear term on the FDE is relatively weak here. Consequently, reasonable approximations can be used to find the solution.

Quasilinear initial value problem For linear differential equations, the FDEs are linear in the unknown quantities. Because of this, we were able, in the previous sections, to find the numerical solutions using efficient techniques for the roots of linear algebraic equations. The presence of the nonlinear term, as for example $\sin\phi(t)$ in Eq. (8-119), changes the situation. Even when the term is small, we cannot simply adapt the methods designed for linear systems to solve the equation without some additional considerations. There are two possible approaches to take. The first is to approximate the nonlinear term. For example, we can calculate the contributions of $\sin\phi(t_k)$ using the best estimate of the value of $\phi(t_k)$ we have for the point. To add a small improvement, we can make a series expansion of the nonlinear term around the best estimate and include in the approximation a term with linear dependence on $\phi(t_k)$. In either case, the FDE is now approximated as a linear one and can therefore be solved using one of the methods discussed earlier for a system of linear algebraic equations.

The second approach is to use an explicit scheme for the FDEs. If, for the moment, we ignore questions concerning stability and convergence, we can in principle propagate the solution by starting from a set of initial conditions. The nonlinear term is not a problem if it involves only the solution at earlier times. Consider Eq. (8-119) as an example. Since it is an initial value problem, we may regard it as an equation to find $\phi(t_{k+1})$ from the values of $\phi(t_k)$ and $\phi(t_{k-1})$. The equation is second order and the two initial values may be taken to be those for $\phi(t_0)$ and $\phi(t_1)$. With these two values as the input, we can obtain $\phi(t_2)$ using Eq. (8-119) for $k = 1$. With $\phi(t_1)$ and $\phi(t_2)$, we can solve for $\phi(t_3)$ by putting $k = 2$, and so on, as we did earlier in Eq. (8-6). Instead of this naive approach, we can use methods such as the Runge-Kutta or extrapolation methods discussed earlier. However, it is difficult to establish the stability of any of these methods for nonlinear ODEs. On the other hand, for many cases, such as the example of Eq. (8-118), we can carry out the actual calculations and find that both methods are able to yield fairly stable solutions.

For a pendulum swinging with finite initial amplitude, the period τ is known from analytical calculations to be

$$\begin{aligned} \tau &= \frac{2}{\omega_0}\int_{-\pi/2}^{+\pi/2} \frac{d\beta}{\sqrt{1-\sin^2\frac{\phi_0}{2}\sin^2\beta}} \\ &= \frac{2\pi}{\omega_0}\left\{1+\frac{1}{4}\sin^2\frac{\phi_0}{2}+\frac{9}{64}\sin^4\frac{\phi_0}{2}+\frac{25}{256}\sin^6\frac{\phi_0}{2}+\cdots\right\} \end{aligned} \tag{8-120}$$

where ϕ_0 is the initial amplitude. We shall make use of this result as a check on the accuracy of our numerical solution for Eq. (8-118). Using either of the methods described above, we can find the solution for the FDE given in Eq. (8-119). To check against Eq. (8-120), we must deduce the period from the numerical results we have

Table 8-1: Period of a pendulum with finite initial amplitude.

Initial amplitude		Period	
Radians	Degrees	Numerical	Series
0.1	5.7	6.287	6.287
0.25	14.3	6.308	6.308
0.5	28.6	6.383	6.382
1.0	57.3	6.700	6.698
1.5	85.9	7.300	7.265

for $\phi(t_k)$. One way is to use inverse interpolation to find the zeros of the oscillatory function and define the period as twice the difference between two consecutive zeros. The Bessel inverse interpolation technique of §3-6 can be easily adapted for this task and the results are shown in Table 8-1 for various initial amplitudes. The values listed in the last column, labeled "series," are calculated using Eq. (8-120) up to the $\sin^6(\phi_0/2)$ term. The small discrepancies for large values of ϕ_0 between the numerical and series results are likely to be coming from the need for additional terms in the series expansion of Eq. (8-120). The comparison provides us with some confidence that the numerical solution to Eq. (8-118) has the required accuracy.

Quasilinear boundary value problem As an example of quasilinear boundary value problems, we shall consider the "kink" solution of the differential equation

$$\frac{d^2\phi}{dx^2} + \phi(x) - \phi^3(x) = 0. \tag{8-121}$$

The equation arises in field theory for a potential of the form

$$V(\phi) = \frac{\lambda}{4}\left\{\phi^2(x) - \frac{\mu^2}{\lambda}\right\}^2.$$

It represents one of the possible extensions we can make on the basic $\phi^2(x)$ potential used as an approximation in Fig. 4-1.

To derive Eq. (8-121), we start with the lagrangian density for a field ϕ that is a function of x and t only:

$$\mathcal{L}(x,t) = \frac{1}{2}\left(\frac{\partial\phi}{\partial t}\right)^2 - \frac{1}{2}\left(\frac{\partial\phi}{\partial x}\right)^2 - V(\phi).$$

From the Lagrange's equation of motion, we obtain the differential equation

$$\frac{\partial^2\phi}{\partial t^2} - \frac{\partial^2\phi}{\partial x^2} = \mu^2\phi(x,t) - \lambda\phi^3(x,t).$$

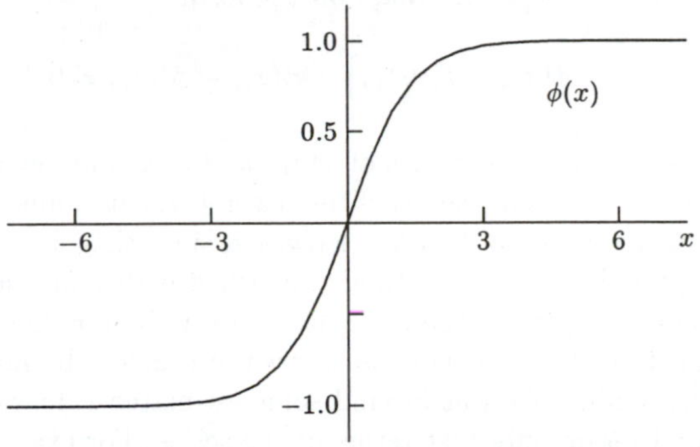

Figure 8-21: A kink solution of the second-order quasilinear ODE Eq. (8-121). Analytically, it has the form $\phi(x) = \tanh(x/\sqrt{2})$.

The constants μ^2 and λ may be absorbed into the definitions for x and $\phi(x)$ by the replacements

$$x \longrightarrow \frac{x}{\mu} \qquad\qquad \phi \longrightarrow \frac{\mu}{\sqrt{\lambda}}\phi.$$

If our interest is in the static solution, we can omit the time dependence and the result is Eq. (8-121). Among other interests, the equation is one of the few whose solitary wave, or soliton, solutions are known analytically. As a differential equation, it is quasilinear, since the nonlinear term is in $\phi(x)$.

We shall be concerned only with a particular solution of Eq. (8-121), generally known as the static "kink" solution. There are several different possible forms of the solution depending on the boundary conditions we wish to impose on the system. The one shown in Fig. 8-21 is obtained with the conditions

$$\phi(-\infty) = -1 \qquad\qquad \phi(+\infty) = +1.$$

Alternatively, we can have an "antikink" solution by imposing, instead, the conditions $\phi(\pm\infty) = \mp 1$. From analytical calculations it is known that the kink solution has the form

$$\phi(x) = \tanh\left(\frac{x - x_0}{\sqrt{2}}\right). \tag{8-122}$$

Here x_0 is the point where the value of $\phi(x)$ changes sign. For our numerical solution, the value of x_0 depends on the initial trial solution used to solve the problem, as the differential equation and the boundary conditions do not contain any explicit dependence on x_0.

Using a mesh of $(N+1)$ points, $x_0, x_1, x_2, \ldots, x_N$, equally spaced along the x-axis, the corresponding FDE for Eq. (8-121) takes on the form

$$\frac{1}{(\Delta x)^2}\big\{\phi(x_{i+1}) - 2\phi(x_i) + \phi(x_{i-1})\big\} + \phi(x_i) - \phi^3(x_i) = 0 \tag{8-123}$$

where Δx is the step size. Since the values of $\phi(x_0)$ and $\phi(x_N)$ are given, it is a two-point boundary value problem. The values of the other $\phi(x_i)$ are found by solving the set of $(N-1)$ equations given by Eq. (8-123). However, since they are no longer linear in $\phi(x_i)$, we cannot apply directly the methods described earlier for linear equations.

There are two ways to overcome the difficulties caused by nonlinear terms in the FDE. The first, as mentioned earlier, is to *linearize* the problem by making a Taylor series expansion of the nonlinear terms around some estimated values of the solution at each mesh point and retain only first terms of the series. For example, the $\phi^3(x_i)$ term in Eq. (8-123) may be expanded as

$$\phi^3(x_i) \approx \tilde{\phi}^3(x_i) + 3\tilde{\phi}^2(x_i)\{\phi(x_i) - \tilde{\phi}(x_i)\} + \cdots \tag{8-124}$$

where $\tilde{\phi}(x_i)$ is the estimated value of $\phi(x_i)$. By truncating the series after the second term, we have a linear approximation for $\phi^3(x_i)$. Initially, the estimate may be obtained from an approximate solution of the original differential equation. Once a first approximation is available, we can improve the results by making use of the approximate solution. This is carried out iteratively until the solution converges, in the same way as we have done on many earlier occasions. However, we shall not pursue this particular method of solution here.

Newton's method to solve the FDE A second approach to solve Eq. (8-123) is Newton's method. As in Eq. (8-124), this method also involves linearizing the problem by approximating the nonlinear term with the first two terms of a Taylor series. To illustrate the differences, consider first the question of finding the roots of an algebraic equation

$$f(s) = 0$$

where $f(s)$ is a nonlinear function of s. Let us assume the root is located at $s = s^{(r)}$. Near the root, we can make a series expansion of the function $f(s)$ around some estimated value $s^{(m)}$

$$f(s) \approx f(s^{(m)}) + \left.\frac{df}{ds}\right|_{s=s^{(m)}}\{s - s^{(m)}\}. \tag{8-125}$$

An improved estimate may be obtained by realizing that, if $s^{(m+1)}$ is the root, then

$$f(s^{(m+1)}) = 0 \approx f(s^{(m)}) + \left.\frac{df}{ds}\right|_{s=s^{(m)}}\{s^{(m+1)} - s^{(m)}\}. \tag{8-126}$$

From this, we obtain the next estimate,

$$s^{(m+1)} = s^{(m)} - \frac{f(s^{(m)})}{\left.\frac{df}{ds}\right|_{s=s^{(m)}}}. \tag{8-127}$$

Since the relation given by Eq. (8-126) is only approximate, we do not expect $s^{(m+1)}$ to be the same as the solution $s^{(r)}$. However, it should represent a better estimate than $s^{(m)}$. The process may be iterated until step k when the value of $f(s^{(k)})$ is sufficiently close to zero for $s^{(k)}$ to be acceptable as the root. Again, we need a starting value $s^{(0)}$, which can be obtained from, for example, an approximate analytical solution to the original differential equation. In general, the convergence of the method is quadratic near the minimum. As a result, the approach is efficient if a good starting value is available.

For any practical applications, we need to generalize the method to a set of n functions $f_1(\boldsymbol{s})$, $f_2(\boldsymbol{s}), \ldots, f_n(\boldsymbol{s})$ that depend on n variables $\boldsymbol{s} = \{s_1, s_2, \ldots, s_n\}$. In the place of Eq. (8-125), we now have the relation

$$\boldsymbol{f}(\boldsymbol{s}) \approx \boldsymbol{f}(\boldsymbol{s}^{(m)}) + \mathcal{J}(\boldsymbol{s}^{(m)})\{\boldsymbol{s} - \boldsymbol{s}^{(m)}\} \tag{8-128}$$

where, in the place of $\frac{df}{ds}|_{s=s^{(m)}}$ in Eq. (8-126), we have the Jacobian matrix

$$\mathcal{J}(\boldsymbol{s}) \equiv \begin{pmatrix} \frac{\partial f_1}{\partial s_1} & \frac{\partial f_1}{\partial s_2} & \cdots & \frac{\partial f_1}{\partial s_n} \\ \frac{\partial f_2}{\partial s_1} & \frac{\partial f_2}{\partial s_2} & \cdots & \frac{\partial f_2}{\partial s_n} \\ \vdots & \vdots & \ddots & \vdots \\ \frac{\partial f_n}{\partial s_1} & \frac{\partial f_n}{\partial s_2} & \cdots & \frac{\partial f_n}{\partial s_n} \end{pmatrix}. \tag{8-129}$$

Similar to Eq. (8-127), we can start from $\boldsymbol{s}^{(m)}$ for the mth-order approximation and find the value for the next order from

$$\boldsymbol{s}^{(m+1)} = \boldsymbol{s}^{(m)} - \boldsymbol{\Delta}^{(m)} \tag{8-130}$$

where $\boldsymbol{\Delta}^{(m)}$ is the solution to the set of equations

$$\mathcal{J}(\boldsymbol{s}^{(m)})\boldsymbol{\Delta}^{(m)} = \boldsymbol{f}(\boldsymbol{s}^{(m)}). \tag{8-131}$$

In general, Jacobian matrices arising from finite difference equations are sparse and may be solved with relative ease.

For our FDE of Eq. (8-123), there are $(N + 1)$ points on the mesh. Since two boundary conditions are supplied by the problem, we are left with $(N - 1)$ unknown quantities, $\phi(x_1)$, $\phi(x_2), \ldots, \phi(x_{N-1})$. It is possible to treat Eq. (8-123) as a set of $(N - 1)$ equations for the $(N - 1)$ unknown $\phi(x_i)$. A typical equation in the set has the form

$$f_i(\boldsymbol{\phi}) = \frac{1}{(\Delta x)^2}\{\phi(x_{i+1}) - 2\phi(x_i) + \phi(x_{i-1})\} + \phi(x_i) - \phi^3(x_i). \tag{8-132}$$

We can use the approximation of Eq. (8-128) to locate the roots using the recursive approach given by Eqs. (8-130) and (8-131). The only nonzero matrix elements of

Box 8-11 Program DM_KINK
Solution of the kink equation by Newton's method

Subprograms used:
- INIT: Construct an equally spaced mesh and an estimate of the solution.
- SLV_TRI: Gaussian elimination for a tridiagonal matrix (cf. §A-3).

Initialization:
Select the error tolerance ϵ, the number of mesh points N, and the maximum number of iterations allowed.

1. Input the range and normalization factor.
2. Use INIT to:
   (a) Construct an equally spaced mesh of $(N+1)$ points.
   (b) Give an initial estimate of the solution as a step function at $x = 0$.
3. Iterate the solution:
   (a) Calculate $f_i(\phi)$ of Eq. (8-132) and elements $\mathcal{T}_{i,i}$ of Eq. (8-133).
   (b) Use SLV_TRI to solve the tridiagonal Jacobian matrix equation Eq. (8-131).
   (c) Update the solution:
      (i) Improve the solution with Eq. (8-130) using the value of $\boldsymbol{\Delta}$ obtained.
      (ii) Estimate the average error $\mathcal{E}$ by averaging over $|\boldsymbol{\Delta}|$.
4. Iterate the solution again if $\mathcal{E}$ is larger than the tolerance of error ϵ.
5. Output the solution if $\mathcal{E}$ is sufficiently small, or terminate the calculation if the maximum number of iterations is exceeded.

the Jacobian matrix are

$$\begin{aligned}\mathcal{J}_{i,i-1} &= \frac{\partial f_i}{\partial\phi(x_{i-1})} = \frac{1}{(\Delta x)^2}\\ \mathcal{J}_{i,i} &= \frac{\partial f_i}{\partial\phi(x_i)} = \frac{-2}{(\Delta x)^2} + 1 - 3\phi^2(x_i)\\ \mathcal{J}_{i,i+1} &= \frac{\partial f_i}{\partial\phi(x_{i+1})} = \frac{1}{(\Delta x)^2}\end{aligned} \tag{8-133}$$

for $i = 1, 2, \ldots, (N-1)$. Since $\phi(x_0)$ and $\phi(x_N)$ are given by the boundary conditions, we have

$$\mathcal{J}_{1,0} = \mathcal{J}_{N-1,N} = 0.$$

The Jacobian matrix is therefore a tridiagonal one of dimension $(N-1)$ and has the same structure as Eq. (8-105). As a result, we can use the Gaussian elimination technique of §A-3 to find the solution.

The procedure to solve the nonlinear FDE of Eq. (8-123) may be summarized in the following way. The starting point is an initial estimate of the values of $\phi(x_i)$ for $i = 1, 2, \ldots, (N-1)$. Using these values, we construct the Jacobian matrix of Eq. (8-133), solve Eq. (8-131) by Gaussian elimination for $\boldsymbol{\Delta}$, and obtain a better set of estimates using Eq. (8-130). These three steps are repeated using the improved estimates until the differences in two successive iterations are smaller than the error that can be tolerated. The procedure is outlined in Box 8-11.

For the calculated results shown in Fig. 8-21, the following step function is used as the initial estimate:

$$\phi^{(0)}(x_i) = \begin{cases} -1 & \text{for } x_i < 0 \\ +1 & \text{otherwise.} \end{cases}$$

This simple form ensures that the point for $\phi(x) = 0$ is located at $x = 0$. By shifting the point where $\phi^{(0)}(x)$ changes from -1 to $+1$ in such a trial solution, we obtain kink solutions corresponding to different values of x_0 in Eq. (8-122).

8-10 Stiffness problems

In §8-2, we encountered the problem of stiffness in the solution for an underdamped harmonic oscillator. A good illustration of the general situation is provided by the second-order equation

$$\frac{d^2\phi}{dt^2} - L^2\phi(t) = 0 \tag{8-134}$$

where L is some large, positive quantity to be specified later. This type of ordinary differential equation appears in a variety of problems. The general solution is

$$\phi(t) = Ae^{-Lt} + Be^{+Lt} \tag{8-135}$$

where A and B are two constants of integration that must be determined from the initial or boundary conditions. For our illustration, we shall take L to be a constant, for example,

$$L^2 = 100$$

and the initial conditions

$$\phi(t=0) = 1 \qquad\qquad \left.\frac{d\phi}{dt}\right|_{t=0} = -L. \tag{8-136}$$

The solution under these conditions is well known:

$$\phi(t) = e^{-10t}. \tag{8-137}$$

The $\exp(+Lt)$ term does not appear here because of the initial conditions. For illustration, let us solve the same differential equation numerically using the method of Euler given in §8-1. As we shall see from the discussion, the difficulties associated with stiffness are not limited to the particular way we solve the problem.

In the same way as we did in going from Eq. (8-4) to (8-5), we can construct an evenly spaced set of points, t_0, t_1, $t_2, \ldots, t_N$, and convert Eq. (8-134) into a finite difference equation of the form

$$\phi(t_{k+1}) = \left\{2 + (\Delta t)^2 L^2\right\}\phi(t_k) - \phi(t_{k-1}).$$

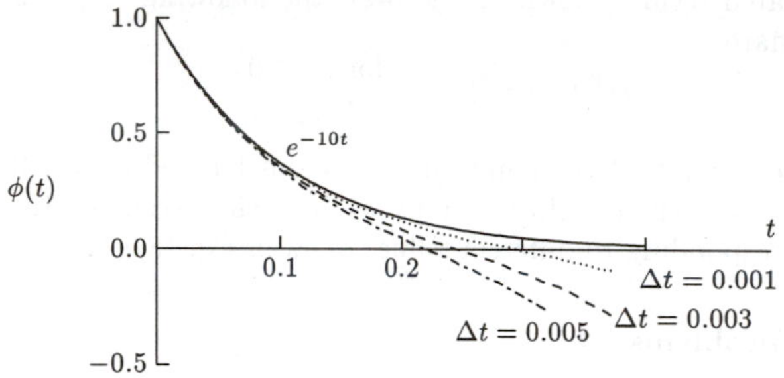

Figure 8-22: Solution of the stiff equation Eq. (8-134). Numerical results obtained with step size of $\Delta t = 0.001$ (dotted curve), 0.003 (dashed curve), and 0.005 (dash-dot curve) are compared with the analytical solution e^{-10t} (solid curve).

The initial conditions of Eq. (8-136) may be introduced into the calculation in the form

$$\phi(t_1) = 1.0 \qquad\qquad \frac{\phi(t_2) - \phi(t_1)}{\Delta t} = -10 \tag{8-138}$$

where we have used a forward difference scheme to approximate the first derivative of $\phi(t)$ at $t = 0$. The calculated results using different step sizes are shown in Fig. 8-22.

It is quite clear by examining the figure that, even for small step sizes, the numerical results deviate fairly early from the exact value of $\exp(-Lt)$, shown as solid curve. Furthermore, the larger the value of Δt, the earlier in time for the departure to become noticeable. The discrepancies from the exact results may be viewed in the following way. The general solution to the differential equation has two terms with different time scales, $\exp(-Lt)$ and $\exp(+Lt)$, as we saw in Eq. (8-135). For our purpose here, let us rewrite it as

$$\phi(t) = e^{-Lt} + \epsilon\, e^{+Lt}$$

by defining ϵ as the relative size of the two terms. For the initial condition of Eq. (8-136), the exact result puts $\epsilon = 0$. However, in a numerical solution, truncation and other errors introduce small admixtures of the unwanted term and $\epsilon \neq 0$. The size of ϵ depends, in part, on the value of Δt used. Regardless of how we may reduce the numerical errors, the unwanted $\exp(+Lt)$ term will eventually dominate the solution once ϵ becomes nonzero at some stage of the calculation, as the function $\exp(+Lt)$ rises rapidly with time. One may argue that, in our illustration, a large part of the error comes from the forward difference approximation used in imposing the initial condition in Eq. (8-138). This is not entirely correct. We can test this point by replacing Eq. (8-138) with $\phi(t_1) = 1$ and $\phi(t_2) = \exp(-10t_2)$ as the starting point of the solution. The difficulty with the numerical solution persists, though with somewhat reduced magnitude.

Solutions of stiffness problem For some cases, methods are available to overcome the stiffness problem and we shall briefly mention two. Let us return to the damped harmonic oscillator example given earlier in Eq. (8-20). We recall that the second-order ODE has the form

$$\frac{d^2\phi}{dt^2} + 2\gamma\frac{d\phi}{dt} + \omega_0^2\phi(t) = 0. \tag{8-139}$$

For the underdamped case, where the problem of stiffness occurs, we can transform the equation into one involving only a single time scale by defining a function $y(t)$ through the relation

$$\phi(t) = e^{-\gamma t}y(t). \tag{8-140}$$

On substituting this form of $\phi(t)$ into Eq. (8-139), we obtain an equation for $y(t)$,

$$\frac{d^2y}{dt^2} + (\omega_0^2 - \gamma^2)y(t) = 0.$$

There is only one time scale here, given by $\sqrt{\omega_0^2 - \gamma^2}$, and we should have no difficulty in solving the equation numerically using any of the techniques discussed earlier for initial value problems.

Transformations similar to that of Eq. (8-140) can often be an effective way to eliminate the stiffness problem. On the other hand, to apply the transformation, we must have beforehand a knowledge of one of the time scales involved. This is not always possible. An alternative for a large variety of problems is to apply the Riccati transformation

$$\frac{d\phi}{dt} = y(t)\phi(t) \tag{8-141}$$

which turns the differential equation for $\phi(t)$ into one for $y(t)$. On differentiating both sides of Eq. (8-141) with respect to t, we obtain

$$\frac{d^2\phi}{dt^2} = \frac{dy}{dt}\phi(t) + y(t)\frac{d\phi}{dt} = \frac{dy}{dt}\phi(t) + y^2(t)\phi(t).$$

Using this, Eq. (8-134) is reduced to a first-order, but quasilinear, equation in $y(t)$:

$$\frac{dy}{dt} + y^2(t) = L^2.$$

This may be solved using one of the methods discussed in the previous section. Through the Riccati transformation, the damped harmonic oscillator equation of Eq. (8-139) may be changed into the form

$$\frac{dy}{dt} + y^2(t) + 2\gamma y(t) = -\omega_0^2.$$

There is no difficulty in solving this quasilinear equation. However, for more complicated systems, the transformation can sometimes lead to equations for which it is difficult to find a solution.

Another way to handle the problem of stiffness is to adopt an implicit method to solve the FDE. All the numerical solutions we have used so far in this section are explicit methods, as $\phi(t_{k+1})$ is given directly in terms of $\phi(t)$ at earlier times. To see why explicit methods can fail, it is instructive to consider a first-order ODE of the form

$$\frac{d\phi}{dt} = \eta\phi(t). \tag{8-142}$$

We shall first take η to be a constant. Using a forward difference,

$$\frac{d\phi}{dt} \rightarrow \frac{\phi(t_{k+1}) - \phi(t_k)}{\Delta t}$$

we can approximate Eq. (8-142) by a finite difference equation of the form

$$\frac{\phi(t_{k+1}) - \phi(t_k)}{\Delta t} = \eta\phi(t_k).$$

This may be put in terms of an algebraic equation,

$$\phi(t_{k+1}) = \{1 + (\Delta t)\eta\}\phi(t_k).$$

If $(\Delta t)\eta < -2$, the value of $\phi(t)$ increases in magnitude and oscillates in sign from one point on the mesh to the next. This is clearly unacceptable as a solution.

On the other hand, if we use a backward difference approximation,

$$\frac{d\phi}{dt} \rightarrow \frac{\phi(t_k) - \phi(t_{k-1})}{\Delta t}$$

we have the result

$$\phi(t_k) = \phi(t_{k-1}) + (\Delta t)\eta\phi(t_k). \tag{8-143}$$

This is an implicit equation for $\phi(t_k)$. For η a constant, we obtain a solution in the form

$$\phi(t_k) = \frac{\phi(t_{k-1})}{1 + \eta\Delta t}$$

which is stable for any step size Δt. In general, implicit methods are better suited for stiff problems.

If η is a function involving $\phi(t)$, we must apply techniques such as Newton's method discussed in the previous section to solve the quasilinear equation corresponding to Eq. (8-143). However, for higher-order equations, it may be easier to find a suitable implicit method to solve the problem instead. There does not seem to be any prospect at the moment for devising a method to handle the stiffness problem in general and, as a result, it remains one of the challenges of numerical methods.

Problems

8-1 A projectile starts from the surface of the earth at $x = x_0$ with an initial velocity v_0 and at an angle of elevation θ_0. If we ignore air resistance, the only force acting on the projectile is gravity along the y-direction. Set up the differential equation describing the trajectory of the projectile assuming that the x-axis is in the horizontal plane. Solve the equation numerically as an initial value problem using v_0 and θ_0 as the initial values. Let x_d be the point where the projectile lands on the surface of the earth again. For a given v_0, plot the value of $d = x_d - x_0$ as a function of θ_0 and demonstrate that the maximum occurs at $\theta_0 = 45°$.

8-2 The projectile motion of Problem 8-1 can be turned into a boundary value problem if we wish the projectile to land at a specific point x_s on the surface of the earth. The solution may be obtained using the same computer program as Problem 8-1 by mimicking one of the shooting methods. Try this approach by varying the angle of elevation θ_0 until the $y = 0$ point occurs at x_t for $|x_s - x_t|$ less than some small value ϵ.

8-3 Rewrite Eq. (8-51) in terms of logarithmic derivatives at $r = r_g$.

8-4 Find the scattering solution of Eq. (8-37) for the $\ell = 0$ modified radial wave function inside a square-well potential using the method of successive overrelaxation (SOR). Explore the dependence of convergence on the relaxation parameter ω and the number of intervals. The shape of the potential well is given by Eq. (8-38). Use $u_0(0) = 0$ and $u_0(r_g) = 1$ as the boundary conditions and $u_0(r) = r/r_g$ as the initial trial solution.

8-5 In the solution of the two-dimensional Poisson equation of §8-6, the potential at the location of a point charge is singular, given roughly by $V(\eta) \propto q/\eta$ where η is the distance to the point charge. What happens to this singularity in the solution to the finite difference equation?

8-6 Write a computer program to calculate the solution for the wave equation of Eq. (8-108) for a stretched string with the end at $x = 0$ fixed and the other end at $x = 1$ moving with a sinusoidal time dependence. The initial condition may be taken to be such that the displacement and velocity of the string are zero everywhere.

8-7 Besides the equation for the kink solution, the static sine-Gordan equation

$$\frac{d^2\phi}{dx^2} - \sin\phi(x) = 0$$

is also a nonlinear ordinary differential equation with known analytical solitary wave solution

$$\phi(x) = 4\tan^{-1}\{\exp(x - x_0)\}.$$

Solve the differential equation numerically using Newton's method and compare the results with the analytical form given above.

8-8 What are the major differences between an eigenvalue problem, such as that given by Eq. (4-67) for a harmonic oscillator potential, and a two-point boundary value problem, such as that represented by Eq. (8-36)? Devise a numerical method to solve the eigenvalue problem of Eq. (4-67) using one of the methods for boundary value problems.

Chapter 9

Finite Element Solution to PDE

In boundary value problems, finite element methods (FEM) are often used instead of the finite difference methods of the previous chapter. Here, the space is divided into a number of small elements, hence the name finite element method. Within each element, the solution is approximated by simple functions, characterized by a few parameters. There are several obvious advantages to such an approach. In the first place, it is easy to handle cases with odd geometrical shapes and such problems occur quite frequently in engineering. Secondly, we have far more flexibility in adjusting the size of individual elements. As a result, it is relatively easy to have finer subdivisions of the space in regions where accuracies are difficult to achieve. As we have seen earlier, this is not easy to do in finite difference methods for partial differential equations.

Finite element analysis has a long tradition and there are large numbers of excellent treatises on the subject, especially in engineering libraries. In this chapter, we shall give an introduction to the subject through a few examples so that the reader can gain a feeling on the subject and, perhaps, even attempt to solve some simple problems. For more details, see, for example, Burnett,[8] Silvester and Ferrari,[50] and Zienkiewicz and Taylor.[60]

9-1 Background

Let us start by examining some of the basic principles of FEM with an example. Although our main interest is in partial differential equations, it is far simpler for our present purpose to use an ordinary differential equation. Consider the following first-order differential equation

$$\frac{d\phi}{dt} + \lambda\phi(t) = 0. \tag{9-1}$$

It is an equation describing exponential decay. For instance, $\phi(t)$ may represent the number of radioactive nuclei in a sample at time t. In this case, the equation describes the decay of the sample with decay constant λ. For the convenience of discussion, we shall restrict our interest in the interval $t = [0, 1]$ and take

$$\lambda = 1.$$

Since it is a first-order ODE, the one boundary condition may be taken as

$$\phi(t = 0) = 1. \tag{9-2}$$

The analytical solution for this case is a familiar one:

$$\phi(t) = e^{-t} = 1 - t + \frac{1}{2!}t^2 - \frac{1}{3!}t^3 + \cdots. \tag{9-3}$$

For the discussion following, it is helpful to think in terms of the series form of the exponential function solution.

Power series solution Let us ignore the analytical solution for the moment and, instead, try to solve Eq. (9-1) using a trial solution consisting of a power series with n terms

$$\tilde{\phi}_n(t) = a_0 + a_1 t + a_2 t^2 + \cdots + a_n t^n. \tag{9-4}$$

The tilde over $\phi(t)$ is to differentiate it from the exact solution, and the subscript n indicates the degrees of freedom of the approximation, given here by the order of the power series used. To satisfy the initial condition of Eq. (9-2), it is necessary that

$$a_0 = 1.$$

The other coefficients a_1, a_2, ..., a_n must be found from the differential equation.

For the convenience of discussion, let us take $n = 2$. In this case, our trial solution assumes the simple form

$$\tilde{\phi}_2(t) = 1 + a_1 t + a_2 t^2. \tag{9-5}$$

Obviously, the result is going to be different from the analytic solution given in Eq. (9-3). However, our interest is only in a restricted domain and, as we shall soon see, it is possible to find a set of values for a_1 and a_2 such that $\tilde{\phi}_2(t)$ represents a very good approximation of the function $\exp(-t)$ within $t = [0, 1]$. For this purpose, we shall treat the coefficients a_1 and a_2 in Eq. (9-5) and, more generally, a_1, a_2, ..., a_n in Eq. (9-4), as parameters, to be adjusted in such a way that $\tilde{\phi}_n(t)$ is as close to the exact solution as possible within the region of interest.

Let us substitute the trial solution Eq. (9-5) into Eq. (9-1). Since it is not the true solution, it is unlikely that $d\tilde{\phi}_2/dt + \tilde{\phi}_2(t)$ vanishes everywhere in $[0, 1]$. We shall define a quantity $R(t; a_1, a_2)$ to measure the difference,

$$R(t; a_1, a_2) \equiv \frac{d\tilde{\phi}_2}{dt} + \tilde{\phi}_2(t) = 1 + (1 + t)a_1 + (2t + t^2)a_2. \tag{9-6}$$

This is known as the *residual.* It is obvious that it is a function of t as well as the parameters a_1 and a_2. More generally, if there are n parameters instead of 2 in Eq. (9-5), the residual may be defined as

$$R(t;\boldsymbol{a}) \equiv \frac{d\tilde{\phi}_n}{dt} + \tilde{\phi}_n(t)$$

where, as before, $\boldsymbol{a} \equiv \{a_1, a_2, \ldots, a_n\}$.

Determination of the parameters To find the values of $\boldsymbol{a}$, we can adjust them such that $\tilde{\phi}_n(t)$ is as close to $\phi(t)$ as possible. In practice, this means to minimize the residual within the domain $t = [0,1]$. There are several ways to achieve this and, in order to illustrate various aspects of the problem, we shall give four of them.

In the collocation method, we require the residual to vanish at n points t_1, $t_2, \ldots$, t_n, within the domain of interest,

$$R(t_i;\boldsymbol{a}) = 0 \qquad \text{for} \quad i = 1, 2, \ldots, n. \tag{9-7}$$

For our $n = 2$ approximation, we can take any two points t_1 and t_2 within $[0,1]$, for example, $t_1 = 1/3$ and $t_2 = 2/3$. This gives us two equations,

$$\begin{aligned} R(t=\tfrac{1}{3};\boldsymbol{a}) &= 1 + \frac{4}{3}a_1 + \frac{7}{9}a_2 = 0 \\ R(t=\tfrac{1}{3};\boldsymbol{a}) &= 1 + \frac{4}{3}a_1 + \frac{7}{9}a_2 = 0. \end{aligned}$$

The roots of this set of equations,

$$a_1 = -\frac{27}{29} \qquad\qquad a_2 = \frac{9}{29}$$

give us

$$\phi_2^{\text{col}}(t) = 1 - \frac{27}{29}t + \frac{9}{29}t^2. \tag{9-8}$$

On the surface, this result appears to be quite different from the exact solution given by Eq. (9-3). However, we see in Fig. 9-1 that the differences are, in fact, quite small within the domain of interest. It is not essential for us to select t_1 and t_2 such that the interval $[0,1]$ is divided into three equal parts. Other choices will result in different values of a_1 and a_2 and, as long as they are reasonable, the solutions are good approximations within the region of interest.

Instead of insisting that the residual vanish at two points in the domain, we can also obtain the necessary equations to solve for a_1 and a_2 by setting the average value of $R(t, \boldsymbol{a})$ in each of two different parts of the domain, Δt_1 and Δt_2, to vanish. That is, we demand

$$\frac{1}{\Delta t_i}\int_{\Delta t_i} R(t;\boldsymbol{a})\,dt = 0 \tag{9-9}$$

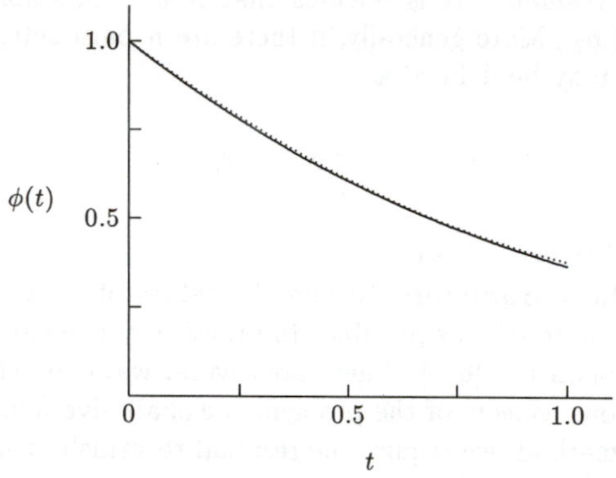

Figure 9-1: Comparison of the $n = 2$ approximation solution to Eq. (9-1) obtained using the collocation method (dotted curve) with the exact result (solid curve).

where the integration is carried out within the subdomain Δt_i. This is known as the subdomain method. For the general case of n parameters, we need n such subdomains or *elements*, $\Delta t_1, \Delta t_2, \ldots, \Delta t_n$. Again, it is not essential that all the elements be of equal size and they may even overlap each other.

For our two-parameter approximation, we shall, for simplicity, take two equal subdomains, $\Delta t_1 = [0, \frac{1}{2}]$ and $\Delta t_2 = [\frac{1}{2}, 1]$. The two average residuals are

$$\begin{aligned} \frac{1}{\Delta t_1}\int_0^{1/2} R(t;\boldsymbol{a})\,dt &= 2\Big\{\frac{1}{2} + \frac{5}{8}a_1 + \frac{7}{24}a_2\Big\} \\ \frac{1}{\Delta t_2}\int_{1/2}^{1} R(t;\boldsymbol{a})\,dt &= 2\Big\{\frac{1}{2} + \frac{7}{8}a_1 + \frac{25}{24}a_2\Big\}. \end{aligned}$$

By requiring both to vanish, we obtain the result,

$$a_1 = -\frac{18}{19} \qquad\qquad a_2 = \frac{6}{19}.$$

Thus

$$\phi_2^{\text{sub}}(t) = 1 - \frac{18}{19}t + \frac{6}{19}t^2 \tag{9-10}$$

is the $n = 2$ approximation solution to Eq. (9-1) using the subdomain method.

A third way to determine $\boldsymbol{a}$ is to make use of least-squares techniques. Using the idea of maximum likelihood discussed in §6-3, the optimum values of $\boldsymbol{a}$ can be obtained by solving the following set of n equations:

$$\frac{\partial}{\partial a_i}\int_{t_b}^{t_d} R^2(t;\boldsymbol{a})\,dt = 2\int_{t_b}^{t_d} R(t;\boldsymbol{a})\frac{\partial R}{\partial a_i}\,dt = 0 \tag{9-11}$$

Table 9-1: Weighting functions in Eq. (9-12) for different finite element methods.

Method		Weighting function
Collocation	$R(t_i; \boldsymbol{a}) = 0$	$W_i(t) = \delta(t_i)$
Subdomain	$\frac{1}{\Delta t_i} \int_{\Delta t_i} R(t; \boldsymbol{a})\, dt = 0$	$W_i(t) = \begin{cases} 1 & t \text{ within } \Delta t_i \\ 0 & \text{otherwise} \end{cases}$
Least-squares	$\int_{t_b}^{t_d} R(t; \boldsymbol{a}) \frac{\partial R}{\partial a_i} dt = 0$	$\frac{\partial}{\partial a_i} R(t; \boldsymbol{a})$
Galerkin	$\int_{t_b}^{t_d} R(t; \boldsymbol{a}) \psi_i(t)\, dt = 0$	$\psi_i(t)$

where $i = 1, 2, \ldots, n$. For our $n = 2$ approximation, the values of a_1 and a_2 produced are different from those found using the subdomain method; however, the calculated results for $\phi(t)$ are very close to each other. Derivation of the necessary formulas and the details of the calculations are left as an exercise.

All three methods are very similar to each other. They belong to a class of more general methods called weighted residual methods. In terms of the integral

$$\int_{t_b}^{t_d} R(t; \boldsymbol{a}) W_i(t)\, dt = 0 \tag{9-12}$$

for $i = 1, 2, \ldots, n$, the only difference between the three methods is in the weighting function $W_i(t)$. In the collocation method, the weighting function is a delta function $\delta(t_i)$. Since, for a delta function,

$$\int_{-\infty}^{\infty} f(x)\, \delta(x - x_0)\, dx = f(x_0)$$

the integral of Eq. (9-12) can be carried out explicitly and the result is Eq. (9-7). The subdomain method, on the other hand, uses a step function as the weight, and the least-squares method, the partial derivative $\partial R/\partial a_i$. All three methods, together with the Galerkin method to be discussed next, are summarized in Table 9-1.

Other forms of the weighting function are possible. In fact, most finite element calculations these days follow the Galerkin method which uses parts of the trial solution themselves as $W_i(t)$. To see this, let us express the trial solution in a more general way. Instead of the power series used in Eq. (9-4), we can expand $\tilde{\phi}_n(t)$ in terms of $(n + 1)$ linearly independent functions $\psi_0(t)$, $\psi_1(t)$, $\psi_2(t), \ldots, \psi_n(t)$. These may be regarded as a set of basis functions with which we can express our trial solution:

$$\tilde{\phi}_n(t) = \psi_0(t) + \sum_{i=1}^{n} a_i \psi_i(t). \tag{9-13}$$

It is convenient to treat the first term $\psi_0(t)$ on the right side slightly differently from the others and consider it as a function which incorporates part of, or all, the initial and boundary conditions of the problem. In this way, it enters $\tilde{\phi}_n(t)$ without having a parameter associated with it. This is equivalent to putting into $\psi_0(t)$ everything that can be fixed by the initial and boundary conditions. For the $n = 2$ trial function in our example of Eq. (9-1), one possible choice is to take

$$\psi_0(t) = 1 \qquad \psi_1(t) = t \qquad \psi_2(t) = t^2. \tag{9-14}$$

A different set of basis functions may actually be better here, but this is not the place to pursue the topic. Our main interest is to see the way that the basis functions $\{\psi_i(t)\}$ may be used as the weighting functions.

The condition of Eq. (9-12) now takes on the form

$$\int_{t_b}^{t_d} R(t; \boldsymbol{a})\psi_i(t)\, dt = 0 \tag{9-15}$$

for $i = 1, 2, \ldots, n$. From these n equations, we obtain a set of values for the n parameters, in the same way as we did in Eqs. (9-7), (9-9), and (9-11) for, respectively, the collocation, subdomain, and least-squares methods.

For our $n = 2$ approximation example, using the basis functions provided by Eq. (9-14), the two Galerkin equations are

$$\begin{aligned}
\int_0^1 \{1 + a_1(1+t) + a_2(2t+t^2)\}t\, dt &= \frac{1}{2} + \frac{5}{6}a_1 + \frac{11}{12}a_2 = 0 \\
\int_0^1 \{1 + a_1(1+t) + a_2(2t+t^2)\}t^2\, dt &= \frac{1}{3} + \frac{7}{12}a_1 + \frac{7}{10}a_2 = 0.
\end{aligned}$$

The values of a_1 and a_2 satisfying this set of algebraic equation are, respectively, $-32/35$ and $2/7$. The resulting solution to the differential equation is then

$$\phi_2^{\mathrm{G}}(t) = 1 - \frac{32}{35}t + \frac{2}{7}t^2. \tag{9-16}$$

Within the domain of interest, the results are very close to those of the exact solution of Eq. (9-4), not too different from those for the collocation method shown in Fig. 9-1.

9-2 Shape functions and finite element approximation

Let us turn our attention to the question of applying finite element methods to solve partial differential equations. In finite difference methods, we have seen in the previous chapter that the first step is to discretize the space into a mesh of nodes. The equivalent step here is to divide the space into small elements. Once this is done, part of the work can be carried out within an element. This is our main function in this section and we shall defer till later the step of joining them together to form the equations for the complete solution.

As illustrations, we shall use examples with two and three independent variables. To simplify the discussion, we shall confine ourselves to shapes with straight lines

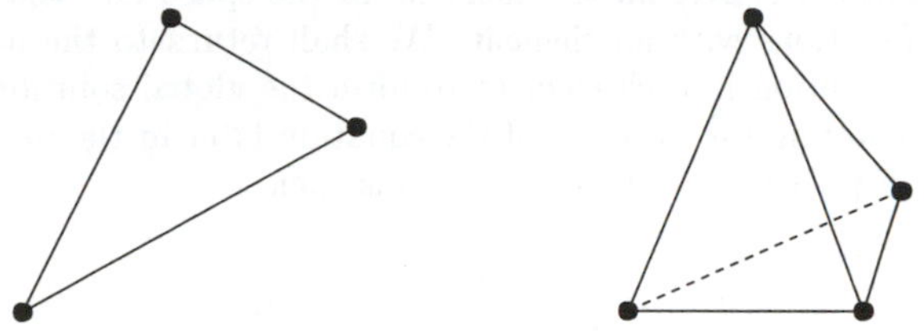

Figure 9-2: Example of finite element shapes: (a) triangle in two dimensions and (b) tetrahedron in three dimensions.

as edges. For such cases, the element in two dimensions is a surface and the most elementary shape is a triangle. Other forms, such as rectangles and polygons, can also be used but we shall not mention them here. In three dimensions, the corresponding quantity is a tetrahedron, as illustrated in Fig. 9-2.

In finite difference methods, one of the keys to better accuracy is to reduce the distance between neighboring points in the space. Here, the corresponding requirement is to take the elements as small as possible. In this way, simple functions, called *shape functions*, may be used to approximate the solution within an element. Let us see what are the simplest forms we can use.

In terms of power series, we find that the minimum order for shape functions is one lower than the order of the differential equation. This can be see from the Galerkin equation Eq. (9-15). As a concrete example, consider the two-dimensional Poisson equation of Eq. (8-78)

$$\frac{\partial^2 V}{dx^2} + \frac{\partial^2 V}{dy^2} = -\rho(x, y). \tag{9-17}$$

Let $\ell_i^{(m)}(x, y)$ represent the ith shape function in element m. The trial solution $\tilde{V}(x, y)$ within this element may be written as

$$V^{(m)}(x, y) = \sum_i v_i^{(m)} \ell_i^{(m)}(x, y)$$

where $v_i^{(m)}$ are unknown coefficients at this stage. Since the shape functions are known functions (and we shall see how to define them later), the solution for $V(x, y)$ in element m is determined once the values of coefficients $v_i^{(m)}$ are found. Let us represent the complete finite element solution in the whole space as

$$\tilde{V}(x, y) = \sum_m \sum_i v_i^{(m)} \ell_i^{(m)}(x, y) \tag{9-18}$$

where the first summation is over all the elements in the space and the second one over all the shapes functions with an element. We shall return to the mechanics of putting together the solution in each element to form the global solution as well as some caution in interpreting the meaning of the equation later in the next section.

The residual for Eq. (9-17) in terms of $\tilde{V}(x,y)$ is then

$$R(x,y;\boldsymbol{v}) = \frac{\partial^2 \tilde{V}}{dx^2} + \frac{\partial^2 \tilde{V}}{dy^2} + \rho(x,y).$$

The Galerkin equation of Eq. (9-15), with the ith shape function as the weight, may now be written in the form

$$\begin{aligned}\int\!\!\int \ell_i^{(m)}(x,y)R(x,y;\boldsymbol{v})\,dxdy &= \int\!\!\int_m \ell_i^{(m)}(x,y)\left\{\frac{\partial^2 \tilde{V}}{dx^2} + \frac{\partial^2 \tilde{V}}{dy^2}\right\} dxdy \\ &\quad + \int\!\!\int_m \ell_i^{(m)}(x,y)\rho(x,y)\,dxdy. \qquad (9\text{-}19)\end{aligned}$$

The surface integrals here need only to be carried out over a single element, as the shape functions $\ell_i^{(m)}(x,y)$ are for element m and vanish everywhere outside. We can integrate the first term on the right side by parts and reduce the order of partial derivatives from two to one. This can be seen by working out the part containing the $\partial^2\tilde{V}/dx^2$ term. Since

$$\frac{\partial}{\partial x}\left\{\ell_i^{(m)}(x,y)\frac{\partial \tilde{V}}{dx}\right\} = \frac{\partial \ell_i^{(m)}}{\partial x}\frac{\partial \tilde{V}}{dx} + \ell_i^{(m)}(x,y)\,\frac{\partial^2 \tilde{V}}{dx^2}$$

we have

$$\begin{aligned}\int\!\!\int_m \ell_i^{(m)}(x,y)\frac{\partial^2 \tilde{V}}{dx^2}\,dxdy &= \int\!\!\int_m \frac{\partial}{\partial x}\left\{\ell_i^{(m)}(x,y)\frac{\partial \tilde{V}}{dx}\right\} dxdy \\ &\quad - \int\!\!\int_m \frac{\partial \ell_i^{(m)}}{\partial x}\frac{\partial \tilde{V}}{dx}\,dxdy. \qquad (9\text{-}20)\end{aligned}$$

The first integral on the right hand side can be transformed into a line integral using the divergence theorem,

$$\int\!\!\int_m \frac{\partial}{\partial x}\left\{\ell_i^{(m)}(x,y)\frac{\partial \tilde{V}}{dx}\right\} dxdy = \oint_m \ell_i^{(m)}(x,y)\,\frac{\partial \tilde{V}}{dx}\hat{\boldsymbol{n}}_x\,dxdy \qquad (9\text{-}21)$$

where $\hat{\boldsymbol{n}}_x$ is the x-component of the normal vector to the boundary of element m. As we shall see in the next section, contributions from this term cancel with similar ones coming from neighboring elements. The only exceptions are elements at the boundary of the whole space. For this reason, we can ignore them for the time being and come back to them in conjunction with the boundary conditions imposed on the problem.

The same integration by parts can be carried out for the $\partial^2\tilde{V}/dy^2$ term as well and Eq. (9-19) reduces to

$$\begin{aligned}\int\int \ell_i^{(m)}(x,y)R(x,y;\boldsymbol{v})\,dxdy &= -\int\int_m \left\{\frac{\partial \ell_i^{(m)}}{\partial x}\frac{\partial \tilde{V}}{dx} + \frac{\partial \ell_i^{(m)}}{\partial y}\frac{\partial \tilde{V}}{dy}\right\} dxdy \\ &\quad + \int\int_m \ell_i^{(m)}(x,y)\rho(x,y)\,dxdy + \text{b.c.} \qquad (9\text{-}22)\end{aligned}$$

where b.c. reminds us of terms involving boundary conditions that must be included at a later stage. Since $\rho(x,y)$ is a known function, the second integral on the right hand side can be carried out explicitly once the shape functions are determined. The first integral can also be written in terms of the shape functions using Eq. (9-18),

$$\begin{aligned}\int\int_m &\left\{\frac{\partial \ell_i^{(m)}}{\partial x}\frac{\partial \tilde{V}}{dx} + \frac{\partial \ell_i^{(m)}}{\partial y}\frac{\partial \tilde{V}}{dy}\right\} dxdy \\ &= \sum_j v_j^{(m)} \int\int_m \left\{\frac{\partial \ell_i^{(m)}}{\partial x}\frac{\partial \ell_j^{(m)}}{dx} + \frac{\partial \ell_i^{(m)}}{\partial y}\frac{\partial \ell_j^{(m)}}{dy}\right\} dxdy. \qquad (9\text{-}23)\end{aligned}$$

We see that the Galerkin equation is now reduced to integrals involving shape functions. The condition that the weighted integrals of the residual must vanish gives us a set of equations for the the unknown coefficients $v_i^{(m)}$.

Let us return now to the question on the form of shape functions. We see from Eq. (9-23) that, for our second-order PDE example, $\ell_i^{(m)}(x,y)$ must be at least linear in x and y; otherwise the derivatives vanish and we cannot construct an equation to solve for $v_i^{(m)}$. The reduction by one from the order of the differential equation comes from integration by parts carried out in Eq. (9-20). Thus, for PDE of order n, partial derivatives up to order $(n-1)$ are involved and the minimum power series order for the shape functions is $(n-1)$. The conclusion is not changed if we have three or higher dimensional PDE. The only difference is that, for three independent variables we have, instead of line integrals in Eq. (9-21), surface integrals whose contributions cancel with those from neighboring elements except those at the boundary.

For our Poisson equation example, the simplest form for shape functions is, therefore, linear and, in two-dimensional space, there are at most three parameters for such a function. For example,

$$\ell_i^{(m)}(x,y) = a_i^{(m)} + b_i^{(m)}x + c_i^{(m)}y. \qquad (9\text{-}24)$$

As a result, we can have a maximum of three linearly independent functions of this form within each element and we shall take them to be $\ell_i^{(m)}(x,y)$ for $i = 1$ to 3. (In three dimensions, the corresponding number is four.) There are several ways to define the values of coefficients $a_i^{(m)}$, $b_i^{(m)}$, and $c_i^{(m)}$ for element m. If we use triangular shapes for our elements, the most convenient way is to take them such that each one of the three linearly independent shape functions is equal to unity at one of three vertices and zero at the other two. That is

$$\ell_i^{(m)}(x_j, y_j) = \begin{cases} 1 & \text{for } i = j \\ 0 & \text{otherwise} \end{cases} \tag{9-25}$$

where (x_j, y_j), for $j = 1$, 2, and 3, are the coordinates for the three vertices of triangular element m. The advantage of this choice is that $v_i^{(m)}$ corresponds to the value for $V(x, y)$ at (x_i, y_i), the location of vertex i for element m, as can be seen from Eq. (9-18).

For this choice, the shape functions $\ell_i^{(m)}(x, y)$ and, hence, the coefficients $a_i^{(m)}$, $b_i^{(m)}$ and $c_i^{(m)}$, are given by

$$\ell_1^{(m)} = \frac{1}{D^{(m)}}\begin{vmatrix} 1 & x & y \\ 1 & x_2 & y_2 \\ 1 & x_3 & y_3 \end{vmatrix} \qquad \ell_2^{(m)} = \frac{1}{D^{(m)}}\begin{vmatrix} 1 & x_1 & y_1 \\ 1 & x & y \\ 1 & x_3 & y_3 \end{vmatrix} \qquad \ell_3^{(m)} = \frac{1}{D^{(m)}}\begin{vmatrix} 1 & x_1 & y_1 \\ 1 & x_2 & y_2 \\ 1 & x & y \end{vmatrix} \tag{9-26}$$

where the denominator is the determinant

$$D^{(m)} = \begin{vmatrix} 1 & x_1 & y_1 \\ 1 & x_2 & y_2 \\ 1 & x_3 & y_3 \end{vmatrix}.$$

Note that both $\ell_i^{(m)}(x, y)$ and D are given in terms of the coordinates of the vertices and are therefore completely determined once element m is specified.

Using the properties of determinants, it is easy to verify that the definition satisfies the condition given by Eq. (9-25). For three-dimensional problems, the determinants are 4×4, rather than 3×3 here. It is also possible to define quadratic and more complicated shape functions in a similar manner and they can be found in most books specializing on finite element methods.

Another good reason to use shape functions defined in the way above is the relation

$$\frac{1}{A^{(m)}}\int\int \left\{\ell_1^{(m)}(x, y)\right\}^p \left\{\ell_2^{(m)}(x, y)\right\}^q \left\{\ell_3^{(m)}(x, y)\right\}^r dxdy = \frac{p!q!r!}{(p + q + r + 2)!} \tag{9-27}$$

where

$$A^{(m)} = \frac{1}{2}D^{(m)}$$

is the area of the triangular element. Since integrals involving products of shape functions form an important part of the calculations within an element, the formula is an useful one. In three dimensional spaces, the volume of an element is given by

$$V^{(m)} = \frac{1}{3!}D^{(m)} = \frac{1}{3!}\begin{vmatrix} 1 & x_1 & y_1 & z_1 \\ 1 & x_2 & y_2 & z_2 \\ 1 & x_3 & y_3 & z_3 \\ 1 & x_4 & y_4 & z_4 \end{vmatrix}$$

where (x_i, y_i, z_i), for $i = 1$ to 4, are the coordinates of the vertices of a tetrahedron. The four shape functions in such an element can be defined in a similar way as Eq. (9-26). Corresponding to Eq. (9-27), we have

$$\frac{1}{V}\int\int\left\{\ell_1^{(m)}\right\}^p\left\{\ell_2^{(m)}\right\}^q\left\{\ell_3^{(m)}\right\}^r\left\{\ell_4^{(m)}\right\}^s dxdydz = \frac{p!q!r!s!}{(p+q+r+s+3)!}. \tag{9-28}$$

Similar relation exists in higher-dimensional space and for higher-order shape functions.

Using shape functions defined by Eq. (9-26), the Galerkin equation for element m and weight $\ell_i^{(m)}(x,y)$ for our simple illustrative example of Eq. (9-17) reduces to

$$\sum_j v_j^{(m)}\left\{b_i^{(m)}b_j^{(m)} + c_i^{(m)}c_j^{(m)}\right\} - \int\int_m \ell_i^{(m)}(x,y)\rho(x,y)\,dxdy = 0. \tag{9-29}$$

The first term involves only coefficients of the shape functions and they are given by the coordinates of the three vertices of the element. The integral in second term can be evaluated, as the source term $\rho(x,y)$ must be provided to us as a part of the statement of the problem. The only unknown quantities are $v_i^{(m)}$. However, we cannot solve for them until we assemble together similar equations for the other elements.

9-3 Assembling contributions from elements

The Galerkin equations for element m given in Eq. (9-29) are not the only ones that $v_j^{(m)}$ must satisfy. This comes from the fact that, for a given j, coefficient $v_j^{(m)}$ is the value of the solution $V(x,y)$ at the location of vertex j for element m. However, vertices in neighboring elements overlap each other, as each node in space is usually the vertex of several elements. This is illustrated schematically in Fig. 9-3 for two neighboring elements. Globally, there are four vertices in this part of the space. Let us assume they are labeled as 1, 2, 3, and 4, as shown in (a). In each one of the two neighboring triangular elements α and β, there are three vertices and they are labeled as α_1, α_2, α_3, and β_1, β_2, β_3, as shown in (b). When we put the weighted residual to zero in element α alone, we have a set of three relations among v_{α_1}, v_{α_2}, and v_{α_3} in the form of Eq. (9-29). Similarly, we have a set of three relations among v_{β_1}, v_{β_2}, and v_{β_3} when we deal with element β. However, since α_1 and β_1 refer to the same point in space, v_{α_1} is in fact the same quantity as v_{β_1}. Hence, the conditions that must be satisfied by v_{α_1} are also the constraints on v_{β_1}. The same restriction applies between α_3 and β_2, as well. This is true for essentially all the vertices since each element is bordered by several others. The only exceptions are a handful of points along the boundary.

We can regard the process of "assembling" the contributions from individual elements to form the global picture for the whole space as a transformation of the vertices from their labeling scheme in individual elements to the global one. This may be illustrated using again the two-element example in Fig. 9-3. If we consider

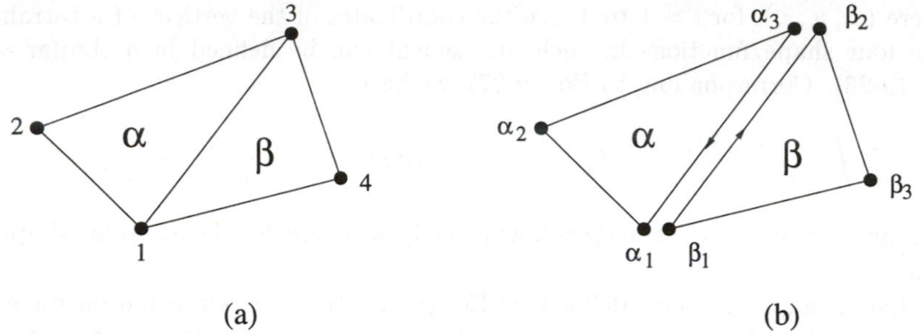

Figure 9-3: Illustration of vertices shared by two neighboring triangular elements.

the labeling scheme in (a) as a column matrix with four elements 1, 2, 3, and 4, and those in (b) as a column matrix with six elements α_1, α_2, α_3, β_1, β_2, and β_3, the relation between them is given by

$$\begin{pmatrix} \alpha_1 \\ \alpha_2 \\ \alpha_3 \\ \beta_1 \\ \beta_2 \\ \beta_3 \end{pmatrix} = \mathcal{T} \begin{pmatrix} 1 \\ 2 \\ 3 \\ 4 \end{pmatrix}. \tag{9-30}$$

The transformation matrix $\mathcal{T}$ is rectangular, with four columns and six rows,

$$\mathcal{T} = \begin{pmatrix} 1 & 0 & 0 & 0 \\ 0 & 1 & 0 & 0 \\ 0 & 0 & 1 & 0 \\ 1 & 0 & 0 & 0 \\ 0 & 0 & 1 & 0 \\ 0 & 0 & 0 & 1 \end{pmatrix}.$$

The three Galerkin equations for element α may be written in matrix form as

$$\begin{pmatrix} S_{11}^{(\alpha)} & S_{12}^{(\alpha)} & S_{13}^{(\alpha)} \\ S_{21}^{(\alpha)} & S_{22}^{(\alpha)} & S_{23}^{(\alpha)} \\ S_{31}^{(\alpha)} & S_{32}^{(\alpha)} & S_{33}^{(\alpha)} \end{pmatrix} \begin{pmatrix} v_{\alpha_1} \\ v_{\alpha_2} \\ v_{\alpha_3} \end{pmatrix} - \begin{pmatrix} g_1^{(\alpha)} \\ g_2^{(\alpha)} \\ g_3^{(\alpha)} \end{pmatrix} = 0.$$

where we have defined

$$S_{i,j}^{(\alpha)} \equiv \left\{ b_i^{(\alpha)} b_j^{(\alpha)} + c_i^{(\alpha)} c_j^{(\alpha)} \right\} A^{(\alpha)} \qquad g_i^{(\alpha)} \equiv \int\int_m \ell_i^{(\alpha)} \rho(x, y)\, dxdy \tag{9-31}$$

and $A^{(\alpha)}$ is the area of element α. Similarly, the three Galerkin equations for element β can be written in an analogous manner. On combining the two elements, we have a six-dimensional matrix equation

$$\begin{pmatrix} S_{11}^{(\alpha)} & S_{12}^{(\alpha)} & S_{13}^{(\alpha)} & 0 & 0 & 0 \\ S_{21}^{(\alpha)} & S_{22}^{(\alpha)} & S_{23}^{(\alpha)} & 0 & 0 & 0 \\ S_{31}^{(\alpha)} & S_{32}^{(\alpha)} & S_{33}^{(\alpha)} & 0 & 0 & 0 \\ 0 & 0 & 0 & S_{11}^{(\beta)} & S_{12}^{(\beta)} & S_{13}^{(\beta)} \\ 0 & 0 & 0 & S_{21}^{(\beta)} & S_{22}^{(\beta)} & S_{23}^{(\beta)} \\ 0 & 0 & 0 & S_{31}^{(\beta)} & S_{32}^{(\beta)} & S_{33}^{(\beta)} \end{pmatrix} \begin{pmatrix} v_{\alpha 1} \\ v_{\alpha 2} \\ v_{\alpha 3} \\ v_{\beta 1} \\ v_{\beta 2} \\ v_{\beta 3} \end{pmatrix} - \begin{pmatrix} g_1^{(\alpha)} \\ g_2^{(\alpha)} \\ g_3^{(\alpha)} \\ g_1^{(\beta)} \\ g_2^{(\beta)} \\ g_3^{(\beta)} \end{pmatrix} = 0. \qquad (9\text{-}32)$$

However, as it stands, this equation is incomplete, since the unknown quantities are not independent of each other.

When we put the two triangular elements together, we have only the four vertices of 1, 2, 3, and 4. Similarly, the unknown quantities are v_1, v_2, v_3, and v_4. In terms of these four unknowns, we see that the correct form of Eq. (9-32) is

$$\begin{pmatrix} S_{11}^{(\alpha)} + S_{11}^{(\beta)} & S_{12}^{(\alpha)} & S_{13}^{(\alpha)} + S_{12}^{(\beta)} & S_{13}^{(\beta)} \\ S_{21}^{(\alpha)} & S_{22}^{(\alpha)} & S_{23}^{(\alpha)} & 0 \\ S_{31}^{(\alpha)} + S_{21}^{(\beta)} & S_{33}^{(\alpha)} & S_{33}^{(\alpha)} + S_{22}^{(\beta)} & S_{23}^{(\beta)} \\ S_{31}^{(\beta)} & 0 & S_{32}^{(\beta)} & S_{33}^{(\beta)} \end{pmatrix} \begin{pmatrix} v_1 \\ v_2 \\ v_3 \\ v_4 \end{pmatrix} - \begin{pmatrix} g_1^{(\alpha)} + g_1^{(\beta)} \\ g_2^{(\alpha)} \\ g_3^{(\alpha)} + g_2^{(\beta)} \\ g_3^{(\beta)} \end{pmatrix} = 0. \qquad (9\text{-}33)$$

The result can also be obtained formally using the transformation matrix $\mathcal{T}$. If we write the six-dimensional matrix with elements $S_{ij}^{(\alpha)}$ and $S_{ij}^{(\beta)}$ symbolically as $\boldsymbol{S}^{(6)}$ and the corresponding four-dimension matrix as $\boldsymbol{S}^{(4)}$, we see that

$$\boldsymbol{S}^{(4)} = \tilde{\mathcal{T}} \boldsymbol{S}^{(6)} \mathcal{T}$$

where $\tilde{\mathcal{T}}$ is the transpose of $\mathcal{T}$.

We see also in Fig. 9-3 a graphical illustration of the cancelation between contributions coming from the line integral in Eq. (9-21) in neighboring elements. By adopting a consistent direction of integration for all the elements, clockwise in Fig. 9-3(*b*), the directions of integration are opposite in neighboring elements along the shared border. As a result, contributions from the line integral of two such elements have opposite signs and their contributions cancel each other.

The assembly of Galerkin equations for two neighboring elements we have shown above as an example must be carried out for all the elements in the space. Consider the simple case of a rectangular space divided into N_x segments along the x direction and N_y segments along the y direction, as shown for example later in Fig. 9-4. (For the example in Fig. 9-5, we have $N_x = N_y = 2$.) The number of small rectangles is $N_x \times N_y$ and the number of triangular elements is twice this number. Apart from boundary conditions, which we shall ignore for the moment, the number of Galerkin equations from all the triangular elements is $6N_xN_y$ and the number of v_i in the solution is $(N_x + 1)(N_y + 1)$, given by the number of mesh points in the space. The assembly step reduces the number of equations from $6N_xN_y$ to the same as the number of unknowns, as we have seen above in the two-element example. Since the number of elements is usually quite large, it is not feasible to carry out the assembly by constructing explicitly the $(N_x+1)(N_y+1)$ by $6N_xN_y$ dimensional transformation matrix $\mathcal{T}$. In practice, the transformation is sufficiently straightforward that the assembly process can be done by inspection. In the next section, we shall see another method by which the assembly step becomes redundant, at least conceptually.

The final algebraic equation we obtain in a finite element calculation has the form

$$\boldsymbol{Sv} = \boldsymbol{g} \tag{9-34}$$

where $\boldsymbol{S}$ is a square matrix consisting of (sums over) $S_{i,j}^{(m)}$, not too different from the form we have seen in Eq. (9-33) above. Similarly, $\boldsymbol{g}$ is a column matrix made of $g_i^{(m)}$. The unknown quantities $\{v_i\}$ are the elements of column matrix $\boldsymbol{v}$ and they represent the values of the solution at the mesh points. The algebraic equation has essentially the same form as the ones we obtained with finite difference methods in the previous chapter and can, therefore, be solved by the same methods.

9-4 Variational approach

There are several advantages in taking a variational approach to finite element calculations, rather than the Galerkin method used in the previous two sections. In the first place, it has a closer contact with the nature of the physical problem. Instead of the differential equation, the starting point is often the lagrangian, or some other conserved quantity in the problem, from which the PDE is derived. In the second place, we do not have to view the steps leading to the algebraic equation Eq. (9-34) as two separate ones, one working within each of the elements and the other in assembling the contributions from the elements.

We have seen that, in a finite element approach to PDE, the solution is expressed in terms of shape functions in each one of the elements. To take advantage of a variational approach, we shall modify Eq. (9-18) slightly and write it as

$$V(x,y) = \sum_m \sum_{i=1}^{3} v_i^{(m)} \ell_i^{(m)}(x,y). \tag{9-35}$$

We should be a little careful in interpreting this equation. If the elements do not overlap each other, there is no ambiguity. However, neighboring elements share nodes, as each one is, in general, the vertex of several elements. As we saw earlier in the simple example of Fig. 9-3, node number 3 is vertex α_3 of element α as well as vertex β_2 of element β. Since, at the vertex of an element, the value of one of the shape functions is unity and the others zero, we have the relation

$$v_{\alpha_3}^{\alpha} = v_{\beta_2}^{\beta} = v_3$$

where $V(x_3, y_3) = v_3$ and (x_3, y_3) is the location of node number 3. As a result, when we carry out the summation over m in Eq. (9-35) for $m = \alpha$ and β, the value of

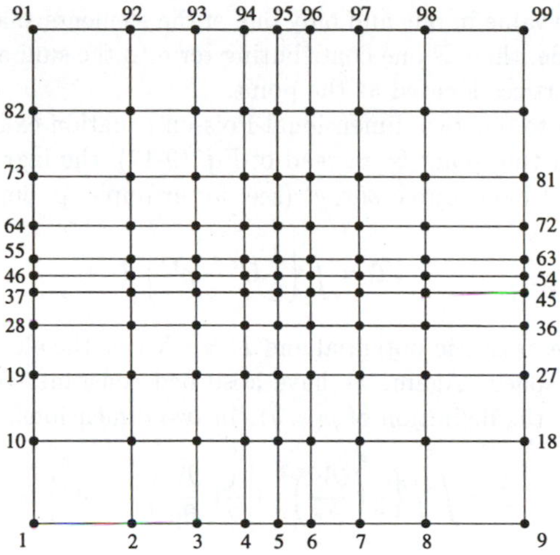

Figure 9-4: Example of node numbering.

$V(x,y)$ at (x_3, y_3) must remain to be v_3. In other words, at nodes, the summation over m must be interpreted such that it is the value in one and only one of the elements sharing the node. Since the value is the same for all the elements sharing the node, it does not matter which element we take.

For the convenience of discussion here, let us use a different method to label the vertices. As we have seen in the previous section, the scheme in Eq. (9-35) is redundant, as the same point in space is referred to differently in different elements (see Fig. 9-3 for example). To resolve this problem we can go back to the labeling scheme used in finite difference methods. One obvious choice is to number the nodes sequentially as done, for example, in Fig. 9-4. Let us represent the value of $V(x,y)$ at node n by the symbol v_n. In other words,

$$v_n = V(x_n, y_n)$$

where (x_n, y_n) are the coordinates of node n in our two-dimensional example. It is easy to see that we can construct a translation table that converts n to the corresponding vertex number in each one of the elements that have node n as one of its vertices. Formally, we can regard n as a function of the element number m and the vertex number i in element m. As a result, Eq. (9-35) may be rewritten in the form

$$V(x,y) = \sum_m \sum_{i=1}^{3} v_{n(m,i)} \ell_i^{(m)}(x,y) \tag{9-36}$$

to emphasize the global nature of the value of $V(x,y)$ at the vertices. Again, we shall interpret the summation over m such that the value of $V(x,y)$ at a node n is v_n, with v_n being the value in one and only one of the elements sharing point n. In other words, at any node, there is one contributing term to the sum among all the elements with one of its vertices located at the point.

Let us return to our two-dimensional Poisson equation example. In a variational approach, our starting point is, instead of Eq. (9-17), the lagrangian for the electrostatic field created by a source $\rho(x, y)$ (see, for example, p. 366 of Goldstein[25]).

$$\mathcal{L} = \int \left\{ \frac{1}{2} \boldsymbol{E}^2 - \rho V \right\} d\tau \tag{9-37}$$

where V is the electrostatic potential and $\boldsymbol{E} = -\nabla V$ is the electric field. The integral is taken over all space. Again, we have absorbed constants such as ϵ_0 and possible factors of 4π into the definition of $\rho(x, y)$. In two dimensions, we have

$$\mathcal{L} = \int\int \left\{ \frac{1}{2}\left(\frac{\partial V}{\partial x}\right)^2 + \frac{1}{2}\left(\frac{\partial V}{\partial y}\right)^2 - \rho V \right\} dxdy. \tag{9-38}$$

One way to see that this is indeed the lagrangian for our example is to derive Eq. (9-17) using the Lagrange's equation which, in this case, takes on the form

$$\frac{\partial}{\partial x}\frac{\delta \mathcal{L}}{\delta\left(\frac{\partial V}{\partial x}\right)} + \frac{\partial}{\partial y}\frac{\delta \mathcal{L}}{\delta\left(\frac{\partial V}{\partial y}\right)} - \frac{\delta \mathcal{L}}{\delta V} = 0. \tag{9-39}$$

Inserting $\mathcal{L}$ of Eq. (9-38), the result is

$$\int\int \left\{ \frac{\partial^2 V}{\partial x^2} + \frac{\partial^2 V}{\partial y^2} + \rho(x, y) \right\} dxdy = 0.$$

We recover Eq. (9-17), as the integrand must vanish everywhere in the space.

For our finite element solution using a variational approach, we shall proceed in a slightly different way. Instead of varying V, as we did in Eq. (9-39), we shall first expand V in terms of v_n using Eq. (9-36) and adopt $\{v_n\}$ as our variational parameters. On substituting Eq. (9-36) into (9-38), the lagrangian takes on the form

$$\begin{aligned}\mathcal{L} \;=\; \int\int \Big\{ & \frac{1}{2}\Big(\sum_m \sum_i v_{n(m,i)} b_i^{(m)}\Big)\Big(\sum_{m'} \sum_{i'} v_{n(m',i')} b_{i'}^{(m')}\Big) \\ & + \frac{1}{2}\Big(\sum_m \sum_i v_{n(m,i)} c_i^{(m)}\Big)\Big(\sum_{m'} \sum_{i'} v_{n(m',i')} c_{i'}^{(m')}\Big) \\ & - \rho(x, y) \sum_m \sum_i v_{n(m,i)} \ell_i^{(m)}(x, y) \Big\} dxdy.\end{aligned}$$

Since the lagrangian is a constant of motion, we have the condition

$$\delta \mathcal{L} = 0.$$

Explicitly for our example, this means

$$\sum_m \sum_i \int\int \Big\{ \Big(\sum_{m'} \sum_{i'} v_{n(m',i')} b_{i'}^{(m')} \Big) b_i^{(m)} + \Big(\sum_{m'} \sum_{i'} v_{n(m',i')} c_{i'}^{(m')} \Big) c_i^{(m)} - \rho(x,y) \ell_i^{(m)}(x,y) \Big\} \delta v_{n(m,i)} \, dxdy = 0. \tag{9-40}$$

For this to be true for arbitrary variations $\delta v_{n(m,i)}$, we obtain the condition

$$\sum_m{}' \sum_i \int\int \Big\{ b_i^{(m)} \sum_{m'} \sum_{i'} v_{n(m',i')} b_{i'}^{(m')} + c_i^{(m)} \sum_{m'} \sum_{i'} v_{n(m',i')} c_{i'}^{(m')} - \rho(x,y) \ell_i^{(m)}(x,y) \Big\} dxdy = 0.$$

The summation over m is now confined to be among elements that include n as one of the vertices and we indicate this restriction using a prime in the summation symbol.

The equation can be further simplified using the fact that the coefficients for the shape functions $a_i^{(m)}$, $b_i^{(m)}$, and $c_i^{(m)}$ vanish outside element m. As a result, the only contribution to the integral comes from those elements with $m' = m$. Furthermore, the integration needs to be taken only over element m. The net result is

$$\sum_m{}' \sum_i \int\int \Big\{ \sum_{i'} v_{n(m,i')} b_{i'}^{(m)} b_i^{(m)} + \sum_{i'} v_{n(m,i')} c_{i'}^{(m)} c_i^{(m)} - \rho(x,y) \ell_i^{(m)}(x,y) \Big\} \, dxdy = 0.$$

This can be put into the form of an algebraic equation for v_n,

$$\sum{}'_m \sum_i \int\int \Big\{ \sum_{i'} b_{i'}^{(m)} b_i^{(m)} + \sum_{i'} c_{i'}^{(m)} c_i^{(m)} \Big\} dxdy \; v_{n(m,i')} = \sum{}'_m \sum_i \int\int \rho(x,y) \ell_i^{(m)}(x,y) \, dxdy. \tag{9-41}$$

This is the same as Eq. (9-34). It differs from Eq. (9-29) as we have already (implicitly) assembled the contributions from all the elements.

9-5 Application to a two-dimensional Poisson equation

Let us carry out a calculation for the two-dimensional Poisson equation of Eq. (9-17) to illustrate some of the considerations in applying a finite element method. For the boundary conditions, we shall take

$$V(x,y) = 0 \qquad \text{for } x = \pm L_x \quad \text{or} \quad y = \pm L_y. \tag{9-42}$$

As the source term, we shall consider

$$\rho(x,y) = \rho_0 \, \delta_{x,0} \, \delta_{y,0}. \tag{9-43}$$

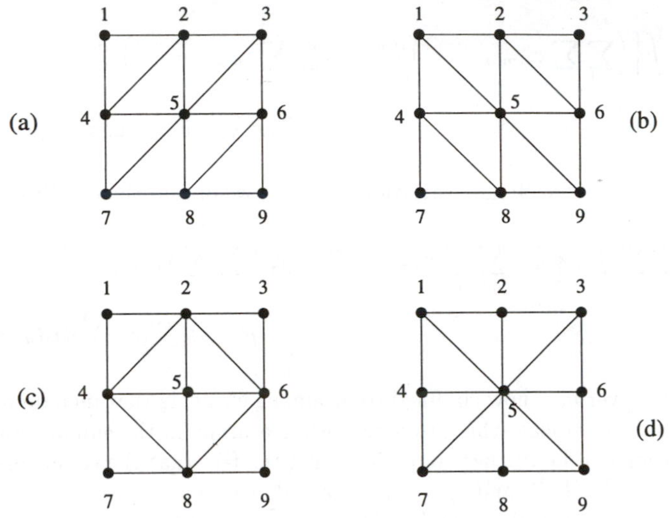

Figure 9-5: Different ways of assigning triangular elements in a two-dimensional space.

This corresponds to the situation of an infinite line of charge in the z-direction in the center of a hollow rectangular conducting pipe. Similar to the example in §8-6, the pipe is grounded, as given by Eq. (9-42), and has dimension $2L_x$ by $2L_y$. Since the problem is invariant along the z-direction, there is no z-dependence in the potential and we have a two-dimensional PDE in (x, y).

Dividing the space into elements For this two-dimensional problem we shall use triangles as the elements. Before going to triangles, it is convenient to reduce our rectangular active space first into small rectangles. This can be done by dividing the line between $x = -L_x$ to $+L_x$ into N_x segments. If we use x_i to represent the location separating two neighboring segments i and $i+1$, we have a total of $(N_x + 1)$ such points located at $(x_0, x_1, \cdots, x_{N_x})$, including the two at the boundary, $x_0 \equiv -L_x$ and $x_{N_x} \equiv +L_x$. Similarly, by defining $y_0 \equiv -L_y$ and $y_{N_y} \equiv +L_y$, the y direction may be divided into N_y segments at $(y_0, y_1, \cdots, y_{N_y})$. Thus the (x, y)-coordinates of a node are (x_{i-1}, y_{j-1}), if it is the ith node (from the left) in the jth row (from the bottom). Note that there is no need to make all the segments the same length. In fact, we shall retain the flexibility to adjust the size of each segment to illustrate later certain aspects of numerical accuracy in the solution.

To get to triangular shapes for our elements, we can divide each small rectangle into two triangles by making a diagonal cut. There are basically two choices here. The first is to make all the diagonal cuts in the same direction, as illustrated in (a) and (b) of Fig. 9-5. The result of adopting this scheme is schematically illustrated in

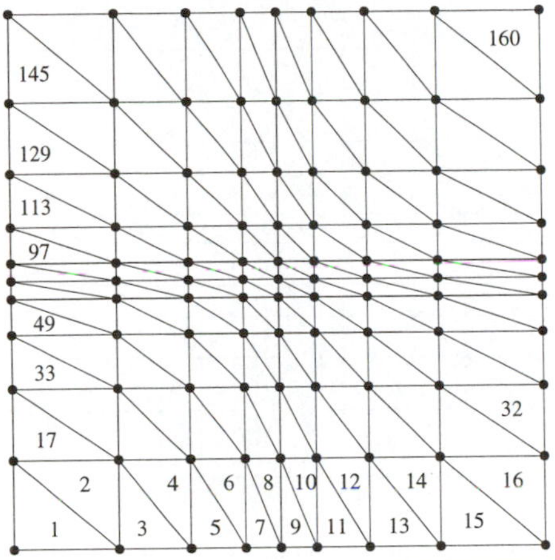

Figure 9-6: Example of numbering the triangular elements.

Fig. 9-6 for a $(N_x, N_y) = (8, 6)$ case. The second choice is to alternate the direction of the diagonal cuts in two adjacent rectangles, as done in (c) and (d) of Fig. 9-5. The result of such a choice for the entire space is given in Fig. 9-7. In both cases, the total number of triangular elements is the same and equals $2N_xN_y$. As we shall see later, the first choice is a little more convenient to carry out the calculations and the second choice results in better accuracy in the final results.

We have a total of $(N_x+1)\times(N_y+1)$ nodes in our space and we shall number them sequentially starting from the bottom left as number 1. The result of a $(N_x, N_y) = (8, 10)$ example was shown earlier in Fig. 9-4. We need also a numbering scheme for the $2N_xN_y$ triangular elements. For our first choice in dividing the rectangles into triangles by taking all the diagonal cuts in the same direction, we have a space schematically represented by Fig. 9-6. Since there are N_x rectangles in each row, we have $2N_x$ triangular elements in a row. If we label the elements sequentially starting from the bottom left, the elements in the lowest row are numbered from 1 to $2N_x$. In the row above, the first one from the left is $2N_x + 1$ and so on.

For the three vertices in each triangle, we shall number them in a clockwise direction, again, starting from the bottom left. That is, we begin with the node with smallest value for the x-coordinate. In cases where two nodes have the same value for the x-coordinate, we begin with the one having the smaller value for the y-coordinate. The net result is that all the odd-numbered triangles have the vertex opposite to the hypotenuse of the right-angle triangles as the first one and all the even numbered ones have the vertex opposite to the hypotenuse as the second one.

This is one of the many possible schemes to number the nodes and any one of them

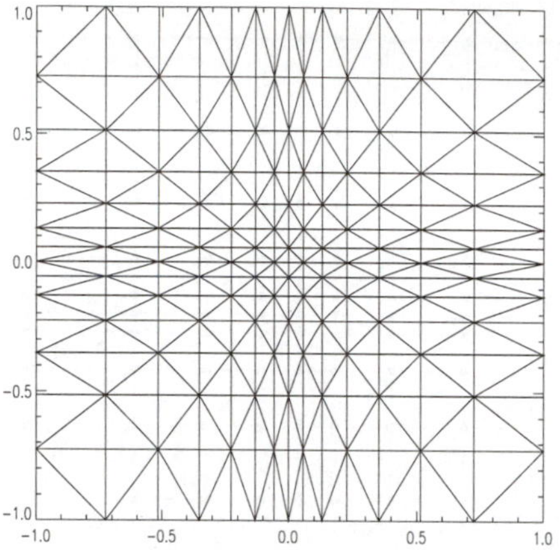

Figure 9-7: Example of constructing two-dimensional finite elements by alternating the direction of diagonal cuts in neighboring elements.

is acceptable so long we can construct a table associating all the vertices in an element with the node numbers in the whole space. In the scheme above, the association is straightforward. For the odd-numbered triangle in the ith rectangle of the jth row, the first vertex is at (x_{i-1}, y_{j-1}) and the node number is $n_1^{(o)} = (j-1) \times (N_x + 1) + i$. The second vertex is at (x_{i-1}, y_j), numbered $n_2^{(o)} = j \times (N_x + 1) + i = n_1^{(o)} + (N_x + 1)$, as it is directly above vertex number 1. The third vertex is at (x_i, y_{j-1}) and numbered $n_3^{(o)} = n_1^{(o)} + 1$. For the even-numbered triangular element in the same rectangle, the three vertices are located at $n_1^{(e)} = n_2^{(o)}$, $n_2^{(e)} = n_2^{(o)} + 1$ at (x_i, y_j), and $n_3^{(e)} = n_3^{(o)}$.

For the second choice, where the diagonal cuts alternate in direction between neighboring rectangles, the association of vertices with node numbers is slightly more complicated. However, there is no basic difficulty. In fact, it is quite easy to construct an association table for any logical choice in assigning the nodes and dividing the active space into elements. Since the way the space is divided into elements can have significant impact on numerical accuracy and computational efficiency, as we shall see later, this is an important step in a finite element calculation.

Shape functions and contributions from individual elements Once the locations of all the vertices in an element are known, we can construct the shape functions by expressing the coefficients $a_i^{(m)}$, $b_i^{(m)}$, and $b_i^{(m)}$, for $i = 1$ to 3, in terms of the coordinates using Eq. (9-26). Since the calculations involve only the values of the coordinates, independent of how the elements are taken, the computer program is not changed if we divide the space in a different way.

With the coefficients of $\ell_i^{(m)}(x, y)$, we can calculate the contributions from each element. For our example, these are $S_{ij}^{(m)}$ and $g_i^{(m)}$ of Eq. (9-31). In our case, $S_{ij}^{(m)}$

is a sum over products of the coefficients of shape functions and are found once the shape functions are known. For the source term $\rho(x, y)$ given by Eq. (9-43), the only nonvanishing $g_i^{(m)}$ are those with $\ell_i^{(m)}(x, y) \neq 0$ at the origin. That is, $g_i^{(m)} \neq 0$ only in those elements involving the origin and the integral can be carried out easily.

Equation for $\boldsymbol{v}_n$ We are now in a position to calculate $\boldsymbol{S}$ and $\boldsymbol{g}$ of Eq. (9-34) for the complete space. Since the matrix elements $S_{nn'}$ and g_n consist of, respectively, sums of $S_{ij}^{(m)}$ and $g_i^{(m)}$ in different elements, the only thing we need to do is to identify the terms that must enter each sum. The principle for carrying out this step has been outlined in §9-3. However, for the actual operation, it may be simpler to follow the variational results of §9-4.

Let us start with the matrix elements of $\boldsymbol{g}$. By comparing Eq. (9-41) with (9-34), we have

$$g_n = \sum_m{}' \sum_i \int \rho(x, y)\, \ell_i^{(m)} dx dy$$

where the prime over the summation reminds us that the sum is restricted to be over those elements that include node n as one of the vertices. To help us in making the selection, we can make use of the diagrams given in Fig. 9-5. For example, consider the $n = 5$ node. For cases (a) and (b), there are six different elements involved. For (c), there are only four, and (d), eight.

The same procedure can be used to find $\boldsymbol{S}$. In terms of matrix elements, Eq. (9-34) may be written as

$$\sum_{n'} S_{nn'} v_{n'} = g_n$$

where n is any node where $V(x, y)$ is not constrained by boundary conditions. Comparing with Eq. (9-41), we see that

$$S_{nn'} = \sum_m{}' \int\int \left\{ b_{i'}^{(m)} b_i^{(m)} + c_{i'}^{(m)} c_i^{(m)} \right\} dx dy$$

with n associated with vertex i', and n' with vertex i, of element m. The summation over m is restricted to those elements having node n as one of its three vertices. Because of this, n' must be a node that is also one of three vertices of the same element (and n' can be the same as n). As a result, $S_{nn'}$ is a sum over the same number of terms as the number of elements sharing both nodes n and n'. For the two-dimensional example we have here, the number is at most two. However, in three and higher dimensions, larger numbers are involved.

From a programming point of view, it may be conceptually simpler to follow the route of constructing first the contributions from each element and then assemble them to form matrices $\boldsymbol{S}$ and $\boldsymbol{g}$. For this purpose, we need a correspondence table that gives the node number n for each one of the three vertices of all the elements. Once this is available, we can go through all the elements one by one and calculate the values of $g_i^{(m)}$ and $S_{ii'}^{(m)}$ for i and i' equal to 1 to 3. The results are then added, respectively, to g_n for n corresponding to i' and $S_{n,n'}$ for n corresponding to i and

n' corresponding to i'. The final result, as expected, is the same as those calculated using the approach described in the previous paragraph.

The dimension of matrices $\boldsymbol{g}$ and $\boldsymbol{S}$ is, in principle, given by the number of degrees of freedom in the problem,

$$n_{\text{free}} = n_{\text{node}} - n_{\text{b.c.}}$$

where, for our two-dimensional example,

$$n_{\text{node}} = (N_x + 1) \times (N_y + 1)$$

is the total number of nodes and $n_{\text{b.c.}}$, the number of nodes with values specified by boundary conditions,

$$n_{\text{b.c.}} = 2(N_x + N_y - 2).$$

In practice, slight modifications can make programming simpler.

The first comes from the fact that our source term is a point charge at the origin. For simplicity, we shall include the origin as one of the nodes. With a true point charge, the potential is infinite where the charge is located. To avoid such a singularity in our solution, we can replace the source term by some fixed value of the potential at the origin. Physically, this means that we are not interested in the potential within some small region around the point charge. As a result, the effect of the source outside this region can be replaced by suitable boundary conditions. This increases $n_{\text{b.c.}}$, the number of points with fixed values, by one.

The second point concerns the question how to treat the points controlled by boundary conditions. Similar to what we have done on many occasions in finite difference calculations, it is simpler to include these points as one of the "unknown" quantities. In other words, we shall make our matrix dimension to be n_{node} instead of n_{free}. To account for the fact these nodes have values that are fixed and should not be changed when we solve Eq. (9-34), we can adopt the following strategy. If node k is such a node, then

$$S_{k,n} = S_{n,k} = \begin{cases} 1 & \text{for } k = n \\ 0 & \text{otherwise.} \end{cases}$$

In addition, we shall include contributions from the line integrals in Eq. (9-22) by incorporating them as a part of g_n. Contributions from the boundary points to the other nodes may be included by redefining the right side to be

$$g_n = g_n - \sum_{k}^{\text{b.c.}} S_{n,k} v_k$$

where the summation is over all nodes with the value of $V(x, y)$ constrained by boundary conditions. In realistic applications, the number of points on the boundary is small compared with n_{freedom}. As a result, the increase in the matrix dimension is a minor price to pay for the ease in programming.

Solve the linear algebraic equation The values of v_n are obtained by solving Eq. (9-34). In principle, this can be done using the method outlined in §5-1. However,

Box 9-1 Program DM_FEM
Finite element calculation of 2d Poisson equation

1. Divide the active space into finite elements.
2. Construct a correspondence table between nodes and vertices.
3. Specify the boundary conditions in terms of the nodes.
4. Build the shape functions in each element.
5. Initialize the calculation by setting $\boldsymbol{S} = 0$ and $\boldsymbol{g} = 0$.
6. Loop through each element and carry out the following calculations:
   (a) Obtain the values for $S_{ij}^{(m)}$ and $g_i^{(m)}$.
   (b) Assemble the contributions into $\boldsymbol{S}$ and $\boldsymbol{g}$:
      (i) Associate vertices i and j with, respectively, nodes n and n'.
      (ii) If both n and n' are not on the boundary, add $S_{ij}^{(m)}$ to $S_{nn'}$ and $g_i^{(m)}$ to g_n.
      (iii) If n is on the boundary, set $S_{nn} = 1$ and g_n boundary condition value.
      (iv) If n not on the boundary and n' is, subtract $S_{ij} * v_{n'}$ from g_n.
6. Solve the linear equations $\boldsymbol{S}\boldsymbol{v} = \boldsymbol{g}$.

in realistic applications, the number of nodes can be large and, as a result, the matrix dimension tends to be high. This causes problems both in the memory required to store the matrix $\boldsymbol{S}$ and in the computer time required to obtain the solution.

We can overcome the difficulty of large dimension by making use of the sparse property of $\boldsymbol{S}$. If we use method (a) or (b) of Fig. 9-5, with the diagonal cuts in all the rectangles in the same direction, it is easy to see that, at most, only five $S_{nn'}$ in a row or a column can be nonzero. If the diagonal cuts are alternating in direction between neighboring rectangles, the corresponding number is either five, if the node is the middle one as in (c) of Fig. 9-5, or nine for the same node in (d). In either case, the total number of nonzero elements in the complete matrix is $\sim 7n_{\text{free}}$. Compared with the total number of possible elements of n_{free}^2, we see that the storage locations required is reduced to $\sim 7/n_{\text{free}}$ if we retain only the nonvanishing elements.

The reduction in storage locations for $\boldsymbol{S}$ is not useful unless we have also a way to solve the linear equations without a lot of additional memory. As we recall from §5-1, the basic process of Gauss-Jordan elimination consists of a series of rotations to reduce all the off-diagonal elements to zero one by one by taking weighted sums of two rows or columns. If we apply this method directly to a sparse matrix, many of the elements that are zero in the original matrix may become nonzero in the intermediate steps of the calculation. As a result, the storage requirement goes back essentially to the original one for a dense matrix and we lose the advantage of the sparse nature of the matrix. There are methods to handle sparse matrices with only a minimum increase in the amount of intermediate storage. The subject belongs to specialized treatment such as Zlatev[61] and we shall not go into them here. Many computer codes have been developed to take advantage of these methods. A good starting point

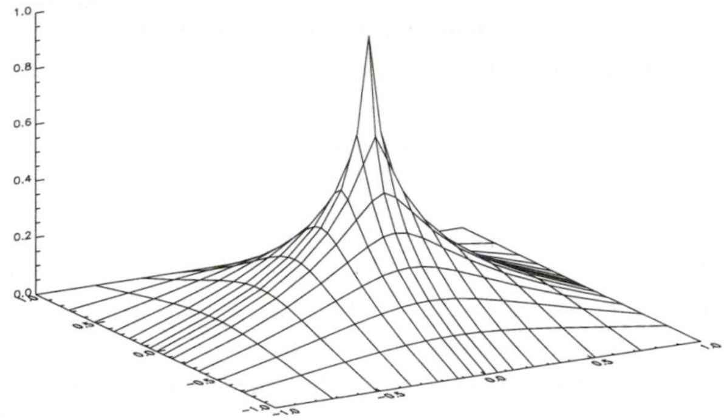

Figure 9-8: Solution of Poisson equation Eq. (9-17) using a $N_x = N_y = 14$ unevenly spaced mesh and boundary conditions given by Eqs. (9-42) and (9-43).

to explore some of these is the netlib.[55]

Once we have the solution for Eq. (9-34), the values of v_n provide us with a solution for $V(x, y)$ at each one of the nodes. The results for $v_n = 1.0$ at the origin, using $N_x = N_y = 14$, is shown in Fig. 9-8 and the steps to carry out the calculations are summarized in Box 9-1. Similar approach can also be used to solve other partial differential equations.

Accuracy considerations Similar to finite difference methods, the accuracy of a finite element calculation depends, in the first place, on the number of nodes taken. However, because of the flexibility offered by the freedom in choosing the elements, we can often make some improvement in a finite element calculation without having to increase the number of nodes. The possibility of unequal spacing between neighboring nodes, mentioned in the introduction of this chapter, is an important asset here, allowing us to put more points where higher accuracies are needed.

A less obvious advantage is the possibility of reducing what may be called "systematic" errors. This may be illustrated in the following way. If we take all the diagonal cuts in the same direction, as in Fig. 9-6, there is the possibility of accumulating the truncation error as we move from one end to the other, either along x- or y-direction. This can be seen, for example, by examining the symmetry, or rather the departures from being completely symmetrical, in the solution on either side of the source point in our two-dimensional Poisson equation example. The remedy is to take advantage of the flexibility in constructing the elements. By alternating the direction of the diagonal cuts, as in Fig. 9-7, the propagation of systematic errors is reduced. This choice is used in obtaining the results shown in Fig. 9-8.

3D space and nonlinear equations For boundary value problems with more than two independent variables, we can proceed essentially in the same way as outlined above for our two-dimensional Poisson equation example. A few special points may be of interests and we shall illustrate them using a three-dimensional space as the example.

The first one concerns the number of nonvanishing elements in a row of our matrix $\boldsymbol{S}$. It is fairly straightforward to make an estimate for it. If linear shape functions are used in the problem, each node can only be "connected" to its neighbor on either side. This means that only a maximum of three nodes are involved in a given line of, for example, fixed y and z. In a layer of, for example, constant z, the corresponding number is nine. This gives us a total of 27 to be the number of possible nonvanishing elements for a given node and, consequently, the maximum number of nonzero elements in a row or column of matrix $\boldsymbol{S}$. Compared with the product $N_xN_yN_z$ for the total number of elements in a row, we see that the matrix $\boldsymbol{S}$ is even more sparse than our two-dimensional example and the use of sparse matrix techniques is even more imperative here.

The second item is the question of numbering the vertices in an element. As we have seen earlier in Eq. (9-21), it is important to ensure contributions from the boundaries of neighboring elements to cancel each other. In two-dimensional cases, this is achieved by numbering the vertices of all triangular elements in the same direction, such as the clockwise direction illustrated in Fig. 9-3. In three dimensions, we need a somewhat more elaborate scheme. This can be achieved by, for example, taking one of the four vertices in a tetrahedron as number 1 and consider it as the top of a (non-symmetrical) pyramid. The three vertices in the base of the pyramid are numbered 2 to 4 in a clockwise direction looking down from the top toward the base. A neighboring tetrahedron can be numbered in the same way by picking a vertex such that, at the surface common to both elements, the surface vectors are pointing at opposite directions.

We have discussed, so far in this chapter, only linear PDEs. For nonlinear problems, one of the methods given in §8-9 must be used first to reduce the equations into linear ones. This can be done either in terms of the original PDE or after the problem is reduced to an algebraic one using finite element. In this way, the problem is not any different from those in finite difference approaches.

As mentioned in the introduction, the subject of finite element methods has a long history and has been developed to a high level of sophistication. In addition to the use of more complicated shape functions than those we have touched upon, there are also the use of graph theory to improve the performance in handling the sparse matrix problem, the use of automatic mesh generation to aid in setting up the problem, and so on. All these topics require treatment far beyond the scope here. The subject of numerical solution to partial differential equation is such an important one in science and engineering that continuous development in the techniques is assured, especially in the area of algorithms for parallel computers.

Problems

9-1 Use a least-squares method to derive the equations satisfied by a_1 and a_2 for an $n = 2$ approximation to the solution of Eq. (9-1). Calculate the values of $\widetilde{\phi}_2(t)$ in the interval $t = [0, 1]$ and compare the results with the exact solution given by the exponential function of Eq. (9-3).

9-2 Solve the differential equation of Eq. (9-1) with three degrees of freedom using the Galerkin method. Calculate the average absolute values of the differences from the $n = 2$ solution of Eq. (9-16) for a number of points in the interval $t = [0, 1]$. Compare the result with the same average difference from the exact solution of Eq. (9-3).

9-3 In cylindrical coordinates, the Laplacian operator has the form

$$\nabla^2\phi(\boldsymbol{r}) = \frac{1}{r}\frac{\partial}{\partial r}\left(r\frac{\partial\phi}{\partial r}\right) + \frac{1}{r}\frac{\partial^2\phi}{\partial\theta^2} + \frac{\partial^2\phi}{\partial z^2}.$$

For an infinitely long line of charge at the center of an equally long, hollow cylinder made of conducting material, the electric field is independent of both θ and z. Use the Galerkin method to solve for the radial dependence of the potential $\phi(r)$ inside the cylinder, assuming that it has a radius of $r_g = 1$ m and is grounded. The linear charge density is $\lambda = 1$ and is confined to a radius $a << r_g$.

9-4 For ordinary differential equations, it is possible to show that both finite element and finite difference methods produce the same algebraic equations. Show this using the prototype given by Eq. (8-57) with the range of interest divided in $(N + 1)$ equal segments, each of size

$$\Delta x = x_{i+1} - x_i$$

and expand the solution in terms of the following shape functions:

$$\ell_i(x) = \begin{cases} \frac{1}{\Delta x}(x - x_{i-1}) & \text{for } x_{i-1} \le x \le x_i \\ \frac{1}{\Delta x}(x_{i+1} - x) & \text{for } x_i \le x \le x_{i+1} \\ 0 & \text{elsewhere} \end{cases}$$

for $i = 1, 2, \ldots, n$. Compare the result with Eq. (8-62).

9-5 Carry out the solution to the two-dimensional Poisson equation of Eq. (9-17) using first the method of Fig. 9-5(a). Check the results by comparing the values on either side of the source point.

9-6 Repeat Problem 9-5 by taking all the diagonal cuts in the opposite direction as in Fig. 9-5(b). Compare the results with those of Problem 9-5.

9-7 Repeat Problem 9-5 by alternating the direction of diagonal cuts as in Fig. 9-6. Compare the results with those obtained in Problems 9-5 and 9-6.

Appendix A

In this appendix, we gather together a few computational techniques that are both useful and interesting but do not fit well in any of the chapters.

A-1 Decomposition into prime numbers

For many calculations, the results appear as ratios of integers. One efficient way of representing such quantities is in terms of a product of prime numbers. Among other advantages, the method offers an easy way to remove common factors between the numerator and the denominator.

Let us use, as illustration, the binomial coefficient

$$\binom{n}{m} = \frac{n!}{m!(n-m)!} \tag{A-1}$$

where the factorial of an integer is defined as the product

$$n! = \prod_{k=1}^{n} k = n(n-1)(n-2)\cdots(2)(1). \tag{A-2}$$

A simple way to calculate a binomial coefficient is to evaluate the numerator and the denominator separately and then take the quotient. This method works only for small values of n and m, as the range of integers we can store in the computer memory is limited. For example, we have seen earlier in §1-2 that the largest integer that can be represented by a four-byte word is less than 13!. On the other hand, the largest binomial coefficient with $n = 13$ is only

$$\binom{13}{6} = 1,716.$$

Obviously, we can take advantage of a method here that is capable of canceling common factors between the numerator and denominator at early stages of the calculation.

A simple method of keeping track of possible common factors among a group of integers is to decompose each into a product of prime numbers, as outlined in

Box A-1 Subroutine PRMDP(NUM,K_LST,ND)
Decomposition of an integer N into a product of prime numbers

Argument list:
- NUM: integer to be decomposed.
- K_LST: Array for the powers of prime number.
- ND: Dimension of K_LST in the calling program.

Initialization:
- (a) Store a list of prime numbers $\{p_i\}$ in the calling program.
- (b) Zero the array K_LST.
- (c) Let N equal to NUM.

1. For $i = 1$ to d, check if N/p_i is an integer:
   - (a) If so,
     - (i) Let $N = N/p_i$,
     - (ii) Increase K_LST(i) by 1.
   - (b) If not, go to the next p_i in the list.
2. Check if $N = 1$.
   - (a) If not, the decomposition failed. Output an error message.
   - (b) If $N = 1$, return K_LST as the list of the powers of prime numbers.

Box A-1. Let p_i for $i = 1$ to d be the first d prime numbers. The value of d depends on the magnitude of the largest integer we wish to decompose. In general, if the next prime number beyond our list of d members is p_{next}, then the largest integer we can decompose using the list of prime numbers is restricted to $n_{\max} < p_{\text{next}}$. Any integer n smaller than p_{next} may be expressed as the product of prime numbers in the following way:

$$n = \prod_{i=1}^{d} p_i^{L_{n,i}} \tag{A-3}$$

where $L_{n,i}$ is a list, or an array, of the powers of all d prime numbers associated with the number n. For example, the number 728 may be expressed as the following product:

$$728 = 2^3 \times 7^1 \times 13^1.$$

This may be represented by an array of d elements in the following way:

$$\{L_{728,i}\} = \{3, 0, 0, 1, 0, 1, 0, 0, \ldots\}$$

indicating that the powers of the (first six) prime numbers, 2, 3, 5, 7, 11, and 13, are, respectively, 3, 0, 0, 1, 0, and 1.

In such a representation, the product of any two integers n and m is given by the sum of their respective prime number arrays $\{L_{n,i}\}$ and $\{L_{m,i}\}$:

$$n \times m = \prod_{i=1}^{d} p_i^{L_{n,i}+L_{m,i}}$$

and the ratio of the same two integers is given by their differences in the powers:

$$\frac{n}{m} = \prod_{i=1}^{d} p_i^{L_{n,i}-L_{m,i}}.$$

If an element in the array becomes negative, the prime number appears in the denominator, with a power equal to the absolute value of the element. For example, the ratio

$$\frac{4356}{728} = \frac{2^2 \times 3^2 \times 11^2}{2^3 \times 7^1 \times 13^1} = \frac{3^2 \times 11^2}{2^1 \times 7^1 \times 13^1}$$

may be represented as $\{-1, 2, 0, -1, 2, -1, 0, 0, \ldots\}$.

A-2 Bit-reversed order

The bit-reversed order is adopted in fast Fourier transform of §3-4 to arrange quantities according to the degree of "evenness." In a binary representation, an even number is one with the last bit 0 and an odd number, the last bit 1. Following this idea, we can judge how "even" an integer is by the number and positions of the 0's in the binary representation. Thus, the integer $b000\cdots000 = 0$ is the most "even." (We have preceded a number with the letter b to indicate that it is written in binary representation, as we did in §1-2.) The next most even integer has a 1 in the first bit and all the rest of the bits are zero: $b100\cdots000 = 2^{\eta-1}$. Here, η is the total number of binary bits required to represent the set of integers under consideration. This is followed by $b010\cdots000 = 2^{\eta-2}$, $b110\cdots000 = 2^{\eta-2} + 2^{\eta-1}$, $b001\cdots000 = 2^{\eta-3}$, and so on. If we reverse the order of the binary bits, the first four members of this list become $b000\cdots000$, $b000\cdots001$, $b000\cdots010$, and $b000\cdots011$, corresponding to the numbers 0, 1, 2, and 3, respectively, in decimal notation. In other words, a list of integer in bit-reversed order is one with members arranged in ascending order if we read the bits in the binary representation backward, as illustrated in Table 3-8 on page 87.

A good way of getting a feeling for the bit-reversed order is to generate such a list. We can start by writing the members in binary representation. If there are $(N+1)$ elements in the list, each element may be represented by $\eta = \log_2(N+1)$ binary bits. In this scheme, the first member in the list is simply $b000\cdots000$, with all bits zero. The second member in the list has the index $b100\cdots000$, with all the bits zero except the first one, the number 1 if we reverse the order of the binary bits. Similarly, the third member in the bit-reversed order list is $b010\cdots000$, the number 2 when the order of bits is reversed. This can go on until we come to the last member, which is $b111\cdots111$ $(= 2^{\eta-1})$, as show in Table 3-8 on page 87.

We now have a natural way to generate a list of 2^η numbers in the bit-reversed order. The first is 0 and this constitutes a group of one element in the list we wish to construct. The next most even number is $M_v = 2^{\eta-1}$. We shall put this number into our list in such a way that the generation process can be put into the form of

Box A-2 Program DM_BTREV
Generate a list of 2^η items in bit-reversed order

1. Input η.
2. Start with one element.
   (a) Set this element to zero and put it into the list.
   (b) Define $M_v = 2^\eta$.
3. For k from 1 to η, carry out the following steps:
   (a) Divide M_v by 2.
   (b) Generate new elements by adding M_v to all the existing elements.
4 Output the list.

an iterative algorithm. For this purpose, we shall put the second member into our list by adding M_v to the first and only member (the number 0) in the list so far. This doubles the length of our list and it has now two elements. We can double the length of the list again by reducing the value of M_v by a factor of 2 and adding it to all existing members in the list we have constructed so far. After repeating this operation η times, the value of M_v is reduced to 1 and the total number of elements in the list becomes 2^η. This completes the bit-reversed order of a list of $(N+1)$ members. The algorithm is outlined in Box A-2.

A-3 Gaussian elimination of a tridiagonal matrix

For a system of linear equations involving a tridiagonal matrix, such as those we encounter in §3-7, 8-7 and 8-9, a very simple way can be used to find the solution. Starting from the prototype of Eq. (3-93)

$$\begin{pmatrix} b_1 & c_1 & 0 & 0 & \dots & 0 \\ a_2 & b_2 & c_2 & 0 & \dots & 0 \\ 0 & a_3 & b_3 & c_3 & \dots & 0 \\ \vdots & \vdots & \vdots & \ddots & \vdots & \vdots \\ 0 & \dots & 0 & a_{N-2} & b_{N-2} & c_{N-2} \\ 0 & \dots & 0 & 0 & a_{N-1} & b_{N-1} \end{pmatrix} \begin{pmatrix} f_1'' \\ f_2'' \\ f_3'' \\ \dots \\ f_{N-2}'' \\ f_{N-1}'' \end{pmatrix} = \begin{pmatrix} d_1 \\ d_2 \\ d_3 \\ \dots \\ d_{N-2} \\ d_{N-1} \end{pmatrix} \tag{A-4}$$

we reduce first the tridiagonal matrix into an upper diagonal one,

$$\begin{pmatrix} b_1' & c_1' & 0 & 0 & \dots & 0 \\ 0 & b_2' & c_2' & 0 & \dots & 0 \\ 0 & 0 & b_3' & c_3' & \dots & 0 \\ \vdots & \vdots & \vdots & \ddots & \vdots & \vdots \\ 0 & \dots & 0 & 0 & b_{N-2}' & c_{N-2}' \\ 0 & \dots & 0 & 0 & 0 & b_{N-1}' \end{pmatrix} \begin{pmatrix} f_1'' \\ f_2'' \\ f_3'' \\ \vdots \\ f_{N-2}'' \\ f_{N-1}'' \end{pmatrix} = \begin{pmatrix} d_1' \\ d_2' \\ d_3' \\ \vdots \\ d_{N-2}' \\ d_{N-1}' \end{pmatrix} . \tag{A-5}$$

The major difference between Eqs. (A-4) and (A-5) is that the subdiagonal elements, that is, the positions that were originally occupied by a_i, for $i = 2$ to $(N-1)$, are now replaced by zeros. If we further normalize all the matrix elements such that all the diagonal ones are unity,

$$b'_i = 1$$

it is easy to see that the elements of the new and old matrices are related in the following way:

$$c'_i = \begin{cases} \dfrac{c_1}{b_1} & \text{for } i = 1 \\ \dfrac{c_i}{b_i - a_i c'_{i-1}} & \text{for } i = 2, 3, \ldots, (N-2) \end{cases} \tag{A-6}$$

$$d'_i = \begin{cases} \dfrac{d_1}{b_1} & \text{for } i = 1 \\ \dfrac{d_i - a_i d'_{i-1}}{b_i - a_i c'_{i-1}} & \text{for } i = 2, 3, \ldots, (N-1). \end{cases}$$

This is a forward substitution process, as the elements are calculated in ascending order according to subscript i.

Next, we can use back substitution to obtain the final solution for the values of f''_i, for $i = 1$ to $(N-1)$. It is convenient to start from the last row. From Eq. (A-5), we have the simple result

$$f''_{N-1} = d'_{N-1} \tag{A-7}$$

since b'_{N-1} is normalized to unity. This gives us the value of f''_{N-1}. The second last row of Eq. (A-5) has the form

$$f''_{N-2} + c'_{N-2} f''_{N-1} = d'_{N-2}.$$

Again, because $b_{N-2} = 1$, we can make use of the value of f''_{N-1} obtained from Eq. (A-7) to solve for f''_{N-2}. The result is

$$f''_{N-2} = d'_{N-2} - c'_{N-2} f''_{N-1}.$$

It is obvious that we can continue this process by moving up one row at a time. The general result is

$$f''_i = d'_i - c'_i f''_{i+1} \qquad \text{for} \quad i = (N-2), (N-3), \ldots, 2, 1. \tag{A-8}$$

In this way, all $(N-1)$ values of f''_i are calculated.

The Gaussian elimination method given here is not the most sophisticated way to solve a matrix equation. As we can see in Chapter 5, other techniques are more convenient to use for the more general cases. The method here has the advantage that it is very simple to implement and works efficiently for cases involving tridiagonal matrices, such as in the cubic spline of §3-7 and in the numerical solution of certain differential equations in Chapter 8.

A-4 Random bit generator

Random bit generator A random bit generator is one that produces a result of either 1 or 0 with equal probability. Similarly to random number generators in general, we start with an integer seed s_i that gives us the first random bit. The main work is in producing the next seed, s_{i+1}, from which we obtain the next random bit. To simplify the discussion, we shall assume that each computer word is 32 bits long, even though the basic method works also for a shorter or longer word length. Among the 32 bits, we shall take an arbitrary one as the random bit of interest to us. Without any loss of generality, we can assume that this is the 19th bit. Again, most other choices among the 32 bits can also serve the purpose, as long as it is not too close to either end of the word.

In the same way as in generating random numbers, the method itself is not concerned with the way the first seed s_i is chosen. However, in practical applications, we need to ensure that it is picked "randomly." If bit 19 of s_i is 1, the random bit given by the generator is 1. Similarly, if bit 19 of s_i is 0, the random bit is 0.

The next seed s_{i+1} is generated in the following manner. If bit 19 of s_i is zero, s_{i+1} is produced by shifting all the bits of s_i to the left by one binary location. This is equivalent to multiplying s_i by 2, except that the first bit of s_i is discarded. In addition to speed, the shift operator also has the advantage that, in contrast to an integer multiply, it does not produce an integer overflow when the first bit of s_i is not zero. The resulting integer seed s_{i+1} is an even number with the last bit zero. This is of no concern to us, since we are only interested in bit 19.

The real work comes when bit 19 of s_i is 1. In this case, a *mask* is used to modify various bits in the seed. In the *primitive polynomial modulo 2* method given in Knuth,[31] the mask is made up of a polynomial consisting of a sum of several terms, each of which is in the form of 2 to an integer power. For our selection of bit 19, a suitable polynomial is

$$m = 2^4 + 2^1 + 2^0.$$

We shall not go into the proof here of why the use of such a mask ensures that a sequence of random bits generated is uncorrelated. The selection of a mask depends also on which bit in the word is used as the random bit. In other words, both the location of the random bit and the mask are an integral part in selecting the polynomial. A list of several choices can be found in Press and others.[44]

One advantage of this method is that we can use the extremely fast Boolean operations AND [IAND(i_1, i_2) in Fortran], OR [IOR(i_1, i_2)], and XOR [IEOR(i_1, i_2)] to carry out all the calculations. Each one of these functions performs a bit by bit comparison between two computer words, i_1 and i_2, and returns the result shown in Table A-1. Where such operations can be used to replace arithmetic calculations, such as integer multiplication and division by 2, there is a gain in speed. In addition, they provide us with the ability to carry out certain manipulations of a computer word not available otherwise.

Table A-1: Boolean operations.

i_1 i_2	0 0 1 1 0 1 0 1	Result
IAND(i_1, i_2)	0 0 0 1	= 1 if both i_1 and i_2 are 1
IOR(i_1, i_2)	0 1 1 1	= 1 if either i_1 or i_2 is 1
IEOR(i_1, i_2)	0 1 1 0	= 1 if either i_1 or i_2 is 1 but not both

In terms of Boolean operations, the steps to obtain a new seed s_{i+1} from the previous one s_i may be summarized as the following:

(a) Let $s' = \mathsf{IEOR}(s_i, m)$. That is, compare all 32 bits of s_i one by one with the corresponding bits in mask m. If both of them are 0 or 1, put the corresponding bit in s' to be 0. If one of them is 1 and the other is 0, put the corresponding bit in s' to be 1.

(b) Shift s' left by 1 bit (equivalent to multiply s' by 2).

(c) Add 1 to s' and store the result as the new seed s_{i+1}.

The results can be checked by applying some of the tests for randomness described in §7-1. A few modifications may be necessary since we are dealing with random bits.

A-5 Reduction of higher-order ODE to first-order

Ordinary differential equation of orders higher than one can always be reduced into a set of coupled first-order ones. For this reason, methods for solving such differential equations are often given only for first-order. Let us demonstrate the reduction using as an example the second-order equation of Eq. (8-4)

$$\frac{d^2\phi}{dt^2} + \omega_0^2\phi(t) = 0. \tag{A-9}$$

By defining

$$y_1(t) = \phi(t) \qquad\qquad y_2(t) = \frac{d\phi}{dt} \tag{A-10}$$

the relation given by Eq. (A-9) can be expressed in terms of the following two equations:

$$\frac{dy_1}{dt} = y_2(t) \qquad\qquad \frac{dy_2}{dt} = -\omega_0^2\, y_1(t). \tag{A-11}$$

For order n, there will be n such equations instead of two for our example.

It is convenient to use a short-hand notation to express the above relations. If we treat the vector $\boldsymbol{y}(t)$ as an array of n functions:

$$\boldsymbol{y}(t) \equiv \{y_1(t), y_2(t), \ldots, y_n(t)\}. \tag{A-12}$$

Eq. (A-11) may be written in the form

$$\frac{d\boldsymbol{y}}{dt} = \boldsymbol{f}(\boldsymbol{y}(t), t) \tag{A-13}$$

where $\boldsymbol{f}(\boldsymbol{y}(t), t) = \{f_1(\boldsymbol{y}(t), t), f_2(\boldsymbol{y}(t), t), \ldots, f_n(\boldsymbol{y}(t), t)\}$. For our $n = 2$ example, we have

$$f_1(\boldsymbol{y}(t), t) = y_2(t) \qquad\qquad f_2(\boldsymbol{y}(t), t) = -\omega_0^2 y_1(t).$$

The meaning of Eq. (A-13) is the same as any relations between vector quantities. That is,

$$\begin{aligned}\frac{dy_1}{dt} &= f_1(y_1(t), y_2(t), \cdots, y_n(t), t)\\ \frac{dy_2}{dt} &= f_2(y_1(t), y_2(t), \cdots, y_n(t), t)\\ \cdots &= \cdots \qquad \cdots \qquad \cdots.\end{aligned}$$

Except where we need the explicit form of each term, we shall make use of the short-hand.

Appendix B
List of Fortran Program Examples

We give here a list of the algorithms used as illustrations for some of the methods discussed. In each case, a computer program is written in Fortran to verify that the methods actually work in practice. The source codes are available from World Scientific:

World Wide Web	http://www.wspc.com.sg/others/software/3365/
ftp	ftp.wspc.com.sg/pub/software/3365/

Name	Box(page)	Short Title
DM_PRIME	1-1 (14)	Generate the first k prime numbers
DM_BNML	1-2 (24)	Binomial coefficients using a table of $\ln(k!)$
TRAPZ	2-1 (33)	Trapezoidal rule integration
DM_SIMPS	2-2 (36)	Simpson's rule integration
DM_QUAD	2-3 (43)	Gauss-Legendre quadrature
DM_MC1D	2-4 (46)	Monte Carlo integration
DM_MC3D	2-5 (51)	Three-dimensional Monte Carlo integration
NVLLE	3-1 (68)	Neville's algorithm for interpolation
RATNV	3-2 (72)	Rational function interpolation using Neville's algorithm
GMMAI	3-3 (75)	Incomplete gamma function, continued fraction approx.
DM_FFT	3-4 (89)	Fast Fourier transform
DM_RCHRD	3-5 (93)	Integration by extrapolation using Richardson's approach
DM_RMBRG	3-6 (95)	Romberg integration using Neville's algorithm
DM_BSNTP	3-8 (106)	Inverse interpolation using Bessel's formula
DM_CBSPL	3-9 (111)	Cubic spline with Gaussian elimination
HRMTE	4-1 (120)	Hermite polynomials coefficients by recursion relation
LGNDR	4-2 (127)	Coefficients of Legendre polynomials using Eq. (4-44)
YLM	4-3 (129)	Associated Legendre polynomial and spherical harmonics
DM_SBES	4-4 (133)	Coefficients of spherical Bessel function
DM_BESP	4-5 (134)	Propagation of the values of spherical Bessel functions
DM_ALAGR	4-6 (145)	Coefficients of associated Laguerre polynomials
GMMA8	4-7 (149)	Gamma function using an eight-term approximation
S_GMMA	4-8 (151)	Incomplete gamma function for small x, series expansion

Name	Box(page)	Short Title
DETRM	5-1 (161)	Determinant value by Gauss-Jordan elimination
LNEQN	5-2 (163)	Solution of linear equations, Gauss-Jordan elimination
MATIV	5-3 (167)	Matrix inversion, Gauss-Jordan elimination
LUDCP	5-4 (173)	LU-decomposition of matrix
FBSBS	5-5 (175)	Forward and back substitution
JCBDG	5-6 (186)	Jacobi diagonalization: real, symmetric matrix
TRIDG	5-7 (197)	Householder tridiagonalization: real, symmetric matrix
BISCT	5-8 (205)	Eigenvalues of tridiagonal matrix by bisection
TRIQL	5-9 (214)	Eigenvalues and eigenvectors: tridiagonal matrix
JNSYM	5-10 (234)	Eigenvalues and eigenvector: real nonsymmetric matrix
LLSQ	6-1 (261)	Linear least-squares fit to a straight line
BETAI	6-2 (270)	Incomplete beta function, continued fraction approximation
MGRS	6-3 (274)	Multiple regression analysis
NLNFT	6-4 (288)	Nonlinear least-squares by Marquardt method
RSHFL	7-1 (299)	Improved random number generator by shuffling
RSUB	7-2 (301)	Subtraction method of random number generation
FQNCY	7-3 (303)	Frequency test of random numbers
RUNUP	7-4 (306)	Run-up test of a sequence of random numbers
DM_PRCOL	7-5 (337)	Percolation on a square lattice
SWEEP	7-6 (345)	One sweep of the lattice with Metropolis algorithm
DM_ISING	7-7 (346)	Two-dimensional Ising model
DM_PTHIT	7-8 (361)	Path integral calc., harmonic oscillator wave function
DM_MNDLB	7-9 (367)	Mandelbrot plot
DM_DLA	7-10 (373)	Simulation of diffusion-limited aggregation
RGKT4	8-1 (386)	Fourth-order Runge-Kutta method
RKODE	8-2 (391)	Runge-Kutta solution of n initial-value, first-order ODEs
MIDPT	8-3 (395)	Modified midpoint method for n first-order ODEs
EXTZS	8-4 (397)	First-order ODEs by extrapolating to zero step size
DM_SHTSQ	8-5 (403)	Shooting method to obtain the scattering solution
MATCH	8-6 (406)	Solution of second-order ODEs by bidirectional shooting
DM_SORWS	8-7 (413)	Successive overrelaxation for boundary value problems
DM_SORPS	8-8 (421)	Solution of partial differential equation using SOR
DM_DFFUS	8-9 (429)	Crank-Nicholson solution of parabolic PDE
DM_WAVE	8-10 (436)	Explicit method to solve the wave equation in (x,t)
DM_KINK	8-11 (444)	Solution of the kink equation by Newton's method
DM_FEM	9-1 (473)	Finite element calculation of 2d Poisson equation
PRMDP	A-1 (478)	Prime number decomposition of an integer
DM_BTREV	A-2 (480)	Generate a list of 2^η items in bit-reversed order

Bibliography

[1] M. Abramowitz and I.A. Stegun. *Handbook of Mathematical Functions.* Dover, New York, 1965.

[2] E. Anderson, Z. Bai, J. Demmel, J Dongarra, J Du Croz, A. Greenbaum, S. Hammarling, A. McKenney, S. Ostrouchov, and Sorensen D. *Lapack Users' Guide.* Society for Industrial and Applied Mathematics, Philadelphia, 1992.

[3] G. Arfken. *Mathematical Methods for Physicists.* Academic Press, San Diego, 1995.

[4] P.R. Bevington. *Data Reduction and Error Analysis for the Physical Sciences.* McGraw-Hill, New York, 1969.

[5] J.M. Borwein and P.B. Borwein. Ramamujan and pi. *Scientific American,* 260:112, February 1988.

[6] H. Bowdler, R.S. Martin, C. Reinsch, and J.H. Wilkinson. The *qr* and *ql* algorithms for symmetric matrices. *Linear Algebra, Handbook for Automatic Computation by Wilkinson, J.H. and Reinsch, C.,* II, 1971.

[7] R.N. Bracewell. The fourier transform. *Scientific American,* 260:86, June 1989.

[8] D.S. Burnett. *Finite Element Analysis, From Concepts to Applications.* Addison-Wesley, Reading, Mass., 1988.

[9] T.M. Cannon and B.R. Hunt. Image processing by computer. *Scientific American,* 245:214, October 1981.

[10] G.J. Chaitin. Randomness and mathematical proof. *Scientific American,* 232:47, May 1975.

[11] T.W. Chiu and T.S. Guu. A shift-register sequence random number generator implemented on the microcomputers with 8088/8086 and 8087. *Comp. Phys. Comm.,* 47:129, 1987.

[12] R. Courant and D. Hilbert. *Methods of Mathematical Physics.* Interscience Publishers, New York, fourth edition, 1995.

[13] H. Cramer. *Mathematical Methods of Statistics.* Princeton University Press, Princeton, N.J., 1946.

[14] J.K. Cullum and R.A. Willoughby. *Lanczos Algorithms for Large Symmetric Eigenvalue Computations*, volume I and II. Birkhäuser, Boston, 1985.

[15] G. Dahlquist and Å. Björck. *Numerical Methods, English translation by N. Anderson.* Prentice Hall, Englewood Cliffs, N.J., 1974.

[16] A.K. Dewdney. Computer recreations. *Scientific American*, 260:108, February 1989.

[17] A.K. Dewdney. Computer recreations. *Scientific American*, 260:125, June 1989.

[18] P. Diaconis and B. Efron. Computer-intensive methods in statistics. *Scientific American*, 248:116, May 1983.

[19] C. Domb and J.L. Lebowitz, editors. *Phase Transitions and Critical Phenomena.* Academic Press, London, 1972-.

[20] P.J. Eberlein and J. Boothroyd. Solution to eigenproblem by a norm reducing jacobi type of method. *Numer. Math.*, 11:1, 1968.

[21] J.F. Fernandez and J. Rivero. Fast alogrithm for random numbers with exponential and normal distributions. *Computers in Physics*, 10:83, 1996.

[22] L. Fox and D.F. Mayers. *Computing Methods for Scientists and Engineers.* Clarendon Press, Oxford, 1968.

[23] I. Galbraith, Y.S. Ching, and E. Abraham. Two-dimensional time-dependent quantum-mechanical scattering event. *Am. J. Phys.*, 52:60, 1984.

[24] A. Goldberg, H.M. Schey, and J.L. Schwartz. Computer generated motion pictures of one-dimensional quantum-mechanical transmission and reflection phenomena. *Am. J. Phys.*, 35:177, 1967.

[25] H. Goldstein. *Classical Mechanics.* Addison-Wesley, Reading, Mass., 1950.

[26] F.B. Hildebrand. *Introduction to Numerical Analysis.* McGraw-Hill, New York, 1956.

[27] J. Hoshen and R. Kopelman. Percolation and cluster distribution: I. cluster multiple labeling technique and critical concentration algorithm. *Phys. Rev.*, B14:3438, 1976.

[28] K. Huang. *Statistical Mechanics.* Wiley, New York, second edition, 1987.

[29] B.D. Keister. Multidimension quadrature algorithms. *Computers in Physics*, 10:119, 1996.

[30] M. Kendall and A. Stuart. *The Advanced Theory of Statistics*, volume I. Macmillan, New York, 1977.

[31] D.E. Knuth. *The Art of Computer Programming*, volume II. Addison-Wesley, Menlo Park, Calif., second edition, 1981.

[32] S.V. Lawande, C.A. Jensen, and H.L. Sahlin. Monte Carlo integration of the Feynman propagator in imaginary time. *Comp. Phys.*, 3:416, 1969.

[33] T.D. Lee. *Particle Physics and Introduction to Field Theory.* Harwood, Chur, Switzerland, 1981.

[34] V.A. Lubimov, E.G. Novikov, V.Z. Nozik, E.F. Tretyakov, and V.S. Kosik. An estimate of the ν_e mass from the β-spectrum of tritium in the valine molecule. *Phys. Lett.*, 94B:266, 1980.

[35] V.A. Lubimov, E.G. Novikov, V.Z. Nozik, E.F. Tretyakov, and V.S. Kosik. An estimate of the ν_e mass from the β-spectrum of tritium in the valine molecule. *Neutrino Mass and Related Topics, Proc. of XVI INS Intl. Symp., Tokyo, 1988, edited by S. Kato and T. Ohshima,*, 1988.

[36] S.K. Ma. *Modern Theory of Critical Phenomena.* Benjamin, Reading, Mass., 1976.

[37] G. Marsaglia, B. Narasimhan, and A. Zaman. A random number generator for pc's. *Comp. Phys. Comm.*, 60:345, 1990.

[38] J. Mathews and R.L. Walker. *Mathematical Methods of Physics.* Benjamin, New York, 1965.

[39] N. Metropolis, A.W. Rosenbluth, M.N. Rosenbluth, and A.H. Teller. Equation of state calculations by fast computing machines. *J. Chem. Phys.*, 21:1087, 1953.

[40] P. Pfeuty and G. Toulouse. *Introduction to the Renormalization Group and to Critical Phenomena, (English translation by G. Barton.* Wiley, London, 1977.

[41] J. Phillips. *The NAG Library: A Beginner's Guide.* Clarendon Press, Oxford, 1986.

[42] C. Pomerance. The search for prime numbers. *Scientific American*, 247:136, December 1982.

[43] W.H. Press and S.A. Teukolsky. Numerical calculation of derivatives. *Comput. Phys.*, 5:68, 1991.

[44] W.H. Press, S.A. Teukolsky, W.T. Vetterling, and B.P. Flannery. *Numerical Recipes, The Art of Scientific Computing.* Cambridge University Press, Cambridge, second edition, 1992.

[45] C. Renfrew. The origins of indo-european languages. *Scintific American*, 261:106, October 1989.

[46] L.H. Ryder. *Quantum Field Theory.* Cambridge University Press, Cambridge, 1985.

[47] L.S. Schulman and P.E. Seiden. Percolation analysis of stochastic models of galactic evolution. *J. Stat. Phys.*, 27:83, 1982.

[48] L.S. Schulman and P.E. Seiden. Percolation and galaxies. *Science*, 233:425, 1986.

[49] G. Sewell. *The Numerical Solution of Ordinary and Partial Differential Equations.* Academic Press, San Diego, Calif., 1988.

[50] P.P. Silvester and R.L. Ferrari. *Finite Elements for Electrical Engineers.* Cambridge University Press, Cambridge, second edition, 1991.

[51] G.D. Smith. *Numerical Solution of Partial Differential Equations: Finite Difference Methods.* Clarendon Press, Oxford, third edition, 1985.

[52] D. Stauffer. *Introduction to Percolation Theory.* Taylor and Francis, London, 1985.

[53] R.M. Steffen. Extranuclear effects on angular correlations of nuclear radiations. *Adv. Phys.*, 4:293, 1955.

[54] J. Stoer and R. Bulirsch. *Introduction to Numerical Analysis, (English translation by R. Bartels, W. Gautschi and C. Witzgall).* Springer-Verlag, New York, 1980.

[55] URL. *http://www.netlib.org.* Anonymous ftp: ftp.netlib.org, Gopher: gopher.netlib.org.

[56] J.H. Wilkinson. *The Algebraic Eigenvalue Problem.* Oxford University Press, Oxford, 1965.

[57] J.H. Wilkinson and C. Reinsch. *Linear Algebra*, volume II of *Handbook for Automatic Computation, ed. by F.L. Bauer and others.* Springer-Verlag, Berlin, 1971.

[58] K.G. Wilson. Confinement of quarks. *Phys. Rev.*, D10:2445, 1974.

[59] S.S.M. Wong. *Introductory Nuclear Physics.* Prentice Hall, Englewood Cliffs, N.J., 1990.

[60] O.C. Zienkiewicz and R.L. Taylor. *The Finite Element Method.* McGraw Hill, Maidenhead, England, fourth edition, 1989.

[61] Z. Zlatev. *Computational Methods for General Sparse Matrices.* Kluwer Academic Publishers, Dordrecht, The Netherlands, 1991.

Index